W9-DHX-838

ANIMAL BEHAVIOR

Readings from
**SCIENTIFIC
AMERICAN**

ANIMAL BEHAVIOR

Selected and Introduced by
Thomas Eisner
Cornell University

Edward O. Wilson
Harvard University

W. H. Freeman and Company
San Francisco

Cover: A spider (*Argiope* sp.) in its web. (Photograph by Thomas Eisner.)

Most of the SCIENTIFIC AMERICAN articles in ANIMAL BEHAVIOR are available as separate Offprints. For a complete list of more than 950 articles now available of Offprints, write to W. H. Freeman and Company, 660 Market Street, San Francisco, California 94104.

Library of Congress Cataloging in Publication Data

Eisner, Thomas, 1929-   comp.
    Animal behavior.

    1. Animals, Habits and behavior of—Addresses, essays, lectures.   I. Wilson, Edward Osborne, 1929-   joint comp.   II. Scientific American.   III. Title.
[DNLM: 1. Ethology—Collected works. QL751 E35a]
QL751.6.E37   591.5   75-2383
ISBN 0-7167-0511-7
ISBN 0-7167-0510-9 pbk.

Printed in the United States of America

9 8 7 6 5 4 3

CONTENTS

Introductory Essay   1

I   ENVIRONMENT TO ACTION

II   THE ADAPTIVENESS OF BEHAVIOR

III   THE ADAPTABILITY OF BEHAVIOR

IV   SOCIAL BEHAVIOR

Note on cross-references: All articles are referred to by title. For those that appear in this book, the page on which the article begins is given. For those that are available as Offprints, but are not included here, the Offprint number is given. For those published by SCIENTIFIC AMERICAN, but not available as Offprints, the month and year of publication are given.

ANIMAL BEHAVIOR

INTRODUCTORY ESSAY

The 1973 Nobel Prize in Physiology and Medicine was given to Karl von Frisch, Konrad Z. Lorenz, and Niko Tinbergen. Students of animal behavior around the world were gratified by this recognition of three of their most eminent colleagues. In the system of reward, status, and prestige that pervades the world of science as surely as it does any other primate society, the discipline of animal behavior had arrived. What had the three accomplished? The citation stated that they were the chief architects of the new science of ethology. This means that they helped to institute the objective study of whole patterns of animal behavior under natural conditions, in ways that emphasized the functions and the evolutionary history of the patterns.

They did more: they showed us how to study animal behavior in the first place. In his celebrated studies of honeybees, stretching unbroken over sixty years, von Frisch used elementary training techniques to demonstrate which environmental stimuli the bees can detect. He presented the stimuli to his insects in conjunction with sugar water and at varying intensities. When the bees later responded to a particular intensity of the stimulus in the absence of sugar water, the intensity was recorded as falling within their range of sensitivity. Methodically and painstakingly, von Frisch and his students plotted the sensory world of this one species of insect. They proved, among other things, that bees have color vision and have the hitherto unsuspected capacity to perceive polarized light and magnetic fields. In the single most surprising discovery, bees were found to communicate the positions of food sources and new hive sites by means of the waggle dance, one of the most complex forms of behavior ever recorded in animals. After the work of von Frisch and his associates, no animal species can ever by viewed the same way again. The honeybee is now known to us as a sophisticated creature, traveling through a complex life cycle enveloped in sensory fields that can be explored and characterized as precisely as the details of its anatomy and physiology.

Lorenz and Tinbergen achieved essentially similar results in the study of vertebrates. It has been Lorenz's central argument that behavioral traits are not the shifting, will-o'-the-wisp phenomena often depicted in the past, but instead are hard, measurable entities that can be described in much the same fashion as skulls, aortas, and other anatomical parts. This theme had been developed in the early years of this century by naturalists such as Oskar Heinroth, Charles Otis Whitman, and William Morton Wheeler, but was then mostly submerged in the 1920's and 1930's by the emphasis on the role of learning and hormones in the modification of behavior. Lorenz refocused the attention of zoologists on the invariant components of behavior. This he accomplished by detailed studies of the behavior of jackdaws, ducks, geese, and other birds, chronicling the step-by-step development of their instinctive repertories. He showed, especially in his famous comparative study of ducks, how evolutionary trends can be deduced from behavioral traits, which can thus be used to arrange species into more natural taxonomic groupings. Lorenz, more than any other modern scientist, launched the new era of careful descriptive work on animal behavior. He identified innate components that are useful for the study of evolution even in the process of learning, by elucidating the phenomenon of imprinting, in which young animals fixate on certain other individuals or objects during a brief "critical period" in their development. In more recent years, Lorenz has created a major controversy with his book *On Aggression* and other writings, which suggest that much of human behavior, no less than that of birds and insects, is stereotyped and the product of an idiosyncratic evolutionary history. He has insisted that aggressiveness is a universal human trait which must be accommodated in social planning and cannot simply be erased by training and the provision of better surroundings. Although many of Lorenz's recent proposals have been strongly disputed, they have at least opened fresh inquiries into the basis of human behavior.

Niko Tinbergen is an associate of Lorenz who shared in the development of vertebrate ethology. But more in the manner of von Frisch, he is an experimentalist who found ways to analyze the behavior patterns observed in free-ranging animals. Tinbergen and his co-workers devised numerous experiments to isolate and to identify the key stimuli, called social releasers, used by animals to communicate with members of their own species. A great deal of complex behavior in fish and birds was found to be "released" by elementary stimuli that could be successfully imitated with crude models held in the experimenter's hands. In various of these experiments, stickleback fish challenged "rivals" consisting solely of a red patch painted on a disk; young game birds performed the characteristic manuevers used to evade hawks when nothing more than a cross was passed over their heads; and wooden "eggs" constructed of various sizes and spotting patterns were retrieved by herring gulls from outside the nest and incubated. Tinbergen played a principal role in opening these and other classes of vertebrate behavior to incisive experimentation. With Lorenz he showed how the new results might be used to gain deeper insights into the organization of the central nervous system. His 1950 book, *The Study of Instinct*, integrating ethology and relating it to the study of the nervous system, has been possibly the single most influential synthesis of animal behavior published in this century.

We have introduced the three Nobel laureates to you as a graphic way to define the essence of the modern study of animal behavior. In the *Scientific American* articles that follow, the main focus is on ethology. Whole patterns of behavior are illustrated in a wide variety of organisms, and the functions of these patterns under natural conditions are investigated. Within ethology, broadly defined, lies the vital and rapidly growing conjunction of neurobiology, ecology, and evolutionary theory. Students of ethology are not concerned with details of the sensory and nervous systems except as they bear directly on the control of whole patterns of behavior. Similarly, ecology and evolutionary theory are considered only as contributing disciplines. The study of animal behavior, in short, is viewed as an intellectually discrete subject, and is taught as much in many departments of biology and psychology in the United States. At the same time, one should keep in mind that the discipline covers two different levels of biological organization, each of which requires a different mode of explanation.

At the lower level, that of the organism, the behavior of the individual is described in terms of the processes of its own nervous system. The organism is broken down into its component sensory and motor elements, but just far enough to adduce a full account of the whole pattern of behavior. The subjects of concern at this level are sensory physiology and integrative neurophysiology, in which the interaction of whole groups of sense cells and nerve cells constitutes the main process. The analysis ordinarily will not go so far as to explore the fine structure of the nerve cell or the biochemistry of the nerve discharge; those subjects belong to "pure" neurobiology and would carry us unnecessarily deep into the third, subcellular level of organization. The higher of the two levels included in our discipline consists of the population. Here are focused the subjects of sociobiology, which treats all aspects of sexual and social behavior, and behavioral ecology, in which behavior is viewed as a device by means of which the organism interacts with all of its environment. At the first, lower level, the organism is analyzed as though it were a machine to be dissected and reassembled. At the second, higher level, behavior is explained in terms of its evolutionary history and the functions it performs in the natural environment.

The two-layered nature of animal-behavior studies is implicit in the sequence of the *Scientific American* articles selected for this volume. We begin with case histories of the most stereotyped forms of behavior, the control mechanisms of which are understood to some extent at the cellular level (Part I: "From Environment to Action"). Then attention shifts to the manner in which behavior evolves, its genetic basis and the highly diverse ways by which it adapts organisms to their environments (Part II: "The Adaptiveness of Behavior"). The articles in Part III ("The Adaptability of Behavior") deal with the methods by which animals alter their responses through learning and hormone-mediated change in order to cope with the inevitable variations encountered in the environments. Finally, Part IV ("Social Behavior") is devoted to communication and the organization of animal societies.

Students in beginning biology courses and general readers who sample the literature of biology are almost invariably attracted to the study of animal behavior. Many find it the most fascinating subject in all of science. In our opinion a straightforward reason for this popularity exists: the essential qualities of man are behavioral, and the remainder of the animal kingdom provides a virtually infinite array of mirrors with which to view fragments of our inner nature, to deduce the evolutionary steps that produced us, and to identify at last, in a scientific sense, the traits that make us truly unique. There is more to the study of animal behavior, of course. It is a foundation piece of ecology, and the physiology peculiar to behavior is more complex and hence more challenging than other kinds of physiology. But ultimately it is the explicit human application, the bridge that is being built between biology and the social sciences, that will make animal behavior of increasing intellectual importance in the years immediately ahead.

Thomas Eisner

Edward O. Wilson

I

ENVIRONMENT TO ACTION

I ENVIRONMENT TO ACTION

INTRODUCTION

Animals are defined to a large extent by their behavior; they are motile creatures that explore and manipulate their environments. Even the most sedentary animals are more active and manipulative than the most motile plants. Hydras search into the surrounding medium with tentacles; clams sweep food-laden currents of water into their mouths with rows of cilia; and parasitic worms creep through the cavities and sway in the body fluid of their hosts. There is a sound ecological reason why animals *must* be active. They are the consumers of energy and not the primary producers, which role is filled by plants and microorganisms. Even when they search out or prey on other animals, they merely act as conduits for energy captured and packaged by the primary producers. One of the few real laws of ecology states that consumers can get only a small fraction of the energy that passes through plants. Under the best of conditions this fraction seldom exceeds 20 per cent, and it is usually much less. Consequently animals must literally glean a living from the environment, patrolling it or sucking parts of it into their alimentary canals in order to secure sufficient quantities of energy. Once this option was chosen in evolution by the earliest animals one billion years ago, the effects spilled into every other aspect of their life. The body was compacted into precise symmetric shapes and powered by masses of muscles; alimentary tracts were assembled for the processing of food in packets; and entire new organs were invented to monitor odor, light, and sound.

And above all there was the nervous system to coordinate behavior. Behavioral responses are not absolutely dependent on a nervous system. The members of colonies of siphonophores, a group that includes the Portuguese man-of-war and certain other jelly-fish-like coelenterates, communicate with one another partly through special cells in the epithelium. They are able to coordinate their movements to such an extent that the whole colony acts like a single organism, or superorganism if you prefer. A few plants "behave" without the benefit of even the rudiments of either a nervous or a muscular system. The Venus flytrap captures insects by suddenly clamping modified leaves together to form a trap. The plant then digests its victims and uses the nitrogen to supplement the meager supplies extracted from the swampy environments in which it lives. The so-called sensitive plants (see Figure I.1) fold and partly retract their leaves when touched, exposing spines that make them a far less attractive target for herbivores browsing in the vicinity. Their action can be considered exactly analogous to the prickling of spines by a hedgehog or porcupine. These plants accomplish their movements by the rapid change of turgor in certain strategically placed cells. In the first article of this *Scientific American* Reader, Joseph J. Maio describes an equally remarkable form of behavior in certain species of soil-dwelling molds. The

Figure I.1. A sensitive plant (*Schrankia microphylla*) responds to manual disturbance by folding its leaflets. The response, which increases exposure of the plant's thorns, may protect it against browsers.

mold filaments grow numerous tiny ring-shaped snares, each composed of just three cells. When a hapless worm sticks its head into one of the rings, the cells suddenly inflate and seize it in a stranglehold.

In spite of these promising "efforts" by a few plants and fungi, the range of behavior in multicellular organisms that lack a nervous system is of necessity strictly limited. We can be reasonably sure that wherever animals have evolved in other parts of the universe, they contain circuits, relays, and control centers that we would instantly classify as a nervous system. And there would be common features in the way these animals scan the environment and respond to it, features analogous to those described below.

Sign stimuli

Certain environmental stimuli are used by animals to trigger and to guide behavior. These stimuli constitute only a minute fraction of those that might be used. They are the ones that are, first, the most relevant to the animal's survival and reproduction and, second, the most predictable and reliable. A bat may wait for the light intensity to fall below 0.5 footcandles to begin its crepuscular flight, and pay no attention to the correlated but far less reliable drop in air temperature. A beetle bites into a leaf because the leaf contains a chemical found only in the species of plant on which its species exists; it ignores the thousands of other chemicals that impinge on its receptors each day. Such key signals are referred to as *sign stimuli*. When the sign stimuli emanate from other members of the same species, and especially when they are used in communication, they are designated *social releasers* or simply *releasers*. Sometimes the term releaser is used in a looser manner to mean the same thing as sign stimulus.

Stimulus filtration

The sense organs of the animal and the afferent nerves that lead in from them have been specifically designed in the course of evolution to select and to identify the sign stimuli. Stimulus filtration provides some of nature's most dramatic examples of specialization: fish that attack upon seeing the flash of a single color; male moths that start to fly after sensing a few hundred molecules of the female scent; beetles that live in burned-over areas and are activated by the far infrared radiation from forest fires. In considering such cases in the *Scientific American* articles in this anthology, the reader should bear in mind the fact that each species molds its perception of the outside world to these basic necessities. German writers have just the right expression for this concept: *Umwelt*, or loosely, the "world around me." Each species has its own *Umwelt*, that peculiar combination of sights, odors, sounds, and other sensory impressions by which it picks its way carefully through the environment. Human beings tend to think of themselves uncritically as fully open, all-perceiving creatures. But of course they are not; they are jacketed by an *Umwelt* not much less restrictive and certainly no less adaptive than that characterizing the great majority of other animals. Human beings, to be more specific, have good vision but are blind to ultraviolet light, which is readily perceived by certain insect species (Figure I.2). Our ears are quite sensitive over a wide range of frequencies, but they are completely deaf to the ultrasonic frequencies used routinely by bats and dolphins. Our sense of taste is mediocre, comparable to that of a honeybee, and both of these are inferior to that of a butterfly. Our sense of smell is still poorer; we have only the most feeble comprehension of the rich world of odors in which our own dogs and other domestic animals live. And finally, we have no comprehension at all of what it must be like to orient to electric and magnetic fields, to polarized light, or to the ripple of waves on the surface of a pond, all of which are stimuli perceived with great clarity by one kind of animal or another.

Figure I.2. To the human eye, the marsh marigold appears as an evenly colored yellow flower. To the ultraviolet-sensitive eye of a pollinator such as the honeybee, however, the flower has a distinct pattern of ultraviolet-absorbent nectar guides (see ultraviolet photograph, below), which direct the insect to the site of food, while incidentally insuring that it will pollinate the flower.

The stereotypy of motor responses

Another human misconception, and one often transferred freely and with even less accuracy to animals, is that of freedom of action. We like to consider ourselves to be at least theoretically capable of a near infinitude of mental images and motor actions. But in fact human beings are severely limited by the way the brain is constructed and by the images that can be gleaned from their narrow *Umwelt*. One *Scientific American* article here has been selected to make this important point. In "Eye Movements and Visual Perception," David Noton and Lawrence Stark describe the amazingly strict and predictable manner in which the eye scans a visual field, forming patterns so divorced from conscious control that they can be revealed only by the use of an ingenious laboratory apparatus.

What is thus true of human beings is even more true of animals. In the narrow, prismatic world of the lower animals, the events of a normal life cycle unfold with the precision of a computer program. Sign stimuli are met with rigid responses, sometimes called *consummatory acts*. These acts are characteristically directed at some particular object in the environment, a quality that distinguishes them from eye blinks, knee jerks, and other simple reflexes. They are also subject to *habituation*. That is, once these acts have been performed, they are less likely to be elicited again, at least for some period of time following the performance. Students of animal behavior find it convenient to recognize the following three phases in the full execution of a stereotyped behavioral act.

1. *Appetitive behavior.*   The animal begins a searching behavior which tends to be persistent but loosely and variably performed. A hungry ant, for example, wanders in wide looping paths away from the entrance of its nest until it encounters food. A male moth flies back and forth across an open field in search of the odor of the female scent during certain hours of the night. (The physiological expression most frequently employed to describe such behavior is "drive," but this word has come into increasing disfavor because of its vagueness and its lack of reference to any particular mechanism within the central nervous system). It is a characteristic of much appetitive behavior that it is autonomous, in the sense that the animal requires no previous familiarity with the environment or even experience in the ultimate behavioral act toward which the search directs it.

2. *The consummatory act.*   This response, typically among the most complex and stereotyped of all behaviors in the animal's repertory, is the climax of the instinct sequence. A prey is discovered, pursued, and killed; a potential mate is recognized, and courted with elaborate displays; a host plant is located and eggs are laid upon it—and so forth through the full life cycle.

3. *Quiescence.*   After the completion of the consummatory act, appetitive behavior is halted and reduced in intensity, and further consummatory acts of that type become less likely. To put the matter in somewhat more physiological language, the response threshold of the behavior is raised: it takes more of the sign stimulus to elicit the consummatory act a second time.

We have noted that a consummatory act is directed typically toward some object. By distinguishing this orientation from the act itself, biologists have greatly simplified the analysis of stereotyped behavior. In the classic example introduced by Konrad Lorenz, graylag geese respond to the placement of one of their eggs outside the nest by tucking the egg beneath the bill and rolling it back into the nest. The tucking motion is defined as the consummatory act; the tilting of the head from side to side to guide the egg toward the nest is the orientation. A male cockroach reacts to the odor of female sex substance by attempting to mate with whatever object produces it. In nature only a female cockroach possesses the scent, but the experimenter can easily transfer it to a glass rod or some other inert substitute (Figure I.3). The motions of copulation are the consummatory act, their direction toward the object, the orientation.

Several basic kinds of orientation are practiced. The most elementary is the *kinesis*, in which locomotion takes no particular direction but speeds up or slows down according to the intensity of the stimulus. As a result the organism either gravitates toward the source of the stimulus or else drifts away from it. A ciliated protozoan exposed to an irritating chemical, for example, increases its random zigzag movements until it reaches a place where the level of the stimulation is less. The woodlouse, a small land-dwelling crustacean, searches for hiding places by an equally simple technique. If the site it occupies starts to dry out, the woodlouse begins to move about a great deal. When a sufficiently moist place is reached, the woodlouse becomes less restless and may cease movement altogether.

A more precise orientation technique is the *taxis* (plural, *taxes*), in which an animal maintains a particular spatial relation to a stimulus source as it moves. When a parasitic water mite (*Unionicola*) swims in search of a clam on which to feed, it orients by means of a positive, phototaxis; that is, it travels toward the strongest available source of light. Since the strongest light in nature almost invariably comes from the sun, the questing mite is drawn into the upper layers of the open water, where it can wander back and forth while encountering a minimum of obstructions. As it passes over a clam buried in the silt below, the mite smells its quarry. This new stimulus reverses the phototaxis from positive to negative. Now the mite swims away from the sun into the darkness of deeper water, where other, probably chemical clues lead it to the clam. Taxes are prevalent throughout the animal kingdom. A geotaxis guides earthworms toward or away from gravity, hence onto the surface of the ground at night or down deeper into the soil in the day. A rheotaxis directs the newly hatched salmon from the spawning site downstream to the sea. Several years later the reversed rheotaxis helps guide the salmon back upstream to spawn. Virtually every kind of physical stimulus to which organisms can react has been used in taxes by one animal species or another. The responses are not always a simple movement toward or away from the stimulus source. One special class of taxes, called menotaxes, consists of movements at a constant angle to the stimulus source. The best known version is the sun-compass orientation. By traveling with its body held at a constant angle to the sun, which is too far away to be approached by any meaningful amount, the animal follows a straight line over the surface of the earth. This is the method used by honeybees to commute back and forth between the hive and flower patches, over distances as great as five kilometers.

The ultimate techniques in orientation use energy emitted by the organism itself. Instead of depending on cues provided by other objects, the organism bounces its own signals off the objects. Almost everyone is familiar with *sonar*, or echolocation, the process by which ships broadcast pulses of sound and record the echoes that come back from submarines, the sea bottom, and other solid objects beneath them. Essentially the same technique is used by bats to pursue moths and other insect prey and to avoid obstacles in their flight paths.

Figure I.3. Males of the cockroach *Periplaneta americana* respond to a glass rod that has been dipped in an extract of the female's sex attractant. The males have converged on the rod, fluttering their wings and raising the tips of their abdomens (see male on upper left), responses that normally initiate the mating act.

The big brown bat (*Eptesicus fuscus*) emits ten clicking sounds a second while flying. These signals are loud in volume, but they are so high in pitch (mostly 50,000 to 100,000 cycles per second) as to be essentially inaudible to human beings. Sonar is also used by porpoises while navigating and hunting in murky water. A second form of emitted-energy orientation, equally sophisticated in design, has been evolved by the electric eels of South America and a variety of other fish that hunt at night or in cloudy water. These creatures maintain electric fields around their bodies, generating the impulses from specially modified muscles located in their tails. When objects intrude in the field, the distortion is recorded by receptor organs located in the head of the fish. (See "Electric Location by Fishes," by H. W. Lissmann).

Biologists have been notably successful in analyzing the behavior of in-

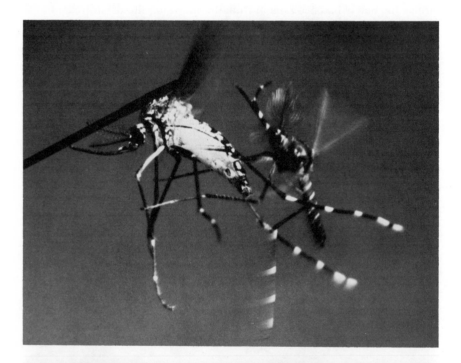

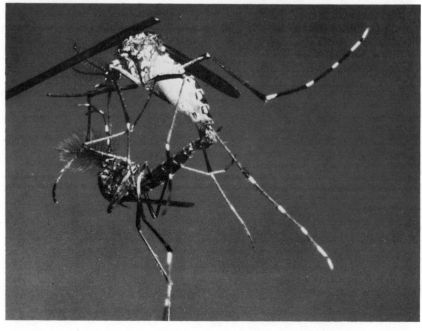

Figure I.4. A tethered female mosquito (*Aedes aegypti*) attracts a male (top), who then mates with her (bottom). The male is lured by the hum of the female's beating wings.

vertebrates and the simpler vertebrate animals, such as fish and birds, in terms of the elementary concepts just reviewed. It is possible, for example, to describe much of the life cycle of an insect as a procession of appetitive behavior and consummatory acts guided by taxes and other basic orientation devices. The adult mosquito provides a striking example (Figure I.4). The female of the yellow-fever mosquito (*Aedes aegypti*) produces a monotonous whining hum with her wings as she flies through the air. Although this sound is irritating to human ears, it is a sexually attractive signal for the male mosquito. The beating of the wings produces sounds with frequencies ranging from 450 to 600 cycles per second. The males fly toward any steady hum between 300 and 800 cycles per second, a response which in nature brings them to the females. The experimenter can induce the same response by striking tuning forks in the 300 to 800 cps range. If these instruments are wrapped in cheesecloth as well, some of the males will even try to mate with them.

Other crucial sign stimuli have been identified in the life cycles of various kinds of mosquitoes. The female locates human beings and other warm-blooded hosts partly by the warmth of their bodies and partly by the odor of lactic acid, one of a host of excretory products that evaporate continuously from the skin. The quantities of lactic acid are too minute to be detected by human beings, but mosquitoes, like many insects, have a much keener sense of smell. In the course of their behavioral evolution they have selected this particular substance as a sign stimulus. When the female alights, she plunges her stylets through the skin like a prospector searching for oil. Sometimes she "sinks a dry hole"—misses the blood vessels altogether—whereupon she pulls out the stylets, shifts position, and tries again. How does she know when the stylets strike "home"? For *Aedes* one of the clues is actually the taste of ATP (adenosine triphosphate) in the blood.

Parallel examples are encountered in the lives of the social insects. The remarkable organization of insect colonies is based to a large extent on *pheromones*, or chemical releasers (Figure I.5). A honeybee alarms her nest-mates by expelling certain secretions from the mandibular glands in the head or from glandular cells around the base of the sting. The odor of these pheromones at first attracts the other bees and then, at close quarters, launches them into an aggressive frenzy. Nestmates are recruited by assembly pheromones that emanate from a gland located on the upper surface of the abdomen. Using a secretion from the mandibular gland, the queen inhibits the workers from rearing new queens. The same pheromone is used in the mating flights to attract males. Other substances distinguish nestmates from the members of other colonies, and various castes and life stages from each other. The glandular apparatus used to achieve these and other social functions, such as nest construction, is complex and extensive.

Students of behavior in birds and mammals, and above all those concerned with the workings of the human mind, tend to regard such representations of the life of mosquitoes and honeybees as an oversimplified account of the nature of animal behavior. They are correct in part. The larger the brain of an animal, the more its behavior can be modified. The repertories of familiar mammalian species such as the laboratory white rat do not unfold with insect-like precision. Hormone levels and learning are decisive determinants, and individuals can easily be induced to develop different patterns of behavior by the appropriate manipulation of these factors. In rhesus monkeys and other monkeys and apes, *socialization* is crucial to normal development. That is, infant and juvenile animals must experience normal relationships with their mothers and other members of the group in order to acquire normal sexual and parental behavior. When deprived of all socialization by being isolated immediately after birth from other members of their own species, males and females cannot copulate properly, and females have great difficulty in rearing their own offspring. Psychologists who extend their studies to animal behavior, especially when they are motivated by a desire to understand human behavior through comparison, quite naturally emphasize such phenomena as socialization and pay close attention to developmental variability as a subject important for its own sake.

Scientific American articles on behavioral modification have been included in later sections of this anthology (see, for example, "How an Instinct Is Learned," by Jack P. Hailman; and "Love in Infant Monkeys," by Harry F. Harlow). However, for two reasons we have stressed the simpler, more stereotyped repertories of invertebrate animals in the introductory section. First, it is time to accept the fact that the overwhelming majority of animals are invertebrates. Of the approximately one million known species, less than 40,000 or 4 per cent, are vertebrates. Invertebrates constitute the bulk of animals both on the land and in the sea. One entomologist has estimated that 10^{18} (one billion billion) insects are alive on earth at any given moment, but even they may well be outnumbered by copepod crustaceans, the dominant invertebrates of the sea. Studies in evolution, ecology, and—yes—behavior are destined to concentrate increasingly on invertebrates in the future; vertebrate parochialism is coming to an end in behavioral biology. The second reason for starting with invertebrates is that these smaller, simpler animals generally provide the best experimental subjects for the complete analysis of whole behavior patterns. As Donald Kennedy points out in "Small Systems of Nerve Cells," the nervous system of a human being is constructed of an almost incomprehensible 10 billion to 100 billion cells. Little wonder that human psychology, and the study of behavior in other higher mammals, is prone to treat the brain as a black box; psychologists often can do no more than measure the input of sensory signals along entire nerves and the output of neuromuscular responses. In sharp contrast, many invertebrate nervous systems contain on the order of a hundred thousand neurons. Although the brain is still a complicated organ, auxiliary control centers much simpler in organization can be distinguished within the ganglia, the knots of associative cells and supporting tissue distributed along the body cavity of the animals. Sometimes the input of information is limited to a small number of sensory cells that can be identified and monitored individually. A few motor systems have been discovered that possess comparable simplicity. It is from such elementary assemblages—which nevertheless control relatively complex behavior—that general principles concerning the internal processes of the black box can be elucidated at the level of the cell. Important advances of course continue to be made with vertebrates, but the clearest and most quickly grasped examples are to be found among the invertebrates, and that is where we shall begin.

Although doing so is by no means essential, the reader is advised to read through the articles in sequence. In the remainder of this introduction and in subsequent sections, we will present a brief discussion of each of the articles in order to help the reader place it within the framework of the whole. Technical points will also be explained in the few cases where authors may have been too brief, and additional references will be provided.

"Predatory Fungi," *by Joseph J. Maio. July 1958.*
This account includes a description of the fungi that strangle nematode worms with their curious three-celled loops. As explained earlier, the response can be regarded as behavior in the broad sense, despite the fact that no nervous system is involved.

"Life and Light," *by George Wald. October 1959.*
The sunlight that falls on the earth is in the "fortunate range" of 300 to 1,100 nanometers. Life would not exist without such electromagnetic energy, and it is conceivable that it would not have evolved unless it were available in

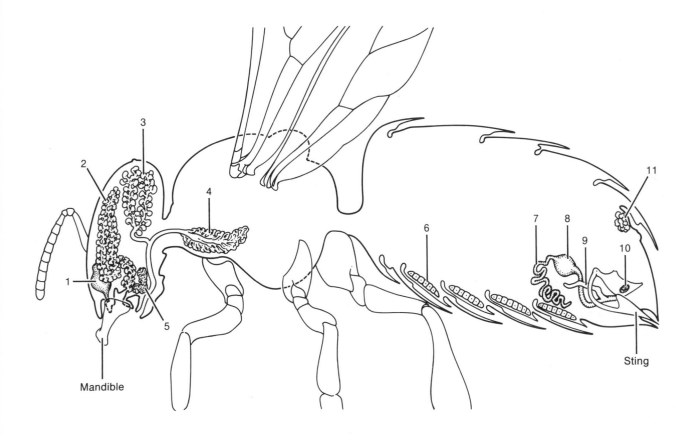

1. Mandibular gland	2-heptanone	Alarm; queen substances (multiple functions in sex and colony control)
2. Hypopharyngeal gland	Royal jelly	Larval food
3. Head labial gland	?	Cleaning, dissolving, digestion (?)
4. Thorax labial gland	?	Cleaning, dissolving, digestion (?)
5. Postgenal gland	?	?
6. Wax gland	Beeswax	Nest construction
7. Poison gland	Venom	Defense
8. Poison-gland reservoir	Venom	Defense
9. Defour's gland	?	?
10. Koschevnikov's gland	?	In queen, attraction of workers
11. Nasanov's gland	Geraniol; citral, nerolic acid	Assembly, orientation to swarm

Figure I.5. The social organization of the honeybee is largely based on secretions from a large, complex glandular system. Many of the chemical products are pheromones, or chemical releasers, used in communication between colony members. Various of the pheromones evoke alarm, attraction, or orientation during swarming; those of the queen both attract males during the mating flight and control worker behavior within the hive.

the narrow segment we refer to as light. Certainly photosynthesis, the principal mode of capturing and packaging energy, could not have originated. We did not select this article for that particular concept, however. More important for our purposes is the fact that once life was coupled to light in such a manner, the particular way that animals could use it for orientation and communication was foreordained. In his distinctively insightful way, Wald shows how retinene and special molecules associated with retinene were selected in the course of evolution of several groups of animals to exploit the properties of light for orientation and communication. The reader is invited to use this case to reflect on the "fitness of the environment."

"Small Systems of Nerve Cells," *by Donald Kennedy. May 1967.*
The sea hare *Aplysia* and the crayfish *Procambarus* have become important organisms for neurobiological and behavior research because of the great simplicity of certain elements of their nervous system and the ease with which these elements can be manipulated in laboratory experiments. For example, complex and specific movements in the tail of the crayfish have been shown to be under the control of individual "command fibers," which are simple nerve cells. The new data summarized in this article have contributed to the realization that far more authority can be invested in simple cells than was previously conceived. They open the possibility that substantial portions of whole behavior patterns can be explained tracing the circuitry of a countable number of these basic units.

"Moths and Ultrasound," *by Kenneth D. Roeder. April 1965.*
Roeder presents here yet another case of the high achievement of a small group of nerve cells in an invertebrate. Only *two* such cells transmit all the information concerning sound in ultrahigh frequencies within each ear of the moth. Thanks to the ingenious field experiments conducted by Roeder and his associates, we also know the ecological significance of the two cells: they detect the sonar signals emitted by bats and allow the moths to escape before being captured and eaten by the bats. No better documentation exists of the highly specific, idiosyncratic relationship between a particular behavior pattern, the nerve cells that mediate it, and the environmental pressure that caused it to evolve in the first place.

"The Sex-Attractant Receptors of Moths," *by Dietrich Schneider. July 1974.*
Schneider here documents another case, with new implications, of the tight relationship between form and function in one of the simplest invertebrate systems. The female sex attractant of the domestic silk moth *Bombyx mori* can be fairly regarded as the classic pheromone. It was the first sex attractant identified chemically, by Adolf F. J. Butenandt and his collaborators in 1959, and it has been subject to more analysis than any other pheromone with the exception of the queen substance of honeybees. As Schneider explains here, the system is precise and specific to an extreme. The male silk moth is virtually a sexual guided missile. Some 50 percent of the odor receptors on his feathery antennae respond only to the sex pheromone. The pheromone itself is one geometric isomer of a particular 16-carbon alcohol: it is "bombykol," or *trans*-10-*cis*-12-hexadecadian-1-ol. The other three hexadecadienols with double bonds at the 10 and 12 carbon positions (but twisted at different angles from bombykol) are far less effective. The male is also extremely sensitive to the pheromone. Only a single molecule is required to activate one of the receptor cells leading to the antennal nerve and hence to the brain. When more than 200 cells are discharged per second, the rate required to overcome the background noise caused by spontaneously firing cells, the male is informed that bombykol is present in the air, and he begins to search for a female. Sexual communication in the silk moth is an example of sensory filter-

ing occurring at the outermost level, in this instance within the odor-receptor cells themselves.

"Eye Movements and Visual Perception," *by David Noton and Lawrence Stark. June 1971.*
We have introduced this article on human behavior in the midst of invertebrate examples to make the point that certain elements of behavior in man are as stereotyped as the consummatory acts of lower animals. When a visual field, such as that provided by a painting or an outdoor vista, is too extensive to be comprehended with a single fixation, the eye travels around it in swift scanning paths that are quite predictable in direction and timing. The eye has a logic of its own; the selection of the key points for fixation, the time spent on each, and the sequence in which they are examined are largely outside the control of the conscious mind.

"The Flight-Control System of the Locust," *by Donald M. Wilson. May 1968.*
With this article by Donald Wilson, whose premature death by a drowning accident soon afterward ended a brilliant career in neurobiology, we shift abruptly from sensory to motor systems. It was the notable achievement of Wilson, Torkel Weis-Fogh of the University of Copenhagen (now of the University of Cambridge, England), and their co-workers to demonstrate a motor-control device in the flight of locusts comparable in simplicity to the sensory systems already elucidated in moths, sea hares, and other invertebrates. The prevailing idea of motor action up to the time of these experiments had been the peripheral-control hypothesis: the perfectly reasonable notion that input from the sense organs is processed by the central nervous system, which then decides on the motor commands to be sent out along the afferent nerves to the appropriate muscles. As muscle contraction takes place, new sensory information is generated. By means of proprioceptors (sense organs responding to pressure), the muscles themselves inform the central nervous system of the actions they have taken, while the change in the body's position causes new signals from the outside environment to reach the sensory system. The result is a feedback loop: a sequence of messages transmitted from sense organs to central nervous system to motor system to sense organs and on around again until the central nervous system "satisfies" itself that the correct action has been completed. Wilson and his associates obtained the surprising result that flight in locusts is partially independent of feedback control. When all the sources of sensory feedback are eliminated in experiments, the wings can continue to beat in a normal rhythm. Flight, in other words, is run by an internal motor that can be turned on or off by signals from the outside but does not require continued instructions to operate. Moreover, to repeat our theme of the simplified modes of invertebrate behavior, the muscles of each wing are activated by fewer than 20 motor-nerve cells.

"Brains and Cocoons," *by William G. Van der Kloot. April 1956.*
The cocoon of a moth is a marvelously engineered apparatus. The patterns by which the grown caterpillar weaves it out of long strands of silk are so precise that in many cases entomologists can tell the species to which the caterpillar belongs by a glance at the form of the cocoon. William G. Van der Kloot and Carroll M. Williams recognized that the cocoon is in fact "frozen" behavior that can be weighted and measured. In the series of experiments described in this article, the two investigators exploited this circumstance to explore some of the circuitry of the insect central nervous system. By damaging selected nerves and portions of the brain, they identified several of the most important pathways that control cocoon spinning. In particular, they observed that the corpora pedunculata of the brain is a principal coordinating center essential to the correct performance of this higly stereotyped behavior.

"The Neurobiology of Cricket Song," *by David Bentley and Ronald R. Hoy. August 1974.*

In this masterful study the authors have brought into play the widest possible range of techniques to analyze a single stereotyped behavior pattern. Cricket song is first examined in terms of its evolution and ecological significance. Then the centers of control in the central nervous system are tracked down with the use of microelectrodes implanted into single nerve cells. Further experiments lead to the discovery of the patterns of activity by which special command fibers cause the wings to scrape against one another in the sound-producing movements. Next the authors trace the development of the command patterns within the nervous system of immature crickets. Finally, they confront the crucial question of whether the songs of crickets are learned or inherited. A meaningful answer is supplied: many kinds of perturbations in the environment, including total isolation during development, fail to deflect the emergence of fully formed song; but the hybridization of different cricket species produces offspring with intermediate song types. Thus Bentley and Hoy establish that the intricate machinery which they have begun to tear down and reconstruct at the level of the cell is indeed programmed genetically in considerable detail, as Konrad Lorenz suggested for much of stereotyped behavior in animals generally.

"How We Control the Contraction of Our Muscles," *by P. A. Merton. May 1972.*

In this article we leap again, as we did earlier in the article by Noton and Stark on eye movements, from insects to man. And as before, the purpose is to illustrate sterotyped components within the most complex behavior patterns. In this article on muscle contraction, Merton introduces the reader to psychophysics, the branch of psychology—by necessity limited to human beings—that considers the relation between conscious mental processes and the interaction of the body with its surrounding environment. We have just seen that the locust flies with an internal motor that is partly independent of sensory input from the environment. Merton here describes similar automatic responses in the more elementary forms of human behavior. The eye is insensitive. It cannot feel movements forced on it from the outside, as when it is displaced slightly by the probe of a finger. We are aware of the voluntary movement of the eye because our brain has instructed it to move and has registered that fact automatically. The image appears to jump when the eye is moved by a finger, as we would expect. But it seems to remain steady when the eye is moved voluntarily, because the brain compensates for the amount of movement it has ordered. The compensation recurs automatically, and it is beyond the power of the conscious mind to alter it. Similarly, the posture of the arm is kept steady by the stretch reflex. When the limb is flexed or extended by the addition of a weight or some other external force, the change is relayed to the spinal cord by impulses that flow from special receptors in the muscle spindles. The reflex operates largely free from intervention by the mind. When the mind wills a change in posture, the order is relayed by means of a strange deception that closely resembles power steering in modern automobiles. The order comes in the form of impulses from the brain to the muscle spindles, making them contract. The contraction activates the stretch reflex, causing the main muscles to contract, just as a mere touch of the steering wheel of an automobile turns on the small servomotor that powers the turning of the main wheels. The human nervous system is filled with such semi-automatic devices that are scarcely more complicated than the basic elements of invertebrate behavior. It is only in the way they are assembled and placed under higher centers of control that the central nervous systems of man and the other highest vertebrates are distinguished from those of simpler animals.

"Annual Biological Clocks," *by Eric T. Pengelley and Sally J. Asmundson. April 1971.*

Stereotypy in behavior does not result entirely from sensory screening, consummatory acts, and semi-automatic motor controls. There also exist internal rhythms that cause the stereotyped behavior to wax and wane in intensity. The most familiar and hitherto best-studied cases are the circadian rhythms, which occur in not completely precise 24-hour cycles (hence the name, from the Latin, *circa*, "about," plus *dies*, "a day"). Female moths, for example, do not emit sex attractants uniformly around the clock. Each species has a sharply defined period, say 10 to 12 P.M. or 4 to 6 A.M., during which the female extrudes her scent glands and "calls" in the males for mating. Circadian rhythms are pervasive in many kinds of behavioral and physiological activities: foraging, egg-laying, metabolism, maintenance of body temperature, growth, susceptibility to disease and poisons, and others. The rhythms appear to be truly endogenous. When organisms are put in closed rooms with constant illumination (or lack of it) and constancy in other environmental factors, the 24-hour cycles run on. In the article reprinted here, Pengelley and Asmundson describe a newly discovered and even more intriguing phenomenon, the circannual rhythm, in which a complete cycle consumes about one year. Ground squirrels can pass in and out of hibernation, birds can run through their fall migratory sequence, and deer can grow and shed antlers, all under constant conditions. Despite a very large amount of research devoted to it, the biochemical nature of the circadian clock has not yet been disclosed. The circannual clock is of course even more of a mystery. A rival hypothesis, championed for many years by Frank A. Brown of Northwestern University, is that the rhythms are really guided by changes in the earth's magnetic field, cosmic rays, or some other subtle geophysical factors. This explanation may seem strange, but it has not been wholly discounted. Whatever the explanation, the widespread existence of such rhythms is a fact that must weigh heavily in every study of behavioral systems.

SUGGESTED ADDITIONAL READING for PART I

Alcock, John. 1974. *Animal Behavior: An Evolutionary Approach.* Sinauer Associates, Sunderland, Mass.

Manning, A. 1972. *An Introduction to Animal Behavior.* Second edition. Addison-Wesley, Reading, Mass. These are probably the best two short textbooks available on basic aspects of animal behavior. They are recommended for further study as the next step beyond, or in conjunction with, this *Scientific American* Reader.

Marler, P. R., and W. J. Hamilton III. 1966. *Mechanisms of Animal Behavior.* John Wiley, New York. The best major textbook on the physiological mechanisms underlying behavior, especially on the more stereotyped patterns stressed here in Part I.

Roeder, Kenneth D. 1967. *Nerve Cells and Insect Behavior.* Revised edition. Harvard University Press. A lucid presentation of highly imaginative research which, although done on insects only, touches upon some of the most fundamental problems of neurobiology and behavior. Wonderful reading!

1   Predatory Fungi

by Joseph J. Maio
July 1958

Plants which trap insects are familiar in the folklore of biology. Less well-known, but more important in the balance of nature, are certain molds which ensnare and consume small animals in the soil

Unless man succeeds in duplicating the process of photosynthesis, it appears that animals will always have to feed upon plants. But the plant world exacts its retribution. A number of plants have turned the tables on the animal kingdom, reversing the roles of predator and prey. These are the plant carnivores—plants that trap and consume living animals. Most famous are the pitcher plant, with its reservoir of digestive fluid in which to drown hapless insects; the sundew, with its fly-paper-like leaves; and Venus's-flytrap, with its snapping jaws. But there are other carnivorous plants of larger significance in the balance of nature. We ought to know them better because they are to be found in great profusion and variety in any pinch of forest soil or garden compost. They are microscopic in size, but just as deadly to their animal prey as the sundew or Venus's-flytrap.

These tiny predators are members of the large group of fungi we call molds. They grow in richly branching networks of filaments visible to the naked eye as hairy or velvety mats. Molds do not engage in photosynthesis. Like most bacteria, they lack chlorophyll and so must

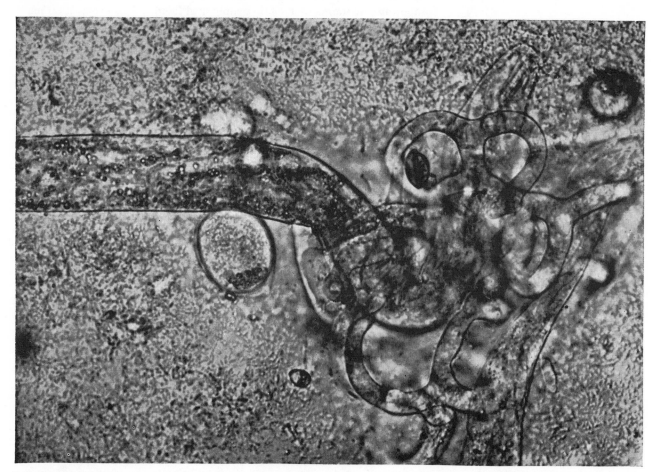

NEMATODE WORM IS TRAPPED by the adhesive fungus *Trichothecium cystosporium* in this photomicrograph by the British biologist C. L. Duddington. The entangling network of the fungus is at right; the body of the worm extends to the left. The oval object below the body of the worm is one of the spores by which the fungus reproduces. At lower right are the remains of another worm.

derive their food from other plants and from animals. Molds have long been familiar as scavengers of dead organisms, promoters of the process of decay. It was not until 1888 that a German mycologist, named Friedrich Wilhelm Zopf, beheld molds in the act of trapping and killing live animals—in this case the larvae of a tiny worm, the wheat-cockle nematode.

The nematodes (eelworms, hookworms and their like) are not the only prey of these animal-eating plants. Their victims run the gamut from the comparatively formidable nematodes down to small crustaceans, rotifers and the lowly amoeba. Charles Drechsler of the U. S. Department of Agriculture, a student of the subject for some 25 years, has identified a large number of carnivorous molds and matched them to their prey. Many are adapted to killing only

one species of animal, and some are equipped with traps and snares which are marvels of genetic resourcefulness. How they evolved their predatory habits and organs remains an evolutionary mystery. These molds belong to quite different species and have in common only their behavior and some similarities of trapping technique. They present a challenging subject for investigation which may throw light on some fundamental questions in biology and may lead also to new methods for control of a number of crop-killing nematodes.

The simplest of the molds have no special organs with which to ensnare their victims. Their filaments, however, secrete a sticky substance which holds fast any small creature that has the misfortune to come in contact with

it. The mold then injects daughter filaments into the body cavity of the victim and digests its contents. Most of the animals caught in this way are rhizopods—sluggish amoebae encased in minute hard shells. Sometimes, however, the big, vigorous soil nematodes are trapped by this elementary means.

More specialized is an unusual water mold, of the genus *Sommerstorffia*, which catches rotifers, its actively swimming prey, with little sticky pegs that branch from its filaments. When a rotifer, browsing among the algae on which this mold grows, takes one of these pegs in its ciliated mouth, it finds itself impaled like a fish on a hook.

Some molds do their trapping in the spore stage. The parent mold produces staggering numbers of sticky spores. When a spore is swallowed by or sticks

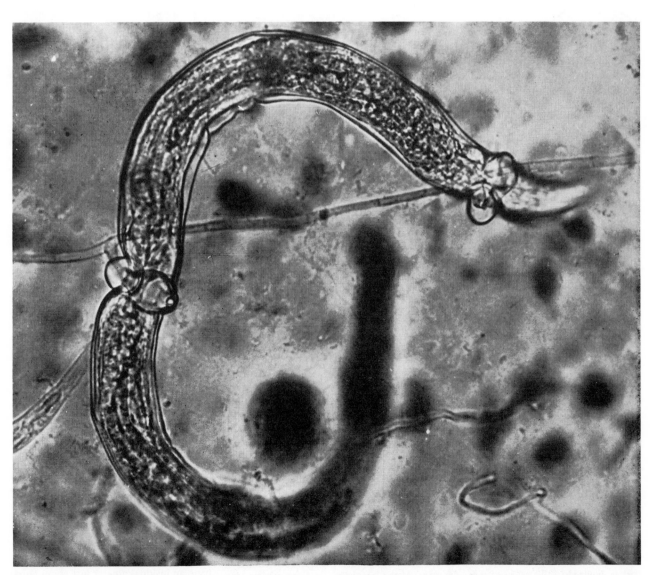

TWO CONSTRICTING RINGS of the fungus *Dactylaria gracilis* grasp another nematode in this photomicrograph by Duddington. The nematode was first caught by the head (*upper right*), and then flicked its body into another ring (*left center*). The rings deeply constrict the body of the worm. The horizontal line above the middle of the picture is a filament to which the rings are attached.

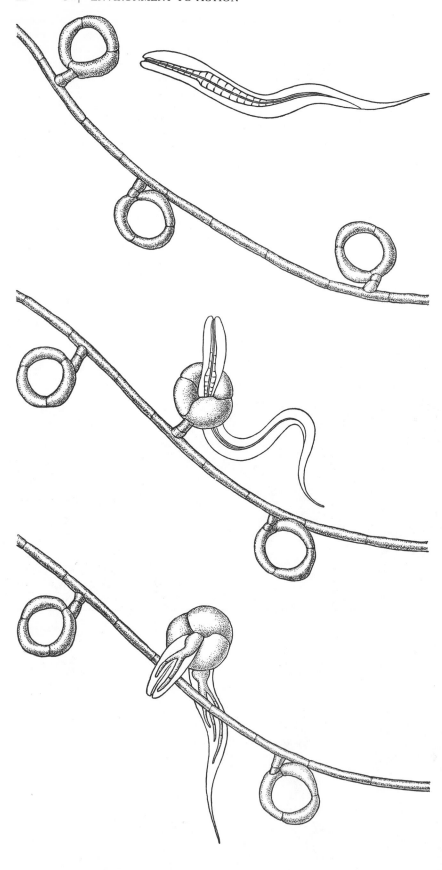

CAPTURE AND CONSUMPTION of a nematode worm by a fungus of the constricting-ring type is depicted in this series of three drawings. At the top the worm approaches a ring attached to a filament of the fungus. In middle the three cells of one ring have expanded to trap the worm. At bottom filaments have branched out of the fungus to digest the worm.

to a passing amoeba or nematode, it germinates in the body of its luckless host and sends forth from the shriveled corpse new filaments and new spores to intercept other victims.

The most remarkable of all killer molds are found among the so-called *Fungi Imperfecti*, or completely asexual fungi. The advanced specialization of these molds is particularly interesting because they are not killers by obligation but can live quite well on decaying organic matter when nematodes, their animal victims, are not available. If nematodes are present, these molds immediately develop highly specialized structures which re-adapt them to a carnivorous way of life. They will do so even if they are merely wetted with water in which nematodes have lived.

One of these molds is *Arthrobotrys oligospora*, the nematode-catching fungus that was first studied by Zopf. When nematodes are available, it develops networks of loops, fused together to form an elaborate nematode trap. An extremely sticky fluid secreted by the mold seems to play an important role in capturing the nematode, which need not even enter the network in order to be held fast. The fluid is so sticky that one-point contact with the network frequently is enough to doom the nematode. In its frenzied struggles to escape, the worm only becomes further entangled in the loops, and finally, after a few hours of exertion, weakens and dies. The destruction these molds can cause in a laboratory culture of nematodes is appalling, particularly from the point of view of the nematode!

Two French biologists, Jean Comandon and Pierre de Fonbrune, have made motion pictures which show that the fungus's secretion of this adhesive substance is accompanied by intense activity in its cells. Material in the cytoplasm of the cell streams toward the point of contact with the worm. The mold may be bringing up reserves of adhesive and digestive enzymes to subdue the nematode; it may also secrete a narcotic or an intoxicant to speed the process.

Even more artfully contrived are the "rabbit snares" employed by some molds. First fully described by Charles Drechsler, these are rings of filament which are attached by short branches to the main filaments, hundreds of them growing on one mold plant. The rings are always formed by three cells and have an inside diameter just about equal to the thickness of a nematode. When a nematode, in its blind wanderings through the soil, has the ill luck to stick its head into one of these rings, the three cells suddenly inflate like a pneumatic tire, gripping it in a stranglehold from which there is no escape.

The rings respond almost instantaneously to the presence of a nematode; in less than one tenth of a second the three cells expand to two or three times their former volume, obliterating the opening of the ring. It is difficult to understand how the delicate filaments can hold the powerfully thrashing worm in so unyielding a grip. Occasionally a muscular worm does escape by breaking the ring off its stalk. But this victory only postpones the inevitable. The ring hangs on like a deadly collar and ultimately generates filaments which invade the worm, kill it and consume it.

We are not yet sure what cellular mechanisms activate these deadly nooses. We know that in the case of the constricting ring of one mold the activating stimulus is the sliding touch of the nematode as it enters the ring. A nematode that touches the outer surface of the ring will not trigger the mechanism. But if the worm passes inside the ring, its doom is certain. This mold, then, exhibits a sharply localized "paratonic" or touch response like that of the Venus's-flytrap.

Perhaps the inflation of the cells is caused by a change in osmotic pressure, resulting in an intake of water either from the environment or from neighboring cells. Or perhaps it results from changes in the colloidal structure of the cell protoplasm. There is a recent report from England that the constricting rings of one species react to acetylcholine—the substance associated with the transmission of impulses across synapses in the animal nervous system!

Some species of molds prey upon root eelworms that infest cereal crops, potatoes and pineapples. This has inspired experiments to use these fungi to control the pests. In one early experiment, conducted in Hawaii by M. B. Linford of the University of Illinois and his associates, a mulch of chopped pineapple tops was added to soil known to harbor the pineapple root-knot eelworm. This mulch produced an increase in the numbers of harmless, free-wandering nematodes which thrive in rich soil. The presence of these decay nematodes stimulated molds in the soil to develop nematode traps, which caught the eelworms as well as the harmless species. A recent experiment in England gives similar promise that the molds may be effective against the cereal root eelworm. Plants protected by stimulated molds showed slight damage compared to the eelworm-ravaged control plants.

Investigators in France have reported an experiment which suggests that molds may be used to control nematode parasites of animals as well as those of plants. Two sheep pens were heavily infested with larvae of a hookworm, closely related to the hookworms of man, which causes severe pulmonary and intestinal damage to sheep. One of the pens was sprinkled with the spores of three molds that employ snares or sticky nets to trap nematodes. Healthy lambs were placed in both pens. After 35 days of exposure the lambs in the pen inoculated with the molds were found free of infection, while those in the control pen showed signs of infestation with the worm.

The carnivorous molds offer many possibilities for future investigation. One subject that needs to be explored is their role in the complex biology of the soil. We would also like to know more about the physiological mechanism that underlies the extraordinary behavior of the nematode "snares." The results of experiments on mold control of nematodes are already encouraging. They suggest that one day these peculiar little plants may perform an even more important role in agriculture than they played in nature, silently and unobtrusively, throughout the millennia before their discovery.

NEMATODE IS INVADED by filaments of a fungus which has trapped it. This photomicrograph was made by David E. Pramer of the New Jersey Agricultural Experiment Station.

2

Life and Light

by George Wald
October 1959

Life depends on a narrow band in the electromagnetic spectrum. This is a consequence of the way in which molecules react to radiation, and must hold true not only on earth but elsewhere in the universe

All life on this planet runs on sunlight, that is, on photosynthesis performed by plants. In this process light supplies the energy to make the organic molecules of which all living things are principally composed. Those plants and animals which are incapable of photosynthesis live as parasites on photosynthetic plants. But light—that form of radiant energy which is visible to the human eye—comprises only a narrow band in the spectrum of the radiant energy that pervades the universe. From gamma rays, which may be only one tenbillionth of a centimeter long, the wavelengths of electromagnetic radiation stretch through the enormous range of 10^{16}—10,000 million million times—up to radio waves, which may be miles in length. The portion of this spectrum that is visible to man is mainly contained between the wavelengths 380 to 760 millimicrons (a millimicron is ten millionths of a centimeter). By using very intense artificial sources one can stretch the limits of human vision somewhat more widely: from about 310 to 1,050 millimicrons. The remarkable fact is that, lying altogether within this slightly wider range of wavelengths, and mainly enclosed between 380 and 760 millimicrons, we also find the vision of all other animals, the bending of plants toward light, the oriented movements of animals toward or away from light and, most important, all types of photosynthesis. This is the domain of photobiology.

Why these wavelengths rather than others? I believe that this choice is dictated by intrinsic factors which involve the general role of energy in chemical reactions, the special role that light energy plays in photochemical reactions, and the nature of the molecules that mediate the utilization of light by living organisms. It is not merely a tautology to say that photobiology requires the particular range of wavelengths we call light. This statement must be as applicable everywhere in the universe as here. Now that many of us are convinced that life exists in many places in the universe (it is hard to see how to avoid this conclusion), we have good reason to believe that everywhere we should find photobiology restricted to about the same range of wavelengths. What sets this range ultimately is not its availability, but its suitability to perform the tasks demanded of it. There cannot be a planet on which photosynthesis or vision occurs in the far infrared or far ultraviolet, because these radiations are not appropriate to perform these functions. It is not the range of available radiation that sets the photobiological domain, but rather the availability of the proper range of wavelengths that decides whether living organisms can develop and light can act upon them in useful ways.

We characterize light by its wave motion, identifying the regions of the spectrum by wavelength or frequency [*see illustration at top of pages 26 and 27*]. But in its interactions with matter—its absorption or emission by atoms and molecules—light also acts as though it were composed of small packets of energy called quanta or photons. These are in fact a class of ultimate particles, like protons and electrons, though they have no electric charge and very little mass. Each photon has the energy content: $E = hc/\lambda$, in which h is Planck's universal constant of action (1.58×10^{-34} calorie seconds), c is the velocity of light (3×10^{10} centimeters per second in empty space) and λ is the wavelength. Thus, while the intensity of light is the rate of delivery of photons, the work that a single photon can do (its energy content) is inversely proportional to its wavelength. With the change in the energy of photons, from one end of the electromagnetic spectrum to the other, their effects upon matter vary widely. For this reason photons of different wavelengths require different instruments to detect them, and the spectrum is divided arbitrarily on this basis into regions called by different names.

In the realm of chemistry the most useful unit for measuring the work that light can do is the "einstein," the energy content of one mole of quanta (6.02×10^{23} quanta). One molecule is excited to enter into a chemical reaction by absorbing one quantum of light; so one mole of molecules can be activated by absorbing one mole of quanta. The energy content of one einstein is equal to 2.854×10^7 gram calories, divided by the wavelength of the photon expressed in millimicrons. With this formula one can easily interconvert wavelength and energy content, and so assess the chemical effectiveness of electromagnetic radiations.

Energy enters chemical reactions in two separate ways: as energy of activation, exciting molecules to react; and as heat of reaction, the change in energy of the system resulting from the reaction. In a reacting system, at any moment, only the small fraction of "hot" molecules react that possess energies equal to or greater than a threshold value called the energy of activation. In ordinary chemical reactions this energy is acquired in collisions with other molecules. In a photochemical reaction the energy of activation is supplied by light. Whether light also does work on the reaction is an entirely separate issue. Sometimes, as in photosynthesis, it does so; at other times, as probably in vision, it seems to do little or no work.

Almost all ordinary ("dark") chemical

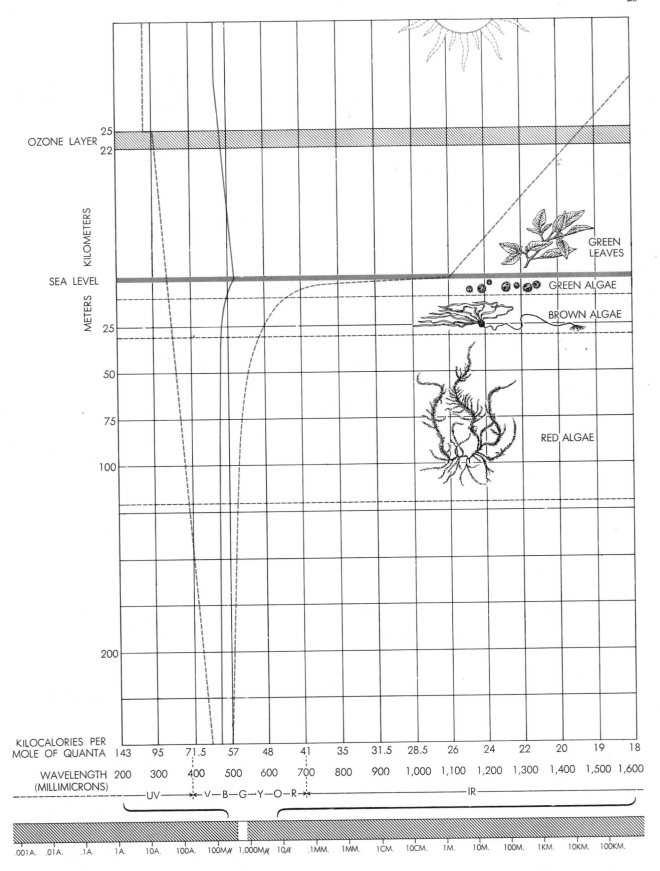

OZONE LAYER

SEA LEVEL

KILOMETERS / METERS

25
22

25

50

75

100

200

GREEN LEAVES

GREEN ALGAE

BROWN ALGAE

RED ALGAE

KILOCALORIES PER MOLE OF QUANTA	143	95	71.5	57	48	41	35	31.5	28.5	26	24	22	20	19	18
WAVELENGTH (MILLIMICRONS)	200	300	400	500	600	700	800	900	1,000	1,100	1,200	1,300	1,400	1,500	1,600

—————— UV ——— ✳ — V — B — G — Y — O — R — ✳ —————————— IR ——————————

.001A. .01A. .1A. 1A. 10A. 100A. 100Mμ 1,000Mμ 10μ .1MM. 1MM. 1CM. 10CM. 1M. 10M. 100M. 1KM. 10KM. 100KM.

SPECTRUM OF SUNLIGHT at the earth's surface is narrowed by atmospheric absorption to the range of wavelengths (from 320 to 1,100 millimicrons) that are effective in photobiological processes. The sunlight reaching the domain of life in the sea is further narrowed by absorption in the sea water. The solid colored line locates the wavelengths of maximum intensity; the broken colored lines, the wavelength-boundaries within which 90 per cent of the solar energy is concentrated at each level in the atmosphere and ocean. The letters above the spectrum of wavelengths at bottom represent ultraviolet (UV), violet (V), blue (B), green (G), yellow (Y), orange (O), red (R) and infrared (IR). Other usages in the chart are explained in the illustration at top of next two pages.

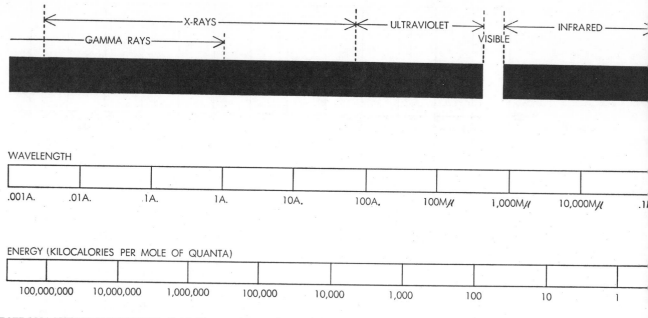

ELECTROMAGNETIC SPECTRUM is divided by man into qualitatively different regions (*top bar*), although the only difference between one kind of radiation and another is difference in wavelength (*middle bar*). From gamma rays, measured here in angstrom units, or hundred millionths of centimeter (A.), through light waves, measured here in millimicrons, or ten millionths of a centi-

reactions involve energies of activation between 15 and 65 kilogram calories (kilocalories) per mole. This is equivalent energetically to radiation of wavelengths between 1,900 and 440 millimicrons. The energies required to break single covalent bonds—a process that, through forming free radicals, can be a potent means of chemical activation—almost all fall between 40 and 90 kilocalories per mole, corresponding to radiation of wavelengths 710 to 320 millimicrons. Finally, there is the excitation of valence electrons to higher orbital levels that activates the reactions classified under the heading of photochemistry; this ordinarily involves energies of about 20 to 100 kilocalories per mole, corresponding to the absorption of light of wavelengths 1,430 to 280 millimicrons. Thus, however one approaches the activation of molecules for chemical reactions, one enters into a range of wavelengths that coincides approximately with the photobiological domain.

Actually photobiology is confined within slightly narrower limits than photochemistry. Radiations below 300 millimicrons (95 kilocalories per mole) are incompatible with the orderly existence of such large, highly organized molecules as proteins and nucleic acids. Both types of molecule consist of long chains of units bound to one another by primary valences. Both types of molecule, however, are held in the delicate and specific configurations upon which their functions in the cell depend by the relatively

weak forces of hydrogen-bonding and van der Waals attraction.

These forces, though individually weak, are cumulative. They hold a molecule together in a specific arrangement, like zippers. Radiation of wavelengths shorter than 300 millimicrons unzips them, opening up long sections of attachment, and permitting the orderly arrangement to become random and chaotic. Hence such radiations denature proteins and depolymerize nucleic acids, with disastrous consequences for the cell. For this reason about 300 millimicrons represents the lower limit of radiations capable of promoting photoreactions, yet compatible with life.

From this point of view we live upon a fortunate planet, because the radiation that is useful in promoting orderly chemical reactions comprises the great bulk of that of our sun. The commonly stated limit of human vision—400 to 700 millimicrons—already includes 41 per cent of the sun's radiant energy before it reaches our atmosphere, and 46 per cent of that arriving at the earth's surface. The entire photobiological range— 300 to 1,100 millimicrons—includes about 75 per cent of the sun's radiant energy, and about 83 per cent of that reaching the earth.

From about 320 to 1,100 millimicrons —virtually the photobiological range— the sun's radiation reaches us with little modification. The atmosphere directly above us causes an attenuation, mainly by scattering rather than absorption of

light, which is negligible at 700 millimicrons and increases exponentially toward shorter wavelengths, so that at 400 millimicrons the radiation is reduced by about half. In the upper atmosphere, however, a layer of ozone, at a height of 22 to 25 kilometers, begins to absorb the sun's radiation strongly at 320 millimicrons, and at 290 millimicrons forms a virtually opaque screen. It is only the presence of this layer of ozone, removing short-wave antibiotic radiation, that makes terrestrial life possible.

At long wavelengths the absorption bands of water vapor cut strongly into the region of solar radiation from 720 to 2,300 millimicrons. Beyond 2,300 millimicrons the infrared radiation is absorbed almost completely by the water vapor, carbon dioxide and ozone of the atmosphere. The sun's radiation, therefore, which starts toward the earth in a band reaching from about 225 to 3,200 millimicrons, with its maximum at about 475 millimicrons, is narrowed by passing through the atmosphere to a range of about 310 to 2,300 millimicrons at the earth's surface.

The differential absorption of light by water confines more sharply the range of illumination that reaches living organisms in the oceans and in fresh water. The infrared is removed almost immediately in the surface layers. Cutting into the visible spectrum, water attenuates very rapidly in succession the red, orange, yellow and green. The short-wavelength limit is also gradually drawn

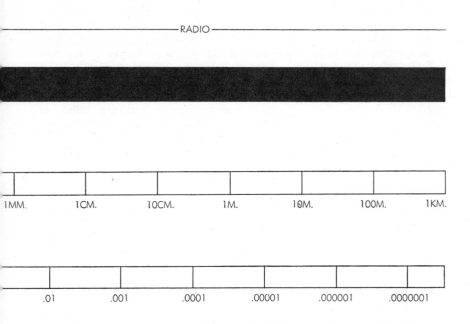

meter (Mμ), the waves range upward in length to the longest radio waves. The difference in wavelength is associated with a decisive difference in the energy conveyed by radiation at each wavelength. This energy content (*bottom bar*) is inversely proportional to wavelength.

matter once every 300 years. All the oxygen in our atmosphere, having been bound by various oxidation processes, is renewed by photosynthesis once in about 2,000 years.

In the original accumulation of this capital of carbon dioxide and oxygen, early in the history of the earth, it is thought that the process of photosynthesis itself profoundly modified the character of the earth's atmosphere and furnished the essential conditions for the efflorescence and evolution of life. Some of the oldest rock formations have lately been discovered to contain recognizable vestiges of living organisms, including what appear to have been photosynthetic forms. So for example iron gunflint cherts found in southern Ontario contain microscopic fossils, among which appear to be colonial forms of blue-green algae. These deposits are estimated to be at least 1.5 billion years old, so that if this identification can be accepted, photosynthesis has existed at least that long on this planet.

It now seems possible that the original development of the use of light by organisms, through the agency of chlorophyll pigments, may have involved not primarily the synthesis of new organic matter, but rather the provision of stores of chemical energy for the cell. A few years ago the process called photosynthetic phosphorylation was discovered, and has since been intensively explored,

in, so that the entire transmitted radiation is narrowed to a band centered at about 475 millimicrons, in the blue.

Photosynthesis

Each year the energy of sunlight, via the process of photosynthesis, fixes nearly 200 billion tons of carbon, taken up in the form of carbon dioxide, in more complex and useful organic molecules: about 20 billion tons on land and almost 10 times this quantity in the upper layers of the ocean. All the carbon dioxide in our atmosphere and all that is dissolved in the waters of the earth passes into this process, and is completely renewed by respiration and the decay of organic

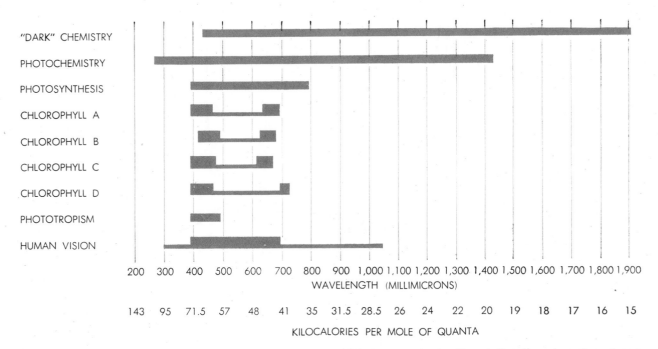

ENERGY CONTENT OF LIGHT is matched to the energy requirements of chemistry and photobiological processes and to the absorption spectra of photoreactive substances. The thicker segments of the bars opposite the chlorophylls indicate the regions of maximum absorption of light in each case, and the thicker segment in the bar opposite human vision indicates the normal boundaries.

mainly by Daniel I. Arnon of the University of California. By a still-unknown mechanism light forms the terminal high-energy phosphate bonds of adenosine triphosphate (ATP), which acts as a principal energy-carrier in the chemistry of the cell. One of the most interesting features of this process is that it is anaerobic; it neither requires nor produces oxygen. At a time when our atmosphere still lacked oxygen, this process could have become an efficient source of ATP. Among the many things ATP does in cells one of the most important is to supply the energy for organic syntheses. This direct trading of

the energy of sunlight for usable chemical energy in the form of ATP would therefore already have had as by-product the synthesis of organic structures. Mechanisms for performing such synthesis directly may have been a later development, leading to photosynthesis proper.

The essence of the photosynthetic process is the use of the energy of light to split water. The hydrogen from the water is used to reduce carbon dioxide or other organic molecules; and, in photosynthesis as performed by algae and higher plants, the oxygen is released into the atmosphere.

We owe our general view of photosynthesis in great part to the work of C. B. van Niel of the Hopkins Marine Station of Stanford University. Van Niel had examined the over-all reactions of photosynthesis in a variety of bacteria. Some of these organisms—green sulfur bacteria—require hydrogen sulfide to perform photosynthesis; van Niel discovered that in this case the net effect of photosynthesis is to split hydrogen sulfide, rendering the hydrogen available to reduce carbon dioxide to sugar, and liberating sulfur rather than oxygen. Still other bacteria—certain nonsulfur purple bacteria, for example—require organic

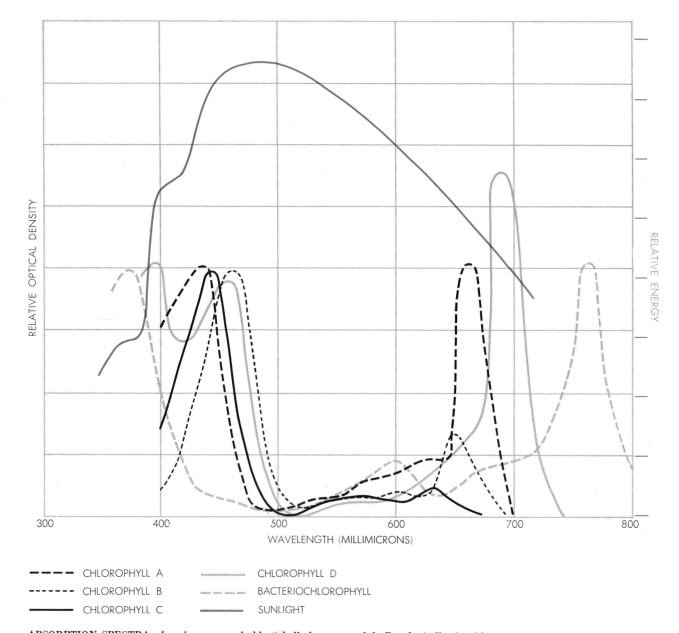

- - - - -	CHLOROPHYLL A
- - - - - -	CHLOROPHYLL B
————	CHLOROPHYLL C
··········	CHLOROPHYLL D
— — —	BACTERIOCHLOROPHYLL
————	SUNLIGHT

ABSORPTION SPECTRA of various types of chlorophyll show the regions of the spectrum in which these substances absorb sunlight most effectively, measured on scale of relative optical density at left. Paradoxically the chlorophylls absorb best at the ends of the spectrum of sunlight, where energy, shown on scale at right, falls off steeply from the maximum around middle of the spectrum.

substances in photosynthesis. Here van Niel found that the effect of photosynthesis is to split hydrogen from these organic molecules to reduce carbon dioxide, liberating in this case neither oxygen nor sulfur but more highly oxidized states of the organic molecules themselves. Finally there are forms of purple bacteria that use molecular hydrogen directly in photosynthesis to reduce carbon dioxide, and liberate no by-product.

The efficiency of photosynthesis in algae and higher green plants is extraordinarily high—just how high is a matter of continuing controversy. The work of reducing one mole of carbon dioxide to the level of carbohydrate is in the neighborhood of 120 kilocalories. This energy requirement, though the exact figure is approximate, cannot be evaded through any choice of mechanism. Thanks to the selective absorption of the green chlorophyll pigment, light is made available for this process in quanta whose energy content is 41 or 42 kilocalories per mole, corresponding to quanta of red light of wavelength about 680 millimicrons. It is apparent, therefore, that several such quanta are required to reduce one molecule of carbon dioxide. If the energy of light were used with perfect efficiency, three quanta might perhaps suffice.

About 35 years ago the great German biochemist Otto Warburg performed experiments which appeared to show that in fact about four quanta of light of any wavelength in the visible spectrum are enough to reduce a molecule of carbon dioxide to carbohydrate. This might have meant an efficiency of about 75 per cent. Later a variety of workers in this country and elsewhere insisted that when such experiments are performed more critically, from eight to 12 quanta are required per molecule of carbon dioxide reduced. This discrepancy led to one of the bitterest controversies in modern science.

Many of us have grown tired of this controversy, which long ago bogged down in technical details and fruitless recriminations. I think it significant, however, that a number of recent, non-Warburgian, investigations have reported quantum demands of about six, and in at least one case the reported demand was as low as five. These numbers represent very high efficiencies (50 to 60 per cent), though not quite as high as Warburg prefers to set them.

Investigators have now turned from the question of efficiency to a more fruitful study of the specific uses to which quanta are put in photosynthesis. This is yielding estimates of quantum demand related to specific mechanisms rather than to controversial details of experimentation.

To reduce one molecule of carbon dioxide requires four hydrogen atoms and apparently three high-energy phosphate bonds of ATP. If we allow one quantum for each hydrogen atom (a point not universally conceded), that yields directly a quantum demand of four. If the ATP can be supplied in other ways, for example by respiration, four may be enough. If, however, light is needed also to supply ATP, by photosynthetic phosphorylation, then more quanta are needed; how many is not yet clear. Yet if one quantum were to generate one phosphate bond, the theoretical quantum-demand of photosynthesis, with all the energy supplied by light, would be four plus three, or seven. That would represent a high order of efficiency in the conversion of the energy of light to the energy of chemical bonds.

It is curious to put this almost obsessive concentration on the efficiency of photosynthesis together with what I think to be one of the most remarkable facts in all biology. Chlorophylls, the pigments universally used in photosynthesis, have absorption properties that seem just the opposite of what is wanted in a photosynthetic pigment. The energy of sunlight as it reaches the surface of the earth forms a broad maximum in the blue-green to green region of the spectrum, falling off at both shorter and longer wavelengths. Yet it is precisely in the blue-green and green, where the energy of sunlight is maximal, that the chlorophylls absorb light most poorly; this, indeed, is the reason for their green color. Where the absorption by chlorophyll is maximal—in widely separated bands in the violet and red—the energy of sunlight has fallen off considerably [see illustration on opposite page].

After perhaps two billion years of selection, involving a process whose efficiency is more important than that of any other process on earth, this seems an extraordinarily poor performance. It is a curious fact to put together with Warburg's comment (at one point in the quantum-demand controversy) that in a perfect nature, photosynthesis also is perfect. I think that the question it raises may be put more usefully as follows: What properties do the chlorophylls have that are so profoundly advantageous for photosynthesis as to override their disadvantageous absorption spectra?

We have the bare beginnings of an answer; it is emerging from a deeper understanding of the mechanism of photosynthesis, in particular as it is expressed in the structure and function of chlorophyll itself. Chlorophyll *a*, the type of chlorophyll principally involved in the photosynthesis of algae and higher plants, owes its color, that is, its capacity for absorbing light, to the possession of a long, regular alternation of single and double bonds, the type of arrangement called a conjugated system [see illustration on next two pages]. All pigments, natural and synthetic, possess such conjugated systems of alternate single and double bonds. The property of such systems that lends them color is the possession of particularly mobile electrons, called pi electrons, which are associated not with single atoms or bonds but with the conjugated system as a whole. It requires relatively little energy to raise a pi electron to a higher level. This small energy-requirement corresponds with the absorption of radiation of relatively long wavelengths, that is, radiation in the visible spectrum; and also with a high probability, and hence a strong intensity, of absorption.

In chlorophyll this conjugated system is turned around upon itself to form a ring of rings, a so-called porphyrin nucleus, and this I think is of extraordinary significance. On the one hand, as the illustration on these two pages shows, it makes possible a large number of rearrangements of the pattern of conjugated single and double bonds in the ring structure. Each such arrangement corresponds to a different way of arranging the external electrons, without moving any of the atoms. The molecule may thus be conceived to resonate among and be a hybrid of all these possible arrangements. In such a structure the pi electrons can not only oscillate, as in a straight-chain conjugated system; they can also circulate.

The many possibilities of resonance, together with the high degree of condensation of the molecule in rings, give the chlorophylls a peculiar rigidity and stability which I think are among the most important features of this type of structure. Indeed, porphyrins are among the most inert and stable molecules in the whole of organic chemistry. Porphyrins, apparently derived from chlorophyll, have been found in petroleum, oil shales and soft coals some 400 million years old.

This directs our attention to special features of chlorophyll, which are directly related to its functions in photosynthesis. One such property is not to utilize the energy it absorbs immediately in

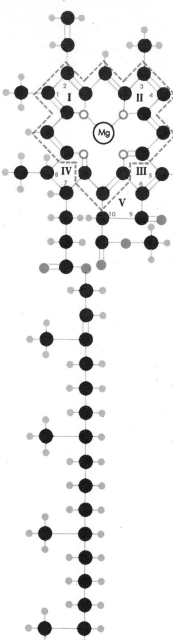

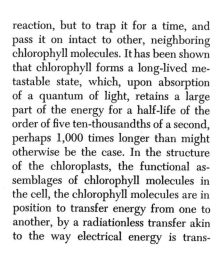

● CARBON   (Mg) MAGNESIUM

● OXYGEN   ○ NITROGEN

● HYDROGEN

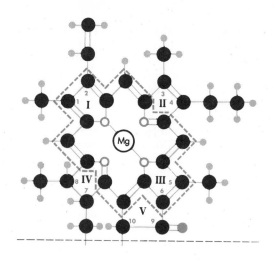

CHLOROPHYLL MOLECULE, diagrammed in its entirety at left, owes its photobiological activity to the rigid and intricate porphyrin structure at top. The arrangement of the bonds in this structure may resonate among the configurations

reaction, but to trap it for a time, and pass it on intact to other, neighboring chlorophyll molecules. It has been shown that chlorophyll forms a long-lived metastable state, which, upon absorption of a quantum of light, retains a large part of the energy for a half-life of the order of five ten-thousandths of a second, perhaps 1,000 times longer than might otherwise be the case. In the structure of the chloroplasts, the functional assemblages of chlorophyll molecules in the cell, the chlorophyll molecules are in position to transfer energy from one to another, by a radiationless transfer akin to the way electrical energy is trans-

ferred in an induction motor. This capacity for transferring the energy about, so that it virtually belongs to a region of the chloroplast rather than to the specific molecule of chlorophyll that first absorbed it, makes possible the high efficiency of photosynthesis. While photosynthesis is proceeding rapidly, many chlorophyll molecules, having just reacted, are still in position to absorb light, but not to utilize it. In this way large amounts of absorbed energy that would otherwise be degraded into heat are retained and passed about intact until used photosynthetically.

One sign of the capacity to retain the energy absorbed as light and pass it on relatively intact is the strong fluorescence exhibited by chlorophyll. This green or blue-green pigment fluoresces red light; and however short the wavelengths that are absorbed—that is, however large the quanta—the same red light is fluoresced, corresponding to quanta of energy content about 40 kilocalories per mole. This is the quantity of energy that is passed from molecule to molecule in the chloroplast and eventually made available for photosynthesis.

The generally inert structure of chlorophyll must somewhere contain a chemically reactive site. Such a site seems to exist in the five-membered carbon ring, usually designated ring V in the structural diagrams. James Franck of the University of Chicago some years ago called attention to the possibility that it is here that the reactivity of chlorophyll is localized. Recent experiments by Wolf

Vishniac and I. A. Rose of Yale University, employing the radioactive isotope of hydrogen (tritium), have shown that chlorophyll, both in the cell and in solution, can take up hydrogen in the light though not in the dark, and can transfer it to the coenzyme triphosphopyridine nucleotide, which appears to be principally responsible for transferring hydrogen in photosynthesis. There is some evidence to support Franck's suggestion that the portion of chlorophyll involved in these processes is the five-membered ring.

Chlorophyll thus possesses a triple combination of capacities: a high receptivity to light, an inertness of structure permitting it to store the energy and relay it to other molecules, and a reactive site equipping it to transfer hydrogen in the critical reaction that ultimately binds hydrogen to carbon in the reduction of carbon dioxide. I would suppose that these properties singled out the chlorophylls for use by organisms in photosynthesis in spite of their disadvantageous absorption spectrum.

Photosynthetic organisms cope with the deficiencies of chlorophyll in a variety of ways. In 1883 the German physiologist T. W. Engelmann pointed out that in the various types of algae other pigments must also function in photosynthesis. Among these are the carotenoid pigments in the green and brown algae, and the phycobilins, phycocyanin and phycoerythrin (related to the animal bile-pigments) in the red and blue-green algae. Engelmann showed that each type

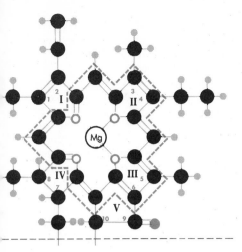

diagrammed in the middle and at the right. These and other possible configurations of the bonds help to make it possible for the chlorophyll molecule to trap and store energy which is conveyed to it by light quanta.

of alga photosynthesizes best in light of the complementary color: green algae in red light, brown algae in green light, red algae in blue light. He pointed out that this is probably the basis of the layering of these types of algae at various depths in the ocean.

All these pigments act, however, by transferring the energy they absorb to one another and eventually to chlorophyll *a;* whatever pigments have absorbed the light, the same red fluorescence of chlorophyll *a* results, with its maximum at about 670 to 690 millimicrons. The end result is therefore always the same: a quantum with an energy content of about 40 kilocalories per mole is made available to chlorophyll *a* for photosynthesis. The accessory pigments, including other varieties of chlorophyll, perform the important function of filling in the hole in the absorption spectrum of chlorophyll.

Still another device helps to compensate for the failure of chlorophyll to absorb green and blue-green light efficiently: On land and in the sea the concentration of chlorophyll and the depth of the absorbing layer are maximized by plant life. As a result chlorophyll absorbs considerable energy even in the wavelengths at which its absorption is weakest. Leaves absorb green light poorly, yet they do absorb a fraction of it. One need only look up from under a tree to see that the cover of superimposed leaves permits virtually no light to get through, green or otherwise. The lower leaves on a tree, though plentifully supplied with chloro-

plasts, may receive too little light to contribute significantly to photosynthesis. By being so profligate with the chlorophylls, plants compensate in large part for the intrinsic absorption deficiencies of this pigment.

Phototropism

The phototropism of plants—their tendency to bend toward the light—is excited by a different region of the spectrum from that involved in photosynthesis. The red wavelengths, which are most effective in photosynthesis, are wholly ineffective in phototropism, which depends upon the violet, blue and green regions of the spectrum. This relationship was first demonstrated early in the 19th century by a worker who reported that when he placed a flask of port wine between a growing plant and the light from a window, the plant grew about as well as before, but no longer bent toward the light. Recently, more precise measurements with monochromatic lights have shown that the phototropism of both molds and higher plants is stimulated only by light of wavelengths shorter than approximately 550 millimicrons, lying almost completely within the blue-green, blue and violet regions of the spectrum.

Phototropism must therefore depend on yellow pigments, because only such pigments absorb exclusively the short wavelengths of the visible spectrum. All types of plant that exhibit phototropism appear to contain such yellow pigments, in the carotenoids. In certain instances the carotenoids are localized specifically in the region of the plant that is phototropically sensitive. The most careful measurements of the effectiveness of various wavelengths of light in stimulating phototropism in molds and higher plants have yielded action spectra which resemble closely the absorption spectra of the carotenoids that are present.

A number of lower invertebrates—for example, hydroids, marine organisms that are attached to the bottom by stalks —bend toward the light by differential growth, just as do plants. The range of wavelengths which stimulate this response is also about the same as that in plants. It appears that here also carotenoids, which are usually present in considerable amount, may be the excitatory agents. Phototactic responses, involving motion of the whole animal toward or away from the light, also abound throughout all groups of invertebrates. Unfortunately no one has yet correlated accurately the action spectra for such re-

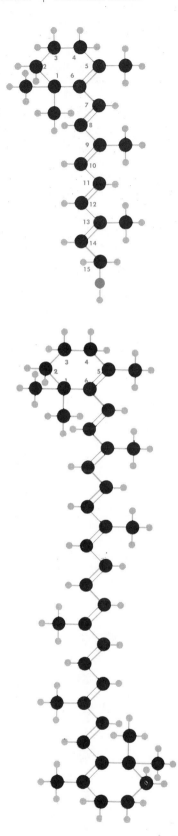

CAROTENE MOLECULE (*bottom*) is probable light-receptor in phototropism and is synthesized by plants. In structure it is a double vitamin A molecule (*top*). Vitamin A, in turn, is precursor of retinine molecule (*illustration on page 32*), which mediates vision.

sponses with the absorption spectra of the pigments that are present, so that no rigorous identification of the excitatory pigments can be made at present. This is a field awaiting investigation.

Vision

Only three of the 11 major phyla of animals have developed well-formed, image-resolving eyes: the arthropods (insects, crabs, spiders), mollusks (octopus, squid) and vertebrates. These three types of eye are entirely independent developments. There is no connection among them, anatomical, embryological or evolutionary. This is an important realization, for it means that three times, in complete independence of one another, animals on this planet have developed image-forming eyes.

It is all the more remarkable for this reason that in all three types of eye the chemistry of the visual process is very nearly the same. In all cases the pigments which absorb the light that stimulates vision are made of vitamin A, in the form of its aldehyde, retinene, joined with specific retinal proteins called opsins. Vitamin A $(C_{20}H_{29}OH)$ has the structure of half a beta-carotene $(C_{40}H_{56})$, with a hydrogen and a hydroxyl radical (OH) added at the broken double bond.

Thus animal vision not only employs substances of the same nature as the carotenoids involved in phototropism of plants; there is also a genetic connection. Animals cannot make vitamin A *de novo*, but derive it from the plant carotenoids

consumed in their diet. All photoreception, from phototropism in lower and higher plants to human vision, thus appears to depend for its light-sensitive pigments upon the carotenoids.

The role of light in vision is fundamentally different from its role in photosynthesis. The point of photosynthesis is to use light to perform chemical work, and the more efficiently this conversion is accomplished, the better the process serves its purpose. The point of vision is excitation; there is no evidence that the light also does work. The nervous structures upon which the light acts, so far as we know, are ready to discharge, having been charged through energy supplied by internal chemical reactions. Light is required only to trigger their responses.

Because this distinction is not always understood, attempts are frequently made to force parallels between vision and photosynthesis. In fact, these processes differ so greatly in their essential natures that no deep parallelism can be expected. The problem of quantum demand, for example, raises entirely different issues in vision as compared with photosynthesis. In photosynthesis one is interested in the minimum number of quanta needed to perform a given chemical task. In vision the problem hinges not on energetic efficiency but on differential sensitivity. The light intensities within which animals must see range from starlight to noonday sunlight; the latter is about a billion times brighter than the former. It is this enormous range of intensities that presents organ-

isms with their fundamental visual problem: how to see at the lowest intensities without having vision obliterated by glare at the highest.

In the wholly dark-adapted state a vertebrate rod, the receptor principally involved in night vision, can respond to the absorption of a single quantum of visible light. To be sure, in the human eye, in which this relationship has been studied most completely, this minimal response of a single rod does not produce a visual sensation. In the dark-adapted state, seeing requires that at least five such events occur almost simultaneously within a small area of retina. This arrangement is probably designed to place the visual response above the "noise level" of the retina. From careful electrophysiological measurements it seems that a retina, even in total darkness, transmits a constant barrage of randomly scattered spontaneous responses to the brain. If the response of a single rod entered consciousness, we should be seeing random points of light flickering over the retina at all times.

The eye's extraordinary sensitivity to light is lost as the brightness of the illumination is increased. The threshold of human vision, which begins at the level of a few quanta in the dark-adapted state, rises as the brightness of the light increases until in bright daylight one million times more light may be needed just to stimulate the eye. But the very low quantum-efficiency in the light-adapted condition nonetheless represents a high visual efficiency.

The statement that the limits of human vision are 380 and 760 millimicrons is actually quite arbitrary. These limits are the wavelengths at which the visual sensitivity has fallen to about a thousandth of its maximum value. Specific investigations have pursued human vision to about 312 millimicrons in the near ultraviolet, and to about 1,050 in the near infrared.

In order to see at 1,050 millimicrons, however, 10,000 million times more light energy is required if cones are being stimulated, and over a million million times more energy if rods are being stimulated. This result came out of measurements made in our laboratory at Harvard University during World War II in association with Donald R. Griffin and Ruth Hubbard. As we exposed our eyes to flashes of light in the neighborhood of 1,000 millimicrons, we could not only see the flash but feel a momentary flush of heat on the cornea of the eye. At about 1,150 millimicrons, just a little farther into the infrared than our ex-

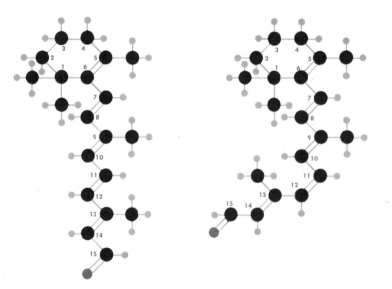

RETINENE MOLECULE is the active agent in the pigments of vision. Upon absorbing energy of light the geometry of the molecule changes from the so-called *cis* arrangement at left to the *trans* arrangement at right. This change in structure triggers process of vision.

periments had taken us, the radiation should have become a better stimulus as heat than as light.

The ultraviolet boundary of human vision, as that of many other vertebrates, raises a special problem. Ordinarily our vision is excluded from the ultraviolet, not primarily because the retina or its visual pigments are insensitive to that portion of the spectrum, but because ultraviolet light is absorbed by the lens of the eye. The human lens is yellow in color and grows more deeply yellow with age. One curious consequence of this arrangement is that persons who have had their lenses removed in the operation for cataract have excellent ultraviolet vision.

One may wonder how it comes about that man and many other vertebrates have been excluded from ultraviolet vision by the yellowness of their lenses. Actually this effect is probably of real advantage. All lens systems made of one material refract shorter wavelengths more strongly than longer wavelengths, and so bring blue light to a shorter focus than red. This phenomenon is known as chromatic aberration, or color error, and even the cheapest cameras are corrected for it. In default of color correction the lens seems to do the next best thing; it eliminates the short wavelengths of the spectrum for which the color error is greatest.

One group of animals, however, makes important use of the ultraviolet in vision. These are the insects. The insect eye is composed of a large number of independent units, the ommatidia, each of which records a point in the object, so that the image as a whole is composed as a mosaic of such points. Projection by a lens plays no part in this system, and chromatic aberration is of no account.

How does it happen that whenever vision has developed on our planet, it has come to the same group of molecules, the A vitamins, to make its light-sensitive pigments? I think that one can include plant phototropism in the same question, and ask how it comes about that all photoreception, animal and plant, employs carotenoids to mediate excitation by light. We have already asked a similar question concerning the chlorophylls and photosynthesis; and what chlorophylls are to photosynthesis, carotenoids are to photoreception.

Both the carotenoids and chlorophylls owe their color to the possession of conjugated systems. In the chlorophylls these are condensed in rings; in the carotenoids they are mainly in straight

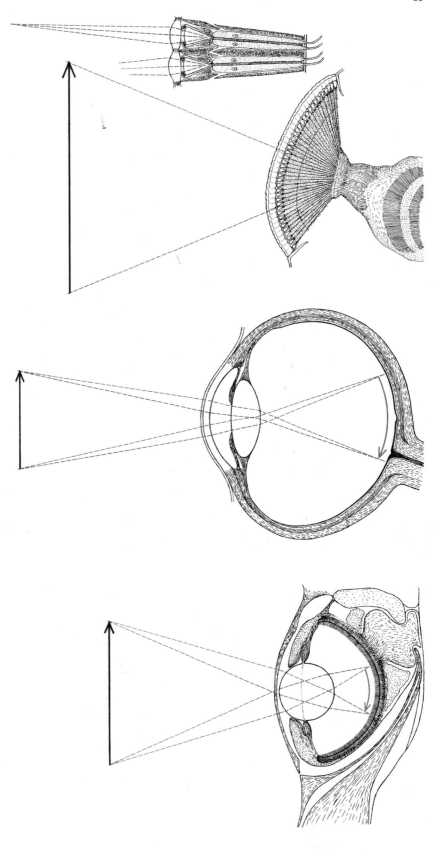

EYES of three kinds have evolved quite independently in three phyla: insects (*top*), vertebrates (*center*) and mollusks (*bottom*). In all three types of eye, however, the chemistry of vision is mediated by retinene derived from the carotenoids synthesized by plants.

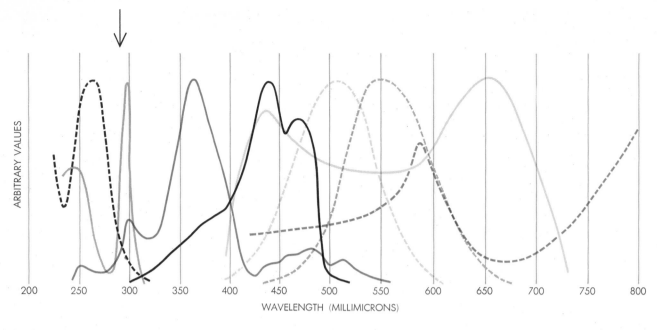

A - - - - -   E ————

B ~~~~~~~   F - - - - -

C ————   G - - - - -

D ————   H - - - - -

PHOTOBIOLOGICAL PROCESSES are activated by different regions of the spectrum: killing of a bacillus (A), sunburn in human skin (B), insect vision (C), phototropism in an oat plant (D), photosynthesis in wheat (E), human "night" vision (F), human "day" vision (G), photosynthesis in a bacterium (H). Arrow at left marks limit of solar short waves.

chains. The chlorophylls fluoresce strongly; the carotenoids, weakly or not at all. Much of the effectiveness of the chlorophylls in photosynthesis is associated with a high capacity for energy transfer; there is as yet no evidence that such energy transfer has a place in vision.

I think that the key to the special position of the carotenoids in photoreception lies in their capacity to change their shapes profoundly on exposure to light. They do this by the process known as *cis-trans* isomerization. Whenever two carbon atoms in a molecule are joined by a single bond, they can rotate more or less freely about this bond, and take all positions with respect to each other. When, however, two carbon atoms are joined by a double bond, this fixes their position with respect to each other. If now another carbon atom is joined to each of this pair, both the new atoms may attach on the same side of the double bond (the *cis* position) or on opposite sides, diagonally (the *trans* position). These are two different structures, each of them stable until activated to undergo transformation—isomerization—into the other.

Carotenoids, possessing as they do long straight chains of conjugated double bonds, can exist in a great variety of such *cis-trans* or geometrical isomers. No other natural pigments approach them in this regard. Porphyrins and other natural pigments may have as

many or more double bonds, but are held in a rigid geometry by being bound in rings.

Cis-trans isomerization involves changes in shape. The all-*trans* molecule is relatively straight, whereas a *cis* linkage at any point in the chain represents a bend. In the composition of living organisms, which depends in large part on the capacity of molecules to fit one another, shape is all-important.

We have learned recently that all the visual pigments known, in both vertebrate and invertebrate eyes, are made with a specifically bent and twisted isomer of retinene. Only this isomer will do because it alone fits the point of attachment on the protein opsin. The intimate union thus made possible between the normally yellow retinene and opsin greatly enhances the color of the retinene, yielding the deep-orange to violet colors of the visual pigments. The only action of light upon a visual pigment is to isomerize—to straighten out—retinene to the all-*trans* configuration. Now it no longer fits opsin, and hence comes away. The deep color of the visual pigment is replaced by the light yellow color of free retinene. This is what is meant by the bleaching of visual pigments by light.

In this succession of processes, however, it is some process associated with the *cis-trans* isomerization that excites vision. The subsequent cleavage of reti-

nene from opsin is much too slow to be responsible for the sensory response. Indeed, in many animals the visual pigments appear hardly to bleach at all. This seems to be the case in all the invertebrate eyes yet examined, in which the entire transformation in light and darkness appears to be restricted to the isomerization of retinene. It seems possible that similar *cis-trans* isomerizations of carotenoid pigments underlie phototropic excitation in plants. Experiments are now in progress in our laboratory to explore this possibility.

Bioluminescence

In addition to responding to light in their various ways, many bacteria, invertebrates and fishes also produce light. All bioluminescent reactions require molecular oxygen; combustions of one kind or another supply the energy that is emitted as light. In photosynthesis light performs organic reductions, releasing oxygen in the process. In bioluminescence the oxidation of organic molecules with molecular oxygen emits light. I used to think that bioluminescence is like vision in reverse; but in fact it is more nearly like photosynthesis in reverse.

What function bioluminescence fulfills in the lives of some of the animals that display it is not yet clear. The flashing of fireflies may act as a signal for integrating their activities, and perhaps as a sexual excitant. What role may be fulfilled by the extraordinary display of red, green and yellow illumination in a railroad worm is altogether conjectural. There is one major situation, however, in which bioluminescence must play an exceedingly important role. This is in the sea, at depths lower than those reached by surface light, and at night at all depths. It would be difficult otherwise to understand how fishes taken from great depths, far below those to which light from the surface can penetrate, frequently have very large eyes. For vision at night or at great depths, it is not necessary that the organisms and objects that are visible themselves be bioluminescent. Bioluminescent bacteria abound in the ocean, and many submerged objects are coated sufficiently with luminous bacteria to be visible to the sensitive eye.

It has lately been discovered that the rod vision of deep-sea fishes is adapted to the wavelengths of surface light that penetrate most deeply into the water: the blue light centered around 475 millimicrons. Furthermore, sensitive new devices for measuring underwater illumination have begun to reveal the remarkable fact that deep-sea bioluminescence may also be most intense at about 475 millimicrons. The same selection of visual pigments that best equips deep-sea fishes to see by light penetrating from the surface seems best adapted to the bioluminescent radiation.

Just as light quanta must be of a certain size to activate or provide the energy for useful chemical reactions, so chemical reactions emit light in the same range of wavelengths. It is for this reason, and no accident, that the range of bioluminescent radiations coincides well with the range of vision and other photobiological processes.

Light and Evolution

The relationship between light and life is in one important sense reciprocal. Over the ages in which sunlight has activated the processes of life, living organisms have modified the terrestrial environment to select those wavelengths of sunlight that are most compatible with those processes. Before life arose, much more of the radiation of the sun reached the surface of the earth than now. We believe this to have been because the atmosphere at that time contained very little oxygen (hence negligible amounts of ozone) and probably very little carbon dioxide. Very much more of the sun's infrared and hard ultraviolet radiation must have reached the surface of

the earth then than now.

Some of the short-wave radiation, operating in lower reaches of the atmosphere and also probably in the surface layers of the seas, must have been important in activating the synthesis and interactions of organic molecules which formed the prelude to the eventual emergence of the first living organisms. These organisms, coming into an anaerobic world, surrounded by the organic matter that had accumulated over the previous ages, must have lived by fermentation, and in this process must have produced as a by-product very large quantities of carbon dioxide. Part of this remained dissolved in the oceans; part entered the atmosphere.

Eventually the availability of large amounts of carbon dioxide, much larger than are in the atmosphere today, made possible the development of the process of photosynthesis. This began to remove carbon dioxide from the atmosphere, fix-

ing it in organic form. Simultaneously, through the most prevalent and familiar form of photosynthesis, it began to produce oxygen, and in this way oxygen first became established in our atmosphere. As oxygen accumulated, the layer of ozone that formed high in the atmosphere—itself a photochemical process—prevented the short-wavelength radiation from the sun from reaching the surface of the earth. This relief from antibiotic radiation permitted living organisms to emerge from the water onto the land.

The presence of oxygen also led to the development of the process of cellular respiration, which involves gas exchanges just the reverse of those of photosynthesis. Eventually respiration and photosynthesis came into approximate balance, as they must have been for some ages past.

One may wonder how much of this history could have occurred in darkness,

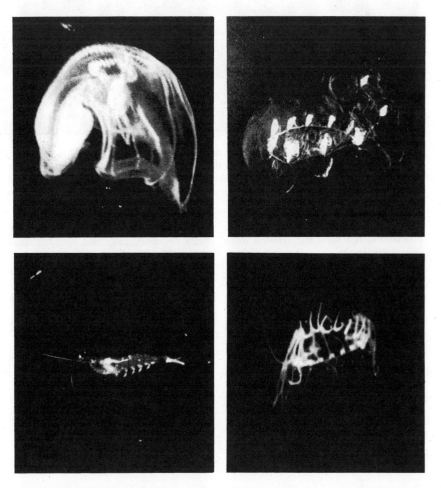

BIOLUMINESCENT CREATURES of the ocean were made to take these photographs of themselves by means of a camera designed by Harold E. Edgerton of Massachusetts Institute of Technology and L. R. Breslau of the Woods Hole Oceanographic Institution. The feeble luminescence of the animals was harnessed to trigger a high-speed electronic flash.

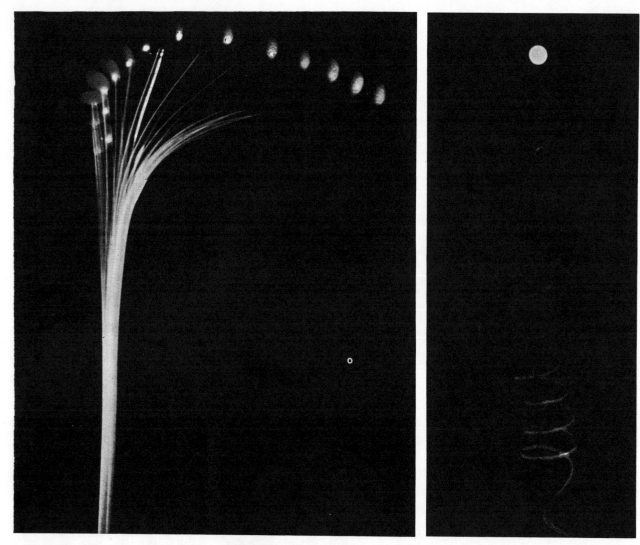

PHOTOTROPISM in the fruiting body of the mold *Phycomyces* is demonstrated in these photographs from the laboratory of Max Delbrück at the California Institute of Technology. The multiple photograph at left, with exposures made at intervals of five minutes, shows the fruiting body growing toward the light source. In the photograph at right, the stalk of the fruiting body has been made to grow in an ascending spiral by placing it on a turntable which revolved once every two hours in the presence of a fixed light-source.

by which I mean not merely the absence of external radiation but a much more specific thing: tne absence of radiation in the range between 300 and 1,100 millimicrons. A planet without this range of radiation would virtually lack photochemistry. It would have a relatively inert surface, upon which organic molecules could accumulate only exceedingly slowly. Granted even enough time for such accumulation, and granted that eventually primitive living organisms might form, what then? They could live for a time on the accumulated organic matter. But without the possibility of photosynthesis how could they ever become independent of this geological heritage and fend for themselves? Inevitably they must eventually consume the organic molecules about them, and with that life must come to an end.

It may form an interesting intellectual exercise to imagine ways in which life might arise, and having arisen might maintain itself, on a dark planet; but I doubt very much that this has ever happened, or that it can happen.

Small Systems of Nerve Cells

by Donald Kennedy
May 1967

*In some invertebrate animals complete behavioral
functions may be controlled by a very few cells. This
makes it possible to trace out the interactions of the
cells and so to investigate nervous integration*

The nervous system of a man comprises between 10 billion and 100 billion cells, and the "lower" mammals men study in an effort to understand their own brains may have two or three billion nerve cells. Even specialized parts of vertebrate nervous systems have an awesome number of elements: the retina of the eye has about 130 million receptor cells and sends more than a million nerve fibers to the brain; a single segment of spinal cord controls the few muscles it operates through several thousand motoneurons, or motor nerve cells, which in turn receive instructions from a larger number of sensory elements.

These vast populations of cells present a formidable challenge to biologists trying to understand how the nervous system works. Since the system is made up of cells, one would like to understand it in terms of cellular activities, and by examining the activity of single nerve cells investigators have been able to learn a great deal about the nature of the nerve impulse and about the generation and transmission of the patterns of impulses that constitute nervous signals. The ultimate object must be, however, to understand not only the activities of single cells but also the rules of their interaction. Since one cannot expect to understand even the most restricted systems by predicting the possible interactions of an inadequately sampled population of cells, mammalian physiologists have devised ingenious ways of circumventing the superabundance of elements, involving particularly biochemical studies of regions of the brain and sophisticated computer analyses of brain waves. The trouble is that most of these methods treat cells as anonymous members of a population rather than as interacting individuals.

Some biologists are taking a different approach, one that retains the individual nerve cell as the focus of attention and yet attempts to deal with groups of them. This approach is made possible by the availability of animals that have fewer nerve cells than any vertebrate and that nonetheless display reasonably complex behavior. The claw-bearing limb of the shore crab, for example, shows impressive coordination and range of movement, made possible by six movable joints with pairs of muscles acting in opposition to each other. As a mechanical device it is in most ways the equal of a mammalian limb, yet the crab operates all this machinery with about two dozen motor nerve cells. A mammal of comparable size would employ several thousand for the analogous purpose.

Nor is such parsimony confined to the motor apparatus: in contrast to the several billion cells of the entire mammalian nervous system, the crab has only half a million or so. The real utility of such systems to the investigator becomes apparent only when one concentrates on the nerve cells belonging to one functional unit such as a reflex pathway, a special sense organ or a particular pattern of behavior. Indeed, the nervous systems of some of the higher invertebrate animals are so economically built that for certain functions one may hope to specify the activity of every individual cell.

This achievement would be a hollow one insofar as learning about mammalian nervous systems is concerned if the additional complexity of more "advanced" systems depended heavily on new capabilities of the individual cells comprising them. All the evidence, however, indicates that the performance limits of the single nerve cell are already reached in relatively simple animals.

The central nervous elements found in the lobster or the sluglike "sea hare" nearly equal those of mammals in size, quantity of input and structural complexity. The marvelous performance of the mammalian brain is, it appears, not so dependent on the individual capabilities of its cells as it is on their greater number and the resulting opportunities for permutation. Therefore an understanding of the connections underlying behavior in a simple system can lead to useful conclusions about the organization of much more complicated ones.

The difficulty lies in the choice of appropriate experimental objects. Ideally one needs a nervous system that produces a reasonably complex repertory of behavior and has only a few cells, each of which can be recognized and located time after time. In certain animals specialized giant cells offer this ready identifiability. The giant axons, or nerve fibers, of the squid and the lobster are an example, and biophysicists have long exploited them for experiments on the properties of nerve-cell membranes. A student of integrative processes in the nervous system needs more; he wants to specify individual properties for each cell in an entire functional assembly.

Two different kinds of nervous system seem particularly promising for this purpose: that of mollusks and that of arthropods. The most thoroughly studied mollusk is the sea hare *Aplysia*, a snail that has only a vestigial shell and leads a somewhat more mobile life than its relatives. Its nervous system is concentrated in a few large ganglia connected by nerve trunks. Several features of the cells are advantageous. The cell bodies are unusually large, some with a diameter of .8 millimeter and the rest well sorted in size below this maximum. They contain

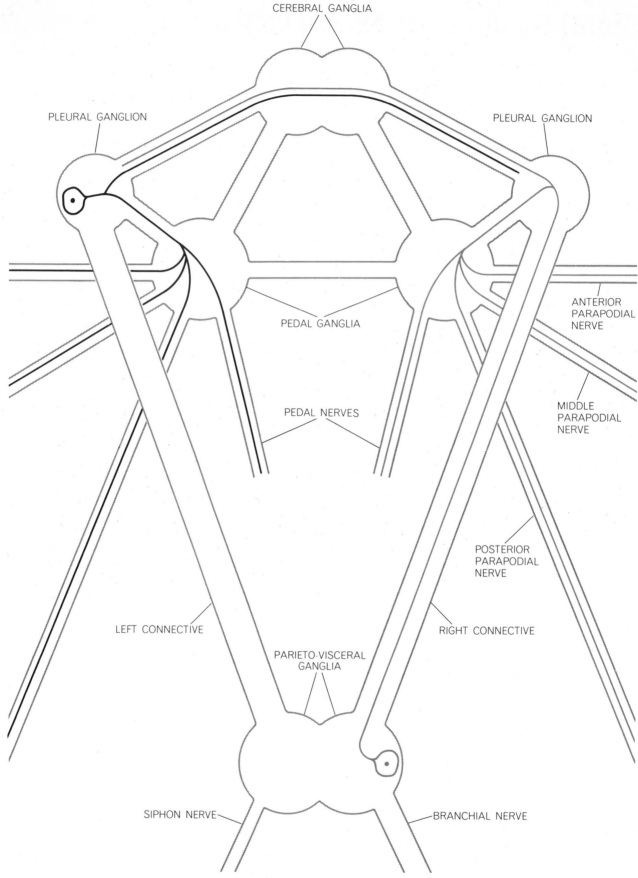

CEREBRAL GANGLIA

PLEURAL GANGLION

PLEURAL GANGLION

PEDAL GANGLIA

ANTERIOR
PARAPODIAL
NERVE

PEDAL NERVES

MIDDLE
PARAPODIAL
NERVE

POSTERIOR
PARAPODIAL
NERVE

LEFT CONNECTIVE

RIGHT CONNECTIVE

PARIETO-VISCERAL
GANGLIA

SIPHON NERVE

BRANCHIAL NERVE

NERVOUS SYSTEM of the "sea hare" *Aplysia*, a mollusk, lends itself to investigation because its cells are few, large and identifiable. Here the entire system is diagrammed and the paths of two nerve cells are shown in color. One of them (*red*) is cell No. 1 in the ganglion illustrated on the opposite page. The routes of a large number of *Aplysia* cells were worked out by L. Tauc of the Centre National de Recherche Scientifique in Paris and G. M. Hughes of Bristol University by recording from individual nerve cell bodies.

a variety of yellow and orange pigments in different proportions. Particular cells are consistent in position from one animal to the next. Several dozen cells can therefore be reliably recognized in different individuals by their size, color and position [*see illustrations on this page*].

Aplysia was first intensively studied in the late 1940's by A. Arvanitaki-Chalazonitis in her laboratory in Monaco. It has since attracted a number of investigators, notably L. Tauc of the Centre National de Recherche Scientifique in Paris, Eric R. Kandel of New York University and Felix Strumwasser of the California Institute of Technology. Tauc and his collaborators, notably G. M. Hughes of Bristol University, have constructed ingenious physiological maps of several *Aplysia* cells [*see illustration on opposite page*]. They accomplished this by inserting a glass microelectrode into a cell body and placing wire electrodes on all the nerve trunks connecting with the ganglion in which the cell body was located. If a particular nerve contained an axon of the cell in which the microelectrode was located, the microelectrode recorded an impulse when that nerve was stimulated with a brief electric shock. The branches of the axons may be arranged in an extremely complex way, but each cell is characterized by a constant arrangement of branches.

Other workers have demonstrated that the connections made in turn by such branches with other nerve cells are also constant. Microelectrodes were inserted in several identified cells, and one microelectrode was used to stimulate the cell it had penetrated while the others recorded impulse activity. A particular cell produced either of two kinds of effect in a nerve cell to which it was connected: an excitation, sometimes strong enough to evoke an impulse in the second cell, or an inhibition, which opposed the discharge of impulses. Strumwasser and Kandel have demonstrated that a single cell can directly excite some cells and inhibit others, and that a certain set of identifiable cells is always excited by a given cell and another set always inhibited.

The ganglia of *Aplysia* have been used in the investigation of two other important problems in nerve physiology: the cellular modifications that take place during "conditioning" and the origin of "discharge rhythms." To investigate the first, Kandel and Tauc recorded from an identifiable cell with a microelectrode while stimulating two nerve trunks containing nerve fibers that excited the recorded cell. A shock delivered to one of

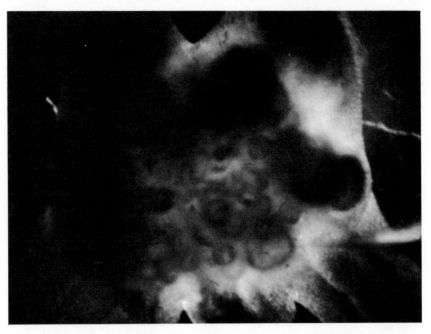

PARIETO-VISCERAL GANGLION of *Aplysia*, photographed unstained by Felix Strumwasser of the California Institute of Technology, measures about 2.5 millimeters across.

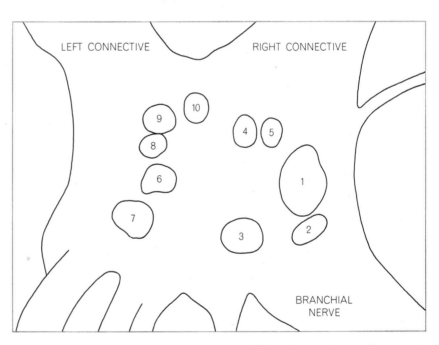

INDIVIDUAL CELLS that are consistently identifiable in the ganglion are diagrammed. Not all are clearly seen in the photograph, in which a microelectrode points to cell No. 3.

these nerves produced strong excitation, indicating that it was the more effective pathway; the other produced only a weak effect. If the shocks to the two pathways were repeatedly paired so that the weak followed the strong at a fixed, short interval, the response to the weak pathway became "conditioned" to the strong excitation; the response increased dramatically and stayed elevated for 15 minutes or more after the conditioning period. If the same number of shocks

was delivered to both pathways for the same period but at random intervals, conditioning did not occur; the response to the weak pathway was not affected. Such systems promise that electrophysiological studies on learning may at last be brought down to the cellular level.

Nerve cells often discharge rhythmically, in bursts that last for a few seconds and are separated by longer intervals. Strumwasser discovered a particularly dramatic instance of rhythmicity

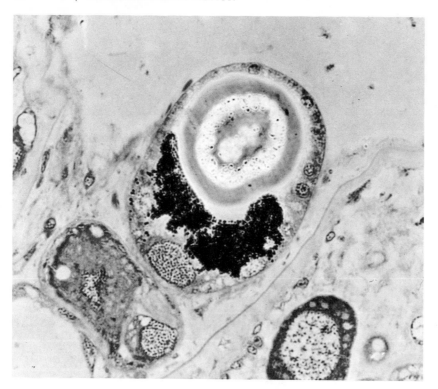

EYE OF *HERMISSENDA* is enlarged about 900 diameters in a photomicrograph made by the late John Barth at Stanford University. The longitudinal section shows the lens (*concentric rings*) and the retina below it. The granular gray structure just below and to the left of the black pigment screening the retina is the nucleus of one of the five receptor cells.

in a certain *Aplysia* cell whose activity—recorded over long periods of time in an isolated ganglion—exhibits a "circadian," or approximately 24-hour, rhythm. Its most active period is a few hours before what would be sunset on a normal light schedule. This inherent periodicity, like that of other biological clocks, is maintained even if such environmental variables as light and temperature are held constant. Strumwasser is now investigating the chemical changes that are correlated with the discharge cycle.

The large size and small number of nerve cells in certain mollusks makes them convenient systems in which to study a number of sensory phenomena. The mollusk *Hermissenda*, for example, has provided information on the nerve-cell interactions that underlie the processing of visual information in a simple "camera" eye: an eye in which a single lens focuses light on a retina of receptor cells. Vertebrate camera eyes have too many cells for such analysis, and so investigators have relied largely on simpler compound eyes, in which a number of independent elements, each with its own lens and sensory cells, send fibers to the central nervous system. Studies of such compound eyes, particularly in the horseshoe crab *Limulus*, revealed the

existence of one especially important process in visual systems: H. K. Hartline and his associates at the Rockefeller Institute found that when a given visual element is illuminated, it not only sends a train of impulses to the central nervous system along its own nerve fiber; it also inhibits the discharge of neighboring elements in the eye. This "lateral inhibition" decreases with distance from the visual element, and it functions to raise the level of contrast at boundaries of stimulus intensity [see "How Cells Receive Stimuli," by William H. Miller, Floyd Ratliff and H. K. Hartline, SCIENTIFIC AMERICAN Offprint 99].

Is lateral inhibition also an essential component of image-formation in other kinds of eye? What is the minimum number of light-receptor cells necessary to form a useful spatial representation of the visual field? What rules are followed in connecting them? These are among the questions investigated in our laboratory at Stanford University, first by the late John Barth and more recently by Michael Dennis, in the course of a study of the eye of *Hermissenda*. The eye measures less than .1 millimeter in its long dimension and consists of a lens, a cup of black pigment to ensure that light enters only from the right direction, and an underlying retina of receptor cells.

The remarkable feature of this miniature camera eye is that its retina consists of only five receptor cells, each one large enough to be penetrated with microelectrodes.

Barth found that cells exposed to light might respond in one of three ways: with an accelerated discharge, with a slower rate of firing followed by an "off" discharge (signaling the cessation of illumination) or with some complex mixture of the two. Dennis' experiments show clearly that this differentiation is not the result of separate classes of receptors. Both the excitatory response and the inhibitory one show similar peaks of sensitivity across the visible spectrum, indicating that they depend on the same light-absorbing pigment. When the activity of a pair (or a trio) of cells is recorded simultaneously, each impulse in one cell is followed by inhibition in the other (or the other two), indicating the presence of cross-connections among them [see *upper illustration on opposite page*]. Since this situation always holds, we conclude that the inhibitory network connects every cell with all four others.

We originally doubted that a mosaic of only five cells could actually form images as larger camera eyes do, but Dennis has demonstrated that the mosaic does indeed have the ability to detect the position of small light sources in the visual field. In one experiment a spot of light five degrees in diameter was moved from right to left and back again on a screen facing the eye [see *lower illustration on opposite page*]. Two of the five receptor cells were impaled with microelectrodes; one, whose activity is shown in the lower trace of each record, responded more strongly when the spot was moved to the left, and held its neighbor, whose activity is shown in the upper trace, under effective inhibitory check. When the light was at the right, the discharge ratio had been such that both cells were firing at almost the same frequency, and it returned to this former value when the spot was moved back to the right. Clearly the relative intensities impinging on the two cells must have been different for the two positions of the light. With the light at the left, the lower cell was more strongly illuminated and consequently fed a stronger inhibition to its neighbor; with the light at the right, the intensities were presumably nearly balanced. In addition, it may be that specific cells display individual personalities even under perfectly homogeneous illumination; those with comparatively little light-sensitive pigment or with relatively strong inhibitory input, for example, would be particularly likely to respond in a predominantly inhibitory fashion and so would be characterized as "off" cells.

As the records show, there is a strong tendency for pairs of cells to fire at the same instant, even when their frequencies of discharge are quite different. This behavior cannot be accounted for on the basis of the inhibitory interaction alone, and it turns out that there is an additional kind of interaction of cells. It is of an excitatory nature, is very brief and is probably mediated by direct electrical connections between two cells. It promotes simultaneous discharge by acting as a trigger for the initiation of impulses in neighboring cells that are nearly ready to fire anyway. The resulting tendency to synchronize may be of value to the region of the brain that receives the visual messages.

These results indicate the technical advantage of small systems of nerve cells to the physiologist. In a larger sense they illustrate how a few elements connected in simple ways can serve an organism remarkably well. It appears that with five receptor cells, a modest optical system and two types of interaction *Hermissenda* can build a crude image-forming system capable of enhancing contrast at boundaries and—at least theoretically—of measuring the speed and direction of moving objects.

The nerve cells of some crustaceans and insects are also easy to recognize individually, and there are relatively few of them. (They are not so spectacularly large as those in *Aplysia*, and instead of recording from the cells with microelectrodes most investigators dissect single axons from the connective nerves that run between central ganglia.) In such nervous systems a very few cells sometimes control a specific, anatomically restricted process. A network of this kind is found, for example, in the hearts of crabs and lobsters, where the beat is triggered and spread by an assembly of only nine to 11 cells embedded in the heart muscle. Some of these cells are "pacemakers," which initiate impulses spontaneously at regular intervals; others are "followers" activated by the pacemakers. Connections among the follower cells marshal their responses into a burst of activity that grows and then subsides until the next pacemaker signals arrive. Activity from the followers also has a subtle feedback effect on the pacemaker frequency. With this system Theodore H. Bullock and his collaborators at the University of California at Los Angeles have pioneered in examining restricted ensembles of nerve cells to establish principles of nervous integration.

Can such analyses be expanded to deal with larger groups of nerve cells that control entire systems of muscles, or even control behavior complex enough to orient an animal in its environment or move the animal through it? They can, provided that the controlling cells are in-

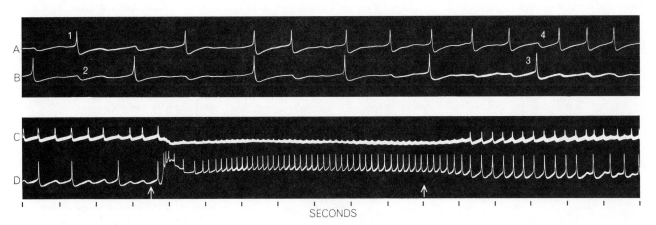

"ON" AND "OFF" responses are seen in records made simultaneously in two *Hermissenda* receptor cells. An impulse from either cell *A* or *B* (*1, 3*) produces an inhibitory hyperpolarization in the other cell (*2, 4*). The illumination of cell *D* (*interval between arrows*) makes it fire faster, inhibiting cell *C*. When the light is turned off, cell *C* is released from inhibition and discharges again.

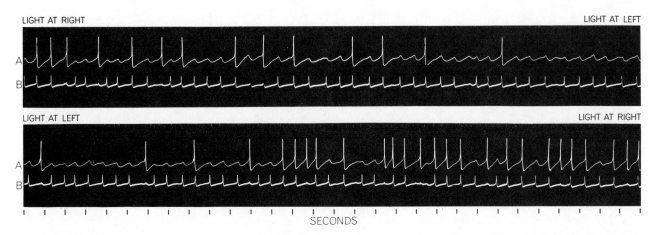

POSITION IS DETECTED by the differential responses of two receptor cells in *Hermissenda*. The discharge rates of cells *A* and *B* were about the same when a light was placed to the right of the field of view. As the light was moved to the left cell *B* discharged relatively more rapidly, effectively inhibiting cell *A*. When the light was moved back again, the former discharge ratio was reestablished.

dividually unique and are identifiable by the experimenter. The feasibility of such studies has been demonstrated primarily by C. A. G. Wiersma and his co-workers at the California Institute of Technology. In a series of investigations spanning the past three decades Wiersma has concerned himself with crustacean muscles and the motor-nerve axons that innervate them. The motor axons—which are remarkably few in number—can be distinguished from one another by their different electrical effects on the muscle and by the kinds of contraction they evoke. Some produce quick twitches by generating large, abrupt depolarizations (reductions in membrane potential that lead to excitation) in the muscle fibers; others make the muscle contract in a more sustained way by producing small and gradually augmenting depolarizations; still others prevent contraction. The number of nerve fibers serving any given muscle is small enough so that one can distinguish the impulses of each one in an electrical record from the entire nerve and correlate these impulses with events in the muscle.

Wiersma has exposed an even more remarkable differentiation of elements in the central nervous system of the crayfish. Single interneurons—nerve cells that collect information from a number of sensory fibers—can be isolated for electrical recording from a part of the central nervous system in the abdomen;

elsewhere such a cell runs its normal course, making connections by branching in each of several different ganglia. Wiersma has prepared maps that give the distribution over the animal's surface of the sensory receptors that will excite each such cell. He has shown that each interneuron is uniquely connected with a set of sensory-nerve fibers, so that a cell with a specific map is always found in the same anatomical location in the central nervous system.

A particular group of touch-sensitive hairs on the back of the fourth abdominal segment, for example, might connect with a dozen different central interneurons. Each interneuron, however, responds to some unique combination of that group of hair receptors and other groups. One interneuron, for instance, might be excited by the fourth-segment group alone; another might be activated by the corresponding group on segment No. 5 as well as by the hairs on segment No. 4; another by those on Nos. 6, 5 and 4; another by those on all segments on one side or on both sides, and so forth. Precise duplication of function is apparently absent; each element encodes a unique spatial combination of sensory inputs [see illustration below].

This specificity suggests that the position and connection pattern of each central nerve cell is precisely determined in the course of differentiation of the nervous system. The reliability of this

mechanism has been demonstrated impressively by Melvin J. Cohen and his colleagues at the University of Oregon, who have analyzed the organization of central ganglia in the cockroach. Cohen has located the cell bodies of specific motoneurons by taking advantage of a striking response shown by such cells when their axons are cut: If a motor nerve is severed at the periphery, even very near the muscle, each cell body supplying an axon in that nerve quickly develops a dense ring around its nucleus. This response, which can be detected with suitable stains, occurs within 12 hours [see illustration on page 45]. The new material comprising the ring has been identified as ribonucleic acid (RNA). Presumably it is required for the protein synthesis associated with regeneration; in any event, the ring provides an unambiguous label for associating a specific central cell with the peripheral destination of its axon.

By cutting individual motor nerves and locating their cell bodies in this way, Cohen has constructed a map that gives the positions for most of the motoneurons in a ganglion. The maps of ganglia from different individuals appear to be almost identical; indeed, specific cells are in nearly perfect register when sections of corresponding ganglia from several animals are superposed. Not only are the cell bodies of motoneurons that serve particular muscles precisely arrayed; they also appear to have rather specific biochemical personalities. Some motoneurons show the ring reaction especially strongly and others show it quite weakly, and these differences are consistently associated with particular cells— as judged by the position of the cell body and the peripheral destination of its axon.

In our own laboratory we have analyzed how the central nervous system of the crayfish deploys a limited array of identifiable nerve cells to control the posture of the abdomen. This structure consists of five segments with joints between them. Its shape is continuously and delicately varied by the action of thin sheets of muscle operating in antagonistic pairs—extensor and flexor—at each of the five joints. Both sets of muscles are symmetrical on each side of the abdomen. The extensors of each half-segment receive six nerve fibers, as do the flexors; the nerve fibers, like the muscles, are repeated almost identically in each segment.

Each individual motor nerve cell can be identified by the size of its impulses

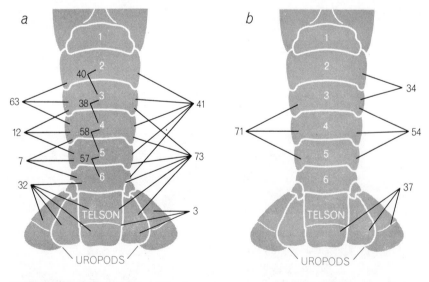

SPECIFIC GROUPS of touch-sensitive hairs on the segments of the crayfish abdomen excite specific interneurons located within the central nervous system. The receptive field of each interneuron listed here (*black numbers*) was mapped by C. A. G. Wiersma's group at Cal Tech. Each responds to stimulation of hair receptors on a unique combination of segments (*white numbers*), as shown by the pointers. The system is bilaterally symmetrical, with an interneuron No. 40, for instance, on each side. Neurons mapped at *a* respond to stimuli delivered to the same side of the animal. Of those mapped at *b*, those listed at the right respond to stimuli on the side opposite them; No. 71 responds to stimuli on either side.

in an electrical record obtained from the nerve bundles that run to flexor or extensor muscles. As one might expect, particular reflexes activate the nerves and muscles in rather stereotyped ways. If, for example, one forcibly flexes the abdomen of a crayfish while holding its thorax clamped and then releases it, the segments extend to approximately their former position. Howard Fields of our laboratory found that this action depends on a pair of receptors that span the dorsal joints in the abdomen. Flexion lengthens the muscle strands associated with these receptor cells, and the cells then discharge impulses that travel toward the central nervous system and activate motoneurons supplying the extensor muscles, which are thereby caused to contract against the imposed load. Such "resistance reflexes" are known in a variety of other systems, including the limbs of mammals and of crabs; in the crayfish they can be studied in a simplified situation, with a single receptor cell and six well-characterized motor cells constituting the entire neural equipment for the reflex loop.

Since only about 120 motoneurons are involved in the regulation of the entire abdomen's position, and since we were able to identify each of them, William H. Evoy and I decided to analyze the central control system for abdominal posture. While recording the motor discharge in several segments at once, we isolated and then stimulated single interneurons located within the central nervous system.

As we had anticipated, most cells had no effect on motor discharge, but we encountered some that regularly released intense, fully coordinated motor-output patterns when we stimulated them with a series of electric shocks. In every case the output was reciprocal: flexors were excited while extensors were inhibited, or vice versa, and the response was similar in several segments. Although equally complex behavior can be produced in many animals by localized stimulation of central nervous structures, it is likely that in most such cases many cells—perhaps thousands—are simultaneously activated by the comparatively gross stimulating procedure that is employed. In the crayfish abdomen, however, we have been able to demonstrate directly that a complete behavioral output can be the result of activity in a single central interneuron. Motor effects produced by stimulation of such central neurons in crayfish had been described earlier by Wiersma and K. Ikeda, who coined the term "command fiber" for those inter-

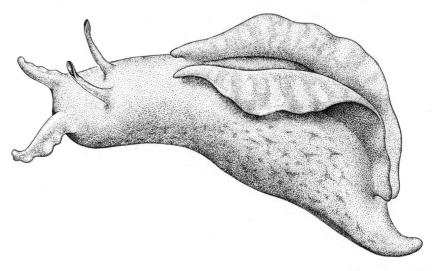

SEA HARE *APLYSIA* is a mollusk with a vestigial shell under the mantle between the winglike parapodia, which are used for locomotion. The animal is about 10 inches long.

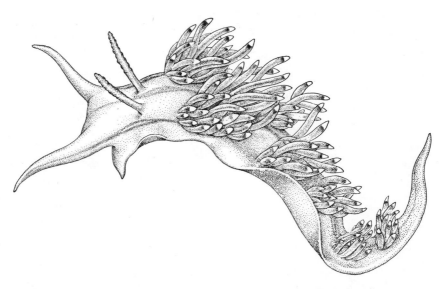

HERMISSENDA, another mollusk, is only one to three inches long. The "camera" eyes are embedded just behind the rhinophores, the two upright stalks near the left end of the animal.

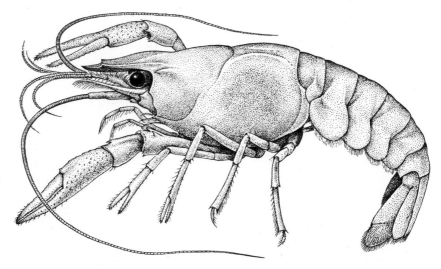

CRAYFISH *Procambarus*, the invertebrate in which complex motor effects produced by single cells are studied by Wiersma and by the author, is from three to five inches long.

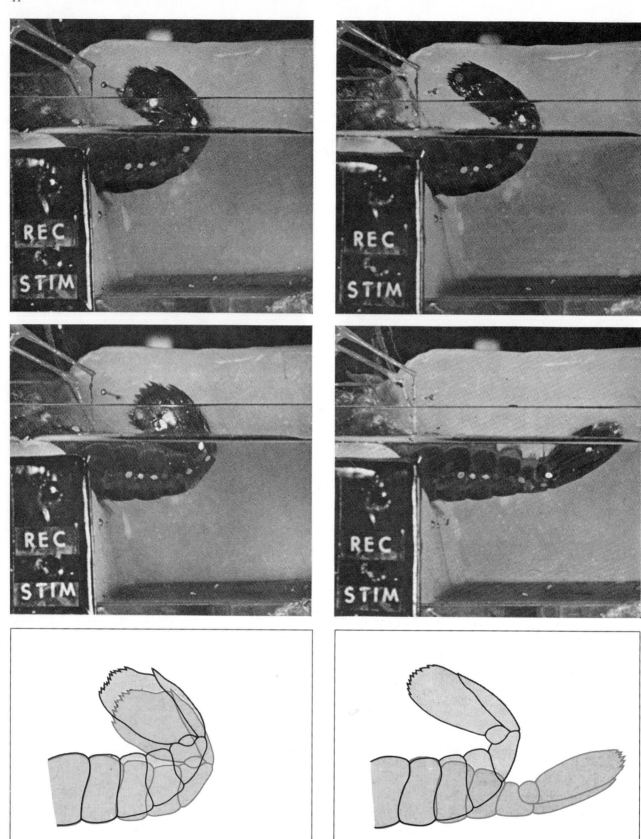

CRAYFISH ABDOMEN responds in complex and specific ways to stimulation of different single cells in the central nervous system. Frames from a motion-picture film made by Benjamin Dane in the author's laboratory record the effects of two different "command fibers"; the drawings specify the initial (gray) and the final (red) positions of the abdomen. One fiber evoked activity primarily in the forward segments (left); the other produced extension in all segments (right). Spots painted on the abdomen facilitated precise measurement of segment angles. Visible effects were confirmed by recording the activity of motor nerve fibers in the first segment.

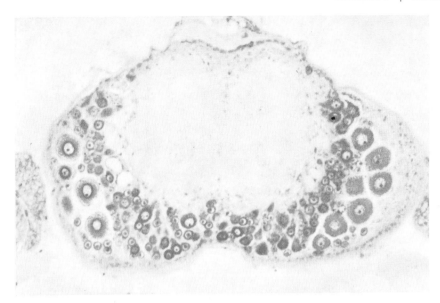

DENSE RINGS of ribonucleic acid surround the nuclei of the cells on one side (*left*) of a stained section of a cockroach ganglion about a millimeter across. Two days before the photograph was made by Melvin J. Cohen of the University of Oregon, all peripheral nerves from one side of this ganglion had been cut. The rings appear in the cells on that side.

neuron axons that produce stereotyped, complex motor effects.

When we compared, in single animals, a series of central nerve cells that produce extension, we found that each produced a special distribution of motor output in different segments. One strongly influenced the last two segments but had a weak effect on segments farther forward; another yielded an almost even balance of output; another activated the forward segments more strongly, and so on. This result led naturally to the inference that individual command elements might each code for a unique combination of segmental actions and thereby produce specific movements or positions. In collaboration with Benjamin Dane, who is now at Tufts University, we filmed the maneuvers of the abdomen that occurred when command fibers were stimulated, while simultaneously recording the discharge of motoneurons supplying muscles in one of the segments. This experiment confirmed the neurophysiological predictions: the films show that two command elements with a generally similar action affect the segments in different ratios and so produce unique abdominal postures [see illustration on opposite page].

These results indicate that the coordination of at least some behavior is based on the intrinsic "wiring" of a nervous center and can be released by single-cell triggers. Such a conclusion would have been more surprising a decade ago than it is now. It was once

thought that complex sequential behavior such as locomotion depended for its organization on a series of instructions fed back to the central nervous system from peripheral sensory cells. Each new act, according to this view, was initiated by proprioceptive sense organs that were excited by the animal's own movement and signaled that one response had been completed; the whole sequence was likened to a chain of reflexes successively activated by the continuous inflow of sensory signals from the periphery. In recent years biologists have become more convinced that the connections among interneurons at the nervous center, selected by evolution and precisely made during the developmental process, contain most of the organization inherent in such behavior, and that sense organs play a role that is more permissive than instructive.

One of the most convincing demonstrations of the importance of these central connections was provided by the experiments of Donald M. Wilson of the University of California at Berkeley on the flight of locusts. This behavior pattern consists of a sequence of actions in the muscles that elevate, depress and twist the wings. Since these muscles are served by only one or two motor nerve cells, one can obtain a record of the activity of each cell by inserting fine wires into the muscles of a relatively intact animal. In this way Wilson and Torkel Weis-Fogh were able to define the impulse sequences, along various motor nerves, characteristic of normal flight.

The pattern can be triggered in several ways. When the insect is stimulated by having air blown on its head, the motor activity characteristic of flight is frequently released, but Wilson has shown that the output pattern has no relation to the timing of the input. Indeed, when he stimulated one of the central connectives electrically at random intervals, the flight motor output still had its normal frequency and pattern.

Associated with the base of each wing of the locust there is a sensory-nerve cell that responds with single impulses or a short burst when the wing comes to the top of its upstroke. It had been supposed that these receptors might be providing phase information about the wingbeat cycle and perhaps initiating activity in the motor nerves controlling the downstroke. Wilson eliminated these receptors and found that although the frequency of flight motor output dropped somewhat, its repetitive character and the sequence of events within a single cycle were unimpaired. This showed that the stretch receptors, although they participate in a reflex that controls frequency, do not provide any information about the phase relations within a single cycle. Since the entire pattern of behavior can be produced by a central nervous system isolated from peripheral structures—as long as enough excitation of some kind is supplied—it must be concluded that all the information for the flight sequence is stored in a set of central connections.

At this point one can only assume that the sequential events of the flight pattern, like the motions that result in postural changes in the crayfish abdomen, are released by activity in central "command" elements. Some single interneurons are known to release similar complex, sequential behavior in the crayfish. One, for example, produces a sequence of flexions that ascends slowly from segment to segment until it reaches the forward end of the abdomen and then begins again at the back. Another causes an intricate series of movements in the appendages of the tail. We wonder how such central command neurons are activated during voluntary movements, whether they are organized entirely in parallel or in part as a hierarchy of related elements, and what kinds of sensory stimulus excite them. While it would be premature to assume that all these questions can be answered or that we can apply the results to other systems, it is the peculiar advantage of small networks of nerve cells that such a prospect does not seem hopeless.

4

Moths and Ultrasound

by Kenneth D. Roeder
April 1965

*Certain moths can hear the ultrasonic cries by which
bats locate their prey. The news is sent from ear to
central nervous system by only two fibers. These can be
tapped and the message decoded*

If an animal is to survive, it must be able to perceive and react to predators or prey. What nerve mechanisms are used when one animal reacts to the presence of another? Those animals that have a central nervous system perceive the outer world through an array of sense organs connected with the brain by many thousands of nerve fibers. Their reactions are expressed as critically timed sequences of nerve impulses traveling along motor nerve fibers to specific muscles. Exactly how the nervous system converts a particular pattern of sensory input into a specific pattern of motor output remains a subject of investigation in many branches of zoology, physiology and psychology.

Even with the best available techniques one can simultaneously follow the traffic of nerve impulses in only five or perhaps 10 of the many thousands of separate nerve fibers connecting a mammalian sense organ with the brain. Trying to learn how information is encoded and reported among all the fibers by following the activity of so few is akin to basing a public opinion poll on one or two interviews. (Following the activity of all the fibers would of course be like sampling public opinion by having the members of the population give their different answers in chorus.) Advances in technique may eventually make it possible to follow the traffic in thousands of fibers; in the meantime much can be learned by studying animals with less profusely innervated sense organs.

With several colleagues and students at Tufts University I have for some time been trying to decode the sensory patterns connecting the ear and central nervous system of certain nocturnal moths that have only two sense cells in each ear. Much of the behavior of these simple invertebrates is built in, not learned, and therefore is quite stereo-typed and stable under experimental conditions. Working with these moths offers another advantage: because they depend on their ears to detect their principal predators, insect-eating bats, we are able to discern in a few cells the nervous mechanisms on which the moth's survival depends.

Insectivorous bats are able to find their prey while flying in complete darkness by emitting a series of ultrasonic cries and locating the direction and distance of sources of echoes. So highly sophisticated is this sonar that it enables the bats to find and capture flying insects smaller than mosquitoes. Some night-flying moths—notably members of the families Noctuidae, Geometridae and Arctiidae—have ears that can detect the bats' ultrasonic cries. When they hear the approach of a bat, these moths take evasive action, abandoning their usual cruising flight to go into sharp dives or erratic loops or to fly at top speed directly away from the source of ultrasound. Asher E. Treat of the College of the City of New York has demonstrated that moths taking evasive action on a bat's approach have a significantly higher chance of survival than those that continue on course.

A moth's ears are located on the sides of the rear part of its thorax and are directed outward and backward into the constriction that separates the thorax and the abdomen [*see top illustration on page 48*]. Each ear is externally visible as a small cavity, and within the cavity is a transparent eardrum. Behind the eardrum is the tympanic air sac; a fine strand of tissue containing the sensory apparatus extends across the air sac from the center of the eardrum to a skeletal support. Two acoustic cells, known as *A* cells, are located within this strand. Each *A* cell sends a fine sensory strand outward to the eardrum and a nerve fiber inward to the skeletal support. The two *A* fibers pass close to a large nonacoustic cell, the *B* cell, and are joined by its nerve fiber. The three fibers continue as the tympanic nerve into the central nervous system of the moth. From the two *A* fibers, then, it is possible—and well within our technical means—to obtain all the information about ultrasound that is transmitted from the moth's ear to its central nervous system.

Nerve impulses in single nerve fibers can be detected as "action potentials," or self-propagating electrical transients, that have a magnitude of a few millivolts and at any one point on the fiber last less than a millisecond. In the moth's *A* fibers action potentials travel from the sense cells to the central nervous system in less than two milliseconds. Action potentials are normally an all-or-nothing phenomenon; once initiated by the sense cell, they travel to the end of the nerve fiber. They can be detected on the outside of the fiber by means of fine electrodes, and they are displayed as "spikes" on the screen of an oscilloscope.

Tympanic-nerve signals are demonstrated in the following way. A moth, for example the adult insect of one of the common cutworms or armyworms, is immobilized on the stage of a microscope. Some of its muscles are dissected away to expose the tympanic nerves at a point outside the central nervous system. Fine silver hooks are placed under one or both nerves, and the pattern of passing action potentials is observed on the oscilloscope. With moths thus prepared we have spent much time in impromptu outdoor laboratories, where the cries of passing bats provided the necessary stimuli.

In order to make precise measurements we needed a controllable source of ultrasonic pulses for purposes of comparison. Such pulses can be generated by electronic gear to approximate natural bat cries in frequency and duration. The natural cries are frequency-modulated: their frequency drops from about 70 kilocycles per second at the beginning of each cry to some 35 kilocycles at the end. Their duration ranges from one to 10 milliseconds, and they are repeated from 10 to 100 times a second. Our artificial stimulus is a facsimile of an average series of bat cries; it is not frequency-modulated, but such modulation is not detected by the moth's ear. Our sound pulses can be accurately graded in intensity by decibel steps; in the sonic range a decibel is roughly equivalent to the barely noticeable difference to human ears in the intensity of two sounds.

By using electronic apparatus to elicit and follow the responses of the A cells we have been able to define the amount of acoustic information avail-

MOTH EVADED BAT by soaring upward just as the bat closed in to capture it. The bat entered the field at right; the path of its flight is the broad white streak across the photograph. The smaller white streak shows the flight of the moth. A tree is in background. The shutter of the camera was left open as contest began. Illumination came from continuous light source below field.

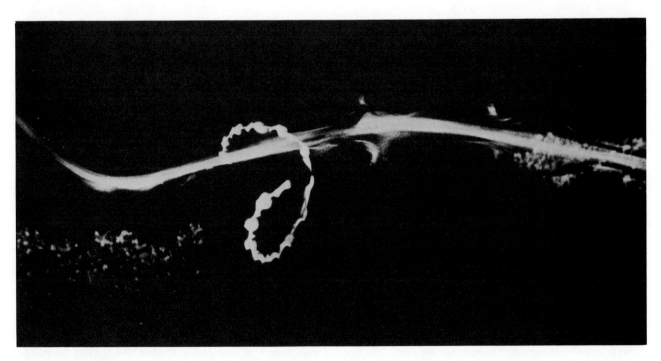

BAT CAPTURED MOTH at point where two white streaks intersect. Small streak shows the flight pattern of the moth. Broad streak shows the flight path of the bat. Both streak photographs were made by Frederic Webster of the Sensory Systems Laboratories.

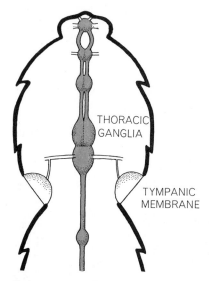

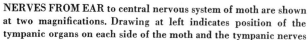

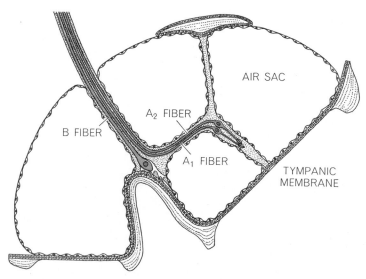

NERVES FROM EAR to central nervous system of moth are shown at two magnifications. Drawing at left indicates position of the tympanic organs on each side of the moth and the tympanic nerves connecting them with the thoracic ganglia. Central nervous system is colored. Drawing at right shows two nerve fibers of the acoustic cells joined by a nonacoustic fiber to form the tympanic nerve.

able to the moth by way of its tympanic nerve. It appears that the tympanic organ is not particularly sensitive; to elicit any response from the A cell requires ultrasound roughly 100 times more intense than sound that can just be heard by human ears. The ear of a moth can nonetheless pick up at distances of more than 100 feet ultrasonic bat cries we cannot hear at all. The reason it cannot detect frequency modulation is simply that it cannot discriminate one frequency from another; it is tone-deaf. It can, however, detect frequencies from 10 kilocycles to well over 100 kilocycles per second, which covers the range of bat cries. Its greatest talents are the detection of pulsed sound—short bursts of sound with intervening silence—and the discrimination of differences in the loudness of sound pulses.

When the ear of a moth is stimulated by the cry of a bat, real or artificial, spikes indicating the activity of the A cell appear on the oscilloscope in various configurations. As the stimulus increases in intensity several changes are apparent. First, the number of A spikes increases. Second, the time interval between the spikes decreases. Third, the spikes that had first appeared only on the record of one A fiber (the "A_1" fiber, which is about 20 decibels more sensitive than the A_2 fiber) now appear on the records of both fibers. Fourth, the greater the intensity of the stimulus, the sooner the A cell generates a spike in response.

The moth's ears transmit to the oscilloscope the same configuration of spikes they transmit normally to the central nervous system, and therein lies our interest. Which of the changes in auditory response to an increasingly in-

tense stimulus actually serve the moth as criteria for determining its behavior under natural conditions? Before we face up to this question let us speculate on the possible significance of these criteria from the viewpoint of the moth. For the moth to rely on the first kind of information—the number of A spikes—might lead it into a fatal error: the long, faint cry of a bat at a distance could be confused with the short, intense cry of a bat closing for the kill. This error could be avoided if the moth used the second kind of information—the interval between spikes—for estimating the loudness of the bat's cry. The third kind of information—the activity of the A_2 fiber —might serve to change an "early warning" message to a "take cover" message. The fourth kind of information—the length of time it takes for a spike to be generated—might provide the moth with

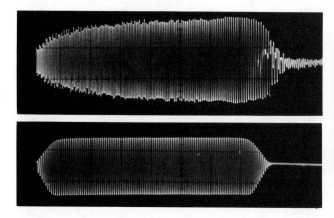

OSCILLOSCOPE TRACES of a real bat cry (top) and a pulse of sound generated electronically (bottom) are compared. The two ultrasonic pulses are of equal duration (length), 2.5 milliseconds, but differ in that the artificial pulse has a uniform frequency.

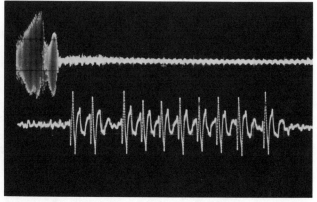

BAT CRY AND MOTH RESPONSE were traced on same oscilloscope from tape recording by Webster. The bat cry, detected by microphone, yielded the pattern at left in top trace. Reaction of the moth's acoustic cells produced the row of spikes at bottom.

the means for locating a cruising bat; for example, if the sound was louder in the moth's left ear than in its right, then *A* spikes would reach the left side of the central nervous system a fraction of a millisecond sooner than the right side.

Speculations of this sort are profitable only if they suggest experiments to prove or disprove them. Our tympanic-nerve studies led to field experiments designed to find out what moths do when they are exposed to batlike sounds from a loudspeaker. In the first such study moths were tracked by streak photography, a technique in which the shutter of a camera is left open as the subject passes by. As free-flying moths approached the area on which our camera was trained they were exposed to a series of ultrasonic pulses.

More than 1,000 tracks were recorded in this way. The moths were of many species; since they were free and going about their natural affairs most of them could not be captured and identified. This was an unavoidable disadvantage; earlier observations of moths captured, identified and then released in an enclosure revealed nothing. The moths were apparently "flying scared" from the beginning, and the ultrasound did not affect their behavior. Hence all comers were tracked in the field.

Because moths of some families lack ears, a certain percentage of the moths failed to react to the loudspeaker. The variety of maneuvers among the moths that did react was quite unpredictable and bewildering [*see illustrations at top of next page*]. Since the evasive behavior presumably evolved for the purpose of bewildering bats, it is hardly surprising that another mammal should find it confusing! The moths that flew close to the loudspeaker and encountered high-intensity ultrasound would maneuver toward the ground either by dropping passively with their wings closed, by power dives, by vertical and horizontal turns and loops or by various combinations of these evasive movements.

One important finding of this field work was that moths cruising at some distance from the loudspeaker would turn and fly at high speed directly away from it. This happened only if the sound the moths encountered was of low intensity. Moths closer to the loudspeaker could be induced to flee only if the signal was made weaker. Moths at about the height of the loudspeaker flew away in the horizontal plane; those above the loudspeaker were observed to turn directly upward

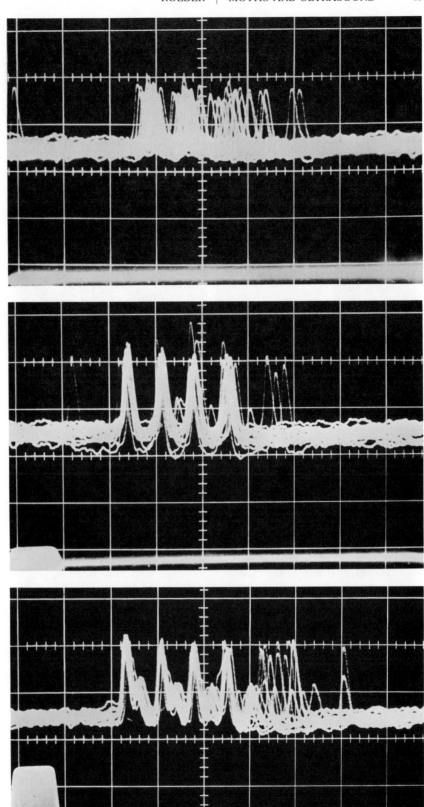

CHANGES ARE REPORTED by moth's tympanic nerve to the oscilloscope as pulses used to simulate bat cries gain intensity. Pulses (*lower trace in each frame*) were at five decibels (*top frame*), 20 (*middle*) and 35 (*bottom*). An increased number of tall spikes appear as intensity of stimulus rises. The time interval between spikes decreases slightly. Smaller spikes from the less sensitive nerve fiber appear at the higher intensities, and the higher the intensity of the stimulus, the sooner (*left on horizontal axis*) the first spike appears.

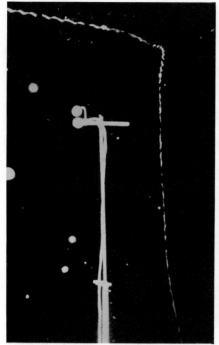 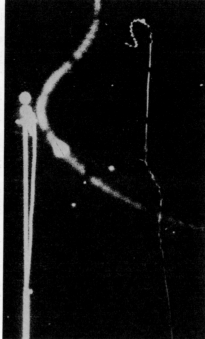 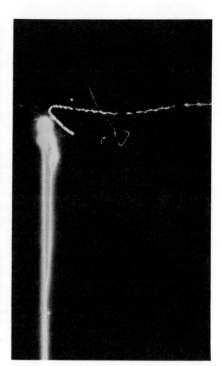

POWER DIVE is taken by moth on hearing simulated bat cry from loudspeaker mounted on thin tower (*left of moth's flight path*).

PASSIVE DROP was executed by another moth, which simply folded its wings. Blur at left and dots were made by other insects.

TURNING AWAY, an evasive action involving directional change, is illustrated. These streak photographs were made by author.

or at other sharp angles. To make such directional responses with only four sensory cells is quite a feat. A horizontal response could be explained on the basis that one ear of the moth detected the sound a bit earlier than the other. It is harder to account for a vertical response, although experiments I shall describe provide a hint.

Our second series of field experiments was conducted in another outdoor laboratory—my backyard. They were designed to determine which of the criteria of intensity encoded in the pattern of *A*-fiber spikes play an important part in determining evasive behavior. The percentage of moths showing "no re-

action," "diving," "looping" and "turning away" was noted when a 50-kilo-cycle signal was pulsed at different rates and when it was produced as a continuous tone. The continuous tone delivers more *A* impulses in a given fraction of a second and therefore should be a more effective stimulus if the number of *A* impulses is important. On the other hand, because the *A* cells, like many other sensory cells, become progressively less sensitive with continued stimulation, the interspike interval lengthens rapidly as continuous-tone stimulation proceeds. When the sound is pulsed, the interspike interval remains short because the *A* cells have had time to regain their sensitivity during the

brief "off" periods. If the spike-generation time—which is associated with difference in the time at which the *A* spike arrives at the nerve centers for each ear—plays an important part in evasive behavior, then continuous tones should be less effective. The difference in arrival time would be detected only once at the beginning of the stimulus; with pulsed sound it would be reiterated with each pulse.

The second series of experiments occupied many lovely but mosquito-ridden summer nights in my garden and provided many thousands of observations. Tabulation of the figures showed that continuous ultrasonic tones were much less effective in producing evasive

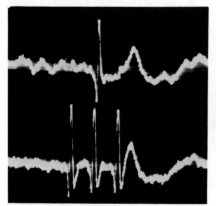 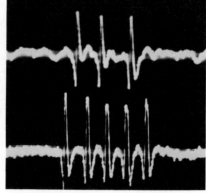 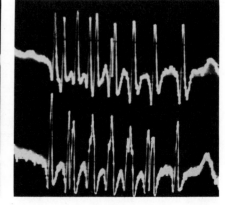

RESPONSE BY BOTH EARS of a moth to an approaching bat was recorded on the oscilloscope and photographed by the author. In trace at left the tympanic nerve from one ear transmits only one spike (*upper curve*) while the nerve from the other ear sends three. As the bat advances, the ratio becomes three to five (*middle*), then 10 to 10 (*right*), suggesting that the bat has flown overhead.

behavior than pulses. The number of nonreacting moths increased threefold, diving occurred only at higher sound intensities and turning away was essentially absent. Only looping seemed to increase slightly.

Ultrasound pulsed between 10 and 30 times a second proved to be more effective than ultrasound pulsed at higher or lower rates. This suggests that diving, and possibly other forms of nondirectional evasive behavior, are triggered in the moth's central nervous system not so much by the number of A impulses delivered over a given period as by short intervals (less than 2.5 milliseconds) between consecutive A impulses. Turning away from the sound source when it is operating at low intensity levels seems to be set off by the reiterated difference in arrival time of the first A impulse in the right and left tympanic nerves.

These conclusions were broad but left unanswered the question: How can a moth equipped only with four A cells orient itself with respect to a sound source in planes that are both vertical and horizontal to its body axis? The search for an answer was undertaken by Roger Payne of Tufts University, assisted by Joshua Wallman, a Harvard undergraduate. They set out to plot the directional capacities of the tympanic organ by moving a loudspeaker at various angles with respect to a captive moth's body axis and registering (through the A_1 fiber) the organ's relative sensitivity to ultrasonic pulses coming from various directions. They took precautions to control acoustic shadows and reflections by mounting the moth and the recording electrodes on a thin steel tower in the center of an echo-free chamber; the effect of the moth's wings on the reception of sound was tested by systematically changing their position during the course of many experiments. A small loudspeaker emitted ultrasonic pulses 10 times a second at a distance of one meter. These sounds were presented to the moths from 36 latitude lines 10 degrees apart.

The response of the A fibers to the ultrasonic pulses was continuously recorded as the loudspeaker was moved. At the same time the intensity of ultrasound emitted by the loudspeaker was regulated so that at any angle it gave rise to the same response. Thus the intensity of the sound pulses was a measure of the moth's acoustic sensitivity. A pen recorder continuously graphed the changing intensity of the ultrasonic pulses against the angle from which

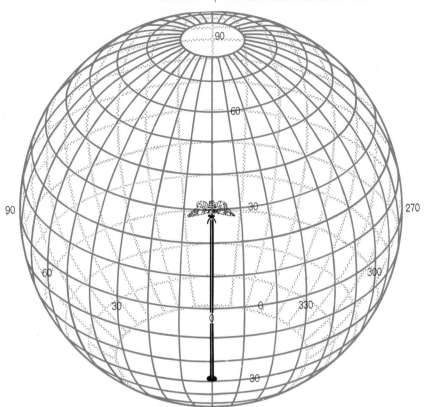

SPHERE OF SENSITIVITY, the range in which a moth with wings in a given position can hear ultrasound coming from various angles, was the subject of a study by Roger Payne of Tufts University and Joshua Wallman, a Harvard undergraduate. Moths with wings in given positions were mounted on a tower in an echo-free chamber. Data were compiled on the moths' sensitivity to ultrasound presented from 36 latitude lines 10 degrees apart.

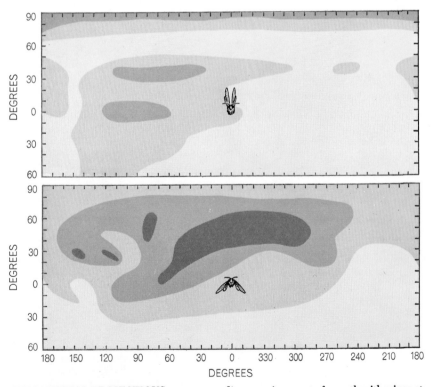

MERCATORIAL PROJECTIONS represent auditory environment of a moth with wings at end of upstroke (*top*) and near end of downstroke (*bottom*). Vertical scale shows rotation of loudspeaker around moth's body in vertical plane; horizontal scale shows rotation in horizontal plane. At top the loudspeaker is above moth; at far right and left, behind it. In Mercatorial projections, distortions are greatest at poles. The lighter the shading at a given angle of incidence, the more sensitive the moth to sound from that angle.

they were presented to the moth. Each chart provided a profile of sensitivity in a certain plane, and the data from it were assembled with those from others to provide a "sphere of sensitivity" for the moth at a given wing position.

This ingenious method made it possible to assemble a large amount of data in a short time. In the case of one moth it was possible to obtain the data for nine spheres of sensitivity (about 5,000 readings), each at a different wing position, before the tympanic nerve of the moth finally stopped transmitting impulses. Two of these spheres, taken from one moth at different wing positions, are presented as Mercatorial projections in the bottom illustration on the preceding page.

It is likely that much of the information contained in the fine detail of such projections is disregarded by a moth flapping its way through the night. Certain general patterns do seem related, however, to the moth's ability to escape a marauding bat. For instance, when the moth's wings are in the upper half of their beat, its acoustic sensitivity is 100 times less at a given point on its side facing away from the source of the sound than at the corresponding point on the side facing toward the source. When flight movements bring the wings below the horizontal plane, sound coming from each side above the moth is in acoustic shadow, and the left-right acoustic asymmetry largely disappears. Moths commonly flap their wings from 30 to 40 times a second. Therefore left-right acoustic asymmetry must alternate with up-down asymmetry at this frequency. A left-right difference in the

A-fiber discharge when the wings are up might give the moth a rough horizontal bearing on the position of a bat with respect to its own line of flight. The absence of a left-right difference and the presence of a similar fluctuation in both left and right tympanic nerves at wingbeat frequency might inform the moth that the bat was above it. If neither variation occurred at the regular wingbeat frequency, it would mean that the bat was below or behind the moth.

This analysis uses terms of precise directionality that idealize the natural situation. A moth certainly does not zoom along on an even keel and a straight course like an airliner. Its flapping progress—even when no threat is imminent—is marked by minor yawing and pitching; its overall course is rare-

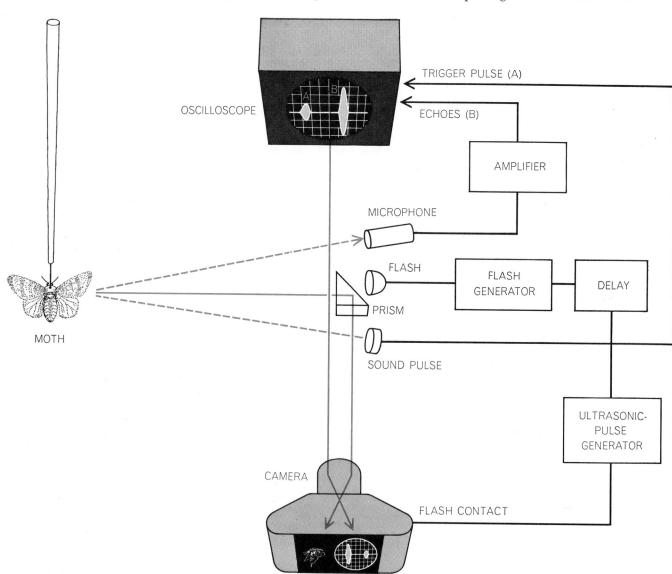

ARTIFICIAL BAT, the electronic device depicted schematically at right, was built by the author to determine at what position with respect to a bat a moth casts its greatest echo. As a moth supported by a wire flapped its wings in stationary flight, a film was made by means of a prism of its motions and of an oscilloscope that showed the pulse generated by the loudspeaker and the echo picked up by the microphone. Each frame of film thus resembled the composite picture of moth and two pulses shown inverted at bottom.

ly straight and commonly consists of large loops and figure eights. Even so, the localization experiments of Payne and Wallman suggest the ways in which a moth receives information that enables it to orient itself in three dimensions with respect to the source of an ultrasonic pulse.

The ability of a moth to perceive and react to a bat is not greatly superior or inferior to the ability of a bat to perceive and react to a moth. Proof of this lies in the evolutionary equality of their natural contest and in the observation of a number of bat-moth confrontations. Donald R. Griffin of Harvard University and Frederic Webster of the Sensory Systems Laboratories have studied in detail the almost unbelievable ability of bats to locate, track and intercept

small flying targets, all on the basis of a string of echoes thrown back from ultrasonic cries. Speaking acoustically, what does a moth "look like" to a bat? Does the prey cast different echoes under different circumstances?

To answer this question I set up a crude artificial bat to pick up echoes from a live moth. The moth was attached to a wire support and induced to flap its wings in stationary flight. A movie camera was pointed at a prism so that half of each frame of film showed an image of the moth and the other half the screen of an oscilloscope. Mounted closely around the prism and directed at the moth from one meter away were a stroboscopic-flash lamp, an ultrasonic loudspeaker and a microphone. Each time the camera shutter opened and exposed a frame of film a

short ultrasonic pulse was sent out by the loudspeaker and the oscilloscope began its sweep. The flash lamp was controlled through a delay circuit to go off the instant the ultrasonic pulse hit the moth, whose visible attitude was thereby frozen on the film. Meanwhile the echo thrown back by the moth while it was in this attitude was picked up by the microphone and finally displayed as a pulse of a certain height on the oscilloscope. All this took place before the camera shutter closed and the film moved on to the next frame. Thus each frame shows the optical and acoustic profiles of the moth from approximately the same angle and at the same instant of its flight. The camera was run at speeds close to the wingbeat frequency of the moth, so that the resulting film presents a regular series of wing positions and the echoes cast by them.

Films made of the same moth flying at different angles to the camera and the sound source show that by far the strongest echo is returned when the moth's wings are at right angles to the recording array [see illustrations at left]. The echo from a moth with its wings in this position is perhaps 100 times stronger than one from a moth with its wings at other angles. Apparently if a bat and a moth were flying horizontal courses at the same altitude, the moth would be in greatest danger of detection if it crossed the path of the approaching bat at right angles. From the bat's viewpoint at this instant the moth must appear to flicker acoustically at its wingbeat frequency. Since the rate at which the bat emits its ultrasonic cries is independent of the moth's wingbeat frequency, the actual sequence of echoes the bat receives must be complicated by the interaction of the two frequencies. Perhaps this enables the bat to discriminate a flapping target, likely to be prey, from inert objects floating in its acoustic field.

The moth has one advantage over the bat: it can detect the bat at a greater range than the bat can detect it. The bat, however, has the advantage of greater speed. This creates a nice problem for a moth that has picked up a bat's cries. If a moth immediately turns and flies directly away from a source of ultrasound, it has a good chance of disappearing from the sonar system of a still-distant bat. If the bat has also detected the moth, and is near enough to receive a continuous signal from its target, turning away on a straight course is a bad tactic because the moth is not likely to outdistance its pursuer. It is then to the moth's advantage to

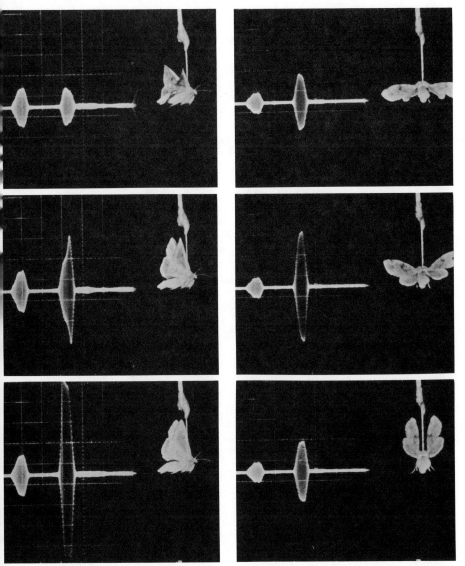

COMPOSITE PHOTOGRAPHS each show an artificial bat's cry (left) and the echo thrown back (middle) by a moth (right). The series of photographs at left is of a moth in stationary flight at right angles to the artificial bat. Those at right are of a moth oriented in flight parallel to the bat. The echo produced in the series of photographs at left is much the larger.

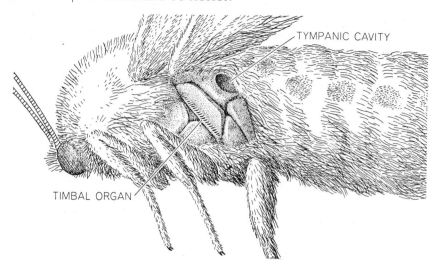

TYMPANIC CAVITY

TIMBAL ORGAN

NOISEMAKING ORGAN possessed by many moths of the family Arctiidae and of other families is a row of fine parallel ridges of cuticle that bend and unbend when a leg muscle contracts and relaxes. This produces a rapid sequence of high-pitched clicks.

go into tight turns, loops and dives, some of which may even take it toward the bat.

In this contest of hide-and-seek it seems much to a moth's advantage to remain as quiet as possible. The sensitive ears of a bat would soon locate a noisy target. It is therefore surprising to find that many members of the moth family Arctiidae (which includes the moths whose caterpillars are known as woolly bears) are capable of generating trains of ultrasonic clicks. David Blest and David Pye of University College London have demonstrated the working of the organ that arctiids use for this purpose.

In noisemaking arctiids the basal joint of the third pair of legs (which roughly corresponds to the hip) bulges outward and overlies an air-filled cavity. The stiff cuticle of this region has a series of fine parallel ridges [*see illustration above*]. Each ridge serves as a timbal that works rather like the familiar toy incorporating a thin strip of spring steel that clicks when it is pressed by the thumb. When one of the moth's leg muscles contracts and relaxes in rapid sequence, it bends and unbends the overlying cuticle, causing the row of timbals to produce rapid sequences of high-pitched clicks. Blest and Pye found that such moths would click when they were handled or poked, that the clicks occurred in short bursts of 1,000 or more per second and that each click contained ultrasonic frequencies within the range of hearing of bats.

My colleagues and I found that certain arctiids common in New England could also be induced to click if they were exposed to a string of ultrasonic pulses while they were suspended in stationary flight. In free flight these moths showed the evasive tactics I have already described. The clicking seems almost equivalent to telling the bat, "Here I am, come and get me." Since such altruism is not characteristic of the relation between predators and prey, there must be another answer.

Dorothy C. Dunning, a graduate student at Tufts, is at present trying to find it. She has already shown that partly tamed bats, trained to catch mealworms that are tossed into the air by a mechanical device, will commonly swerve away from their target if they hear tape-recorded arctiid clicks just before the moment of contact. Other ultrasounds, such as tape-recorded bat cries and "white" noise (noise of all frequencies), have relatively little effect on the bats' feeding behavior; the tossed mealworms are caught in midair and eaten. Thus the clicks made by arctiids seem to be heeded by bats as a warning rather than as an invitation. But a warning against what?

One of the pleasant things about scientific investigation is that the last logbook entry always ends with a question. In fact, the questions proliferate more rapidly than the answers and often carry one along unexpected paths. I suggested at the beginning of this article that it is my intention to trace the nervous mechanisms involved in the evasive behavior of moths. By defining the information conveyed by the acoustic cells I have only solved the least complex half of that broad problem. As I embark on the second half of the investigation, I hope it will lead up as many diverting side alleys as the study of the moth's acoustic system has.

5

The Sex-Attractant Receptor of Moths

by Dietrich Schneider
July 1974

The sex attractant of the female silk moth is detected by an array of receptors on the feathery antennae of the male. A nerve impulse in a receptor cell can be triggered by one molecule of attractant

All sensory systems, each with its specific peripheral receptor cells and its integrating neurons in the brain, are designed to form a biologically relevant image of the outer environment of the organism. In many animals chemical signals play a major and often decisive role as a means of communication. The domesticated silk moth, for example, cannot fly and therefore the male cannot readily scout the terrain in search of a mate. It does, however, have two featherlike antennae that are finely tuned to detect a certain chemical compound: the scent emitted by the female. So sensitive are the male's antennae that one molecule of the vaporous sex attractant will trigger a nerve impulse in a receptor cell. When approximately 200 nerve impulses have been generated in the span of a second, a message is received in the moth's brain and it moves upwind to claim its mate.

Chemical compounds such as the sex attractant of the silk moth, which are secreted by an animal and elicit a specific kind of behavior in animals of the same species, are called pheromones [see "Pheromones," by Edward O. Wilson; SCIENTIFIC AMERICAN Offprint 157]. A major difficulty in studying the olfactory system in most animals is the enormous variety of chemical compounds that elicit a response. The situation is different in the olfactory pheromone-receptor system of the silk moth. Here we have found a rather simple olfactory system where many receptor cells respond identically to only one compound.

In the past 20 years interest in pheromones has grown steadily, and the number of identified pheromones has rapidly increased. This holds particularly for the pheromones of insects because here the investigator's curiosity is augmented by the hope of putting pheromones to work as lures in the control of insect pests

[see "Insect Attractants," by Martin Jacobson and Morton Beroza; SCIENTIFIC AMERICAN Offprint 189]. In most cases sex-attractant pheromones are produced and emitted by the female in order to lure a mate. Sex attractants are widespread in the insect world, but they are also found in other animal classes, as is well known to every owner of a female cat or dog in heat.

The impressive attractiveness of a virgin female moth to its male partners was described as early as the 18th century by naturalists such as René Antoine Ferchault de Réaumur, F. Ch. Lesser and August Johann Rösel von Rosenhof. At the beginning of the 20th century the psychiatrist and entomologist Auguste Forel reported that when some wild European female silk moths emerged from the pupa in his studio in Lausanne, they attracted large numbers of male moths to the windows (along with a large number of *Gassenbuben,* or street urchins, who were attracted by the spectacular assembly of moths).

Although most of the early observers of the sexual attraction of male moths to females agreed that an odorous signal was involved, some were in doubt because of certain puzzling facts. Not only was the human nose unable to detect the alluring odor but also it was hardly conceivable that the amount of odorant that could be produced by the female would be able to lure the males from a distance of at least a kilometer. The critical experiment demonstrating that the attraction was definitely based on odor was described in 1879 by Jean Henri Fabre in his famous *Souvenirs Entomologiques.* When Fabre picked up a female moth and put her under a glass hood, male moths that flew into his house paid no attention to her but went to the place where she had been sitting a short time

earlier. Although Fabre realized that the female's scent was guiding the male at short range, he still believed unknown radiations from the female lured and guided the male from greater distances.

Other experiments with male moths clearly showed that the presumed olfactory faculty of these insects is localized in the antennae, since the males did not react to the "calling" female after their antennae had been removed or covered with varnish. This was the state of knowledge about these moth sex attractants until about 20 years ago. Then advances in chemical, physiological and histological methods led to the identification of the chemical nature of the luring substance and to the definite proof that the receptors are on the male moth's antennae.

The first sex-attractant pheromone to be chemically identified was the lure substance of the female of the commercial silk moth *Bombyx mori.* Adolf F. J. Butenandt and his co-workers at the Max Planck Institute for Biochemistry in Tübingen (and later in Munich) reported in 1959 that the attractant has a chain of 16 carbon atoms and is a doubly unsaturated fatty alcohol (later identified as *trans*-10-*cis*-12-hexadecadien-1-ol). They named the compound bombykol. Their success was the outcome of long years of difficult analytical work, and their choice of *Bombyx mori* was a wise one. This moth is bred for silk production in many parts of the world, so that it is possible to obtain large quantities of the female glands that manufacture the compound. In order to extract 12 milligrams of pure bombykol the biochemists needed the glands of half a million moths.

In the early 1950's, when only enriched extracts from female *Bombyx* glands were available, I met with my biochemical colleagues in Tübingen and was challenged by the problems of olfactory perception in the silk moth. I thought that insight into the highly specific olfactory function in this animal might lead to a better understanding of the still unknown mechanisms of olfactory perception in general. My research began as a one-man enterprise but later involved my students and associates. Members of the silk moth research team in my laboratory at the Max Planck Institute for Behavioral Physiology in Seewiesen are the biologists K.-E. Kaissling, E. Kramer, E. Priesner and R. A. Steinbrecht and the chemist G. Kasang. My recent studies of the gypsy moth and the nun moth were done in collaboration with the biophysicist W. A. Kafka. In our research we hoped to approach an answer to questions about the threshold and the dose-response functions of the odor receptors, the specificity of the receptors, the mechanism of odor-molecule capture, the mechanism of stimulus transduction and the fate of the odor molecule after it has transferred its information.

The peripheral part of any sensory cell reacts to an adequate stimulus (chemicals, light, mechanical displacement or temperature) with a temporary change in the electric charge of its membrane. This response is called the receptor potential. It can be recorded from whole sense organs, provided that the sensory cells are lined up rather like an array of interconnected electric batteries. In a silk-moth antenna simultaneous receptor potentials of many olfactory cells can be recorded simply by mounting the antenna between two electrodes connected to an amplifier and a recording instrument [see illustration below]. The antenna is stimulated by putting an odor source into a glass tube and blowing air through the tube onto the antenna. We tested the response of the male silk-moth antenna to various concentrations of bombykol, of steric isomers of the compound and of homologous fatty alcohols. Although all these compounds generated a response of the olfactory cells, none was nearly as effective as the natural material. We also found that the dose-response curve of this olfactory system covers a wide range of stimulus intensities, as do the response curves of the visual and auditory systems [see bottom illustration on page 58]. Interestingly, the antenna of the female silk moth does not respond to bombykol.

Microelectrode probes revealed that only those receptor cells connected to the specialized long hairs of the antennae respond to bombykol and its isomers. The amplitude of the discharge of the receptor cells increases with increasing concentration of the stimulating compound. This receptor potential generates a series of nerve impulses that travel to the olfactory center of the brain. The frequency of these nerve impulses depends on the amplitude of the receptor potential.

We then asked the following key questions: How many bombykol molecules are required to elicit the male's behavior response and what is the minimum number of molecules a cell needs to generate a nerve impulse? Before we could deal with these questions, however, we had to collect information on the amount of bombykol in the stimulating airstream and on the number of molecules being adsorbed on the antenna and on the hairs. For this purpose we resynthetized the pheromone in a form that incorporated tritium, the radioactive isotope of hydrogen. The measurements we then made with the tritium-labeled bombykol yielded surprising results: more than 25 percent of the bombykol is

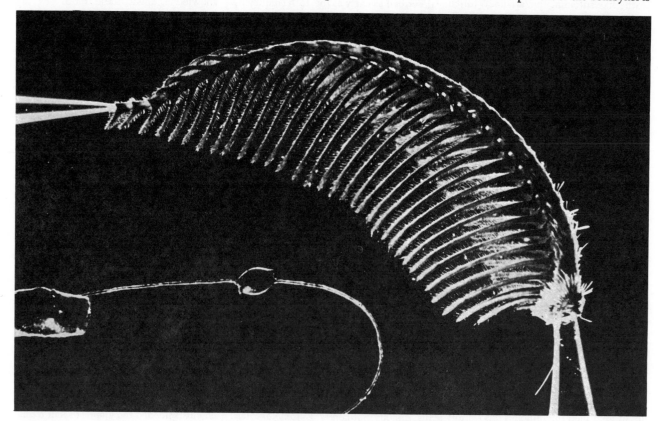

ANTENNA OF THE MALE SILK MOTH has some 17,000 long odor-receptor hairs on its branches. The electrophysiological response to an odor can be measured by mounting an isolated antenna between two glass-capillary electrodes (top left and bottom right) and passing over the antenna a stream of air containing the odor. The oscillographic record of the changes of electric potential in the antenna is called an electroantennogram. The wire loop at lower left holds a thermistor that measures the airflow past the antenna. The photograph was provided by K.-E. Kaissling of the Max Planck Institute for Behavioral Physiology in Seewiesen.

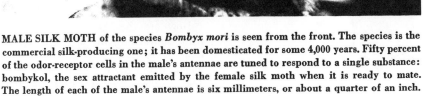

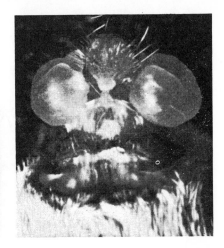

MALE SILK MOTH of the species *Bombyx mori* is seen from the front. The species is the commercial silk-producing one; it has been domesticated for some 4,000 years. Fifty percent of the odor-receptor cells in the male's antennae are tuned to respond to a single substance: bombykol, the sex attractant emitted by the female silk moth when it is ready to mate. The length of each of the male's antennae is six millimeters, or about a quarter of an inch.

TIP OF THE ABDOMEN of the female silk moth holds a pair of glands, the *sacculi laterales,* that contain about one microgram of the sex attractant bombykol. The glands shown here are in an expanded active state.

filtered out of the airstream when it hits the antenna.

We next conducted behavior tests and recorded electrophysiological signals from single odor-receptor cells in *Bombyx* males. When the male silk moth senses the sex attractant, it responds by fluttering its wings. A barely noticeable but nonetheless significant response is observed when the stimulating airstream contains about 1,000 bombykol molecules per cubic centimeter. Such a stimulus is produced by an odor source of only 3×10^{-6} microgram of bombykol. Within a second, which is somewhat more than the insect's reaction time, approximately 300 of the odor molecules are adsorbed on the 17,000 sensory hairs of the antenna. Each hair is innervated by the dendrites, or fiber endings, of two bombykol receptor cells. In this situation Poisson probability statistics, which mathematicians use to distinguish between different kinds of random events, enable us to predict that during that one second only two of the hairs receive double hits of bombykol molecules. The rest

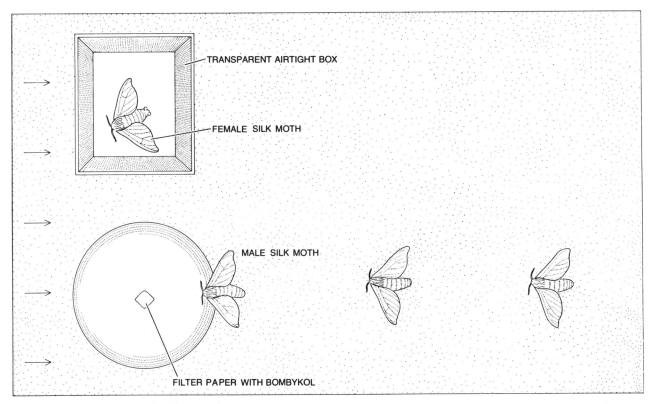

EXPERIMENTAL ARRANGEMENT for testing the sexual attractiveness of an odor is depicted. Male silk moths are placed downwind from a fan; the direction of the airflow is indicated by the arrows. The male moth cannot fly and is always sexually responsive. A female silk moth is placed in an airtight transparent box. A small piece of filter paper soaked with bombykol is put on a glass dish near the female moth. When the males sense the sex attractant, they move upwind to the source of the odor and not to the visible female. Bending of the abdomen of the male moth is a copulatory movement regularly observed with strong stimulation.

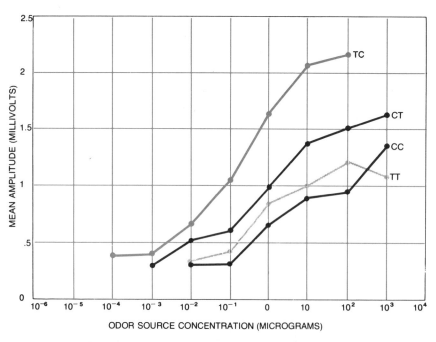

a

b

● CARBON
○ OXYGEN
• HYDROGEN

TRANS
CIS

TRANS
CIS

c

CIS
CIS

d

TRANS TRANS

BOMBYKOL is an unsaturated fatty alcohol with the chemical designation *trans*-10-*cis*-12-hexadecadien-1-ol (*a*). It is the *trans-cis* isomer of the compound. The three other geometrical isomers, the *cis-trans* (*b*), the *cis-cis* (*c*) and the *trans-trans* (*d*) are much less effective.

ELECTROANTENNOGRAM RESPONSE of the male silk moth to different concentrations of bombykol and its isomers is shown. When the concentration of the odor is less than .01 microgram, there is little or no difference in the antennal response to each isomer. At higher concentrations bombykol (*TC*) gives rise to a much greater response than the *cis-trans* (*CT*), *cis-cis* (*CC*) and *trans-trans* (*TT*) isomers do. In all cases the responses to very low odor concentrations are not significantly different from antennal responses to pure air.

receive single hits. This observation already made it highly probable that the receptor cells are activated by single bombykol molecules.

We now recorded olfactory-nerve impulses from hundreds of single bombykol receptor cells. The stimulating technique was essentially the same as the one we had used during the behavior tests, which enabled us to compare the two types of experiment. With a bombykol-source load of 10^{-4} microgram, one in three hairs adsorbs somewhat more than one molecule per second and fires about one impulse. Both of these values are averages.

How are the impulses distributed? After subtracting the spontaneous activity of the cells we plotted the experimental data against the Poisson curves. The responses were separately plotted for the cells that reacted with one impulse or more, two impulses or more and 10 impulses or more. The one-impulse curve and the two-impulse curve exactly fitted the "one hit" and "two hit" Poisson probability curves, whereas the 10-impulse curve did not correspond at all to the 10-hit probability curve [*see illustration on page 62*]. This shows that the small impulse numbers near the response threshold are randomly distributed, as one would expect to be the case with the stimulating molecules also. Behavior and single-cell responses therefore allow us to state that a single nerve impulse is generated when one molecule of bombykol hits a receptor and that two impulses are generated when two molecules hit. The transduction of hits into impulses is obviously more complex with higher hit rates per second.

One difficulty with the radioactivity measurements on which the calibrations of these experiments are based was the limited sensitivity of even the best measuring instruments available. The detection limit for tritium-labeled bombykol with these devices is 3×10^{-8} microgram, or about 10^8 molecules. That is enough for the measurement of the bombykol sources, but much smaller amounts are adsorbed on the antenna. Extrapolation is necessary to determine what the threshold amounts are. Our extrapolation is based on the assumption that the constant relation between the odor-source load and bombykol adsorption that is found in the range of the stronger stimuli holds true for the low rates near threshold.

We are now fairly certain that near the threshold level the receptors actually count the stimulating molecules. The cell can be said to be so sensitive that it reacts to single quanta of odor. A comparably high sensitivity has been found in the rod cells of the vertebrate retina, which respond to a single quantum of light.

If the receptor cells are already responding to a single pheromone molecule, why must 200 cells be activated to make the male moth respond with its typical vibrations? The answer is that all the bombykol receptor cells of the antenna spontaneously fire about 1,600 impulses per second in the resting state, and the threshold signal of 200 impulses is necessary to overcome this background noise. Information theory requires that a detectable signal be greater than three times the square root of the noise. Since here the noise is about 1,600 impulses per second, a meaningful signal must be greater than $3 \times \sqrt{1,600}$, which is equal to 120. The threshold signal of 200 impulses is well above the required minimum, so that the signal is sufficient to tell the moth's brain: "There is bombykol in the air."

The next questions we asked were: Why is the *Bombyx* antenna so effective in filtering the bombykol molecules out of the air? How do the molecules, after adsorption on the hair surface, find their way to the receptor-cell dendrite? Our thinking and experimentation on these lines have been strongly influenced and guided by Gerold Adam and Max Delbrück of the California Institute of Technology, who clearly outlined for us the physical principles that must govern these processes. Adam and Delbrück predicted that the antenna must be an optimal sieve for molecules because of the spacing and arrangement of the receptor hairs. The width of the mesh represented by the hairs is small enough so that the molecules of an odorant, because of their fast thermal movements, cannot pass through the hairs without coming in contact with them and being preferentially adsorbed. By measuring the adsorption of bombykol by the whole antenna and the adsorption by individual hairs that had been shaved from the antenna, we found that more than 80 percent of the molecules are adsorbed on the hairs. This result was significant because the total surface area of the hairs is less than 13 percent of the total surface area of the antenna.

If a receptor cell on a hair is to be activated by one or two molecules of bombykol, the molecules presumably need to diffuse to the receptor-cell dendrite from the surface of the hair. In order to demonstrate the validity of the assumption that there is such a two-dimensional diffusion of bombykol on the surface of the olfactory hair, it was necessary to analyze the structure of the hair. Such a hair, along with its olfactory receptor cells and auxiliary cells, is called a sensillum. The hair is part of the cuticle, which is the tough outer lining of the moth's body. The striking feature of the hairs is that they are perforated by pores connected to fine tubules. In some cases these tubules extend right down to the surface membrane of one of the receptor-cell dendrites. The pores and the tubules can be invaded by test

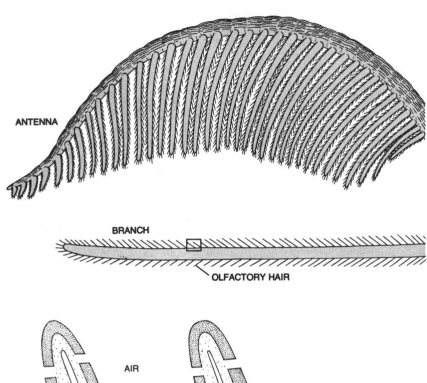

ANTENNA

BRANCH

OLFACTORY HAIR

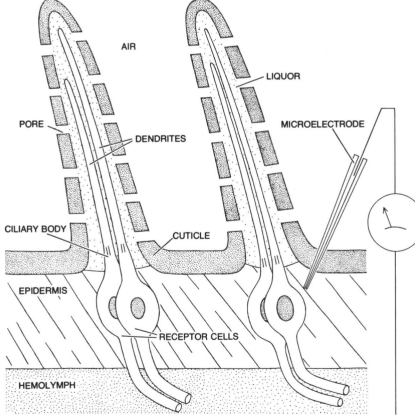

AIR

LIQUOR

PORE

DENDRITES

MICROELECTRODE

CILIARY BODY

CUTICLE

EPIDERMIS

RECEPTOR CELLS

HEMOLYMPH

OLFACTORY RECEPTOR SYSTEM of the male silk moth is shown at increasing magnifications, first the entire antenna (*top*), then a single branch of the antenna (*middle*) and finally a schematic longitudinal section of two olfactory hairs (*bottom*). Each hair is innervated by dendrites of two receptor cells. Molecules of bombykol diffuse through the pore openings in the hair and give rise to an electrical change in the membrane of the receptor cell. Nerve impulses are recorded from a microelectrode inserted into the base of the hair.

substances from the outside, as K. D. Ernst has shown in our laboratory, but the content of the tubules is unknown [*see illustrations on opposite page*].

These observations enabled us to construct the following model: The odor molecules are adsorbed on the hair surface, diffuse to the pores and from there through the tubules to the receptor-cell dendrite, where they elicit the receptor potential. We calculated the diffusion time of bombykol on the hairs and found it to be well within the electrophysiologically determined response time. Although the adsorption of bombykol on the hair surface is an established fact and diffusion along the surface of the hair is highly probable, we have not yet been able to follow the pheromone into the pores and tubules. The spatial resolution of the currently available autoradiographic methods is not high enough to locate the bombykol molecule in these conduits.

The process by which stimulus energy is transduced into receptor excitation is only partly understood for any sensory receptor cell, but with some reasonable assumptions we can at least outline it for the *Bombyx* odor-receptor cell. Our working hypothesis is that the bombykol molecule interacts with an acceptor in the membrane of the receptor-cell dendrite. The acceptor could function as a gating device, controlling the flow of ions through the membrane and thus the distribution of electric charges inside and outside the cell. The gating might be achieved by conformational changes in the molecular structure of the acceptor when it adsorbs the bombykol molecule.

Although the gating mechanism of the acceptor is definitely speculative, we can make some predictions about the properties of the active site of the acceptor. Fortunately we have quantitative information on a number of compounds that are effective in stimulating odor-receptor cells in other insects. When we looked into the physical properties of those molecules that activate a given type of cell, we were led to the conclusion that what happens in the process is not chemical bonding but weak physical interaction. We assume that the binding site and its reaction partner, the odor molecule, are complementary. On this basis the molecular specificity of the binding site can be deduced from the relative effectiveness of various stimulating molecules. Our observation that even the synthetic steric isomers of bombykol were from 100 to 1,000 times less effective than the natural pheromone indicates that the selective binding capacity of the bombykol acceptors is very high. The answer to the often-raised question "What makes a molecule an odor?" is on this functional level "The binding properties of the acceptor."

What is the fate of the odor molecule after it is bonded to the acceptor site in the receptor-cell membrane? With such a sensitive system it is unlikely that after the information transfer the odor molecule has a chance to activate the cell again. On the other hand, the odor molecule should not stay on the receptor site for too long, because the cell must be freed to be able to respond to another stimulus. We do not know as yet how the odor molecule is removed. One possible mechanism would be that it is immediately metabolized after the interaction. We have found that there is such a mechanism available, but it is neither specific enough nor fast enough to be directly involved in the transduction process.

Three years ago Morton Beroza of the U.S. Department of Agriculture asked us if we would be interested in extending our investigations to the gypsy moth, *Porthetria dispar*. His group had just successfully identified the lure pheromone of the female of this insect. It is *cis*-7,8-epoxy-2-methyloctadecane, dubbed "disparlure." As a result of the work of Beroza and his colleagues the compound was available in synthetic form. They had also succeeded in synthesizing 50 related epoxides, some with a different carbon-chain length, a shifted epoxy bridge (an oxygen attached to two carbons in the chain) and/or a shifted methyl group (CH_3). Starting a fruitful collaboration, we first repeated the measurement of the antennal response that had been made by our American colleagues and the experiments they had conducted with gypsy moths in the field. We also recorded the responses of single pheromone-receptor cells.

Our experiments clearly showed that disparlure is a more potent sex lure than

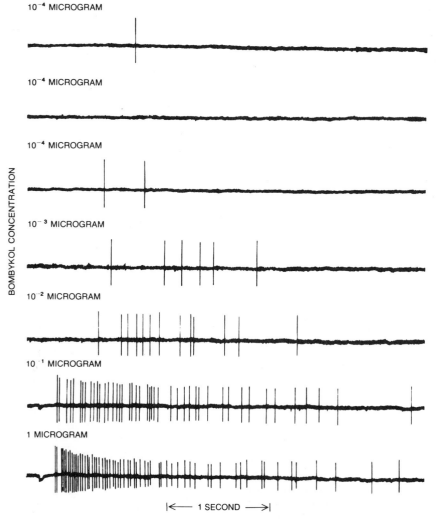

10^{-4} MICROGRAM

10^{-4} MICROGRAM

10^{-4} MICROGRAM

10^{-3} MICROGRAM

10^{-2} MICROGRAM

10^{-1} MICROGRAM

1 MICROGRAM

BOMBYKOL CONCENTRATION

|←——— 1 SECOND ———→|

NERVE IMPULSES generated in a bombykol receptor cell increase in frequency as the concentration of odor increases. A concentration of 10^{-4} microgram of bombykol on odor source gives rise to one or two impulses or none. Recordings were made by E. Priesner.

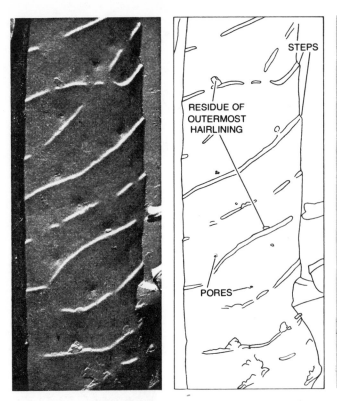

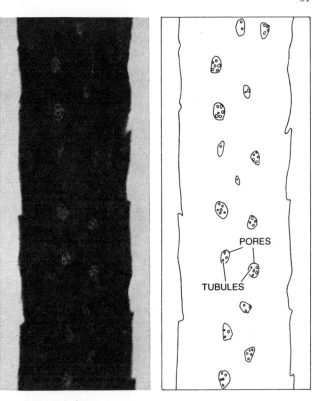

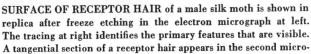

SURFACE OF RECEPTOR HAIR of a male silk moth is shown in replica after freeze etching in the electron micrograph at left. The tracing at right identifies the primary features that are visible. A tangential section of a receptor hair appears in the second micro- graph. The pores are shown in cross section. Inside the pores tubules that extend to the dendrite of the bombykol receptor cell can be seen. The electron micrographs were made by R. A. Steinbrecht of the Max Planck Institute for Behavioral Physiology.

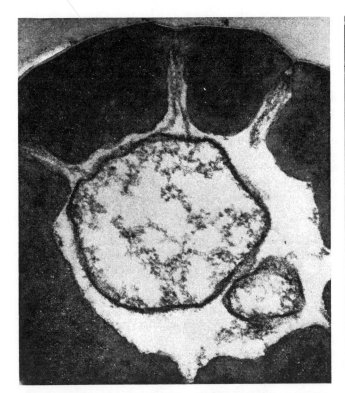

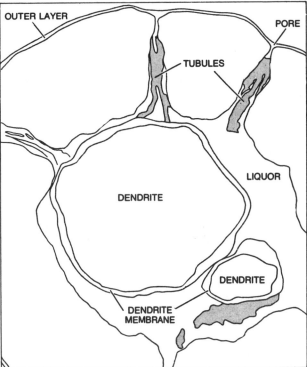

CROSS SECTION OF BOMBYKOL RECEPTOR HAIR is shown in this electron micrograph made by Steinbrecht. The cross section is through the apex of the hair. The tracing at right identifies certain of the hair's features. Some of the tubules in the pores reach to the outer membrane of the dendrite of the bombykol receptor cell. The outer pore openings are covered by several layers of a different electron density. Although the chemical composition of these layers is unknown, they are probably lipophilic, or fat-loving. Layers of such a composition would allow the fatty-alcohol molecules of bombykol to diffuse into the pore openings more readily.

any of the related compounds. The two compounds that elicited the next-largest responses are 20 and 100 times less effective. The male antennae of the gypsy moth and those of the silk moth are very much alike, so that there is little doubt that in principle the mechanisms by which the odor molecules are captured, are transferred and elicit electrical responses are identical in the two species. The details, however, have yet to be worked out.

The gypsy moth is a destructive forest pest in large parts of the eastern U.S. and in some parts of Europe (whence it was introduced to New England in 1869). The closest kin of this species is the nun moth, *Porthetria monacha*, which is found in the forests of central and northern Europe. We extended our studies to the nun moth, whose female sex attractant is not yet known. In our electrophysiological recordings we found that the nun moth had the same preference pattern for disparlure and related compounds as the gypsy moth has. The parallelism in their responses strongly suggests that the two species use the same compound as a sex attractant. Such a lack of species specificity in a sex attractant seems not to be an exception with species of the same genus but rather the rule.

Field experiments conducted earlier by H. Schönherr of the University of Freiburg, in which disparlure was successfully used to trap nun moths, also suggest that disparlure is probably the attractant for this species. In collaboration with R. Lange and F. Schwarz of the same institution, we have continued these studies by comparing disparlure as a bait with some of its related compounds. Again disparlure was by far the most effective attractant.

For two species to have the same attracting pheromone would not present any problem if the species were widely separated ecologically. For the nun moth and the gypsy moth, which live in close proximity in some parts of Europe, it could cause confusion. Possible mechanisms that would prevent uneconomical cross-attraction or even hybridization are differences in the rhythm of daily activity of the species, differences in general and sexual behavior and even a morphological incompatibility for copulation.

When Butenandt started the chemical analysis of the silk moth's sex attractant in the 1930's, he was not just interested in the composition of this enigmatic compound. He was already thinking of the possibility that such substances could be synthesized and used to lure pest in-

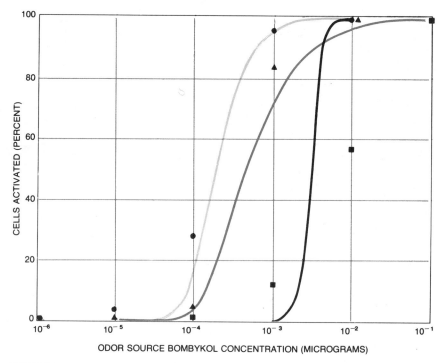

NUMBER OF CELLS that responded with one or more impulses (*circles*), two or more impulses (*triangles*) and 10 or more impulses (*squares*) were plotted against the concentration of bombykol. The data for the one-impulse response exactly fitted the Poisson probability curve for a random one-hit process (*darker color*). The data for the two-impulse responses fitted the two-hit curve (*lighter color*). The data for 10 or more impulses did not fit the 10-hit probability curve (*black*). These results, together with measurements of the adsorption of radioactively labeled bombykol on antennae, indicate that single impulses are generated by single molecules of bombykol and double impulses by two molecules.

sects into traps with high specificity. He and others hoped that this stratagem could be used to avoid the dangerous side effects of the generalized chemical insecticides. Although the need for species-specific biological pest controls is now even more pressing than it was 40 years ago, the trapping of insects (or perhaps the confusion of insect sexual behavior) with pheromones does not appear to be the panacea. Nonetheless, in a number of cases the luring of pest insects with pheromone traps has proved to be useful in predicting the outbreak of a large infestation and thus in timing, calibrating and eventually reducing the application of generalized insecticides. The U.S. Department of Agriculture still hopes to be able to use the gypsy moth's sex attractant to prevent or even halt the insect's dangerous rate of expansion.

Successful pheromone research and promising field trapping has also been conducted with leaf-roller moths by Wendell L. Roelofs and his colleagues at the State Agricultural Experiment Station in Geneva, N.Y. They found that a mixture of pheromonal components will specifically attract several of these fruit tree pests. Another pheromone control method has been examined by H. H.

Shorey and his colleagues at the University of California at Riverside. They evaporated a synthetic sex attractant of the cabbage looper in cabbage fields and the attractant of the pink bollworm moth in cotton fields. Under favorable conditions and with high doses of the attractant they produced an impressive degree of confusion among the males and a high percentage of unfertilized females.

In the future pheromones will probably become a component of a concerted system of biological control methods against pests that threaten crops and human health. Biologists the world over are searching for the Achilles' heel in the life of many pest species. My own view is that in many of these investigations a detailed analysis of the physiology, behavior and ecology of the insect involved is neglected. To be sure, behavior and ecology are complex and not easily studied expressions of the phenomenon called life. They nonetheless deserve the best effort of investigators not only for economic reasons such as the control of insect pests but also for a more fundamental reason: to better learn how we humans can survive in an ecologically balanced world together with our fellow organisms.

6

Eye Movements and Visual Perception

by David Noton and Lawrence Stark
June 1971

Recordings of the points inspected in the scanning of a picture and of the path the eyes follow in the inspection provide clues to the process whereby the brain perceives and recognizes objects

The eyes are the most active of all human sense organs. Other sensory receptors, such as the ears, accept rather passively whatever signals come their way, but the eyes are continually moving as they scan and inspect the details of the visual world. The movements of the eyes play an important role in visual perception, and analyzing them can reveal a great deal about the process of perception.

We have recently been recording the eye movements of human subjects as they first inspected unfamiliar objects and then later recognized them. In essence we found that every person has a characteristic way of looking at an object that is familiar to him. For each object he has a preferred path that his eyes tend to follow when he inspects or recognizes the object. Our results suggest a new hypothesis about visual learning and recognition. Before describing and explaining our experiments more fully we shall set the stage by outlining some earlier experiments that have aided the interpretation of our results.

Eye movements are necessary for a physiological reason: detailed visual information can be obtained only through the fovea, the small central area of the retina that has the highest concentration of photoreceptors. Therefore the eyes must move in order to provide information about objects that are to be inspected in any detail (except when the object is quite small in terms of the angle it subtends in the visual field). The eye-movement muscles, under the control of the brain, aim the eyes at points of interest [see "Control Mechanisms of the Eye," by Derek H. Fender, SCIENTIFIC AMERICAN, July 1964, and "Movements of the Eye," by E. Llewellyn Thomas, SCIENTIFIC AMERICAN Offprint 516].

During normal viewing of stationary objects the eyes alternate between fixations, when they are aimed at a fixed point in the visual field, and rapid movements called saccades. Each saccade leads to a new fixation on a different point in the visual field. Typically there are two or three saccades per second. The movements are so fast that they occupy only about 10 percent of the viewing time.

Visual learning and recognition involve storing and retrieving memories. By way of the lens, the retina and the optic nerve, nerve cells in the visual cortex of the brain are activated and an image of the object being viewed is formed there. (The image is of course in the form of neural activity and is quite unlike the retinal image of the object.) The memory system of the brain must contain an internal representation of every object that is to be recognized. Learning or becoming familiar with an object is the process of constructing this representation. Recognition of an object when it is encountered again is the process of matching it with its internal representation in the memory system.

A certain amount of controversy surrounds the question of whether visual recognition is a parallel, one-step process or a serial, step-by-step one. Psychologists of the Gestalt school have maintained that objects are recognized as wholes, without any need for analysis into component parts. This argument implies that the internal representation of each object is a unitary whole that is matched with the object in a single operation. More recently other psychologists have proposed that the internal representation is a piecemeal affair—an assemblage of parts or features. During recognition the features are matched serially with the features of the object step by step. Successful matching of all the features completes recognition.

The serial-recognition hypothesis is supported mainly by the results of experiments that measure the time taken by a subject to recognize different objects. Typically the subject scans an array of objects (usually abstract figures) looking for a previously memorized "target" object. The time he spends considering each object (either recognizing it as a target object or rejecting it as being different) is measured. That time is normally quite short, but it can be measured in various ways with adequate accuracy. Each object is small enough to be recognized with a single fixation, so that eye movements do not contribute to the time spent on recognition.

Experiments of this kind yield two general results. First, it is found that on the average the subject takes longer to recognize a target object than he does to reject a nontarget object. That is the result to be expected if objects are recognized serially, feature by feature. When an object is compared mentally with the internal representation of the target object, a nontarget object will fail to match some feature of the internal representation and will be rejected without further checking of features, whereas target objects will be checked on all features. The result seems inconsistent with the Gestalt hypothesis of a holistic internal representation matched with the object in a single operation. Presumably in such an operation the subject would take no longer to recognize an object than he would to reject it.

A second result is obtained by varying the complexity of the memorized target object. It is found that the subject takes longer to recognize complex target objects than to recognize simple ones. This result too is consistent with the serial-recognition hypothesis, since more features must be checked in the more complex object. By the same token the result also appears to be inconsistent with the Gestalt hypothesis.

It would be incorrect to give the impression that the serial nature of object recognition is firmly established to the exclusion of the unitary concept advanced by Gestalt psychologists. They have shown convincingly that there is indeed some "primitive unity" to an object, so that the object can often be singled out as a separate entity even before true recognition begins. Moreover, some of the recognition-time experiments described above provide evidence, at least with very simple objects, that as an object becomes well known its internal representation becomes more holistic and the recognition process correspondingly becomes more parallel. Nonetheless, the weight of evidence seems to support the serial hypothesis, at least for objects that are not notably simple and familiar.

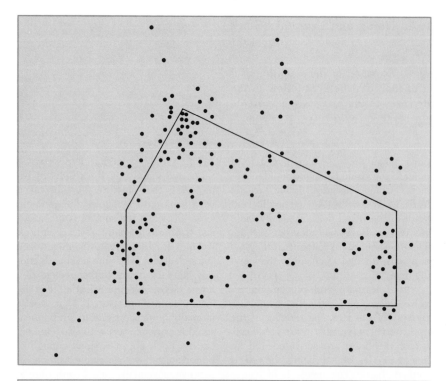

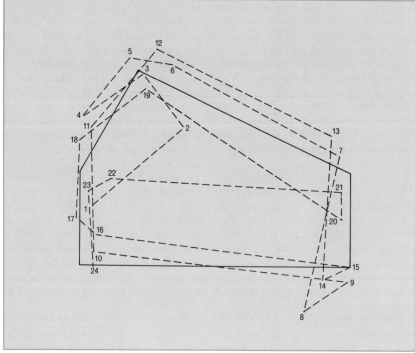

IMPORTANCE OF ANGLES as features that the brain employs in memorizing and recognizing an object was apparent in experiments by Leonard Zusne and Kenneth M. Michels at Purdue University. They recorded fixations while subjects looked at drawings of polygons for eight seconds. At top is one of the polygons; the dots indicate the fixations of seven subjects. Sequence of fixations by one subject in an eight-second viewing appears at bottom.

If the internal representation of an object in memory is an assemblage of features, two questions naturally suggest themselves. First, what are these features, that is, what components of an object does the brain select as the key items for identifying the object? Second, how are such features integrated and related to one another to form the complete internal representation of the object? The study of eye movements during visual perception yields considerable evidence on these two points.

In experiments relating to the first question the general approach is to present to a subject a picture or another object that is sufficiently large and close to the eyes so that it cannot all be registered on the foveas in one fixation. For example, a picture 35 centimeters wide and 100 centimeters from the eyes subtends a horizontal angle of 20 degrees at each eye—roughly the angle subtended by a page of this magazine held at arm's length. This is far wider than the one to two degrees of visual field that are brought to focus on the fovea.

Under these conditions the subject must move his eyes and look around the picture, fixating each part he wants to see clearly. The assumption is that he looks mainly at the parts of the picture he regards as being its features; they are the parts that hold for him the most information about the picture. Features are tentatively located by peripheral vision and then fixated directly for detailed inspection. (It is important to note that in these experiments and in the others we shall describe the subject is given only general instructions, such as "Just look at the pictures," or even no instructions at all. More specific instructions, requiring him to inspect and describe some specific aspect of the picture, usually result in appropriately directed fixations, as might be expected.)

When subjects freely view simple pictures, such as line drawings, under these conditions, it is found that their fixations tend to cluster around the angles of the picture. For example, Leonard Zusne and Kenneth M. Michels performed an experiment of this type at Purdue University, using as pictures line drawings of simple polygons [see *illustration on opposite page*]. From the fixations made by their subjects in viewing such figures it is clear that the angles of the drawings attracted the eyes most strongly.

Our tentative conclusion is that, at least with such line drawings, the angles are the principal features the brain employs to store and recognize the drawing. Certainly angles would be an efficient choice for features. In 1954 Fred Attneave III of the University of Oregon pointed out that the most informative parts of a line drawing are the angles and sharp curves. To illustrate his argument he presented a picture that was obtained by selecting the 38 points of greatest curvature in a picture of a sleeping cat and joining the points with straight lines [*see illustration above*]. The result is clearly recognizable.

Additional evidence that angles and sharp curves are features has come from electrophysiologists who have investigated the activity of individual brain cells. For example, in the late 1950's Jerome Y. Lettvin, H. R. Maturana, W. S. McCulloch and W. H. Pitts of the Massachusetts Institute of Technology found angle-detecting neurons in the frog's retina. More recently David H. Hubel and Torsten N. Wiesel of the Harvard Medical School have extended this result to cats and monkeys (whose angle-detecting cells are in the visual cortex rather than the retina). And recordings obtained from the human visual cortex by Elwin Marg of the University of California at Berkeley give preliminary indications that these results can be extended to man.

Somewhat analogous results have been obtained with pictures more complex than simple line drawings. It is not surprising that in such cases the features are also more complex. As a result no formal description of them has been achieved. Again, however, high information content seems to be the criterion. Norman H. Mackworth and A. J. Morandi made a series of recordings at Harvard University of fixations by subjects viewing two complex photographs. They concluded that the fixations were concentrated on unpredictable or unusual details, in particular on unpredictable contours. An unpredictable contour is one that changes direction rapidly and irregularly and therefore has a high information content.

We conclude, then, that angles and other informative details are the features selected by the brain for remembering and recognizing an object. The next question concerns how these

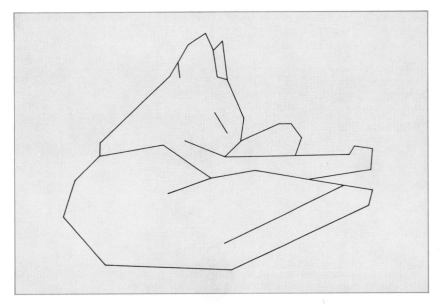

SHARP CURVES are also important as features for visual identification, as shown by Fred Attneave III of the University of Oregon in a picture made by selecting the 38 points of greatest curvature in a picture of a sleeping cat and joining them with straight lines, thus eliminating all other curves. The result is still easily recognizable, suggesting that points of sharp curvature provide highly useful information to the brain in visual perception.

features are integrated by the brain into a whole—the internal representation—so that one sees the object as a whole, as an object rather than an unconnected sequence of features. Once again useful evidence comes from recordings of eye movements. Just as study of the locations of fixations indicated the probable nature of the features, so analysis of the order of fixations suggests a for-

mat for the interconnection of features into the overall internal representation.

The illustration at left shows the fixations made by a subject while viewing a photograph of a bust of the Egyptian queen Nefertiti. It is one of a series of recordings made by Alfred L. Yarbus of the Institute for Problems of Information Transmission of the Academy of Sciences of the U.S.S.R. The illustration

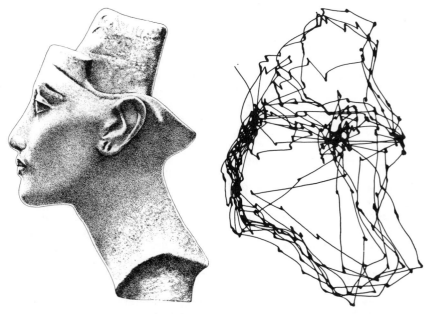

REGULARITIES OF EYE MOVEMENT appear in a recording of a subject viewing a photograph of a bust of Queen Nefertiti. At left is a drawing of what the subject saw; at right are his eye movements as recorded by Alfred L. Yarbus of the Institute for Problems of Information Transmission in Moscow. The eyes seem to visit the features of the head cyclically, following fairly regular pathways, rather than crisscrossing the picture at random.

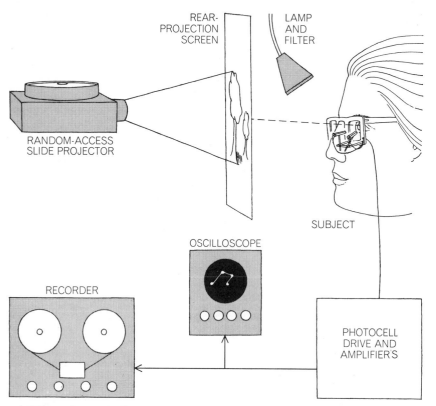

REAR-PROJECTION SCREEN

LAMP AND FILTER

RANDOM-ACCESS SLIDE PROJECTOR

OSCILLOSCOPE

SUBJECT

RECORDER

PHOTOCELL DRIVE AND AMPLIFIER'S

EXPERIMENTAL PROCEDURE employed by the authors is depicted schematically. The subject viewed pictures displayed on a rear-projection screen by a random-access slide projector. Diffuse infrared light was shined on his eyes; his eye movements were recorded by photocells, mounted on a spectacle frame, that detected reflections of the infrared light from one eyeball. Eye movements were displayed on oscilloscope and also recorded on tape.

image on a screen that was fully exposed to the ordinary light in the laboratory. In this way we produced an image of low visibility and could be sure that the subject would have to look directly (foveally) at each feature that interested him, thus revealing to our recording equipment the locus of his attention.

Our initial results amply confirmed the previous impression of cycles of eye movements. We found that when a subject viewed a picture under these conditions, his eyes usually scanned it following—intermittently but repeatedly —a fixed path, which we have termed his "scan path" for that picture [see illustration on opposite page]. The occurrences of the scan path were separated by periods in which the fixations were ordered in a less regular manner.

Each scan path was characteristic of a given subject viewing a given picture. A subject had a different scan path for every picture he viewed, and for a given picture each subject had a different scan path. A typical scan path for our pictures consisted of about 10 fixations and lasted for from three to five seconds. Scan paths usually occupied from 25 to 35 percent of the subject's viewing time, the rest being devoted to less regular eye movements.

It must be added that scan paths were not always observed. Certain pictures (one of a telephone, for example) seemed often not to provoke a repetitive response, although no definite common characteristic could be discerned in such pictures. The commonest reaction, however, was to exhibit a scan path. It was interesting now for us to refer back to the earlier recordings by Zusne and Michels, where we observed scan paths that had previously passed unnoticed. For instance, in the illustration on page 115 fixations No. 4 through No. 11 and No. 11 through No. 18 appear to be two occurrences of a scan path. They are identical, even to the inclusion of the small reverse movement in the lower right-hand corner of the figure.

This demonstration of the existence of scan paths strengthened and clarified our ideas about visual perception. In accordance with the serial hypothesis, we assume that the internal representation of an object in the memory system is an assemblage of features. To this we add a crucial hypothesis: that the features are assembled in a format we have termed a "feature ring" [see illustration on page 68]. The ring is a sequence of sensory and motor memory traces, alternately recording a feature of

shows clearly an important aspect of eye movement during visual perception, namely that the order of the fixations is by no means random. The lines representing the saccades form broad bands from point to point and do not crisscross the picture at random as would be expected if the eyes visited the different features repetitively in a random order. It appears that fixation on any one feature, such as Nefertiti's eye, is usually followed by fixation on the same next feature, such as her mouth. The overall record seems to indicate a series of cycles; in each cycle the eyes visit the main features of the picture, following rather regular pathways from feature to feature.

Recently at the University of California at Berkeley we have developed a hypothesis about visual perception that predicts and explains this apparent regularity of eye movement. Essentially we propose that in the internal representation or memory of the picture the features are linked together in sequence by the memory of the eye movement required to look from one feature to the next. Thus the eyes would tend to move from feature to feature in a fixed order, scanning the picture.

Most of Yarbus' recordings are summaries of many fixations and do not contain complete information on the ordering of the fixations. Thus the regularities of eye movements predicted by our hypothesis could not be definitely confirmed from his data. To eliminate this constraint and to subject our hypothesis to a more specific test we recently made a new series of recordings of eye movements during visual perception.

Our subjects viewed line drawings of simple objects and abstract symbols as we measured their eye movements (using photocells to determine the movements of the "white" of the eye) and recorded them on magnetic tape [see illustration above]. We thereby obtained a permanent record of the order of fixations made by the subjects and could play it back later at a lower speed, analyzing it at length for cycles and other regularities of movement. As in the earlier experiments, the drawings were fairly large and close to the subject's eyes, a typical drawing subtending about 20 degrees at the eye. In addition we drew the pictures with quite thin lines and displayed them with an underpowered slide projector, throwing a dim

the object and the eye movement required to reach the next feature. The feature ring establishes a fixed ordering of features and eye movements, corresponding to a scan path on the object.

Our hypothesis states that as a subject views an object for the first time and becomes familiar with it he scans it with his eyes and develops a scan path for it. During this time he lays down the memory traces of the feature ring, which records both the sensory activity and the motor activity. When he subsequently encounters the same object again, he recognizes it by matching it with the feature ring, which is its internal representation in his memory. Matching consists in verifying the successive features and carrying out the intervening eye

movements, as directed by the feature ring.

This hypothesis not only offers a plausible format for the internal representation of objects—a format consistent with the existence of scan paths—but also has certain other attractive features. For example, it enables us to draw an interesting analogy between perception and behavior, in which both are seen to involve the alternation of sensory and motor activity. In the case of behavior, such as the performance of a learned sequence of activities, the sensing of a situation alternates with motor activity designed to bring about an expected new situation. In the case of perception (or, more specifically, recognition) of an object the verification of features alternates with

movement of the eyes to the expected new feature.

The feature-ring hypothesis also makes a verifiable prediction concerning eye movements during recognition: The successive eye movements and feature verifications, being directed by the feature ring, should trace out the same scan path that was established for the object during the initial viewing. Confirmation of the prediction would further strengthen the case for the hypothesis. Since the prediction is subject to experimental confirmation we designed an experiment to test it.

The experiment had two phases, which we called the learning phase and the recognition phase. (We did not, of

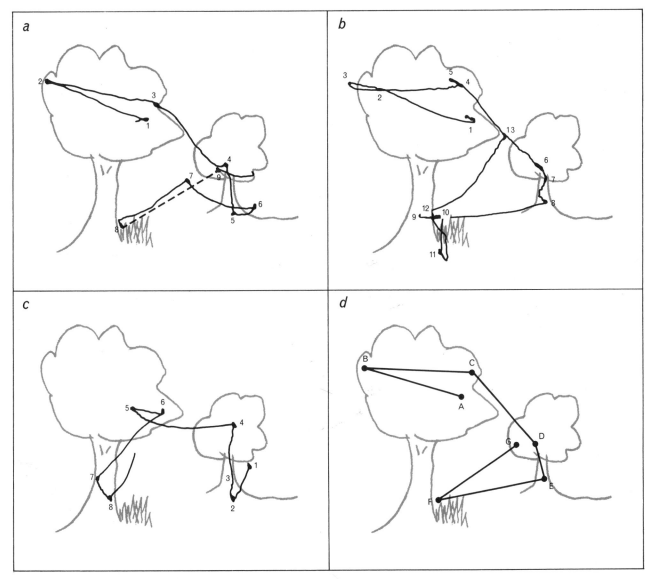

REGULAR PATTERN of eye movement by a given subject viewing a given picture was termed the subject's "scan path" for that picture. Two of five observed occurrences of one subject's scan path as he looked at a simple drawing of trees for 75 seconds are shown here (a, b). The dotted line between fixations 8 and 9 of a indicates that the recording of this saccade was interrupted by a blink. Less regular eye movements made between these appearances of the scan path are at c. Subject's scan path is idealized at d.

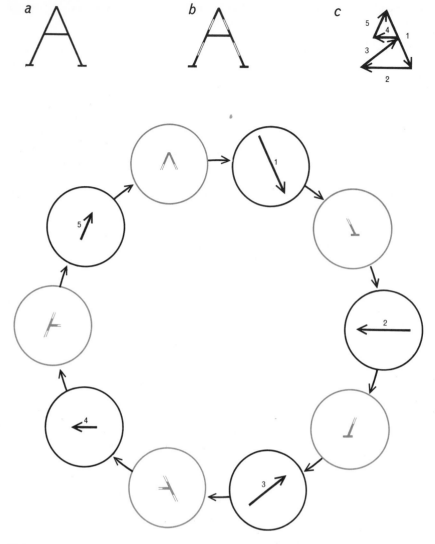

FEATURE RING is proposed by the authors as a format for the internal representation of an object. The object (*a*) is identified by its principal features (*b*) and is represented in the memory by them and by the recollection of the scan path (*c*) whereby they were viewed. The feature ring therefore consists of sensory memory traces (*color*) recording the features and motor memory traces (*black*) of the eye movements from one feature to the next.

ing-phase occurrences of the scan path; in the recognition phase he was matching the feature ring with the picture, following the scan path dictated by the feature ring.

An additional result of this experiment was to demonstrate that different subjects have different scan paths for a given picture and, conversely, that a given subject has different scan paths for different pictures [*see illustration on page 71*]. These findings help to discount certain alternative explanations that might be advanced to account for the occurrence of scan paths. The fact that a subject has quite different scan paths for different pictures suggests that the scan paths are not the result of some fixed habit of eye movement, such as reading Chinese vertically, brought to each picture but rather that they come from a more specific source, such as learned feature rings. Similarly, the differences among subjects in scan paths used for a given picture suggest that the scan paths do not result from peripheral feature detectors that control eye movements independent of the recognition process, since these detectors might be expected to operate in much the same way in all subjects.

Although the results of the second experiment provided considerable support for our ideas on visual perception, certain things remain unexplained. For example, sometimes no scan path was observed during the learning phase. Even when we did find a scan path, it did not always reappear in the recognition phase. On the average the appropriate scan path appeared in about 65 percent of the recognition-phase viewings. This is a rather strong result in view of the many possible paths around each picture, but it leaves 35 percent of the viewings, when no scan path appeared, in need of explanation.

Probably the basic idea of the feature ring needs elaboration. If provision were made for memory traces recording other eye movements between features not adjacent in the ring, and if the original ring represented the preferred and habitual order of processing rather than the inevitable order, the occasional substitution of an abnormal order for the

course, use any such suggestive terms in briefing the subjects; as before, they were simply told to look at the pictures.) In the learning phase the subject viewed five pictures he had not seen before, each for 20 seconds. The pictures and viewing conditions were similar to those of the first experiment. For the recognition phase, which followed immediately, the five pictures were mixed with five others the subject had not seen. This was to make the recognition task less easy. The set of 10 pictures was then presented to the subject three times in random order; he had five seconds to look at each picture. Eye movements were recorded during both the learning phase and the recognition phase.

When we analyzed the recordings, we were pleased to find that to a large

extent our predictions were confirmed. Scan paths appeared in the subject's eye movements during the learning phase, and during the recognition phase his first few eye movements on viewing a picture (presumably during the time he was recognizing it) usually followed the same scan path he had established for that picture during the learning phase [*see illustration on following page*]. In terms of our hypothesis the subject was forming a feature ring during the learn-

RECURRENCE OF SCAN PATH during recognition of an object is predicted by the feature-ring hypothesis. A subject viewed the adaptation of Klee's drawing (*a*). A scan path appeared while he was familiarizing himself with the picture (*b*, *c*). It also appeared (*d*, *e*) during the recognition phase each time he identified the picture as he viewed a sequence of familiar and unfamiliar scenes depicted in similar drawings. This particular experimental subject's scan path for this particular picture is presented in idealized form at *f*.

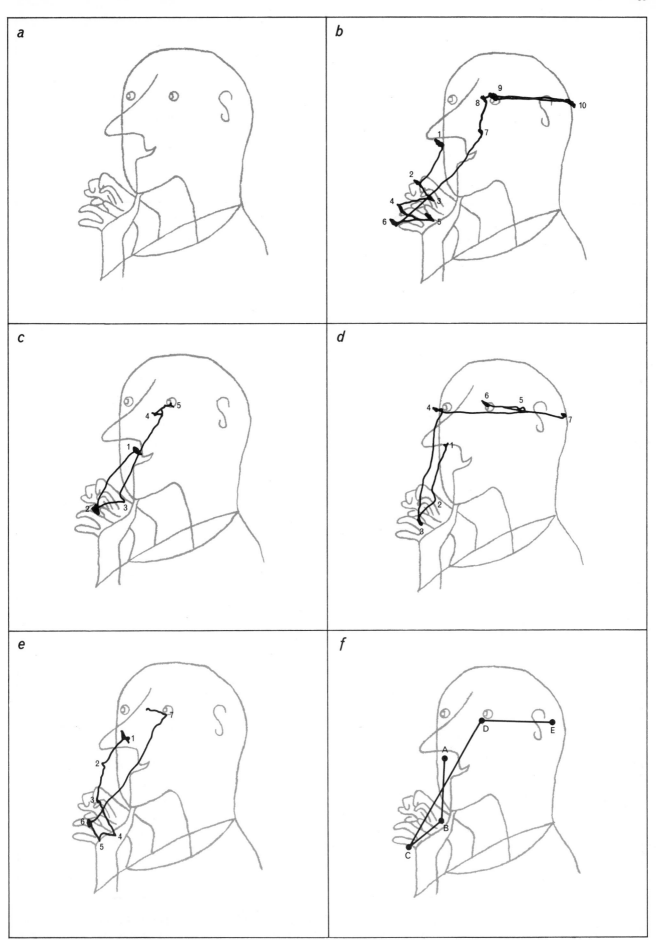

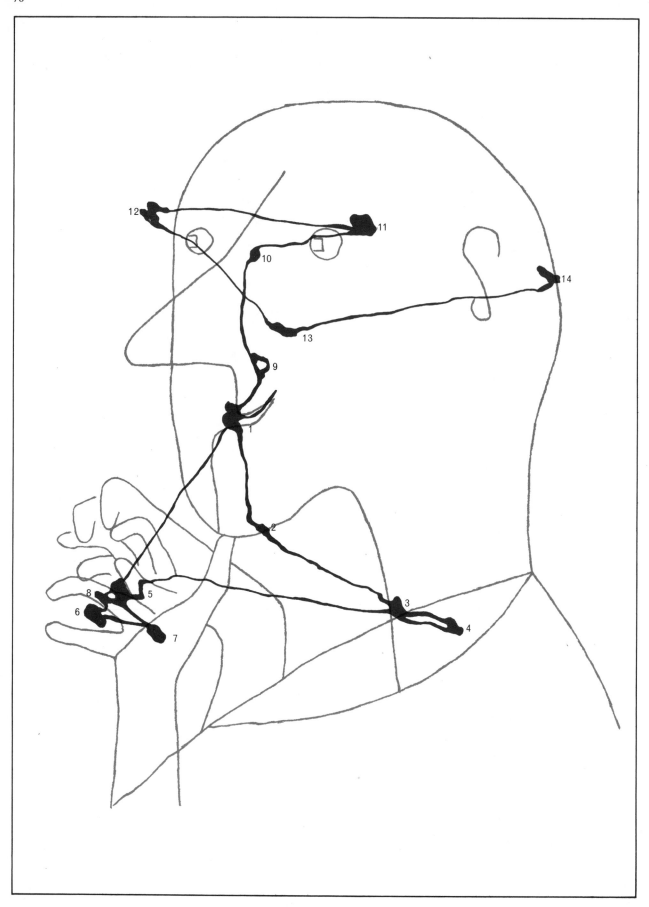

EYE MOVEMENTS made by a subject viewing for the first time a drawing adapted from Paul Klee's "Old Man Figuring" appear in black. Numbers show the order of the subject's visual fixations on the picture during part of a 20-second viewing. Lines between them represent saccades, or rapid movements of eyes from one fixation to the next. Saccades occupy about 10 percent of viewing time.

scan path would be explained [*see top illustration on following page*].

It must also be remembered that the eye-movement recordings in our experiments were made while the subjects viewed pictures that were rather large and close to their eyes, forcing them to look around in the picture to see its features clearly. In the more normal viewing situation, with a picture or an object small enough to be wholly visible with a single fixation, no eye movements are necessary for recognition. We assume

that in such a case the steps in perception are parallel up to the point where an image of the object is formed in the visual cortex and that thereafter (as would seem evident from the experiments on recognition time) the matching of the image and the internal representation is carried out serially, feature by feature. Now, however, we must postulate instead of eye movements from feature to feature a sequence of internal shifts of attention, processing the features serially and following the scan

path dictated by the feature ring. Thus each motor memory trace in the feature ring records a shift of attention that can be executed either externally, as an eye movement, or internally, depending on the extent of the shift required.

In this connection several recordings made by Lloyd Kaufman and Whitman Richards at M.I.T. are of interest. Their subjects viewed simple figures, such as a drawing of a cube, that could be taken in with a single fixation. At 10 randomly chosen moments the subject was asked

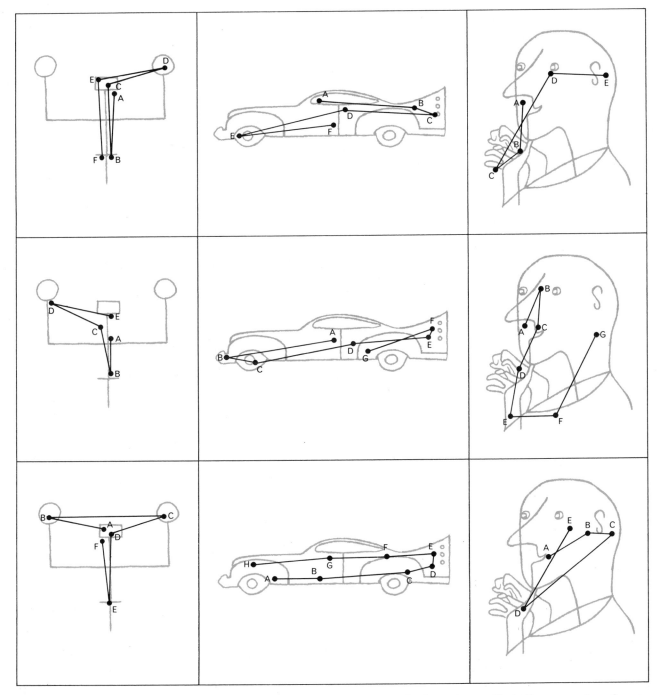

VARIETY IN SCAN PATHS is shown for three subjects and three pictures. Each horizontal row depicts the scan paths used by one subject for the three pictures. Vertically one sees how the scan paths of the three subjects for any one picture also varied widely.

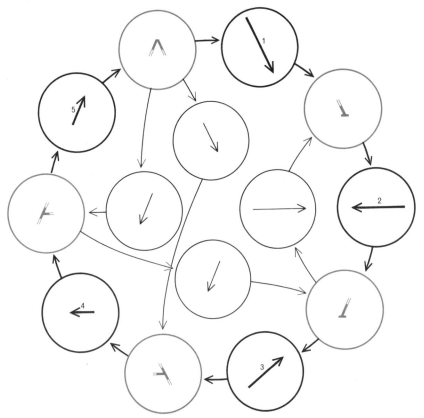

MODIFIED FEATURE RING takes into account less regular eye movements that do not conform to scan path. Several movements, which appeared in 35 percent of recognition viewings, are in center of this ring. Outside ring, consisting of sensory (*color*) and motor memory traces (*black*), represents scan path and remains preferred order of processing.

to indicate where he thought he was looking. His answer presumably showed what part of the picture he was attending to visually. His actual fixation point was then recorded at another 10 randomly selected moments [*see the illustration below*]. The results suggest that the subject's attention moved around the picture but his fixation remained fairly steady near the center of the picture. This finding is consistent with the view that smaller objects too

are processed serially, by internal shifts of attention, even though little or no eye movement is involved.

It is important to note, however, that neither these results nor ours prove that recognition of objects and pictures is necessarily a serial process under normal conditions, when the object is not so large and close as to force serial processing by eye movements. The experiments on recognition time support the serial hypothesis, but it cannot yet be regard-

ed as being conclusively established. In our experiments we provided a situation that forced the subject to view and recognize pictures serially with eye movements, thus revealing the order of feature processing, and we assumed that the results would be relevant to recognition under more normal conditions. Our results suggest a more detailed explanation of serial processing—the feature ring producing the scan path—but this explanation remains conditional on the serial hypothesis.

In sum, we believe the experimental results so far obtained support three main conclusions concerning the visual recognition of objects and pictures. First, the internal representation or memory of an object is a piecemeal affair: an assemblage of features or, more strictly, of memory traces of features; during recognition the internal representation is matched serially with the object, feature by feature. Second, the features of an object are the parts of it (such as the angles and curves of line drawings) that yield the most information. Third, the memory traces recording the features are assembled into the complete internal representation by being connected by other memory traces that record the shifts of attention required to pass from feature to feature, either with eye movements or with internal shifts of attention; the attention shifts connect the features in a preferred order, forming a feature ring and resulting in a scan path, which is usually followed when verifying the features during recognition.

Clearly these conclusions indicate a distinctly serial conception of visual learning and recognition. In the trend to look toward serial concepts to advance the understanding of visual perception one can note the influence of current work in computerized pattern recognition, where the serial approach has long been favored. Indeed, computer and information-processing concepts, usually serial in nature, are having an increasing influence on brain research in general.

Our own thoughts on visual recognition offer a case in point. We have developed them simultaneously with an analogous system for computerized pattern recognition. Although the system has not been implemented in working form, a somewhat similar scheme is being used in the visual-recognition system of a robot being developed by a group at the Stanford Research Institute. We believe this fruitful interaction between biology and engineering can be expected to continue, to the enrichment of both.

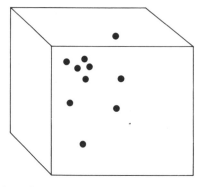

INTERNAL SHIFTS OF ATTENTION apparently replace eye movements in processing of objects small enough to be viewed with single fixation. A subject's attention, represented by statements of where he thought he was looking, moved around picture (*left*), whereas measured fixation point (*right*) remained relatively stationary. Illustration is based on work by Lloyd Kaufman and Whitman Richards at the Massachusetts Institute of Technology.

7

The Flight-Control System of the Locust

by Donald M. Wilson

May 1968

*Groups of nerve cells controlling such activities as
locomotion are regulated not only by simple reflex
mechanisms but also by behavior patterns apparently
coded genetically in the central nervous system*

Physicists can properly be concerned with atoms and subatomic particles as being important in themselves, but biologists often study simple or primitive structures with the long-range hope of understanding the workings of the most complex organisms, including man. Studies of viruses and bacteria made it possible to understand the basic molecular mechanisms that we believe control the heredity of all living things. Adopting a similar approach, investigators concerned with the mechanisms of behavior have turned their attention to the nervous systems of lower animals, and to isolated parts of such systems, in the hope of discovering the physiological mechanisms by which behavior is controlled.

One way to approach the study of behavior mechanisms is to ask: Where does the information come from that is needed to coordinate the observable activities of the nervous system? We know that certain behavior patterns are inherited.

This means that some of the informational input must be directly coded in the genetic material and therefore has an origin that is remote in time. Nonetheless, probably all behavior patterns depend to some degree on information supplied directly by the environment by way of the sense organs. Behavior that is largely triggered and coordinated by the nervous input of the moment is commonly called reflex behavior. Much of neurophysiological research has been directed at the analysis of reflex behavior mechanisms. Recent work makes it clear, however, that whole programs for the control of patterns of animal activity can be stored within the central nervous system [see the article "Small Systems of Nerve Cells," by Donald Kennedy, beginning on page 37]. Apparently these inherited nervous programs do not require much special input information for their expression.

I should point out here that whereas there is now general agreement among

biologists that many aspects of animal behavior are under genetic control, it is not easy to show in particular cases that a kind of behavior is inherited and not learned. I believe, however, that this is a reasonable assumption for the cases to be discussed in this article, namely flight and walking by arthropods (insects, crustaceans and other animals with an external skeleton).

The studies I shall describe were begun as part of an effort to demonstrate how several reflexes could be coordinated into an entire behavior pattern. Until recently it was thought by most students of simple behavior such as locomotion that much of the patterning of the nervous command that sets the muscles into rhythmic movement flowed rather directly from information in the immediately preceding sensory input. Each phase of movement was assumed to be triggered by a particular pattern of input from various receptors. According to this hypothesis, known as the peripheral-

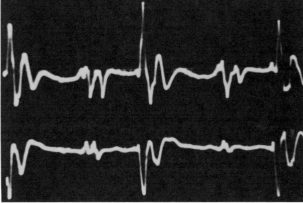

LOCUST WING position and wing-muscle action potentials were
recorded in synchronous photographs. The flash that illuminates
the locust (*left*) is triggered by the first muscle potential (*at left on
oscilloscope trace*). The wing motion is traced by spots of white
paint on each wing tip that reflect room light through the open

shutter. The trace at the top shows three "doublet" firings of downstroke muscles controlling the forewing; the bottom trace shows
similar firings for the hindwing. The smaller potentials visible between the large doublets are from elevator muscles more remote
from the electrodes. The oscilloscope traces span 100 milliseconds.

control hypothesis, locomotion might begin because of a signal from external sense organs such as the eye or from brain centers, but thereafter a cyclic reflex process kept it properly timed.

This cyclic reflex could be imagined to operate as follows. An initiating input causes motor nerve impulses to travel to certain muscles, and the muscles cause a movement. The movement is sensed by position or movement receptors within the body (proprioceptors), which send impulses back to the central nervous system. This proprioceptive feedback initiates activity in another set of muscles, perhaps muscles that are antagonists of the first set. The sequence of motor outputs and feedbacks is connected so that it is closed on itself and cyclic activity results. Clearly such a system depends on a well-planned (probably inherited) set of connections among the many parts involved; thus both the central nervous system and its peripheral extensions (the

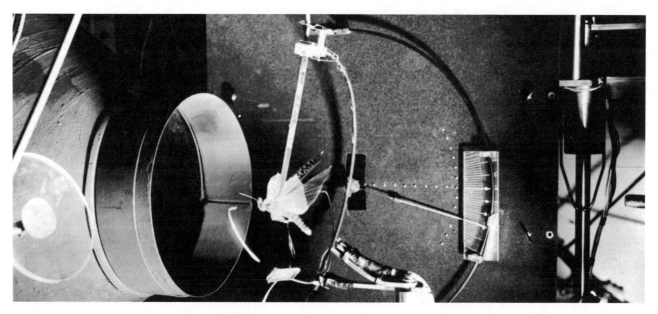

NERVE AND MUSCLE impulses were recorded during flight with this equipment. The locust is flying, suspended at the end of a pendulum, at the mouth of a wind tunnel. The scale at the right registers the insect's angle of pitch. The wires lead to amplifiers.

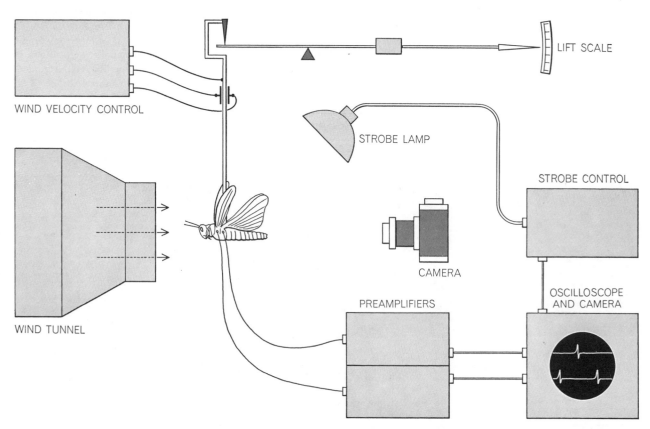

EXPERIMENTAL SETUP is diagrammed. The motion of the pendulum controls the wind-tunnel blower, so that the insect can fly at its desired wind speed. Muscle or nerve impulses are displayed on the oscilloscope, which in turn controls the stroboscopic flash lamp.

muscular and sensory structures) are crucial to the basic operation of the system.

An alternative hypothesis, known as the central-control hypothesis, suggests that the output pattern of motor nerve impulses controlling locomotion can be generated by the central nervous system alone, without proprioceptive feedback. This hypothesis has received much support from studies of embryological development. Only a few zoophysiologists have favored it, however, because the existence of proprioceptive reflexes had been clearly demonstrated. It seemed that if such reflexes exist, they must operate.

Proprioceptive reflexes certainly play an important role in the maintenance of posture. I suspect that this may be their basic and primitive function. In many animals—insects and man included—proprioceptive reflexes help to maintain a given body position against the force of gravity. A simple example was described by Gernot Wendler of the Max Planck Institute for the Physiology of Behavior in Germany. The stick insect, named for its appearance, stands so that its opposed legs form a flattened "M" [see illustration at right]. Sensory hairs are bent in proportion to the angle of the leg joints. The hairs send messages to the central nervous ganglia, concentrated groups of nerve cells and their fibrous branches that act as relay and coordinating centers. If too many impulses from the hairs are received, motor nerve cells are excited that cause muscles to contract, thereby moving the joint in the direction that decreases the sensory discharge. Thus the feedback is negative, and it results in the equilibration of a certain position.

If a weight is placed on the back of the insect, one would expect the greater force to bend the leg joints. Instead the proprioceptive feedback loop adjusts muscle tension to compensate for the extra load. The body position remains approximately constant, unless the weight is more than the muscles can bear. If the hair organs are destroyed, the feedback loop is opened and the body sags in relation to the weight, as one would expect in an uncontrolled system.

If the leg reflexes of arthropods are studied under dynamic rather than static conditions, one finds also that they are similar to the reflexes of vertebrates. When a leg of an animal is pushed and pulled rhythmically, the muscles respond reflexively with an output at the same frequency. At high frequencies of movement the reflex system cannot keep up and the output force developed by the muscles lags behind the input move-

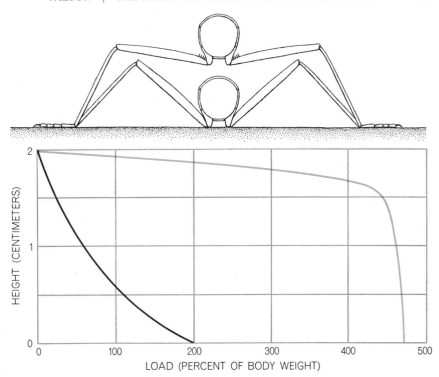

STICK INSECT keeps its body height nearly constant as the load on its leg muscles changes. Hairs at the first leg joint sense the angle of the joint and a reflex loop maintains the angle until the animal is overloaded (*colored curve*). If the sensory hairs are damaged, the reflex loop is opened and the body sags quickly as more weight is added to it (*black curve*).

ment. The peripheral-control hypothesis postulates an interaction of similar reflexes in each of the animal's legs. If one leg is commanded to lift, the others must bear more weight and postural reflexes presumably produce the increased muscle force that is needed. If any part oscillates, other parts oscillate too, perhaps in other phase relations. Although no such total system has been analyzed, one can imagine a sequence of reflex relationships that could coordinate all the legs into a smooth gait.

Against this background I shall describe the work on the nervous control of flight in locusts I began in the laboratory of Torkel Weis-Fogh at the University of Copenhagen in 1959. Weis-Fogh and his associates had already investigated many aspects of the mechanisms of insect flight, including the sense organs and their role in the initiation and maintenance of flight [see "The Flight of Locusts," by Torkel Weis-Fogh; SCIENTIFIC AMERICAN, March, 1956]. These studies, and the general climate of opinion among physiologists, tended to support a peripheral-control hypothesis based on reflexes. To test this hypothesis I set out to analyze the details of the reflex mechanisms.

An important consideration in the early phases of the work was how to study nervous activities in a small, rap-

idly moving animal. This was accomplished by having locusts fly in front of a wind tunnel while they were suspended on a pendulum that served as the arm of an extremely sensitive double-throw switch. The switch operated relays that controlled the blower of the tunnel, so that whenever the insect flew forward, the wind velocity increased and vice versa. Thus the insect chose its preferred wind speed, but it stood approximately still in space. Other devices measured aerodynamic lift and body and wing positions; wires that terminated in the muscles or on nerves conducted electrical impulses to amplifying and recording apparatus.

Early in the program of research we found that fewer than 20 motor nerve cells control the muscles of each wing, and that we could record from any of the motor units controlled by these cells during normal flight. We drew up a table showing when each motor unit was activated for various sets of aerodynamic conditions. The results of this rather tedious work were not very exciting but did provide a necessary base for further investigation. Moreover, I think we can say that these results constitute one of the first and most complete descriptions of the activity of a whole animal analyzed in terms of the activities of single motor nerve cells. In brief, we found that the output pattern consists of nearly synchronous impulses in two small populations of cooperating motor units, with activity alternating between antagonistic sets of muscles, the muscles that elevate the wings and the muscles that depress them. Each muscle unit normally receives one or two impulses per wingbeat or no impulse at all. The variation in the number of excitatory impulses sent to the different muscles serves to control flight power and direction.

We also found it possible to record from the sensory nerves that innervate, or carry signals to, the wings. These nerves conduct proprioceptive signals from receptors in the wing veins and in the wing hinge. The receptors in the wing veins register the upward force, or lift, on the wing; the receptors in the wing hinge indicate wing position and movement in relation to the body. These sensory inputs occur at particular phases of the wing stroke. The lift receptors usually discharge during the middle of the downstroke; each wing-hinge proprioceptor is a stretch receptor that discharges one, two or several impulses toward the end of the upstroke [*see illustration on page 78*].

Everything I have described so far about the motor output and sensory input of an insect in flight is consistent with the peripheral-control hypothesis. Motor impulses cause the movements the receptors register. According to the hypothesis the sensory feedback should trigger a new round of output. Does this actually happen?

A useful test of feedback-loop function is to open the loop. This we did simply by cutting or damaging the sense organs or sensory nerves that provide the feedback. Cutting the sensory nerve carrying the information about lift forces caused little change in the basic pattern of motor output, although it did affect the insect's ability to make certain maneuvers. On the other hand, burning the stretch receptors that measure wing position and angular velocity always resulted in a drastic reduction in wingbeat frequency. These proprioceptors provide the only input we could discover that had such an effect. Most important of all, we found that, even when we eliminated all sources of sensory feedback, the wings could be kept beating in a normal phase pattern, although at a somewhat reduced frequency, simply by stimulating the central nervous system with random electrical impulses.

From these studies we must conclude that the flight-control system of the locust is not adequately explained by the peripheral-control hypothesis and patterned feedback. Instead we find that the coordinated action of locust flight muscles depends on a pattern-generating system that is built into the central nervous system and can be turned on by an unpatterned input. This is a significant finding because it suggests that the networks within the nerve ganglia are endowed through genetic and developmental processes with the information needed to produce an important pattern of behavior and that proprioceptive reflexes are not major contributors of coordinating information.

Erik Gettrup and I were particularly curious to learn how the wing-hinge proprioceptor, a stretch receptor, helped to control the frequency of wingbeat. When Gettrup analyzed the response of this receptor to various wing movements, he found that to some degree it signaled to the central nervous system information on wing position, wingbeat amplitude and wingbeat frequency. We then cut out the four stretch receptors so that the wingbeat frequency was reduced to about half the normal frequency and artificially stimulated the stumps of the stretch receptors in an attempt to restore normal function. Under these conditions we found that electrical stimulation of the stumps could raise the frequency of wingbeat no matter what input pattern we used. Although the normal input

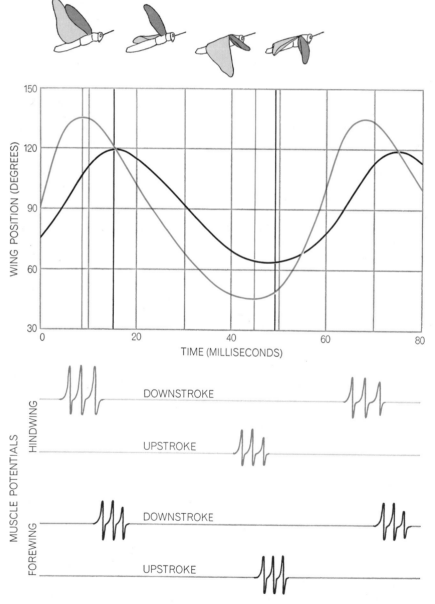

MUSCLE-POTENTIAL RECORDS are summarized in relation to wing positions in a flying locust. The curves (*top*) show the angular position (90 degrees is horizontal) of the hindwings (*color*) and forewings (*black*). The four simulated traces at the bottom show how the downstroke and upstroke muscles respectively fire at the high and low point for each wing.

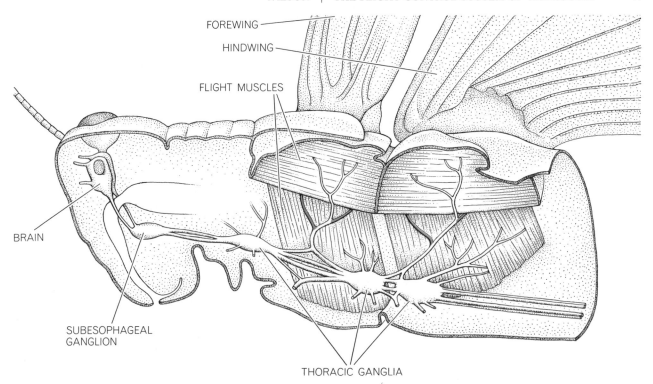

NERVES AND MUSCLES controlling flight in the locust are shown in simplified form. The central nervous system includes the brain and the various ganglia. From the thoracic ganglia, motor nerves lead to the wing's upstroke muscles (*vertical fibers*) and the downstroke muscles (*horizontal fibers*) above them. There are also sensory nerves (*color*) that sense wing position and aerodynamic forces.

from the stretch receptor arrives at a definite, regular time with respect to the wingbeat cycle, in our artificially stimulated preparations the effect was the same no matter what the phase of the input was.

We also found that the response to the input took quite a long time to develop. When the stimulator was turned on, the motor output frequency would increase gradually over about 20 to 40 wingbeat cycles. Hence it appears that the ganglion averages the input from the four stretch receptors (and other inputs too) over a rather long time interval compared with the wingbeat period, and that this averaged level of excitation controls wingbeat frequency. In establishing this average of the input most of the detailed information about wing position, frequency and amplitude is lost or discarded, with the result that no reflection of the detailed input pattern is found in the motor output pattern. We must therefore conclude that the input turns on a central pattern generator and regulates its average level of activity, but that it does not determine the main features of the pattern it produces. These features are apparently genetically programmed into the central network.

If that is so, why does the locust even have a stretch reflex to control wingbeat frequency? If the entire ordered pattern needed to activate flight muscles can be coded within the ganglion, why not also include the code for wingbeat frequency? The answer to this dual question can probably be found in mechanical considerations. The wings, muscles and skeleton of the flight system of the locust form a mechanically resonant system—a system with a preferred frequency at which conversion of muscular work to aerodynamic power is most efficient. This frequency is a function of the insect's size. It seems likely that even insects with the same genetic makeup may reach different sizes because of different environmental conditions during egg production and development. Hence each adult insect must be able to measure its own size, as it were, to find the best wingbeat frequency. This measurement may be provided by the stretch reflex, automatically regulating the wingbeat frequency to the mechanically resonant one.

What kind of pattern-generating nerve network is contained in the ganglia? We do not know as yet. Nonetheless, a plausible model can be suggested. The arguments leading to this model are not rigorous and the evidence in its favor is not overwhelming, but it is always useful to have a working hypothesis as a guide in planning future experiments. Also, it seems worthwhile to present a hypothesis of how a simply structured network might produce a special temporally patterned output when it is excited by an unpatterned input.

When neurophysiologists find a system in which there is alternating action between two sets of antagonistic muscles, they tend to visualize a controlling nerve network in which there is reciprocal inhibition between the two sets of nerve cells [see top illustration on page 79]. Such a network can turn on one or both sets at first, but one soon dominates and the other is silenced. When the dominant set finally slows down from fatigue, the inhibiting signal it sends to the silent set also decreases, with the result that the silent set turns on. It then inhibits the first set. This reciprocating action is analogous to the action of an electronic flip-flop circuit; timing cues are not needed in the input. The information required for the generation of the output pattern is contained largely in the structure of the network and not in the input, which only sets the average level of activity. A nerve network that acts in this way can consist of as few as two cells or be made up of two populations of cells in which there is some mechanism to keep the cooperating units working together.

In the locust flight-control system several tens of motor nerve cells work together in each of the two main sets. The individual nerve cells within each set

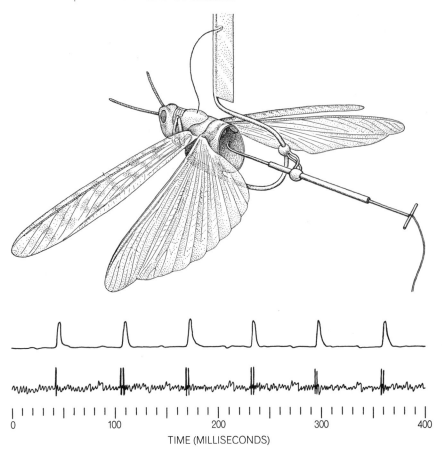

TIME (MILLISECONDS)

SENSORY DISCHARGES in nerves from the wing and wing hinge are recorded with wires manipulated into the largely eviscerated thoracic cavity of a locust. The top record is of downstroke muscle potentials, which are repeating at the wingbeat frequency. The bottom record is of a sensory (stretch) receptor from one wing, firing one or two times per wingbeat.

seem to share some excitatory interconnections. These not only provide the coupling that keeps the set working efficiently but also have a further effect of some importance. Strong positive coupling between nerve cells can result in positive feedback "runaway"; the network, once it is activated, produces a heavy burst of near-maximum activity until it is fatigued [*see middle illustration on opposite page*] and then turns off altogether until it recovers. A network of this kind can also produce sustained oscillations consisting of successive bursts of activity alternating with periods of silence without any patterned input. Thus either reciprocal inhibition or mutual excitation can give rise to the general type of burst pattern seen in locust motor units. Both mechanisms have been demonstrated in various behavior-control systems. It is likely that both are working in the locust and that these two mechanisms, as well as others, converge to produce a pattern of greater stability than might otherwise be achieved.

In summary, the model suggests that each group of cooperating nerve cells is

mutually excitatory, so that the units of each group tend to fire together and produce bursts of activity even when their input is steady. In addition the two sets are connected to each other by inhibitory linkages that set the two populations into alternation. Notice that in these hypothetical networks the temporal pattern of activity is due to the network structure, not to the pattern of the input. Even the silent network stores most of the information needed to produce the output pattern.

The locust flight-control system consists in large part of a particular kind of circuit built into the central ganglia. Some other locomotory systems seem much more influenced by reflex inputs. As an example of such a system I shall describe briefly the walking pattern of the tarantula spider. There is much variation in the relative timing of the eight legs of this animal, but on the average the legs exhibit what is called a diagonal rhythm. Opposite legs of one segment alternate and adjacent legs on one side alternate so that diagonal pairs of legs are in step [*see illustration on page 80*].

The tarantula can lose several of its legs and still walk. Suppose the first and third pairs of legs are amputated. If the spider's legs were coordinated by means of a simple preprogrammed circuit like the one controlling the locust's flight muscles, one would expect the spider to move the remaining two legs on one side in step with each other and out of step with the legs on the other side. A four-legged spider that did this would fall over. In actuality the spider adjusts relations between the remaining legs to achieve the diagonal rhythm. Other combinations of amputations give rise to other adaptations that also maintain the mechanically more stable diagonal rhythm.

Thus it appears that the pattern of coordination does depend on input from the legs. One can advance a possible explanation. Each leg is either driven by a purely central nervous oscillator or each leg and its portion of ganglion forms an oscillating reflex feedback loop. Suppose the several oscillators are negatively coupled. A pair of matched negatively coupled oscillators will operate out of phase. If the nearest leg oscillators are negatively coupled more strongly than the ones farther apart, the normal diagonal rhythm will result. For example, if left leg 1 has a strong tendency to alternate with right leg 1 and right leg 1 alternates with right leg 2, then left leg 1 must operate synchronously with right leg 2, to which it is more weakly connected. Now if some of the oscillators are turned off by amputating legs, so that either the postulated oscillatory feedback loop is broken or the postulated central oscillator receives insufficient excitatory input, new patterns of leg movements will appear that will always exhibit a diagonal rhythm.

The real nature of the oscillators involved in the leg rhythms is not known. These results and the postulated model nonetheless illustrate how sensory feedback could be used by the nervous system in such a way that the animal could adjust to genetically unpredictable conditions of the body or environment without recourse to learning mechanisms. Could this be the role of reflexes in general? We have seen that in the locust flight system much information for pattern generation is centrally stored—presumably having been provided genetically—and that the reflexes do seem to supply only information that could not have been known genetically.

A way to describe the two general models of muscle-control systems has been suggested by Graham Hoyle of the University of Oregon. He calls the cen-

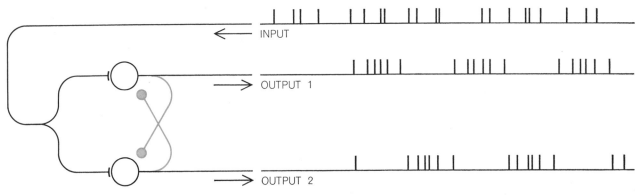

CROSS INHIBITION is one kind of interaction between nerve cells. The cells are connected in such a way that impulses from one inhibit the other (*color*). This can cause a pattern of alternating bursts, each cell firing (inhibiting the other) until fatigued. The hypothetical network shows how an unpatterned input can be transformed into a patterned output by structurally coded information.

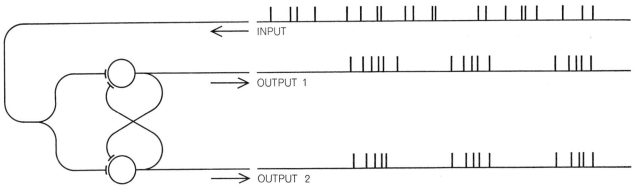

IN CROSS EXCITATION the output from each cell excites the other. This makes for approximate synchrony. There may also be a positive feedback "runaway" until fatigue causes deceleration or a pause; once rested, the network begins another accelerating burst.

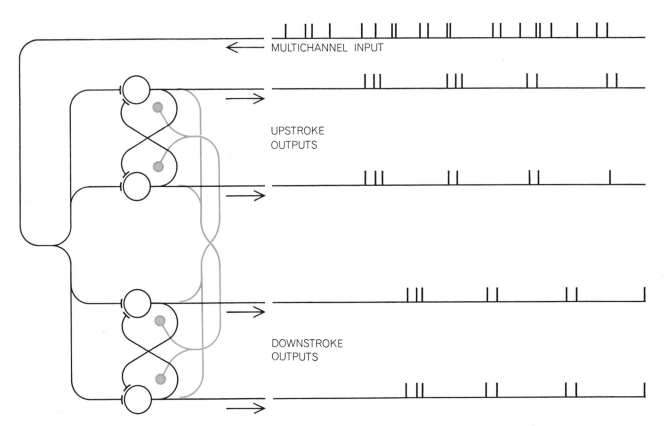

HYPOTHETICAL NETWORK of nerve cells in the locust might involve two cell populations, an upstroke group (*top*) and a downstroke group (*bottom*). Cells *within* a group excite one another but there are inhibitory connections *between* groups (*color*). The inhibition keeps the activity of one group out of phase with that of the other, so that upstroke and downstroke muscles alternate.

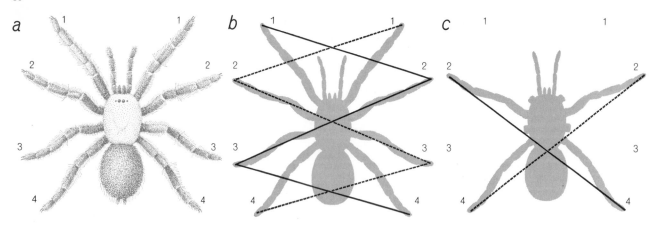

LEGS OF TARANTULA (*a*) move in diagonally arranged groups of four (*b*). The diagonal pattern persists if some legs are removed (*c*), suggesting that proprioceptive reflexes set the pattern or that changed input alters inherent central nervous system activity.

trally stored system of pattern generation a motor-tape system. In such a system a preprogrammed "motor score" plays in a stereotyped manner whenever it is excited. For a system in which reflexes significantly modify the behavioral sequences, Hoyle suggests the term "sensory tape" system. I prefer the term "sensory template." Such systems have a preprogrammed input requirement that can be achieved by various outputs. If the output at any moment results in an input that does not match the template, the mismatch results in a changed output pattern until the difference disappears. Such a goal-oriented feedback system can adapt to unexpected environments or to bodily damage. Spider and insect walking patterns show a kind of plasticity in which new movement patterns can compensate for the loss of limbs. In the past this kind of plasticity has often been interpreted as evidence for the reflex control of locomotory pattern. The locust flight-control system, on the other hand, certainly has a motor score that is not organized as a set of reflexes. Can a motor-score system show the plastic behavior usually associated with reflex, or sensory-template, systems?

For several years we thought that the locust flight-control system was relatively unplastic. We knew about the control of wingbeat frequency by the stretch reflex and about other reflexes, for example a reflex that tends to keep the body's angle of pitch constant. When I had made recordings of nerve impulses to some important control muscles before and after cutting out other muscles or whole wings, however, I had not found differences in the motor output. I therefore concluded that the flight-control system was not capable of a wide range of adaptive behavior. On this basis one

would predict that a damaged locust could not fly, or could fly only in circles.

The locust has four wings. Recently I cut whole wings from several locusts, threw the locusts into the air and found to my surprise that they flew quite well. The flying locust shows just as much ability to adapt to the loss of a limb as a walking insect or spider does. For the crippled locust to fly it must significantly change its motor output pattern. Why did the locust not show this change in the experiments in which I made recordings of nerve impulses to its muscles?

In all the laboratory experiments the locust was approximately fixed in space. If the insect made a motor error, it might sense the error proprioceptively, but it could not receive a feedback signal from the environment around it indicating that it was off course. The free-flying locust has at least two extraproprioceptive sources of feedback about its locomotory progress: signals from its visual system and signals from directionally sensitive hairs on its head that respond to the flow of air. Either or both of these extraproprioceptive sources can tell the locust that it is turning in flight. In the free-flying locust the signals are involved in a negative feedback control that tends to keep the animal flying straight in spite of functional or anatomical errors in the insect's basic motor system. When animals are studied in the laboratory under conditions that do allow motor output errors or anatomical damage to produce turning motions, then compensatory changes in the motor output pattern are observed provided only that the appropriate sensory structures are intact.

The locust flight-control system consists in part of a built-in motor score, but it also shows the adaptability expected of a reflex, or sensory-template, type of control. From these observations on plasticity in the locust flight system one

can see that there may be no such thing as a pure motor-tape or a pure sensory-template system. Many behavior systems probably have some features of each.

What we are striving for in studies such as the ones I have reported here is a way of understanding the functioning of networks of nerve cells that control animal behavior. Neurophysiologists have already acquired wide knowledge about single nerve cells—how their impulses code messages and how the synapses transmit and integrate the messages. Much is also known about the electrical behavior and chemistry of large masses of nerve cells in the brains of animals. The intermediate level, involving networks of tens or hundreds of nerve cells, remains little explored. This is an area in which many neurobiologists will probably be working in the next few years. I suspect that it is an area in which important problems are ripe for solution.

I shall conclude with a few remarks on the unraveling of the mechanisms of genetically coded behavior. As I see it, there are two major stages in the readout of genetically coded behavioral information. The first stage is the general process of development of bodily form, including the detailed form of the networks of the central nervous system and the form of peripheral body parts, such as muscles and sense organs that are involved in the reflexes. This stage of the genetic readout is not limited to neurobiology, of course. It is a stage that will probably be analyzed largely at the molecular level. The second stage involves a problem that is primarily neurobiological: How can information that is coded in the grosser level of nervous-system structure, in the shapes of whole nerve cells and networks of nerve cells, be translated into temporal sequences of behavior?

Brains and Cocoons

by William G. Van der Kloot
April 1956

*The fluffy shroud of the silkworm represents a map of
its spinning movements. This pattern may be altered
by brain surgery, shedding light on the relationship
between the nervous system and behavior*

Every silkworm that is to become a moth must first spin itself a cocoon. This is an intricate process. Just how intricate can be seen by looking at the cross-sectioned cocoon in the photograph below. The caterpillar first spun a thin, tough, densely woven outer envelope. Then it laid down a loose cushion of silk. Inside this it enclosed itself in a second thin, tough envelope. There the caterpillar rests, metamor-

phosed to the pupal stage. In building the cocoon the caterpillar extruded more than a mile of silk.

The intricately woven cocoon gives us a three-dimensional map of the caterpillar's behavior. It thereby presents a visible record of the operations of the insect's nervous system. As such it serves as a convenient and relatively simple model for analyzing nervous systems in general. Beavers build dams, and men

may build cathedrals. Silkworm cocoons, beaver dams and cathedrals all arise basically from the action of units which are much the same in each case—the nerve cells. We know that all nerve cells function on the same elementary "all or nothing" principle. A nerve cell either fires an impulse or it remains dormant; the impulse either excites or inhibits a neighboring cell. It is the patterned firing of thousands and millions of nerve

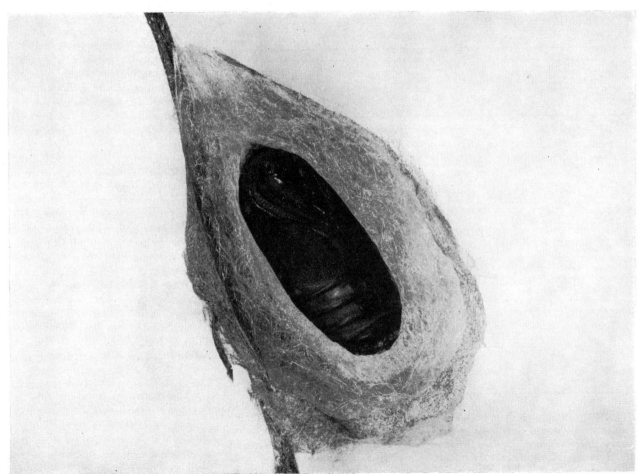

COCOON of the *Cecropia* silkworm is cut in half to show its inner construction. In the middle of the cocoon is the pupa of Cecropia.

The cocoon consists of three layers: a thin outer envelope, a thin inner envelope and a thick, loose cushion of silk between them.

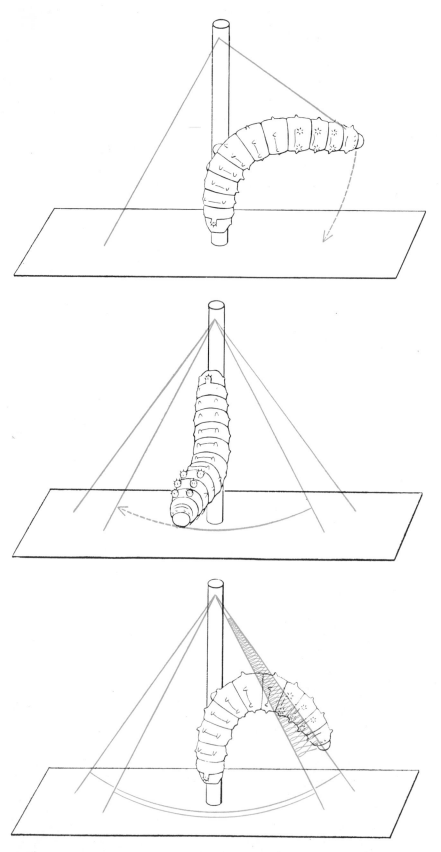

cells that determines specific behavior. To understand how behavior is directed and controlled we must look into the circuitry by which a creature's nerve cells are hooked up in its central nervous system.

The building of cocoons provides an excellent approach to the investigation of nervous system circuitry. Although the silkworm's circuitry is sufficiently baffling, it is far simpler than the humblest vertebrate brain. Cocoon building is an inborn behavior pattern, dependably repeated by each individual without prior learning. The *Cecropia* silkworm invariably builds the same kind of cocoon; other caterpillars build other kinds. And the ways in which the silkworm's behavior may be modified by experiments are recorded visibly in its spinning patterns. These conveniences led Carroll M. Williams and me to select cocoon construction for study in the Biological Laboratories at Harvard.

We provide our silkworms with a spinning platform which consists of a dowel mounted perpendicularly in the center of a board. In this controlled situation they give us a uniform performance. Two patterns of movement shape the cocoon. The silkworm climbs up the dowel until its hind end is just above the board. In this position it stretches the forepart of its body upward and fastens the beginning of its silk thread to a point near the top of the dowel. Then it bends its forebody downward and to the side, paying out the line of silk as it moves, until its head touches the board. The caterpillar fastens the silk thread there, stretches upward again, glues the thread again to the dowel, and bends down to a different point on the board. The sequence of movements is repeated again and again. The silkworm gradually shrouds itself with a cone of silk.

Periodically during spinning its behavior changes. The silkworm reverses its position on the dowel so that its hind end is up and its head toward the board. It now swings the forebody first to one side and then to the other, spinning a relatively flat sheet of silk on the board. This will be the bottom of the cocoon. In nature the caterpillar rarely finds a flat surface beneath it, and the bottom is usually rounded. The shape of its cocoon, in short, is influenced by the environment: the silkworm must find points on a surface (or the webbing of its own silk) to attach the thread. If it does not make contact with some surface at the end of a movement, the silkworm will sway about until a contact is made. Where points of silk attachment are limi-

MOVEMENTS OF A SILKWORM spinning its cocoon are shown in these drawings. The movements are executed on a dowel mounted on a board in the laboratory. At first the silkworm climbs the dowel, fastens a thread to it and bends downward to attach the thread to the board (*top drawing*). Then the silkworm faces downward and spins the floor of the cocoon (*middle drawing*). When the floor and the framework of the cone are complete, the silkworm weaves a fabric between the threads of the framework with figure 8 movements (*bottom*). The silkworm is unable to make these movements near the apex of the cone.

ted, the caterpillar goes through much wasted swaying motion.

When the silkworm has built the skeleton of its cocoon, it weaves the fabric of the envelopes with a third movement. Rocking its head back and forth in a figure 8 pattern, it connects the silk threads already laid down—adding, so to speak, the woof to the warp. At the apex of the cone, however, the space between the silk threads is too confined for figure 8 movements, and the top of the cocoon is therefore only loosely woven—so loosely that a pencil can be slipped through the fabric. This feature, though not the result of conscious foresight, is important in the life of the silkworm. It provides an escape hatch through which the mature moth will make its exit to the world outside.

Now these three movements, which normally produce a cocoon, will produce strange and distorted variants when the silkworm is restricted in some way in the laboratory. For example, we tried the experiment of tying the caterpillar to the dowel with its head toward the board, so that it could not reverse its body and face upward. The animal began by stretching its forebody to the farthest point on the board it could reach and fastening the thread there. Then it bent the forebody toward the dowel, but since it could not reach upward, it had to fasten the thread at a point below or alongside its body. The result was that, though the silkworm succeeded in spinning a cone, it was left outside its cocoon [photograph above].

The structure had the usual two layers, corresponding to the outer and inner envelopes of the normal cocoon. The question arises: What sort of stimulus causes the spinning of the second layer? Evidently it is not contact with the completed first layer, because the insect was not enclosed in it.

Normally the silkworm invests 60 per cent of its silk in the construction of the outer envelope. We wondered what a silkworm would construct if we could make it spend the initial 60 per cent of its silk in making a sheet instead of an envelope. To do this we had to provide the silkworm with a two-dimensional environment. Accordingly we inserted it in a large inflated balloon. In this endless two-dimensional world the insect could find no points in the third dimension for attachment of its silk. It had to spin the silk as a sheet spread out on the inner surface of the balloon. When a caterpillar had spun 60 per cent of its silk in this fashion, we restored it to the three-dimensional world. It proceeded

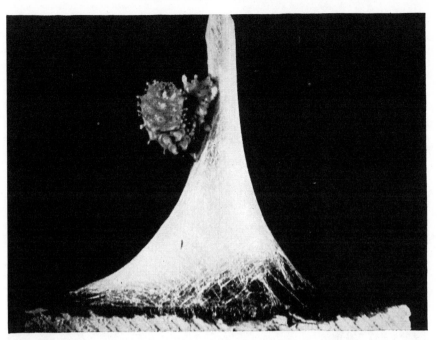

EXPERIMENT with one silkworm consisted in tying its tail to the dowel so that it could only face downward. The silkworm was able to spin the cocoon, but remained outside it.

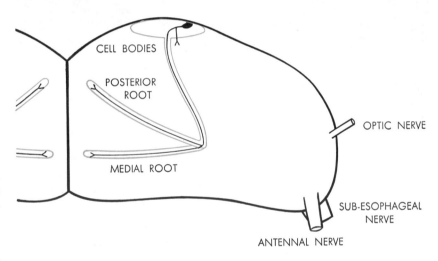

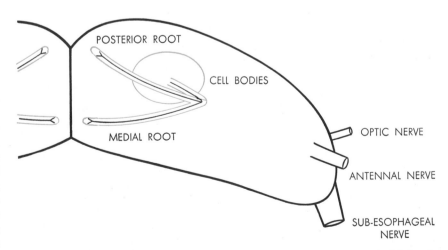

BRAIN of the silkworm is depicted in these schematic drawings. At the top the left hemisphere of the brain is seen from above. Its front faces down. At the bottom the hemisphere is seen from the front. The corpus pedunculatum, the message center of the brain, is in color.

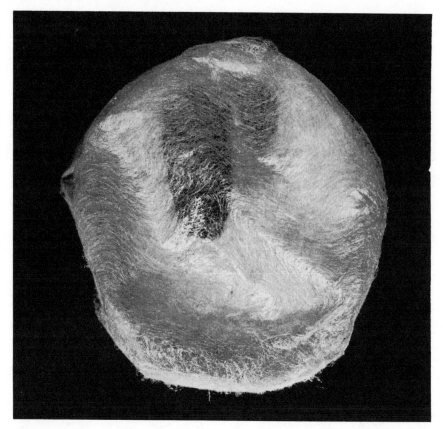

DAMAGED BRAIN caused another silkworm to spin its cocoon in this pattern. Placed in a cylindrical cardboard container, the silkworm covered the floor of the container with silk. Visible inside the silk sheet of the cocoon is the pupa and the shed skin of the silkworm.

as if it had made an outer envelope and built what corresponded to an inner one.

We concluded that the starting of the second layer must be controlled by sensory messages from the silk-forming gland, signaling when 60 per cent of the material has been spent. Further experiments showed that the silkworms actually measure the length of silk spun, "record" this information and finally act on the accumulated data.

Sensory messages from the spinning apparatus are important in other ways. By means of a simple device in our laboratory we are able to record the gross spinning movements of a silkworm even when it is not actually building a cocoon. When we plug the insect's silk outlet with paraffin, it performs the usual ordered pattern of movements—it "spins an imaginary cocoon." But when we remove the silk glands surgically, the movements become disorderly. Apparently control of the movements to build a cocoon depends on a constant background of messages from the silk gland to the central nervous system.

Our task now was to get inside the silkworm to see if we could find out how its central nervous system translates the inflowing sensory barrage into an outflow of coordinated motor signals. We conceived this to be essentially a geographical question: What areas of the central nervous system perform this sensory-to-motor translation?

A first look inside the head of a silkworm raises doubt whether its brain is capable of conducting such a sophisticated operation. One is reminded of the thesis of the 18th-century naturalist and classifier Carolus Linnaeus, who held that insects have no brains at all: "Organs of sense: tongue, eyes, antennae on the head, brain lacking, ears lacking, nostrils lacking." The silkworm does have a brain, but it is not much more than a millimeter across and weighs but one or two milligrams. The structure gets something of the appearance of a brain from a constriction down the middle which divides it into right and left hemispheres [see diagram on page 83]. Two nerves, one from each hemisphere, connect the brain to the rest of the central nervous system. They run downward around the esophagus and to a mass of nerve cells called the sub-esophageal ganglion. From here a pair of nerves extends down the length of the animal's abdomen. This is the ventral nerve cord, the insect's equivalent of the vertebrate spinal cord. Periodically the ventral

nerve cord branches off small packets of nerve cells. These ganglia send nerves to the body muscles and receive nerves from sense organs nearby.

Does the brain coordinate the spinning movements or can this function be performed by the ventral cord alone? To answer this question we cut the cord. We found that even when the cut is made close to the head, the muscles at the front end of the silkworm that are still commanded by the brain drag the rest of the body through the movements of spinning. As long as a small fraction of the body is directed by the brain, the silkworm tries gamely to shape a normal cocoon.

We get a very different result when we remove the brain. The caterpillar retains coordination; it can crawl or climb. But it does not attempt to spin a cocoon. This experiment showed that the brain must be the control center for cocoon construction.

In the next experiment we disconnected one of the hemispheres of the brain from the central nervous system by cutting the sub-esophageal nerve. The silkworms spun, but in a disorganized way. They crawled about laying down silk as a sheet over every surface they encountered [see photograph at left]. A similar derangement resulted when a hemisphere itself was cut through. Again the silkworms lined the inner surface of the container with a sheet of silk.

We went on to a more precise exploration of the brain. When a hemisphere was cut along a line close to the midline, the silkworms retained part of their normal behavior: they spun two layers of silk, one above the other, corresponding to the outer and inner envelopes of the normal cocoon. When we cut the brain through right on the midline, the animals spun perfect cocoons. Thus it was clear that nerve tracts which pass across the midline from hemisphere to hemisphere were not an essential part of the neural apparatus for spinning.

These results hinted that the control center could be located precisely somewhere in the hemispheres of the minute brain. To pinpoint that center, cuts through the brain, even with microscissors, were unavailing. We turned to burning tiny parts of the brain with high-frequency electric currents, which permitted us to destroy pieces of brain less than one twentieth of a millimeter in diameter.

We learned with this technique that more than half of the brain tissue was not involved in cocoon construction. The silkworm's spinning behavior was not affected by destruction of the receiving areas for nerves from the eyes and antennae or of regions containing tangles of association fibers. Eventually we located the critical areas around a pair of structures known as the corpora pedunculata. Even slight damage to these brain regions produced profound aberrations of spinning behavior.

The corpora pedunculata have long been familiar to insect neurologists. Félix Dujardin, a French zoologist, discovered these structures over a century ago in the course of an investigation which upset Linnaeus' contention that insects have no brains. He found that the structures attain maximum development in the social insects (bees and ants) and he called them "the seat of intellectual faculties."

There is a corpus pedunculatum ("body with stem") in each hemisphere of an insect brain; it can be seen even on external inspection. The "body" is a clump of nerve cells, and the "stem" is a cable which carries the cells' long filamentary axons to other centers in the brain. In the silkworm the structures are so small that it must be admitted the insect is an intellectual pauper. The stems follow a peculiar course. From the cell body, located at the rear near the top of the brain, each stem extends forward and downward into the hemisphere and later forks in a complex manner. Near the cell body each axon in the stem gives off one or more short branches, which make contact with axons coming from the sensory centers of the brain. A little further along, the stem forks into two roots, one running to the midline of the brain, the other running upward and back toward the cell body. At the end of the latter root, the axons carried by the stem make contact with motor pathways in the brain.

This circuitry suggests that the corpora pedunculata function as central message centers in the insect brain. Experiments strongly sustain the deduction. When the cell bodies of either corpus pedunculatum are destroyed, the caterpillar spins only one flat sheet. The same result in behavior occurs when we burn away the two roots running to the midline of the brain. If we leave these intact, however, and destroy only the roots going to the rear of the brain, caterpillars spin two layers. Thus it appears that the midline roots control the division of the silk into layers. The silkworms whose circuits were destroyed only at some points could perform all the motions of spinning, but they were unable to coordinate the individual motions into the building of a cocoon.

Anatomical studies suggest that the cells of the corpora pedunculata may be fired by sensory stimuli originating both inside and outside the body, and that the messages may interact, so that each pattern of stimulation causes a particular pattern of firing in the circuit. The resulting impulses pass down the stem and roots to excite the motor pathways and induce a patterned series of movements.

There is another possible circuit. Some of the axons that go to the roots do not connect with motor centers but instead return to the sensory centers. In the sensory center they may fire nerve fibers going to the input axons on the stem. This would provide a pathway by which some of the output of the circuit would be returned as input. We may deduce, in short, a feedback loop [see diagram above right]. It will be recalled

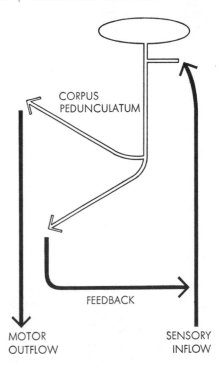

RELATIONSHIP between the corpus pedunculatum (*open lines*) and functions of the silkworm nervous system are outlined.

that feedback circuits of similar design are the essence of mechanical and electronic control mechanisms.

The circuitry of the silkworm brain appears to be an ideally simple testing ground for investigating animal coordination. In most nervous systems integration occurs in incredibly tangled masses of nerve cells. The corpus pedunculatum presents the experimenter with an accessible bundle of oriented fibers, now known to be essential for complex behavior.

9 The Neurobiology of Cricket Song

by David Bentley and Ronald R. Hoy
August 1974

The song pattern of each cricket species is stored in its genes. The songs are thus clues to the links among genetic information, development, the organization of the nervous system and behavior

Nowadays few people live where they can hear crickets singing on summer evenings, but makers of motion pictures know that nothing evokes a bucolic mood more effectively than the sound of crickets chirping in the background. Cricket song, which may seem like a random sequence of chirps and trills, is actually a communication system of considerable complexity. As Richard D. Alexander of the University of Michigan and others have shown, each species of cricket has a distinctive repertory of several songs that have evolved to transmit messages of behavioral importance.

From the work of many laboratories, including our own at the University of California at Berkeley and at Cornell University, it is known that the nerve networks needed for generating the songs of crickets are closely allied to those involved in flight. Both systems develop in stages as the cricket passes through its growth cycle from the larva to the adult. Through breeding experiments it has also been shown that the distinctive song patterns of various species are not learned behavior but are encoded in the insect's genes. Thus we have found that hybrids produced by mating different species exhibit song patterns that are intermediate between those of the parents. Progress is also being made in discovering the genes that control the song patterns and finding where they are located among the cricket's chromosomes. Ultimately it should be possible to piece together all the links in the chain from a sequence of chemical units in the DNA of the cricket's genes to the specification of a nerve network that leads to the production of a distinctive song. Cricket song is only one example of animal behavior that has a large genetically determined component.

The commonest cricket song is the calling song, which is sung by the males to guide sexually receptive females to the singer's burrow. The calling songs of different species that mature in the same area at the same time of year are always distinct. Confusion would result if all the males sang the same song. Once the male and the female have found each other, a new song, the courtship song, facilitates copulation. Following the transfer of the spermatophore, or sperm sac, the male may sing a postcopulatory song (known to French investigators as the "triumphal song"), which may help to maintain proximity between the partners. When a male cricket invades the territory of another male, a vigorous fight frequently results. (Cricket fights are even a traditional sport in the Far East.) Fighting is accompanied by the aggressive, or rivalry, song, which is sung by both combatants in an intense encounter and is nearly always sung by the winner. Thus the communication system consists of transmitters, which are always males, and receivers, which are both males and females, together with a variety of signals that convey information.

How is information encoded in the song? Each song is a sequence of sound pulses. The cricket generates a pulse by scissoring its fore wings once, drawing a scraper on one wing across a toothed file on the other [*see illustration on opposite page*]. This produces a remarkably pure tone. Features of the song theoretically capable of carrying information include the pitch, the relative amplitude of the pulses and their pattern in time. The temporal pattern has been shown to be the critical factor. Among different species the pattern varies from simple trills to complex sequences of chirps and trills with different intervals.

Fascinating as cricket communication may be to students of animal behavior, why should it interest the authors, who are neurobiologists? The answer is that one hopes the analysis of simple nerve networks associated with simple behavior will provide a foundation for the future understanding of more complex networks and more complex behavior, up to and including the behavior of our own species. Just as investigation of the bacterium *Escherichia coli* and the fruit fly *Drosophila* has been fundamental to current knowledge of molecular biology and genetics, we hope that through the study of cricket singing and similar simple behavioral systems some doors in neurobiology that have so far been closed will now be opened.

Throughout the animal kingdom neurons, or nerve cells, are basically similar in operation. Furthermore, both in vertebrates and in higher invertebrates some assemblies of neurons are organized in characteristic ways to facilitate the execution of sensory or motor tasks. The nervous systems of invertebrates are more amenable to analysis, however, because they generate simpler and more stereotyped behavior, because the neurons are fewer and larger and because many neurons are uniquely identifiable. The last feature means that individual neurons can be repeatedly "queried" by the investigator, whereas their vertebrate counterparts are nameless faces that can be polled once but then become lost in the crowd. Repeatability is so important to progress that investigation of invertebrate networks has proved a very powerful technique in analyzing nervous systems [see the article "Small Systems of Nerve Cells," by Donald Kennedy, beginning on page 37]. The special appeal of crickets lies in the access they provide to a broad range of problems.

The cricket nervous system is a chain of 10 ganglia, or knots of neurons: two in the head, three in the thorax and five in the abdomen. Each ganglion consists of a cortex, or outer layer, of nerve-cell bodies that surrounds a dense feltwork of nerve fibers called the neuropile. The interactions between neurons that control behavior take place within the neuropile. These interactions produce trains of impulses that are conducted along axons either through connectives to other ganglia or through nerves that run to muscles and other peripheral structures. Information from sensory neurons lying in the peripheral parts of the cricket's body is conducted along axons through nerve trunks into the ganglia.

Where in this simple nervous system are the song patterns generated? Franz Huber, the founding father of cricket neurobiology, who is now at the Max Planck Institute for Behavioral Physiology at Seewiesen in Germany, demonstrated with the aid of his students Wolfram Kutsch, Ditmar Otto and Dieter Möss that only the two thoracic ganglia nearest the cricket's head ganglia are necessary for singing. This left open the question of what elements of the pattern are generated within the central nervous system and what elements rely on sensory feedback from the periphery. The experiments of the late Donald M. Wilson of Stanford University on locust flight suggested that virtually the entire pattern might be produced by neural circuits within the ganglia [see the article "The Flight-Control System of the Locust," by Donald M. Wilson, beginning on page 73]. This view was strongly supported by Huber's group in studies in which they observed the effect on the cricket's song patterns when the peripheral sensory system was either heavily loaded or acutely deprived. Neither condition had any significant effect on the song pattern.

One of us (Bentley) then began studying the song-production mechanism with the aid of microelectrodes implanted in various cells of the cricket's nervous system. This work was begun in Huber's laboratory and continued at Berkeley. In an early group of experiments the cricket's thoracic ganglia were completely isolated from sensory timing cues by the cutting of the peripheral nerves. Recordings from identified motor neurons showed that the cricket's nervous system could still produce a motor pattern practically indistinguishable from the normal calling-song pattern. This implied that the two anterior thoracic ganglia must

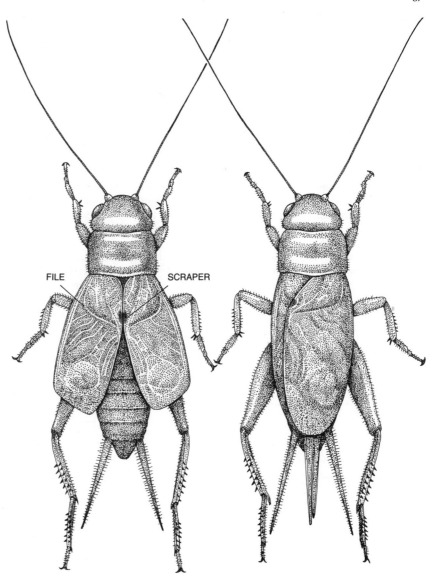

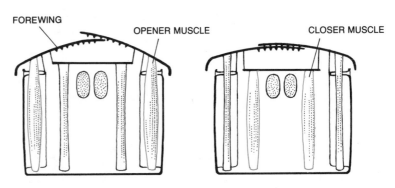

CRICKET SONG IS PRODUCED by specialized structures that are activated when the cricket closes its wings. The upper pair of diagrams show the wings moving from an open position (*left*) to a closed position (*right*). The lower pair of diagrams are simplified cross sections of the same positions as viewed from the front. When the wings are closing, a reinforced segment of cuticle (the scraper) on the edge of one wing bumps across a series of teeth, or ridges (the file), on the underside of the other wing. Both wings are similarly equipped, so that it does not matter which wing is on top and which on the bottom. The movement of the scraper across the file causes the wing to vibrate at about 5,000 cycles per second, producing a remarkably pure tone. Each closure of the wings produces a single sound pulse that lasts roughly 25 milliseconds, or about 125 cycles. The actual linkage between muscle contraction and wing movement involves a complex set of sclerites (small pieces of cuticle hinged to other pieces) that connect the wings to the thorax and muscles.

contain a network of nerve cells that are responsible for generating the calling-song pattern and that they are remarkably independent of sensory input.

The next task was to try to identify among the 1,000 or more nerve cells in each ganglion the neurons concerned with singing. During singing the wings of the cricket are operated by a small set of powerful "twitch" muscles. Each muscle is driven either by a single "fast" motor neuron or by up to five such neurons. The arrival of a nerve impulse at the bundle of muscle fibers that are innervated by a particular motor neuron results in a large action potential, or electrical impulse, in the muscle. Thus action potentials in a muscle unit are a direct one-to-one monitor of impulses in the corresponding motor neuron [*see illustration at left below*].

Fine insulated wire electrodes, which

will record these muscle action potentials, can be inserted through tiny holes drilled in the cricket's exoskeleton and implanted in each unit of any selected muscle. Many such electrodes can be implanted in an animal without interfering with its normal behavior. Therefore by successively implanting each muscle any behavior can be characterized in terms of which motor neurons are active, of the sequential impulse pattern in each neuron and of the relative timing of discharge in the different neurons. With this technique the motor neurons involved in the cricket's calling song were discovered and labeled according to the muscle units the neurons innervate.

Within the ganglion the same neurons can be found and identified with the aid of intracellular microelectrodes, ultrafine glass pipettes through which dye can be injected. The tip of the electrode is driv-

en into the neuron's cell body or one of its fibers. Once embedded the electrode can be used in either of two ways, passively or actively. When it is used passively, the electrode records the electrical activity of the cell and the synaptic inputs to the cell, either excitatory or inhibitory. When it is used actively, the electrode conveys current into the cell, making it possible to analyze the electrical activity within the cell and to stimulate the cell, revealing its effects on other cells [*see illustration at right below*]. When dye is injected into a cell through a micropipette, the dye permeates the cell body and all its fibers, revealing a structure that can be examined by either optical or electron microscopy.

With these techniques it has been possible to identify the activity patterns of the neurons involved in cricket singing, to learn their characteristic morphology

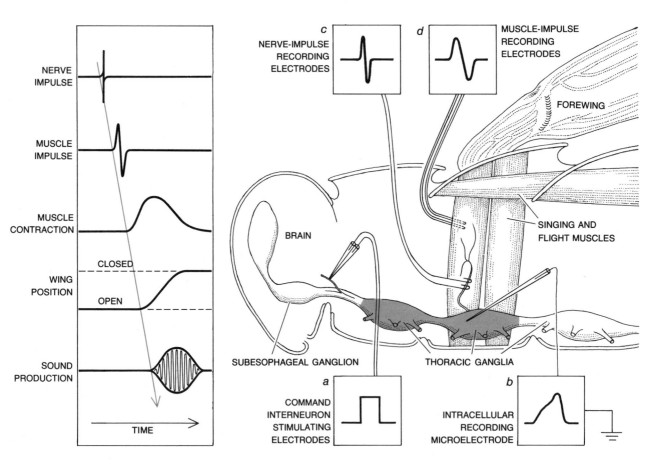

NEURAL EVENTS leading to a single pulse of the cricket song begin with the firing of a motor neuron (*top trace*) in one of the cricket's thoracic ganglia (*location "b" in illustration at right*). The arrival of the nerve impulse triggers a muscle impulse, or action potential (*location "d" at right*), that contracts one of the wing-closing muscles. The closing of the wing causes the scraper on one wing to rub across the file underneath the other wing, producing a sound pulse.

NEUROPHYSIOLOGY OF CRICKET SONG is studied with the aid of implanted microelectrodes that enable investigators to either trace or actively elicit the flow of nerve impulses that result in a sequence of chirps and trills. The cricket's central nervous system consists of a chain of 10 ganglia, of which only five, including the "brain" ganglion, are shown here. The two ganglia in color are sufficient for generating normal song. The song pattern can be elicited by stimulating specific "command" interneurons that lie in the nerve bundle connecting the two head ganglia to the thoracic ganglion (*a*). The command interneurons turn on the thoracic interneurons and motor neurons that generate song. Impulses from motor neurons are conducted along peripheral nerves to the song muscles. Vertical muscle fibers shown close the wings to produce a sound pulse and also to elevate the wings in flight. Horizontal muscle fibers open the wings in singing and depress the wings in flight.

and to discover how they are connected to other cells, particularly how motor neurons are connected to muscle units. In this way every fast motor neuron involved in singing was located, filled with dye and mapped within the ganglion.

What, then, are the characteristics and interconnections of the motor neurons that underlie the generation of cricket-song patterns? Are interneurons, or intermediate neurons, involved, and if so, what is their role? Like any motor pattern, the activity underlying the calling song requires a sequential timing of impulses in individual nerve cells and a coordinated timing of discharges in particular cell populations. The population of motor neurons involved in singing falls into two groups of synchronously firing cells that alternate with each other to open and close the wings. Intracellular recordings indicate that some synchronously firing motor neurons are coupled in such a way that an impulse in one greatly increases the probability of an impulse in its neighbor. The alternate firing of antagonistic motor neurons is established by the powerful inhibition of one group, the wing-closing motor neurons, during discharge of the other group, the wing-opening motor neurons. The wing-closing cells fire immediately after this inhibition, with the result that there is a characteristic spacing of impulses in the two groups [see lower two illustrations at right].

It has now been shown that interneurons help to establish the sequential timing of the song patterns. Even when crickets chirp normally, gaps appear in the song from time to time. The gaps do not, however, upset the established rhythm: the timing of the chirps continues as if there had been no gap [see second illustration from top at right]. This suggests that some internal oscillator, or "clock," within the ganglion has continued to run undisturbed in spite of the missing chirp. It has now been established that the motor neurons themselves are not part of the neuronal oscillator. During gaps not only do the motor neurons fail to fire but also the input signal that drives them is missing. Moreover, anomalous extra chirps that occur from time to time have no effect on the basic rhythm. One must conclude that the timing of the chirps is established by elements higher up in the cricket's nervous system, evidently interneurons whose output signal does not go directly to the motor neurons.

Although the interneuronal song oscillator is quite insensitive to influences

STIMULATION OF COMMAND INTERNEURONS (*top trace*) causes motor neurons to discharge (*bottom trace*) in a typical calling-song pattern. Stimulus site is labeled *a* and recording site *c* in illustration at right on the opposite page. Motor neurons fire as long as command interneurons are stimulated; arrow indicates several minutes of continuous firing.

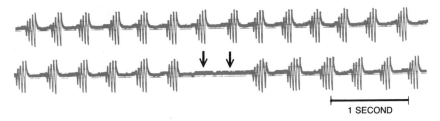

CHIRP REGULARITY is shown in this sequence of 26 consecutive chirps, recorded inside a motor neuron in the second thoracic ganglion (*site "b" in illustration at right on opposite page*). Each chirp consists of four or five sound pulses, marked by oscillations of intracellular potential. Continuous record is here divided into two rows and aligned to show how rhythm persists in spite of two skipped chirps (*arrows*). Song resumes after the gap at exactly the right time, indicating existence of a "rhythm keeper" higher in the nervous system.

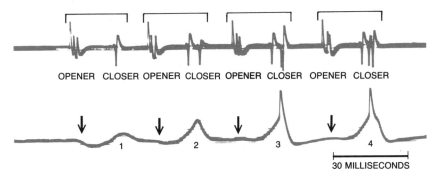

CHIRP CONSISTING OF FOUR RAPID SOUNDS is produced by alternate firing of wing-opening motor neurons and wing-closing motor neurons. The top trace shows nerve impulses arriving at their respective muscles (*site "c" in illustration at right on opposite page*). The bottom trace shows a recording from the larger of two closing motor neurons involved in producing the top trace. Immediately after the discharge of wing-opening motor neurons the closing motor neuron is inhibited from firing (*arrows*). Following inhibition the closing motor neuron is excited but first two excitations (*1, 2*) are below threshold.

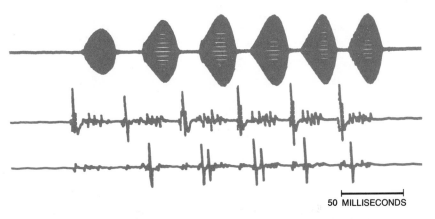

RELATION OF CHIRP SOUNDS TO MUSCLE IMPULSES is shown in simultaneous recordings. The top trace depicts the amplitude of the emitted sounds. The other two traces represent the impulses of a wing-opening muscle (*middle trace*) and of a wing-closing muscle (*bottom trace*). Muscle impulses were recorded at site *d* in illustration at right on opposite page. Simultaneous contraction of several wing-closing muscles produces sound pulse.

FLIGHT-MUSCLE SYSTEM can be studied by tethering crickets in a low-speed wind tunnel. Flight employs many of the same muscles and motor neurons used in singing. The neuronal system involved in song generation appears to be actively suppressed until the nymph, or young cricket, undergoes a final molt and reaches adulthood. This is not the case with flight behavior, which can be elicited and studied in the wind tunnel when nymphs are still four molts away from adulthood. Fine wire electrodes can be implanted in tethered insects.

from outside the cricket's central nervous system, it can be manipulated internally. Huber and his colleagues have shown that singing behavior can be elicited either by electrical stimulation of the cricket's brain or by making small lesions in the brain. This indicates that interneurons running from the brain to the thoracic ganglia are capable of activating the song network. Cells of similar capability have been found in other simple nervous systems and are termed command interneurons.

At Berkeley one of us (Bentley) has been able to locate the axon of a command interneuron in the bundle of about 10,000 nerve fibers that runs between the cricket's brain and the first thoracic ganglion. The command interneurons are always found in the same location.

FRUIT FLY AND NEWLY HATCHED CRICKET are about the same size. A common fruit fly, *Drosophila melanogaster*, is at left; a "wild type" (genetically typical) nymph of the species *Teleogryllus oceanicus* is in the middle. Mass screening of cricket nymphs for interesting mutations can be done immediately after hatching, before the first molt, when they are termed first instars. Nymphs molt 10 times on their way to adulthood. The first-instar nymph on the right is a mutant that lacks certain sensory hairs on rear-end antennae.

They are bilaterally symmetrical, and when they are electrically stimulated at an appropriate frequency, they cause the song network to generate a perfectly normal calling-song pattern [*see top illustration on preceding page*]. One can show that a single command interneuron suffices to elicit the song pattern. There appear to be no conceptual or technical barriers to learning much more about how this hierarchically organized neural subsystem operates.

Many important questions are presented by the appearance in the adult cricket of a neuronal network that will generate a behavior pattern as precise as cricket song with such reliability. When are the neurons built? When do the cells become physiologically mature and what kind of electrical activity do they exhibit before reaching maturity? When are the functional connections that coordinate activity of the cells established? Is the network assembled before or after the cricket becomes an adult? If it is after, does perfection of the pattern depend on acoustical feedback, that is, on the cricket's hearing its first attempts at singing and then making corrections? Some answers to these questions have been found for singing and also for the closely related behavior of flight.

The development of the cricket proceeds without any dramatic metamorphosis. The female deposits her eggs singly in the soil. Following embryonic development the eggs hatch into miniature nymphs about the size of a fruit fly, which conspicuously lack wings, reproductive structures and associated behavior. During postembryonic development, which is several times as long as embryonic development, the nymphs pass through 10 instars, or developmental stages, separated by molts. With each successive molt the nymphs increase in size and resemblance to adults except for certain structures such as the wings and the ovipositor: the tube through which the female lays her eggs. These structures are not fully developed until the final molt to adulthood.

The male cricket normally begins to call about a week after its final molt. Nymphal crickets never attempt to sing, even if they are placed in a situation that would stimulate singing in the adult. For example, nymphs are strongly aggressive in competition for food, but they do not move their small wing pads in the pattern of aggressive song. Either the neural circuits that mediate singing are not yet laid down or, if they are, they must be actively suppressed. When we made lesions in the brain of last-instar nymphs in an area that would evoke singing in the adult, the wing pads finally moved in a pattern resembling song. To determine whether or not the motor pattern was the same as the one that gives rise to the calling song, we recorded muscle action potentials from identified motor units and compared the impulse pattern with the pattern the same unit would be expected to generate during the calling song of an adult [*see bottom illustration at right*].

Since the strength of the argument depends on the predictability of the adult motor pattern, this kind of analysis is possible only in animals, such as the cricket, that show highly stereotyped forms of behavior. Several conclusions can be drawn from this study: (1) the neuronal network for the calling song is completed in the nymph, (2) the assembly of the network does not depend on acoustical feedback and (3) song patterns are not prematurely activated in nymphs because of active inhibition originating in the brain.

We have not followed the maturation of the song networks in detail because the brains of the younger nymphs are so small that it is difficult to make the lesions required to elicit singing behavior. This is not the case, however, with the closely related behavior of flight. The highly invariant, rhythmic motor pattern of flight is similar to singing in that it involves the same set of muscles and the same motor neurons to operate the fore wings and also their homologues that operate the hind wings. The motor pattern consists in the alternate firing of elevator (upstroke) motor units and depressor (downstroke) motor units, with the hind-wing segments leading their fore-wing counterparts by about a third of a wing-stroke cycle.

In nymphs there does not seem to be any suppression of the neuronal network for flight. As a result one can induce nymphs of very early instars to attempt flying by suspending them in a small wind tunnel. Electrophysiological recordings are made from identified motor units, and their performance is evaluated by comparing their pattern with the pattern the same unit would make in the adult.

We find the first definite signs of the motor pattern of flight in nymphs of the seventh instar. Removing the nymph from contact with the ground and suspending it in a wind tunnel is sufficient to induce some flight motor neurons to discharge a few impulses at frequencies

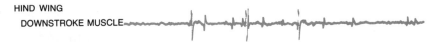

SEVENTH-INSTAR NYMPH

HIND WING
DOWNSTROKE MUSCLE

EIGHTH-INSTAR NYMPH

HIND WING
DOWNSTROKE MUSCLE

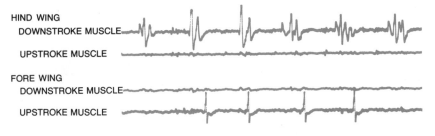

NINTH-INSTAR NYMPH

HIND WING
DOWNSTROKE MUSCLE

UPSTROKE MUSCLE

FORE WING
DOWNSTROKE MUSCLE

UPSTROKE MUSCLE

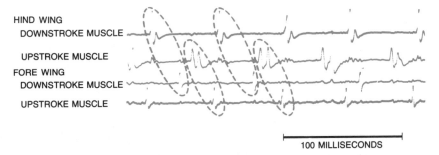

10TH-INSTAR NYMPH

HIND WING
DOWNSTROKE MUSCLE

UPSTROKE MUSCLE

FORE WING
DOWNSTROKE MUSCLE

UPSTROKE MUSCLE

100 MILLISECONDS

ASSEMBLY OF NERVE NETWORK FOR FLIGHT is completed during the last third of the cricket's larval life. The records show muscle-impulse patterns produced by tethered nymphs trying to fly in a wind tunnel. In seventh-instar nymphs the pattern is only fragmentary. In later instars new muscles come into play and the pattern becomes stronger. In the last stage before adulthood muscles are properly coordinated: upstroke and downstroke units alternate and the hind wing leads the fore wing (*indicated by broken lines*).

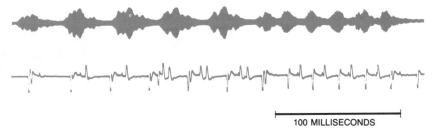

100 MILLISECONDS

ASSEMBLY OF NERVE NETWORK FOR SINGING can be demonstrated by making lesions in the brain of 10th-instar cricket nymphs. Before the final molt to adulthood cricket nymphs do not attempt to sing. Certain lesions, however, that elicit singing in adult crickets also elicit the calling-song motor pattern in nymphs. The top trace is the calling-song sound-pulse pattern of an adult of the species *T. commodus*. The bottom trace shows the closely parallel activity of wing-opening muscles (*downgoing impulses*) and wing-closing muscles (*upgoing impulses*) elicited by brain lesions in a 10th-instar nymph of the same species.

approaching the normal rate of the wing stroke [*see top illustration on preceding page*]. During subsequent development the performance improves in several respects: first, there are more impulses per burst, corresponding to wing strokes of greater amplitude; second, there are more bursts per response, corresponding to more wing strokes and longer flights, and third, additional motor neurons are recruited into the pattern. As with singing, the neuronal network involved in flight seems to be fully assembled by the last instar, although the overall frequency of its oscillatory behavior does not reach normal speed until after the molt to adulthood.

How does the neuronal network develop structurally before its physiological activation begins? The information we have suggests that it develops as follows. Cell bodies and peripheral axons (long fibers, one from each cell body) arise while the cricket is still in the embryo stage. The richer growth of dendrites (short fibers) within the ganglia may come during the first third of postembryonic development. By filling identical neurons with dye one can show that by the sixth instar, halfway through postembryonic development, the major branching network has been completed. In the next instar the first signs of physiological activity in adult patterns can be

detected. During the last third of postembryonic development the sequential firing pattern steadily improves and precise coordination with other neurons is achieved. This last step may reflect the actual establishment of synapses, or connections, between neurons. By the last instar the neuronal network is fully assembled and potentially operative, although it may be suppressed by inhibition from the brain. Thus immediately after the molt to adulthood the nervous system of the cricket is ready to generate both flight and the calling song.

It is well established that in order to attract females of the same species

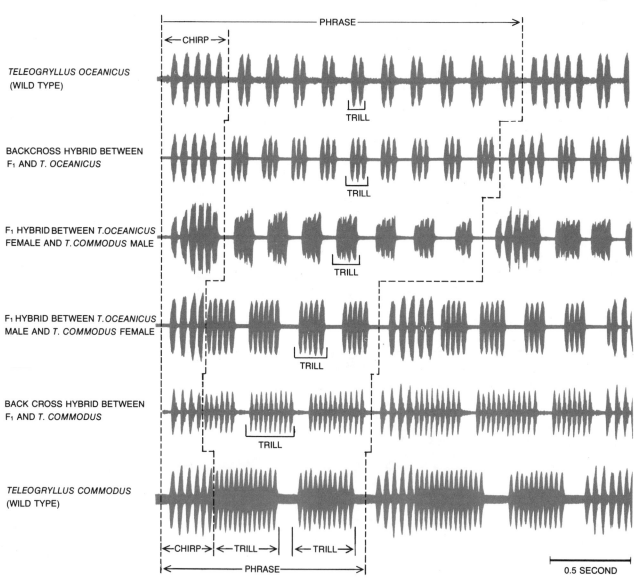

SONG PATTERNS OF HYBRID CRICKETS shift systematically in proportion to the ratios of the different wild-type genes inherited by the individual. These records show the song patterns of two cricket species and their hybrids. The records are aligned so that a complete phrase of the song starts at the left; each phrase consists of a chirp followed by two or more trills. The phrase of *T. oceanicus* (*top*) is not only much longer than the phrase of *T. commodus* (*bottom*) but also distinctively different. In each of the last three patterns the first trill is fused to the chirp. The F_1, or first-generation, hybrids of these two species produced the third and fourth traces. Again they are quite different depending on which species served as the male parent and which as the female parent. The second and fifth traces were produced by backcrosses between the two different F_1 hybrids and members of the parental species.

successfully the male cricket must broadcast a very precise message. What is the source of the information underlying this precision? We have seen that neither motor practice of the songs nor acoustical feedback is required. The correct pattern arises from the neural properties and neural connections established during development. How do they become properly established? One hypothesis invokes the environmental information available to cricket nymphs during their development. For example, some songbirds have been shown to remember song patterns heard during their adolescence and to defer the use of the information until the following year in the songs they sing as adults. The main alternative hypothesis is that all the necessary information for cricket singing is stored genetically and is read out in the form of neuronal structures during the course of development.

These hypotheses can be tested by changing either the environmental input or the genetic one during development and observing the effects. We raised crickets under a variety of environmental conditions, including different regimes of population density, diet, temperature, cycles of light and darkness, and of course acoustical experience. Some crickets heard no songs, some heard only songs of their own species and some heard only songs of another species. In every case male crickets that had reached maturity produced the calling song characteristic of their own species. This indicates that environmental information is not utilized in the determination of the song pattern.

What would happen, however, if a wild-type (genetically "normal") male of one species were mated with a female of another? What song would the hybrid male offspring of such a union sing? Since these particular hybrids are fertile, we were able to backcross them with individuals of the parental species. From such genetic manipulations we learned that each genotype (that is, each hybrid, backcross or other mixture) gives rise to a unique calling song and that all individuals of each genotype sing the same song. Even more remarkable, the song patterns shift systematically according to the proportion of the wild-type genes carried by the male cricket [see illustration on opposite page]. One can only conclude that the information specifying song patterns is encoded in the genes.

In order to give rise to the song-generating neuronal network the information coded in the genes must be read out

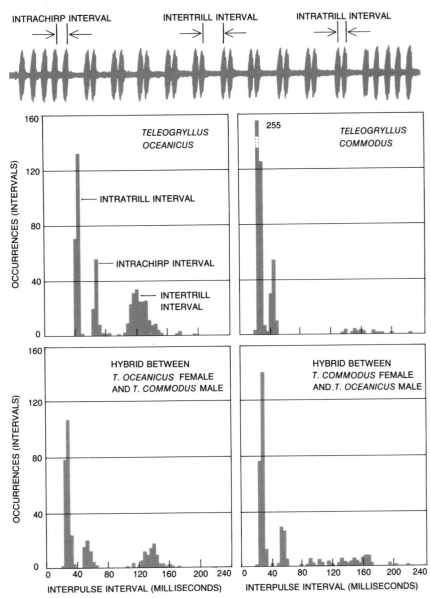

PRECISION OF SONG PATTERN becomes evident when the intervals between sound pulses in the calling song are measured for several hundred pulses. The two top histograms show the intervals in the calling songs of wild-type *T. oceanicus* (left) and *T. commodus* (right), which also appear in the illustration on the opposite page. When hybrids are made between these two species (bottom), their intertrill intervals resemble those of the species that served as the maternal parent: *T. oceanicus* on the left, *T. commodus* on the right. This shows that the genes influencing the intertrill interval are on the X, or sex, chromosome.

during development by a series of complex and subtle interactions between the cricket's environment and the genes of the cells involved. It seems, however, that the range of possible products of this interaction is stringently limited. If development is successful, the calling song of the adult is a very accurate reflection of the genotype.

How many genes are involved and where are they located on the cricket's 15 chromosomes? A start toward answering these questions can be made by examining the pattern of inheritance

of features of the calling song. If a particular feature (such as the number of chirps or the interval between chirps) were determined by a single gene that was dominant over the corresponding gene in another species, the feature should be transmitted unchanged to the first-generation (F_1) hybrids between the species.

When we examined 18 features in the calling songs of two cricket species (genus *Teleogryllus*) and their F_1 hybrids, we found no evidence that the features involved the dominance of single genes. If a character were controlled by a

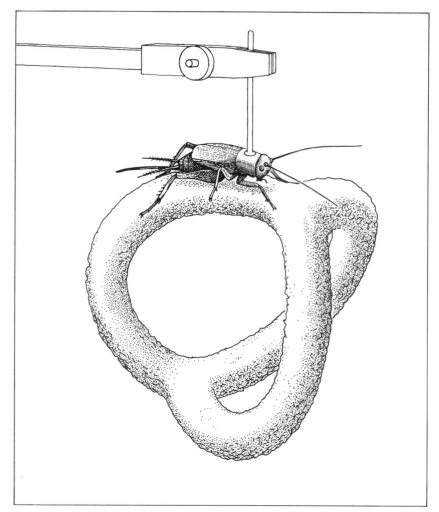

TETHERED FEMALE CRICKET WALKING ON Y MAZE reveals her degree of preference for calling songs of males of different species by the choice she makes when she comes to a fork in the Y. The cricket actually holds the featherweight Styrofoam maze as she travels along it. One loudspeaker is located on the left side of the cricket and another on the right side. In each test the cricket hears only one male calling song, played 40 times through one speaker or the other in random order. The frequency with which the cricket turns toward the song is taken as an index of its attractiveness (*see illustration on opposite page*).

the *X* chromosome and that other elements are not [*see illustration on preceding page*]. The genetic system that specifies the neuronal network accounting for cricket song is therefore a complex one, involving multiple chromosomes as well as multiple genes.

So far we have written exclusively about the transmitter in the cricket-song communication system. What about the receiver? A good deal has been learned about the way the female responds to the male's song. For example, Thomas J. Walker of the University of Florida and others have studied the selective responsiveness of females to song patterns. The female's orientation to a sound source and attraction to it have been investigated recently at Berkeley by Rodney K. Murphey and Malcolm Zaretsky. Zaretsky and John Stout in Huber's laboratory have begun to identify and characterize the sensory interneurons involved in responding to the song and "recognizing" it.

A fundamental problem common to the analysis of all animal communication systems is how the timing and the synchronous evolution of the transmitter and receiver are maintained. Found everywhere on the globe, crickets are classified into about 3,000 species whose song patterns have diverged widely in the course of evolution. How has it happened that in each case the receiver has changed along with the transmitter, so that the female still responds selectively to the call of a male of the same species?

One of us (Hoy), in collaboration with Robert Paul, has been studying this problem, first at the State University of New York at Stony Brook and recently at Cornell. Experiments were designed to quantify the ability of female crickets to detect and select calling songs of their own species, to determine the role of the genotype in that selectivity and to explore the relation between the genetic systems that control song transmission and song reception.

In these experiments a female cricket is suspended by her thorax and allowed to hold a hollow sphere of Styrofoam cut in the form of a continuous Y-shaped maze [*see illustration on this page*]. As she "walks" on the maze, which she is actually holding, she periodically comes to junctions that call for a decision to turn right or left. On each side of her there is a small loudspeaker through which different song patterns can be played. In each experiment the walking female is required to make 40 choices, 20 while the song is played through one speaker and 20 while it is played through

single nondominant gene, the crossing of an *F*₁ individual with a wild-type individual of the parental species should give rise to two distinct classes of backcross offspring, one like the parent and one like the *F*₁ individual. The more genes there are that influence a character, the broader and smoother is the distribution of types produced by backcrossing. Our analysis of many backcrosses has failed to reveal any examples of a simple bimodal distribution, which would indicate single-gene control of some characteristic of cricket song. Therefore we conclude that many genes are required to specify the neuronal network responsible for song production.

Genes that influence a particular characteristic or behavior, such as cricket song, can be localized on a specific chromosome, provided that they are on the *X*

chromosome. Female crickets, like the females of many other species, have two *X* chromosomes (*XX*), whereas male crickets have only one *X* chromosome (and lack the *Y* chromosome found in many other animals). As a result two types of cross can be made between two species, one using males from species *A* and females from species *B* and the other using males from species *B* and females from species *A*. Male offspring from these crosses will be genetically alike except that they will have *X* chromosomes from different maternal parents. Thus differences in the songs of the two types of male can be attributed to genes located on the *X* chromosome. Analysis of hybrid calling songs reveals that certain elements of the song pattern (for example the interval between trills) do appear to be controlled by genes on

the other. The number of decisions made to turn toward the source of the sound divided by the total number of choices is taken as an index of the "attractiveness" of the song. Each female is tested only once and is presented with only one song. The tests clearly establish that female crickets prefer the calling song of the males of their own species.

The role of the genes in establishing this preference can be investigated by manipulating the genotype. In a typical study crickets of two different species are mated and the hybrid F_1 females are tested with songs of three types: the calling songs of the males of each parental species and the calling song produced by the females' F_1 hybrid brothers. Surprisingly, the hybrid females prove to be attracted to the songs of their brothers much more than to the songs of either parental species [see illustration below]. This result demonstrates that genetically shared information specifies the pattern of song recognition as well as the pattern or song production. Moreover, it suggests that similar genetic systems could be involved in encoding information for constructing either a neuronal network that will respond to a specific song pattern or a network that will produce a specific song pattern. Indeed, there is the fascinating possibility that some of the same genes are involved in both systems. Such an assemblage of genes would be a fail-safe means of ensuring the synchronous evolution of the transmitter and the receiver.

The experiments described above firmly establish the link between genetically stored information and the cricket's nervous system, but how does the first control the design of the second? What kind of information about the structure, the physiology and the connectivity of neurons is stored, and how is it read out? Two quite different strategies for approaching the problem immediately suggest themselves and are currently being pursued by one of us (Bentley) at Berkeley. The first is to focus attention on single neurons and ask what features of the nervous system are actually under genetic control. The second is to concentrate on single genes and ask what a particular gene contributes to the design of the nervous system.

The first question can be tackled straightforwardly by crossing different species and hybrid individuals, thereby constructing cricket nervous systems according to different genetic blueprints. Then by examining identifiable homologous neurons in the different systems one can determine what is different about these neurons. The firing pattern during the calling song of two particular neurons in five different genotypes has now been examined: two wild types, the F_1 cross and two backcrosses. Not surprisingly, one finds that the song precisely reflects the firing pattern of the motor neurons, that neurons of each genotype fire in a distinctive pattern and finally that very small differences in pattern can be genetically specified. For example, motor neurons usually fire only once for each pulse of the trill sound. One wild-type cricket, Teleogryllus oceanicus, has trills consisting of two short pulses, whereas the backcross between the F_1

hybrid and T. oceanicus has three-pulse trills [see top illustration on next page]. This means that the actual difference in firing patterns of the responsible motor neurons in the two genotypes is only a single impulse. It is a remarkable example of fine genetic control.

Experiments are now in progress to discover why the neurons of the backcross fire three times rather than twice. One possibility is that there is a difference in how the neurons are excited by command interneurons. If one artificially stimulates the appropriate interneurons in the connective bundle between the cricket's brain and its thorax, one finds that only a slight increase in the rate of stimulation is needed to change the firing rate of the motor neurons involved in the calling song from a two-pulse pattern to a three-pulse one [see bottom illustration on next page]. This result suggests that the effect of the genetic change could be a similar increase in the firing rate of the command interneurons or perhaps an increased efficiency in the transmission of impulses at the synapses. There are many other possibilities, but these experiments at the very least show that it is possible to get at the heart of the pattern-generating mechanism and directly test the effect of genetic manipulation.

The alternative strategy of examining the role of single genes is also being investigated at Berkeley. One begins this process by accumulating a "stable" of organisms with mutations in a single gene. The mutations can be induced by exposing the organisms to X rays or to

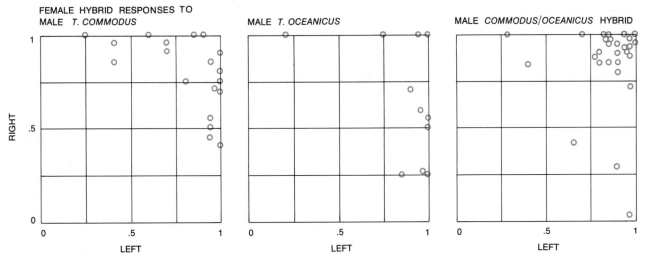

FEMALE HYBRID RESPONSES TO MALE T. COMMODUS

MALE T. OCEANICUS

MALE COMMODUS/OCEANICUS HYBRID

FEMALES' PREFERENCE FOR MALE SONGS is plotted in scattergrams. Each female is required to make 40 choices while walking on a Y maze, 20 while the sound is played through one speaker and 20 while the sound is played through the other. Each open circle represents a 40-choice test. If the female always turned toward the song of a particular male whether played on her left or her right, the song would score 1 on both axes and an open circle would be placed in the extreme upper right corner. The females whose preferences are plotted here were all hybrid offspring of a T. oceanicus female and a T. commodus male. The three scattergrams show the females' relative preference for the song of a T. commodus male (left), the song of a T. oceanicus male (middle) and the song of a hybrid male whose genes closely resemble those of the females being tested (right). Females clearly prefer their "brother's" song.

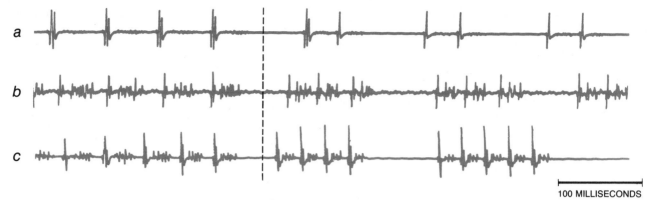

a

b

c

100 MILLISECONDS

NEURAL ACTIVITY UNDERLYING CALLING SONG is patterned according to different genetic specifications, as illustrated by these three records of the firing of a particular identified motor neuron in three different crickets during the calling song. The crickets that produced the traces are of the same genetic type as those that produced the first three sound traces shown in the illustration on page 8: a wild-type *T. oceanicus* (*a*), the *F*$_1$ hybrid between a *T. oceanicus* female and a *T. commodus* male (*c*) and the backcross between the *F*$_1$ hybrid and *T. oceanicus* (*b*). The vertical broken line marks the end of a chirp and the start of a trill. It is evident that genetically stored information exerts extraordinary control over the output of the nervous system. Thus the backcross (*b*) has a three-pulse trill whereas its wild-type parent has a two-pulse trill and its *F*$_1$ hybrid parent has a trill of four or five pulses.

mutagenic chemical compounds. When crickets are raised at elevated temperatures (about 35 degrees Celsius, or 95 degrees Fahrenheit), they have a generation time of about six weeks. One female can produce as many as 2,000 eggs, and the resulting nymphs exhibit a wide diversity of behavioral characteristics. The mass screening of mutants for interesting behavior patterns can be done with first-instar nymphs that are no larger than fruit flies and are just as plentiful [*see bottom illustration on page 90*]. When an interesting mutant is found, it can be grown to an adult 1,000 times larger than a fruit fly in which single nerve cells are readily accessible for study.

The first behavior selected for screening is the evasive leap elicited by the stimulation of the cricket's cerci, which may be called rear-end antennae. The neuronal circuits involved in this response had previously been analyzed by John M. Palka and John Edwards of the University of Washington and by Murphey, working at the University of Iowa. In the screening program at Berkeley two mutants with an abnormal leap response have been isolated and established in breeding lines. In one case the mutant gene is a recessive gene on the *X* chromosome; in the other the gene is a dominant gene on one of the autosomal chromosomes (chromosomes other than the *X* sex chromosomes).

The mutants exhibit a selective loss of a single class of receptor hairs on the cerci. Each hair activates a single sensory neuron that has a direct synaptic connection with certain large, identified interneurons. The mutants are being studied to see whether or not the sensory neurons are also affected by the mutation and whether or not the structure of the interneurons has been changed by the lack of normal input from the sensory neurons. The ease with which such mutants can be isolated encourages hope that the single-gene strategy will ultimately be successful in the analysis of cricket song.

The promise of the approach we have taken in our study of cricket behavior lies in combining several levels of analysis in a single animal. This allows for a powerful infusion of techniques between levels, for example asking developmental questions by means of single-gene mutations or genetic questions by means of single-neuron recordings. An important feature of this approach is that it offers some relief from an affliction of neurobiology that has been called the chimera problem: the accumulation of volumes of data on different aspects of very different creatures. The cricket work links the several levels of analysis in a unitary system and provides a high degree of confidence on how biological integration is achieved. We view the cricket as a kind of decathlon performer in neurobiology: it may not excel at any one thing, but it can be counted on for a sound performance in every event.

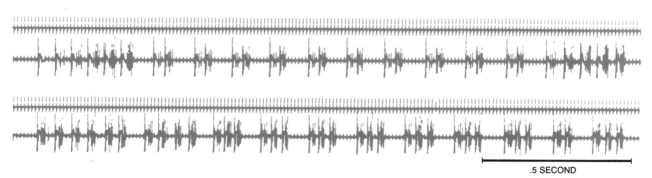

.5 SECOND

CHANGE FROM TWO-PULSE TO THREE-PULSE TRILL can be artificially evoked in the calling-song pattern of a wild-type *T. oceanicus* cricket by changing the firing rate of command interneurons. In each pair of records the top trace shows the stimulus applied to the commond interneuron and the bottom trace shows the impulse pattern of song motor neurons. In the bottom pair of traces the firing rate of the command interneuron has been increased about 10 percent, causing a shift in the motor pattern from the wild-type two-pulse trill to the three-pulse trill characteristic of the backcross whose trill appears in trace *b* in the illustration at the top of the page. Thus the command interneuron may be the neural element responsible for the difference in song patterns of the two genotypes.

How We Control the Contraction of Our Muscles

by P. A. Merton
May 1972

*Voluntary muscular movements are driven by a
servomechanism similar in many respects to the
automatic feedback system employed to control
power-assisted steering in an automobile*

Psychophysics is the branch of experimental science that deals with the relation between conscious mental events and physical events within and without the body. Most psychophysics is sensory psychophysics, which deals with the relation between a physical stimulus and the resulting sensation experienced by the subject. The object of sensory-psychophysical experiments is to gain understanding of the physiological mechanisms that lie between the stimulus and the sensation, and to be able to draw inferences about what goes on inside a sense organ, a nerve or the brain. Measurements of subjective sensory thresholds in any sensory mode (tactile, visual, auditory or whatever), perceptions of color matches and judgments of the pitch of a note or the direction of a sound are examples of sensory-psychophysical observations. Sensory psychophysics is an old and highly respectable subject. In the hands of such investigators as Thomas Young, Jan Purkinje, Hermann von Helmholtz, James Clerk Maxwell, Lord Rayleigh and their modern successors it has told us a great deal about vision, hearing and other senses. Young's celebrated three-color theory of color vision, published in 1802, was formulated entirely on psychophysical evidence and is the basis of modern color photography and color television.

The other branch of psychophysics, motor psychophysics, does not have these credentials. It deals with the reciprocal problem, the relation between a conscious effort of will and the resulting physical movement of the body. It is just as important to know how we move as how we feel, but on the motor side much less has been achieved, partly, I suspect, because physiologists for metaphysical reasons feel that conscious volition is a faintly disreputable thing for them to have dealings with.

In sensory psychophysics it is easy to find illustrative examples of sensory phenomena that have an analytical character, that is, examples that provide some insight into sensory mechanisms, but on the motor side it is not so easy. I can think of one striking instance. A motor psychophysical fact of immense everyday importance is the individuality of a person's signature. Whenever Mr. X makes the appropriate volitional effort and signs his name, it always comes out the same (or enough so to be recognizable) and different from what anyone can write if he tries to write the same name. This is not an analytical observation; it is just a mysterious physiological fact, which we take for granted because we are so familiar with it. What does tell us something, however, is the further observation that if Mr. X takes a piece of chalk and signs his name in large letters on a blackboard, it again comes out the same. The muscles used are different but the individuality remains. From this observation we learn something about the organization of the motor system.

In this article evidence from both branches of psychophysics is taken into account, but the main object is to redress the balance in favor of the motor side. In more concrete terms we ask: What has been learned by making observations on voluntary movements in man about the physiological mechanisms that make our muscles do what we expect of them? Not, of course, very much. The title of this article is somewhat pre-tentious, as titles will be. There are a few definite phenomena to describe. With them we reach a new point of view, from which I hope we can see a general line of advance. I shall stick to simple movements and not come close to explaining the individuality of handwriting. (That subject was introduced partly to advertise the fact that sensory physiologists do not have all the glamor problems.) It will be useful to start by drawing an analogy between the human body and an automobile.

In the old days the steering wheel of a motorcar was directly connected to the road wheels by a series of levers and linkages, and the brake pedal similarly applied pressure directly to the brake shoes. On coming to a hill a gearshift could be moved to engage a suitable pair of gears to climb the hill with.

Today, in order to enable the driver, no matter how frail, to control a massive vehicle with the flick of a wrist or ankle, sophisticated mechanisms have been developed to assist with steering, braking and gear-shifting. All these mechanisms have devices (sensors, we may call them) that measure some physical variable (for example brake pressure or engine revolutions) and use the "feedback" information from them to control the mechanism that assists the driver. Let us concentrate on the mechanism that assists with steering. In its essentials it works as follows. Each position of the steering wheel corresponds to a certain angle of the front road wheels that the driver would like them to assume with respect to the fore-and-aft axis of the chassis. A sensor at the bottom of the steering column detects the difference between this "demanded" position and the actual position of the road wheels. Signals from the sensor, called the misalignment detector, are used to turn on a small servomotor (from the Latin *servus*, meaning slave), which turns the road wheels in such a direction as to cancel the misalignment. Thus the road wheels are made to point in the direction the driver wants, without his having to exert himself. As he turns the steering wheel the road wheels follow automatically.

Such is power-assisted steering. An engineer calls it a follow-up servomechanism. An important point to note is that, the function of the device being to help the driver automatically, he does not want to be bothered with the details of its operation; in particular he would only be distracted from his task of keeping his eyes on the road to see where to steer if signals from the sensor were relayed to him. They give information that is relevant only to the functioning of what ought to be a completely subservient mechanism, and they should remain private to that mechanism.

Power-assisted steering relieves the driver of physical effort only; other such devices relieve him of mental effort too. The automatic transmission, for example, does away with the need to decide when to change gear, as well as the need to perform the change. In an aircraft the automatic pilot does everything and leaves the human pilot completely free.

In the human body there are numerous automatic feedback mechanisms of this kind controlling physiological functions without any mental effort on our part. For instance, the blood pressure

and the output of the heart are controlled so as to suit the current needs of the body; we are quite unaware of the functioning of these systems and of the signals from the pressure sensors in the walls of the arteries and elsewhere that are a part of them.

Such mechanisms are commonplace physiology; they are in the textbooks for medical students and nurses. When we come to muscle, however, the situation is different. To return to our analogy, in the case of the automobile we know what we want to control—direction, speed or retardation—and the problem is to design servomechanisms to help the driver, with appropriate sensors in each instance. The signals from the sensors are just part of the engineering technology, and so we do not display them on dials on the dashboard. They would only put the driver off. In the human machine we have muscles to control. How do we do it? Do the orders to contract go directly from the brain? Presumably not, since on examination it appears that muscles, like the automobile, are equipped with sensors of their own, of whose signals the owner of the muscles, like the owner of the automobile, remains unaware. Presumably, like the sensors in the automobile, they are taking part in automatic mechanisms that assist the subject in controlling his muscles. What are they helping to control? Muscle tension perhaps? It could be; some of them measure tension. Length? Others of them respond to changes in muscle length. A combina-

tion of tension and length? Sometimes tension and sometimes length? Now we see the nature of the problem. It is the inverse of the automobile designer's. We are presented with the sensors and we have to discover what the mechanism they are part of was designed to do. What precisely do we ask of our muscles that they need these confidential sensors to make them do it? It is by no means obvious.

Having thus briefly sketched the picture, let me now go into the physiology in more detail. It falls into two sections. The first presents the evidence that muscles incorporate sensory receptors of whose signals we are not consciously aware; the second discusses what is known of the mechanisms in which they take part.

In the 18th century the great Swiss physiologist Albrecht von Haller established for the first time that the internal organs of the body, such as the heart, the stomach and the brain, are in general insensitive to the kind of stimuli that are so readily felt by the skin: pricking, pinching, cutting, burning and so forth. It is this fact that enables surgeons to perform operations on, say, the brain substance with only local anesthesia around the incision. In his studies of muscle Haller found that stretching a muscle by pulling gently on the tendon exposed in a wound in a human subject did not cause sensations of either movement or tension. (Pulling hard, however, is painful.) Reflecting on Haller's observations, one can perceive that the viscera and the muscles are really in different categories. It is not at all surprising, when one comes to think of it, that the liver should be insensitive to cutting with a knife or burning with a cigarette; such stimuli would be so rare without the animal's getting an earlier and more effective warning from the abdominal skin that to develop a system to report them would give the animal a negligible evolutionary advantage, whereas sensitivity to mechanical contacts, which the skin preeminently possesses, would very likely be a positive disadvantage. Imagine what life would be like if throughout it one were as vividly aware of the beat of one's heart as the surgeon who puts a finger on it exposed during an operation! With muscle, however, it is quite otherwise. It might be useful for us to be conscious of how extended our muscles are at any moment, since that determines the position of our limbs, and also to know their rate of shortening or elongation and the tension in them. If we are to be-

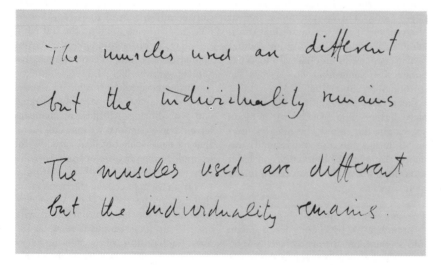

STRIKING EXAMPLE of a simple experimental observation that provides some insight into the organization of the motor psychophysical system is represented by these two handwritten versions of a sentence taken from the text of this article. The sentence was written large on a wall with a felt-tipped pen (*top*) and small on a piece of paper with a fine mapping pen (*bottom*). The writing on the wall is about 10 times larger. The large writing was done by movements of the wrist, elbow and shoulder, whereas the small writing used muscles in the hand itself. Nevertheless, the character of the writing is the same in both cases.

lieve Haller, this is just what they do not tell us. For this and other reasons that will shortly emerge we find that, whereas the insentience of the viscera has long been received as physiological dogma with a status comparable to the circulation of the blood, the insentience of muscle has often been called in question and probably cannot be regarded as universally accepted even today.

For me the question was both raised and answered the day I read the arguments of Helmholtz, published in 1867 in his *Handbook of Physiological Optics*. Helmholtz reached the same conclusion as Haller by experiments with the eye, which have the merit that anyone can repeat them and convince himself of the facts. Helmholtz starts with the familiar observation that if one takes hold of the skin at the outer corner of the eyelids and jerks it sideways, the eye itself is moved and what one sees with that eye appears to jump about. On the other hand, we know that if one moves one's eyes voluntarily, the scene one is looking at does not appear to jump. Helmholtz argues as follows. In both cases the image of the external world moves over the retina as the eye moves. When one moves one's eyes actively, by voluntary effort, one allows for the eye movement and does not interpret the movement of the image on the retina as signifying a movement of the external world. When the eyes are moved passively by an external pull, however, one interprets what one sees as if the eye had remained still. The movement of the image on the retina is assumed to be due to a movement of the external world and not to a movement of the eye. Hence we only know in which direction our eyes are pointing when we move them voluntarily, and this must be because we make an unconscious estimate of the effort put into moving them. (We have a "sense of effort.") Sense organs in the eye muscles (or elsewhere around the eye, if there are any) do not tell us which way our eyes are pointing, because when the eyes are moved passively, we do not seem to know they have moved.

This argument, as it stands, is not conclusive, because when the eyelids are pulled, the sense organs in the eye muscles or elsewhere might not be excited in the same manner as when the eye is turned normally by the contraction of its muscles. The apparent movement of the external world during a passive movement of the eye might therefore be due to a misjudgment of the eye's direction rather than to a complete ignorance of its movement.

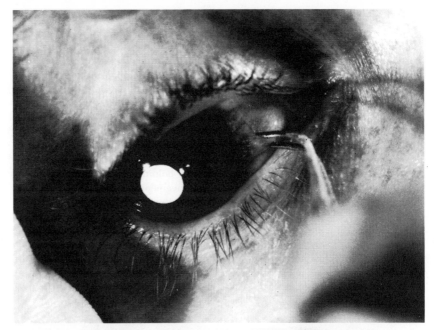

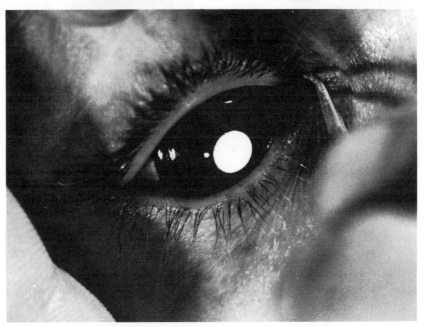

INSENTIENCE OF EYE MUSCLES was demonstrated a few years ago by means of an ingenious experiment devised by G. S. Brindley, now at the Maudsley Hospital in London. In these photographs, made in the course of the experiment, Brindley is manipulating the author's eye with forceps to test whether, after blinding it with a black cap, there was any awareness of passive movements. There was not. The white spot on the cap is to give an indication of eye position. The eye and the lids were treated with local anesthetic.

This objection, as Helmholtz argues, can be answered by considering afterimages. If one stares fixedly at a bright light for 15 to 30 seconds (please, not the sun!), then on looking elsewhere an afterimage of the bright light is perceived and persists for a minute or so. When an object is fixated steadily, the afterimage likewise stays still, but when the gaze is shifted, the afterimage also moves. This, of course, refers to active voluntary eye movements. In passive movements quite the opposite is found. No matter how hard one pulls on the eyelids the afterimage appears to remain completely stationary. In order to be certain of this phenomenon it is necessary to view the afterimage against a featureless background, such as a sheet of plain paper held close to the eye; otherwise the concomitant apparent jerking around of external objects may make the judgment difficult. Hence during passive movements we interpret what we see precisely as if the eye had not moved at all. It is not a matter of a quantitative misjudgment. The reader is encouraged to repeat for himself these crucial observations and reflect on the compelling conclusions Helmholtz drew from them.

A few years ago my friend G. S. Brindley (now at the Maudsley Hospital in London), who has a genius for settling or eliminating argument by incisive experiment, proposed that we confirm Helmholtz directly by blinding an eye with a black cap on the cornea (the eye's transparent front surface) and then moving the eye around with forceps to see if the subject could feel the movement. (Pain was prevented by instilling gener-

ous quantities of local-anesthetic eye drops.) The test proved that subjects are quite unaware of large passive rotations of the eye in its socket of 30 degrees or more; they do not know the eye is being manipulated at all unless the forceps happen to touch the eyelid. Another important point was that if the subject was invited to voluntarily move his eyeball while the forceps were gripping it, he was unable to tell whether the experimenter holding the forceps was allowing the movement to take place or was preventing the eye from moving.

The unequivocal conclusion of all these experiments is that we have no sense organs in the eye muscles or near them that tell us which way our eyes are pointing. We normally know which way

we are looking, but only because an internal "sense of effort" gives us an estimate of how much we have exerted our eye muscles. If voluntary movements are artificially impeded, or if passive movements are imposed, we absolutely do not know what is going on—unless we can see and reason back from the visual illusions we receive.

So much for the eyes. In the limbs the same facts are less easily demonstrated. To use Haller's method with patients whose tendons have been exposed under local anesthetic in the course of an orthopedic operation is one possibility, but it does not satisfy the powerful compulsion that all investigators in sensory physiology have to try it for themselves. A paper on visual illusions in which the author had not experienced the phenomena himself is almost unthinkable, and rightly so. What better way could he have of satisfying himself that they were correctly reported? Hence it is desirable to find a method for studying muscular sensibility in ordinary limb muscles of healthy subjects. The difficulty, of course, is to devise a way of stretching a muscle without the subject's knowing what is being done, since he can feel pressure on the skin or the movement of a joint. Local anesthesia of an extremity provides an answer. Investigators have variously injected local anesthetic around the joint at the base of the big toe or at the base of a finger, or have anesthetized the entire hand by cutting off the blood supply with a pneumatic tourniquet around the wrist for about 90 minutes. Movement of an anesthetized digit then stretches the muscles that move it, which lie above the anesthetized region. My collaborators and I use the top joint of the thumb, which has the advantage that only one muscle (lying well up in the forearm) flexes it, whereas the joints of the fingers are operated by more than one muscle, some in the hand and some in the forearm. Thus when the thumb is anesthetized by a tourniquet at the wrist, voluntary movements of the top joint are unimpaired in strength. We have also used injection of local anesthetic around the base of the thumb.

The uniform result of numerous experiments is that, with an adequate depth of anesthesia, the subject (whose eyes are shut) cannot tell in what position the experimenter is holding the top of his thumb, or whether he is bending it backward and forward. This is true only provided that the movement is not rapid and that the thumb is not forcibly extended or flexed at the limits of its range of movement. It is also the case

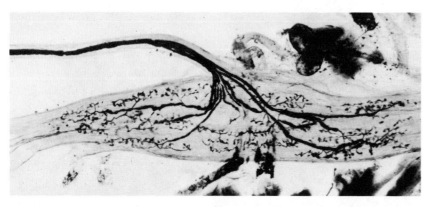

TENDON ORGAN contains sense endings that signal to the nervous system the tension in the part of the muscle in which they lie. A typical location of a tendon organ is shown in the diagram on page 101. The single sensory nerve fiber that services the tendon organ has been made to appear black in this photograph by means of a special silver stain. The nerve fiber divides many times, terminating in very fine branches with knobs at the ends. These structures, in some unknown way, sense the deformation produced by tension and cause nerve impulses to be sent up the sensory fiber at a rate that is determined by the tension. This tendon organ was dissected out of the leg muscle of a cat; it is about half a millimeter long. Surrounding one end are the remains of muscle fibers. Both photographs on this page were made by Colin Smith, Michael Stacey and David Barker of the University of Durham.

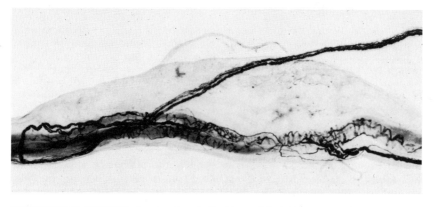

EQUATORIAL REGION of a muscle spindle dissected from the leg muscle of a rabbit appears in this photomicrograph; the part shown is about a millimeter long. Again the nerve fibers and nerve endings have been stained with a silver stain, making it possible to distinguish clearly the equatorial capsule, the intrafusal muscle fibers and the sensory endings wrapped around them. The nerve ending to the right is a primary ending; its sensory nerve fiber enters from lower right. The other ending is a secondary ending; its nerve fiber enters from upper right. The finer nerve fibers are part of the motor nervous system.

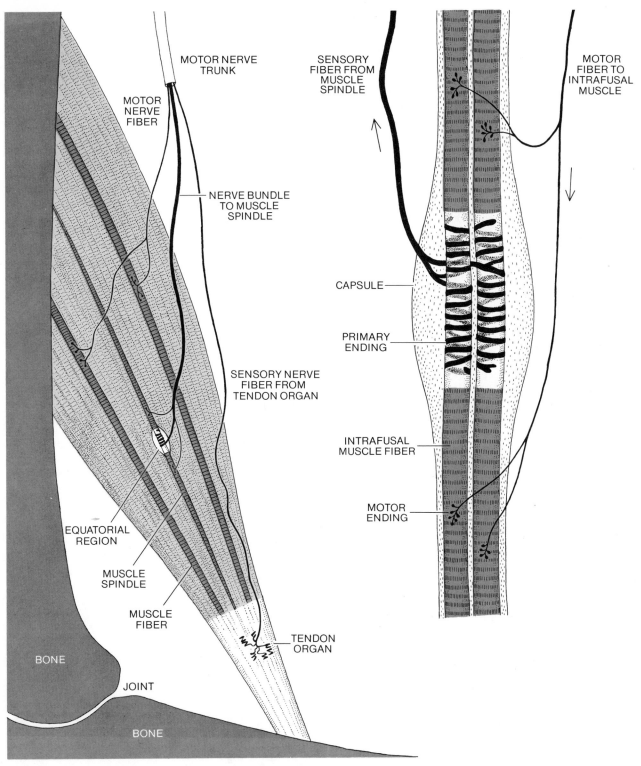

MOTOR NERVE
TRUNK

MOTOR
NERVE
FIBER

NERVE BUNDLE
TO MUSCLE
SPINDLE

SENSORY NERVE
FIBER FROM
TENDON ORGAN

EQUATORIAL
REGION

MUSCLE
SPINDLE

MUSCLE
FIBER

TENDON
ORGAN

BONE

JOINT

BONE

SENSORY
FIBER FROM
MUSCLE
SPINDLE

MOTOR
FIBER TO
INTRAFUSAL
MUSCLE

CAPSULE

PRIMARY
ENDING

INTRAFUSAL
MUSCLE FIBER

MOTOR
ENDING

ARRANGEMENT OF SENSE ORGANS in a typical muscle is indicated in these simplified diagrams. The proportions in the diagram at left are highly distorted. A real muscle fiber is only about a tenth of a millimeter in diameter, but it is often several centimeters long. A muscle spindle is somewhat thinner; it consists of even finer specialized structures called intrafusal muscle fibers. Only two ordinary muscle fibers and one spindle are depicted in detail; a real muscle may contain tens of thousands of muscle fibers and hundreds of spindles. The diagram at right gives an enlarged view of the equatorial region of a muscle spindle. Wrapped around the intrafusal muscle fibers are the terminations of the sensory nerve fiber; the function of these sense endings is to respond to mechanical deformation by causing nerve impulses to be sent up the sensory nerve. In the equatorial region the cross striations, which are an indication of the presence of a contractile mechanism within the fiber, are absent. Hence when the intrafusal fibers contract, this region is extended and excites the sensory endings, just as if the region had been extended by stretching the entire muscle and the spindle within it. In this diagram only two intrafusal fibers are shown; a real spindle often has half a dozen or more. Moreover, intrafusal fibers come in two distinct varieties, only one of which is shown here. Another complication is the fact that there are three distinct kinds of motor nerve to the intrafusal fibers. In a real spindle the equatorial region is also much longer than depicted here. Photomicrographs showing the innervation of a tendon organ and the equatorial region of a muscle spindle appear on page 100.

that if the subject attempts to flex his thumb, he cannot tell whether he has been successful, or whether the experimenter has prevented it from moving. Thus with skin and joint sensation eliminated the thumb behaves just like the eye. Muscle *is* insentient.

I have already argued that one would not on general grounds expect the liver, say, to have sensibility like the skin's. Indeed, if one looks at the liver through a microscope, it has none of the elaborate apparatus of sensibility seen in the skin—no network of branching nerve fibers ending in a variety of characteristic sensitive structures: the sense organs. The same goes for other viscera. Muscles are not so obliging. They are supposed to be insentient, but when we look inside them, they turn out to be full of sense organs, and very fine sense organs at that. The principal kind, the muscle spindles, are the most elaborate sensory structures in the body outside of the eyes and ears. This deep paradox (for which the reader has already been prepared) is at the back of everything in this article. All the essential facts that create it have been known since 1894, when Sir Charles Sherrington proved conclusively that there were nerve fibers going to the muscle spindles that be-

longed to the body's system of sensory nerves, and hence established that the muscle spindles were sense organs. Unfortunately in those distant days Sherrington was insensitive to the class distinction between the information on the road sign that tells the driver to turn right and the information from the sensors in his power-assisted steering gear that enables him to do so effortlessly (between, one might say, the different types of information required by the legislature and the executive). He allowed himself to be persuaded that Helmholtz had been wrong and that his own discovery showed that muscles were sentient after all.

Sherrington had thus taken the view that in effect there was no paradox, and his influence was so immense that it was 60 years before the true situation was at last clearly perceived. By this time the paradox had much less impact, since physiologists had discovered many of the facts about the muscle spindle needed for its resolution. Before going on to these facts I should finish the present story.

In the past few years the paradox has been given a further twist. Several groups of workers on both sides of the Atlantic, whose members are too numerous to name individually, have found

that signals from muscle sense organs find their way to the cerebral cortex. It seems that they get to the cortex but we remain unconscious of them. This is very surprising. No one imagines for a moment that we do not make use of all the information our eyes send to the cerebral cortex to build up the picture of the outside world we consciously perceive, and I am sure that a few years ago any ordinary physiologist would have been prepared to extend this point of view to sensory information of any kind that could be shown to get to the cortex.

The evidence for what I have just said is not complete. The animal most resembling man in which signals from muscle sense organs have been shown to reach the cortex is the baboon. It seems unlikely that they do not reach the cortex in man, and equally unlikely that a baboon should be conscious of the signals from its muscles when a man is not. A strong hint also comes from the cat. John E. Swett and C. M. Bourassa of the Upstate Medical Center of the State University of New York showed that muscle sense organs send signals to the cat's cerebral cortex, but unlike signals from the skin (or for that matter from the eyes or ears) they cannot be used to set up a conditioned reflex. Without ex-

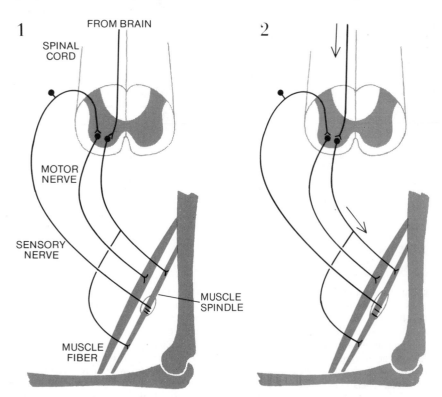
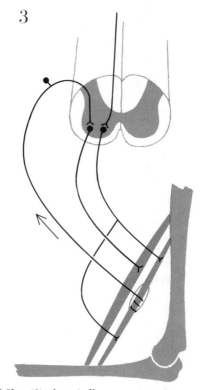

SERVOMECHANISM involved in the control of voluntary muscular contractions is shown here. The basic diagram (*1*) is the same as it is in the illustration of the stretch reflex, but with provision made for signals from the brain to cause the muscle spindle to contract by way of a special motor nerve fiber. When a signal is

transmitted along this special fiber (*2*), the spindle contracts, exciting the spindle sensory ending, just as if the spindle had been stretched. Consequently a contraction of the main muscle is excited by way of the stretch-reflex pathway (*3, 4*). In a real muscle this picture is further complicated by the existence of a direct pathway

plaining what is meant by this fact in detail one can say that it strongly suggests the cat is not conscious of the signals from its muscles.

The first part of this article was intended to introduce the reader to the idea that muscle organs function at a subconscious level in a purely subservient role. Like the perfect servant, they work so unobtrusively that we are unconscious of them, but the findings about cortical projection begin to strain the analogy. The eccentric 18th-century scientist Henry Cavendish reportedly dismissed any servant he caught sight of. He wrote down what food he wanted and it was put out for him. It would have been going too far to expect the butler to wait on him at table without betraying his presence, but that is what the muscle sense organs seem to manage to do!

Scarcely less remarkable than the mere existence of the muscle spindles is the fact that they (the most important of the two kinds of muscle sense organ) are themselves contractile. This is a unique property among sensory structures. That was perfectly clear to Sherrington in 1894, but it still remains one of the most challenging observations in the physiology of the motor system;

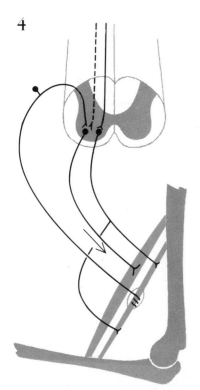

4

(*broken line in diagram 4*) **from the brain to the main motor nerve cells. In the power-steering analogy this pathway corresponds to a direct connection between the steering wheel and the road wheels of an automobile.**

even if the interpretations to be put forward later in this article are on the right lines, it is most improbable that they are more than one facet of the truth.

Muscle spindles (they are called spindles because they are long and thin and have pointed ends) consist of a bundle of modified muscle fibers, the intrafusal muscle fibers (from the Latin *fusus*, meaning spindle), with the sensory nerve fibers wrapped around a short specialized region somewhere near the middle of their length. The stimulus that excites a muscle spindle is the stretching of this specialized sensory region. Now, as I have said, the muscle spindles are contractile. They are not, however, equally contractile along their entire length; the contractile apparatus fades out in the sensory region, and the middle of the sensory region, where the sense endings connected to the largest nerve fibers lie, probably does not contract at all. When the spindle contracts, these sense endings (known as the primary endings) are stretched by the contraction of the remainder of the spindle and discharge nerve impulses.

The next point to observe is that the muscle spindles lie among the ordinary muscle fibers (the much larger red stringy structures, visible to the unaided eye, that actually do the work) and share their attachments to bone or tendon. Hence they change length as the main muscle fibers change length. If a contraction of a muscle spindle, which excites its primary ending, is succeeded by an equal contraction of the main muscle, the stretch will be taken off the sensory region and the ending will be silenced. The spindle primary, in fact, is sensitive to the difference in length between the spindle and the main muscle fibers; it is a misalignment detector. It discharges if contraction of the spindle is not matched by contraction of the main muscle, or, vice versa, if extension of the main muscle is not accompanied by relaxation of the spindle. There is no obligation for the muscle spindles and the main muscle to contract and relax together, because the motor nerve fibers that run to them and carry the nerve impulses from the central nervous system that cause them to contract are largely separate. The spindles could therefore be activated while the main muscle remained passive, and vice versa.

Having seen the circumstances under which nerve impulses are discharged by the spindle primary endings, the next question is: What do these impulses do when they reach the central nervous system? Their best-established

function is to excite an automatic contraction—the stretch reflex—in the muscle from which they come. This they do, at least in part, by impinging directly on those nerve cells in the spinal cord that give rise to the motor nerve fibers to the muscle in question.

The most familiar manifestation of the stretch reflex is the knee jerk, widely used in medicine to test the state of the nervous pathways concerned. A physician strikes the tendon below the knee-cap with a rubber hammer, and in a healthy subject the muscles that straighten the knee briefly contract involuntarily. The effect of striking the tendon is slightly and suddenly to stretch these muscles, and so to excite their muscle spindles. The tendon itself has no part in the sensory mechanism. The tendon jerk is quite transient, but under suitable circumstances a slower, sustained extension of a muscle will result in a sustained reflex contraction. If the reaction in a patient who is otherwise relaxed is exaggerated, the limb is said to be "spastic," that is, affected by spasm.

Human muscles in general can be shown to be under the influence of the stretch reflex when they are engaged in steady contractions of a voluntary nature. The main evidence for this is that if a subject is invited, say, to flex his elbow steadily against a load, it is found that a sudden unexpected increase in the load, which causes his elbow to extend, calls up a larger contraction of his biceps muscle, and conversely a decrease in load causes a relaxation. Electrical recording methods reveal that these reactions begin so soon (within about a twentieth of a second) that they must be automatic, reflex responses.

It has been realized for half a century that the stretch reflex confers valuable self-regulating properties on a muscle, causing it automatically to adjust to changes in load, without any need for the orders that the brain sends down to be altered. Everyone believes the reason the horse does not sag at the knees when Douglas Fairbanks leaps from the castle parapet onto its back is that the horse's leg muscles immediately respond to the extra strain by way of their stretch reflexes. If this interpretation is correct, we have one answer to the question: What does the horse expect of its muscles? In this situation it expects them not only to exert enough force to support its body weight but also to adjust automatically to extra weight. Clearly what the horse really wants is for the length of the muscles to be kept roughly constant so that posture is maintained. The stretch reflex can achieve this result

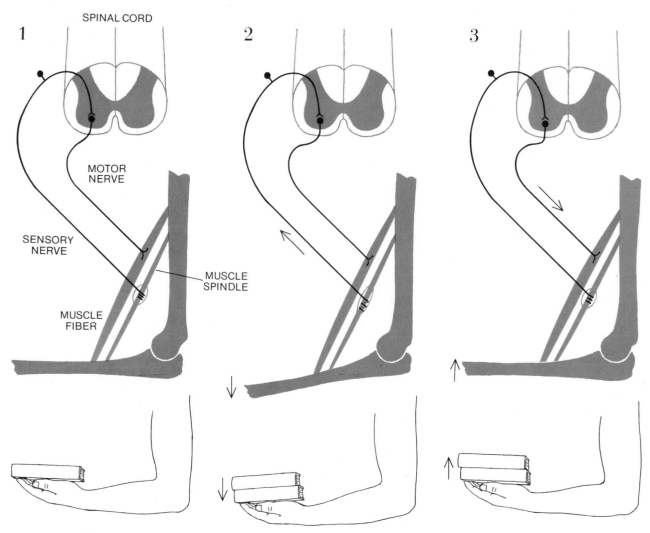

SPINAL CORD

1   2   3

MOTOR
NERVE

SENSORY
NERVE

MUSCLE
SPINDLE

MUSCLE
FIBER

STRETCH REFLEX is mediated by the nervous mechanism depicted in this highly schematic illustration. A muscle is under the influence of the stretch reflex when it is engaged in a steady contraction of a voluntary nature, as when a person's elbow is flexed steadily against a load (1). A sudden unexpected increase in the load (2) stretches the muscle, causing the sense ending on the muscle spindle to send nerve impulses to the spinal cord (*upward arrow*), where they impinge on a motor nerve cell at a synapse and excite it. As a result motor impulses are sent back down to the muscle (*downward arrow*), where they cause it to contract (3). More complicated nervous pathways than the one shown may also be involved in the stretch reflex. Any real muscle is, of course, supplied with many motor nerve fibers and spindles. In addition the synaptic connections to even a single motor nerve cell are multiple.

for the horse because it is based on a sensor—the muscle spindle—that measures length, or, to be more exact, differences in length.

What happens when it is desired that the muscles should execute a movement, not merely maintain a stationary posture or some other steady contraction? The obvious trick is to cause the spindles to contract at the desired rate so that the sensory endings on the spindles will be excited if the main muscle does not itself keep up with the spindles, that is, does not contract at the desired rate. In this way the advantages of automatic compensation for changes of load by means of the stretch reflex could be retained during active shortening. Contraction of the spindles would in effect drive the main muscle by means of the

stretch reflex, turning on more contraction if an unexpected obstruction were met with, or if the rate of shortening for any other reason fell behind, and, vice versa, damping down contraction automatically if the load unexpectedly diminished, or if for some other reason the movement undesirably accelerated. Within the past year C. D. Marsden, H. B. Morton and I have obtained direct evidence that this kind of rapid, reflex compensation does in fact occur during voluntary movements in man.

In this mode of operation the stretch reflex, as the reader will have perceived, functions as a follow-up servomechanism, closely analogous to power-assisted steering in an automobile. Contraction of the spindle corresponds to turning the steering wheel, shortening of the main muscle to turning of the road

wheels, with the spindle sensory ending acting as the misalignment detector. The subject can demand of his muscles either a certain limb position or a certain rate of change of limb position, and within limits (limits not yet known in quantitative terms) his demands will be automatically met by his muscle servo.

That, in brief outline, is as far as we have gone in understanding how, when we make a voluntary effort, the muscle sense organs act at a subconscious level to ensure that our muscles do what we expect of them. Many facts have had to be left out and without doubt many more remain to be discovered. To attempt any account at this stage requires a certain presumption. I can only hope that when the whole truth emerges, it will prove to be an extension and not a contradiction of the story I have told here.

Annual Biological Clocks

by Eric T. Pengelley and Sally J. Asmundson
April 1971

Many organisms have a built-in circadian, or daily, clock. It has now been demonstrated that some also have a circannual, or yearly, clock that operates even when environmental signals are eliminated

As the days become shorter and the temperature drops, birds migrate to a warmer climate, plants become dormant and hibernating animals become fatter and go into their winter sleep. By the same token, the lengthening day and rising temperature in spring bring on the return migration of birds, the budding of plants, arousal from hibernation and (for many animals) the season of copulation and breeding.

The timing of these yearly events is so obviously related to the passing seasons that until recently it was assumed that environmental changes directly supplied the cues for changes in plant and animal behavior. With the increased interest in studying biological phenomena under controlled laboratory conditions (a strategy that has opened many new doors for 20th-century biology) some surprising discoveries have emerged concerning the cyclic behavior of plants and animals. One major finding is that many organisms show cyclic behavior even when the physical environment is kept constant.

A hint of this interesting phenomenon was first reported more than two centuries ago by Jean Baptiste Dortous de Mairan, who noticed that certain plants, although cultivated in total darkness, daily bent their leaves as if toward the sun. Charles Darwin later noted similar phenomena and discussed them in his book *The Power of Movement in Plants*. The matter was not brought under close study, however, until this century, when investigators began to observe that many plants and animals, including man, have daily rhythms of behavior or physiology that seem to be endogenous, that is, generated internally. It is now known that organisms can keep their behavior and physiology in time with the changing environment in three ways. They can respond directly to changes in the environment; they can be internally programmed to respond in a specific way at a specific time regardless of the environmental cues, and both responses can be combined.

In 1959 Franz Halberg of the University of Minnesota gave these rhythms the name "circadian," from the Latin *circa* (about) and *dies* (day). The term "about" is crucial; the rhythm, if it is actually endogenous, never has an exact 24-hour period. A rhythm with such an exact period would have to be attributed to some geophysical signal or stimulus, even though it was unknown, that coincided with the length of the day. A rhythm that only approximates the day's length, on the other hand, can reasonably be ascribed to an internal "clock" of some kind in the organism. To be sure, the clock must have had its evolutionary origin in conditions related to the length of the day, but the fact that it now maintains its own period strongly indicates that the clock is endogenous in the individual plant or animal.

The existence of endogenous daily clocks in certain plants and animals has been confirmed by three criteria: (1) they keep their own time, not exactly corresponding to the day's length; (2) the rhythm they control is not synchronous with any daily environmental signal such as a change in light or temperature, and (3) the period of the rhythm is not affected by the level of the ambient temperature in the organism's environment. It is now well established that the daily rhythmic behavior of many organisms throughout the biological world is under the control of an endogenous daily clock, usually in combination with external factors such as the daily variation in the intensity of light.

The existence of a daily clock led to speculation about the possibility of other endogenous rhythms, particularly an annual rhythm. Obviously it is much more difficult to demonstrate the existence of such a clock, since the task requires a long-term, year-by-year study. Two French investigators, Jacques Benoit and Ivan Assenmacher, seem to have been the first to undertake such an investigation. They selected for study the breeding cycle of the male domestic duck. Keeping the ducks under constant lighting conditions (that is, lighting unchanged in intensity and in the length of the "day"), they measured the activity of the birds' testes over the year. Unfortunately the results were inconclusive.

The first clear indication of the existence of an annual clock was discovered almost by accident in a different investigation by the late Kenneth C. Fisher and the senior author of this article (Pengelley), who was then a student of Professor Fisher's at the University of Toronto. Fisher was studying the phenomenon of hibernation, using as a subject the golden-mantled ground squirrel (*Citellus lateralis*), an inhabitant of the higher reaches of the Rocky Mountains [see "The Adjustable Brain of Hibernators," by N. Mrosovsky; SCIENTIFIC AMERICAN Offprint 513]. The animal was housed in a small, windowless room at a constant temperature of zero degrees Celsius and on a schedule of 12 hours of artificial light each day. It had an unlimited supply of food and water.

Placed in this room in late August, the animal ate, drank and behaved normally until October, keeping active and maintaining its body temperature at a normal 37 degrees C. in the chamber's freezing climate. In October, as might be expected in natural surroundings at that time of year, the squirrel stopped eating and

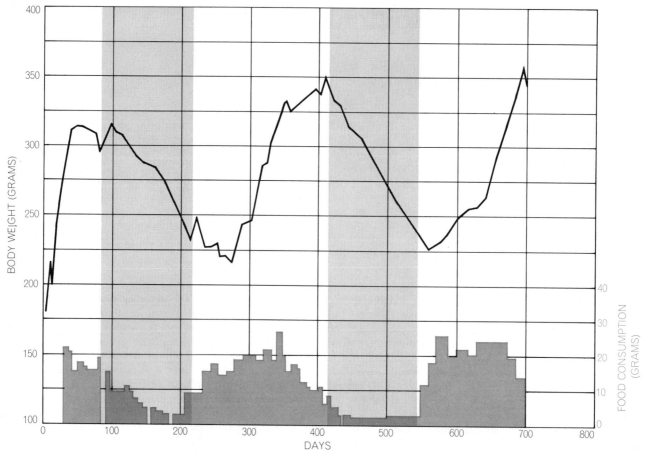

BIOLOGICAL CLOCK of the ground squirrel is responsible for a roughly annual cycle comprised of a wakeful period and a period of hibernation, as this 23-month record of one squirrel indicates. The animal was kept in a room at a temperature of 22 degrees C. (72 degrees F.) that was illuminated 12 hours each day. Although free from outside stimuli, the animal hibernated regularly (*gray bands*). During hibernation it lost weight steadily even though food was available and food consumption continued (*color*).

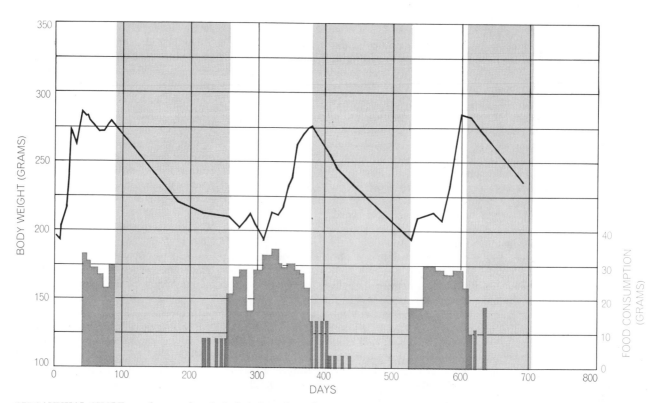

CIRCANNUAL CYCLE was shown to be relatively independent of temperature by a second 23-month study. A ground squirrel was exposed to the same 12-hour day but its room temperature was held at freezing point. No change in hibernation pattern occurred.

drinking and went into hibernation, allowing its body temperature to drop to about one degree C. Except for occasional arousals it remained in hibernation until the following April, when it again became active, raised its body temperature to 37 degrees C. and resumed its usual eating and drinking. In September the animal stopped eating and drinking and entered hibernation without prompting by any discernible signal or change of circumstances.

The experiment was repeated with large numbers of ground squirrels. These animals were exposed to two different ambient temperatures: zero degrees and 22 degrees C. (normal room temperature). Regardless of the ambient temperature, the animals first increased their food consumption and gained weight, then went into hibernation, then awoke to feeding and activity again for a few months and then again lapsed into hibernation in a more or less fixed time cycle. The period of each complete cycle was a little less than a year [*see illustrations on opposite page*]. The animals' behavior fulfilled the criteria for the existence of an endogenous annual clock: the period of the rhythm was not exactly a year, the rhythm was not synchronous with any periodic external signal, and it was not appreciably affected by the ambient temperature. Following Halberg's terminology, this rhythm has been called "circannual."

The authors of this article have gone on to further experiments, with support from the National Science Foundation, and we now have data on the persistence of the circannual rhythm of consecutive hibernation and activity over an observation period of nearly four years. Even more remarkable, the animals show an annual cycle of alternating gain and loss of body weight even when they do not hibernate! We undertook the experiment of keeping the ground squirrel at an ambient temperature of 35 degrees C. (95 degrees Fahrenheit), which is close to the animal's normal body temperature. At this environmental temperature the ground squirrel cannot hibernate. Nevertheless, with plenty of food and water available, the animals reduced their food and water consumption and lost weight during the assumed "winter" and then resumed feeding and gained weight in the "spring" [*see top illustration on next page*]. There could hardly be a more convincing demonstration of the existence of an internal clock operating independently of the environmental conditions. Quite evidently an annual cycle of feeding and fasting is

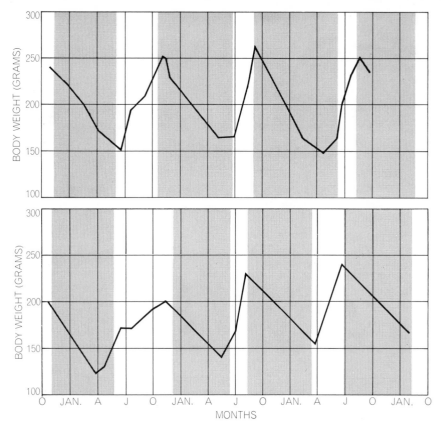

RECORDS OVER FOUR YEARS provide further evidence that the circannual rhythm of ground-squirrel hibernation is not influenced by the normal range of temperatures. The gray bands indicate hibernation. In one instance (*top*) the room temperature was 12 degrees C. (53.6 degrees F.); in the other (*bottom*) the room was just above freezing point.

also programmed in the animal. An experiment conducted by H. Craig Heller and Thomas L. Poulson at Yale University gave even more specific evidence on this question. They rationed the food and water supply of ground squirrels so that the animals lost a great deal of weight just before the usual time of entering hibernation. Then in December the experimenters gave the squirrels all the food and water they might want; the animals, however, took only enough to raise their weight to the usual level for that time of the year, in other words, to their "programmed" weight.

How is the endogenous clock established? Is it created by conditioning or by early imprinting on the nervous system during exposure to the environment soon after birth? Looking into this question, we raised laboratory-born ground squirrels under constant conditions of light and temperature and then placed them while they were still young in a chamber kept dark and at a temperature of three degrees C. These animals still showed the annual rhythm of alternating activity and hibernation in several instances extending over three yearly cycles. The experiment clearly indicated that the rhythm is determined genetical-

ly, not by conditioning; the clock is already set and wound at birth. The annual cycle of one squirrel ranged between 324 and 329 days as measured from the start of one hibernation period to the start of the next. This is the period of the clock under "free-running" conditions, that is, in a constant environment without modification by changing external stimuli.

The discovery of a circannual clock in hibernators has of course been followed up with investigations of other animals marked by conspicuous annual changes in behavior or physiology. The migration of birds, perhaps the most widely and indefatigably studied of all cyclic behavior in the animal world, is a particularly inviting subject for a "clock" investigation. The search for the triggering stimulus that sets off bird migration has ranged over a great diversity of possible factors. Nearly half a century ago William Rowan of the University of Alberta pioneered a laboratory attack on the problem. Working with migratory finches of the genus *Junco,* he found that by manipulating the daily rhythm of light and dark he could influence the rate of growth of the gonads, which in turn

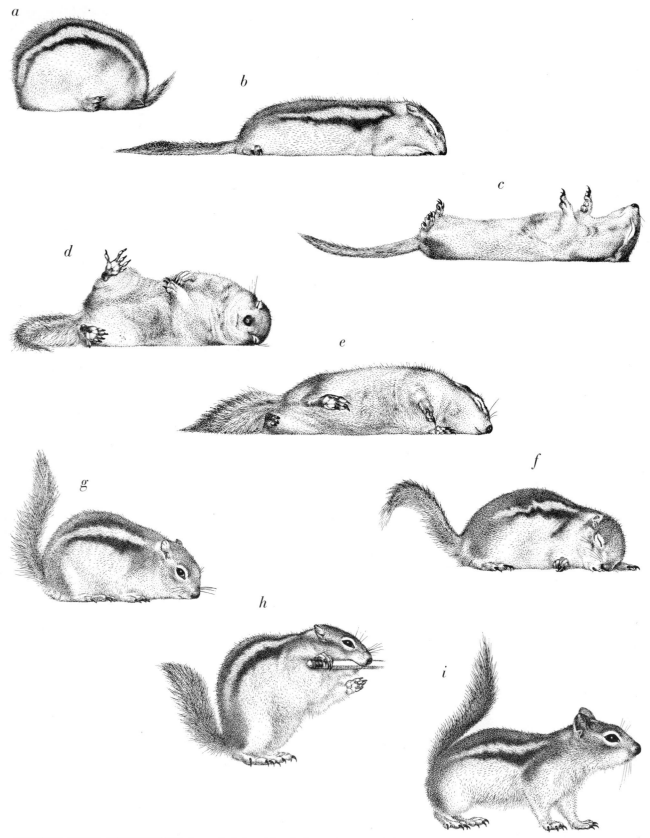

a

b

c

d

e

f

g

h

i

AROUSAL FROM HIBERNATION is marked by rising body temperature in the golden-mantled ground squirrel (*Citellus lateralis*). The ground squirrel was housed in a windowless room that was chilled to the freezing point and illuminated for 12 hours each day. Although it had unlimited food and water, the ground squirrel lost weight and entered hibernation as if it were in the wild. The observations recorded in these drawings were made during one of the animal's periodic arousals, as its temperature rose within two hours from near freezing (1.7 degrees Celsius) to normal (37 degrees C., or 98.6 degrees Fahrenheit). The drawings at top and bottom show it at onset and completion of arousal. The others show it at a number of intermediate temperatures: *b*, three degrees C.; *c*, four degrees C.; *d*, 14.5 degrees C.; *e*, 17.5 degrees C.; *f*, 20 degrees C.; *g*, 26 degrees C., and *h*, 35 degrees C., near normal.

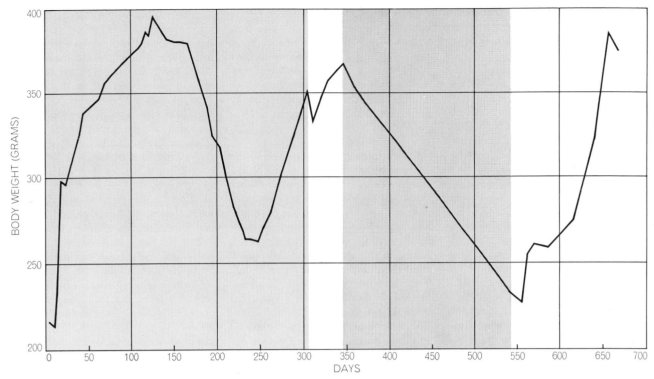

DEPRIVED OF HIBERNATION (*left*) by confinement in a hot room, one ground squirrel still lost weight in circannual fashion although it was awake and had unlimited access to food. The ani-

mal was kept at 35 degrees C. (95 degrees F.) for 302 days (*color*). It did not hibernate but nonetheless lost weight in the way that it did during hibernation (*gray band*) when shifted to a cold room.

acted to bring on the characteristic restlessness and activity that initiated the birds' migration. Rowan's identification of the length of day as a factor influencing the migratory urge has since been well demonstrated, but Rowan and others felt that this could not be the only factor. It remained for Eberhard Gwinner of the Max Planck Institute at Erling-Andechs to discover that some migratory birds, like hibernators, also possess a circannual clock.

Gwinner recently performed a systematic experiment with willow warblers and wood warblers, which breed in cen-

tral and northern Europe, migrate in the fall to central and southern Africa and return to Europe in the spring. Taking fledglings from the nest in spring, he divided them into four groups. Two groups were kept near Munich, one group indoors at a constant temperature of 21 degrees C. and on a daily schedule of 12 hours of light and 12 hours of darkness, the other group in the natural environment, with its variations in temperature and day length. The other two groups were flown to the birds' wintering area in central Africa; there they were sequestered under the same two sets of

conditions, one group under the constant regime and the other exposed outdoors to the natural day length and temperature of the region.

The behavior of the four groups was studied in terms of signs of *Zugunruhe*, or migratory urge, as shown by night activity and by the molt of feathers (which normally occurs during the winter after migration to the wintering area). Gwinner found that the year's cycle was much the same in all four groups: the birds, whether they were on a constant or a natural regime and whether they were kept in the northern or the southern area,

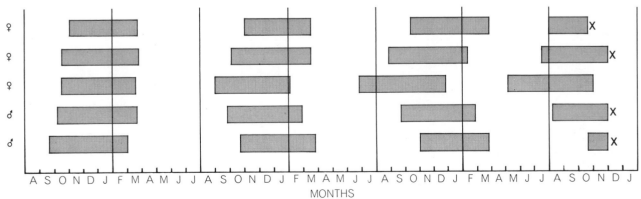

HEREDITARY CHARACTERISTIC, rather than "imprinting" in infancy, is shown experimentally to be the source of the circannual rhythm in ground squirrels. Three females and two males were isolated from external stimuli at birth and then kept continuously in

total darkness at near-freezing temperature. Although deprived of a normal environment, all five animals nonetheless hibernated successively ("*X*" *indicates death*). A near-uniform length of cycle is evidenced by the 313-, 314- and 310-day record of one female.

tended to show a migratory urge in the spring and the fall and to molt in the intervening periods [*see illustration on this page*]. Even when birds were flown to the wintering area in Africa at the height of their fall migratory restlessness they continued in this restless condition just as long as the birds that were not taken to the wintering grounds. In short, differences in environmental conditions did not profoundly affect the birds' basic rhythm of migratory urge and molt. Gwinner concluded that the warblers' rhythm is determined primarily by an endogenous timing mechanism. He has since established that the rhythm is maintained for at least two years regardless of the environmental conditions to which the birds are exposed.

Gwinner noted that the seasonal timing of the events in the birds' rhythm depended on their date of birth. For birds hatched late in the year the dates of migratory urge and molt came later than for those hatched early. This can be taken as further evidence that the timing mechanism is endogenous. Such a mechanism can be seen to have at least two benefits for survival of the species. A programmed duration of the migratory urge helps to ensure that the migrating birds will arrive at a suitable habitat under any conditions; they need only fly at

a given speed and on a defined course until the program runs out. The temporal variation of the individual birds' migrating "seasons" (depending on their time of birth) also spreads out the migrations so that any temporary obstacles or hazards along the flight course or at the destination will involve only part of the population.

Another interesting yearly cycle is the annual growth and shedding of antlers by deer. Richard J. Goss of Brown University has looked into the possibility of an endogenous clock for this phenomenon. In the Northern Hemisphere a deer, having shed its antlers in the winter, ordinarily is in velvet (growing a replacement set of antlers) from spring to fall. The cycle tends to be somewhat irregular in time. It has long been known that tropical deer transported to zoos in the Temperate Zone persist in the same kind of annual cycle they showed in their native habitat, in spite of the difference in the day-length pattern. Goss performed an extensive experiment with the sika deer (a native of Japan) in Massachusetts. A control group was kept outdoors under natural conditions; these deer were in velvet from May through September. The experimental deer were placed in light-tight barns and given a

constant photoperiod of different lengths; some had eight hours of artificial light per day, some 16 hours, some were kept in continuous darkness and some had a "day" evenly divided between 12 hours of light and 12 of dark. In this experiment, extending over three years, all but one of the groups showed essentially the same annual cycle, with irregular timing, of antler growth. Only the deer on the day evenly divided between light and dark failed to replace their antlers each year. The results with the Japanese deer clearly indicated the existence of a circannual clock, probably phased in this case by the rhythm of light and dark.

Perhaps the clearest known demonstration of the control of antler growth by an endogenous clock has been given by a totally blind elk that has been under observation for six years at Colorado State University. Notwithstanding the absence of any light cues, this animal has shed and regenerated its antlers on schedule every year.

There is now evidence that the possession of a circannual clock is by no means confined to certain vertebrates. The Yale investigators Poulson and Thomas C. Jegla have discovered such a clock in a small, translucent crayfish, *Orconectes pellucidus*. In nature this animal lives in an environment almost devoid of seasonal changes; it dwells in dark caves where there is little variation in temperature. Poulson and Jegla collected the animals from a cave in Kentucky and placed them in laboratory aquariums kept in total darkness at a constant temperature of 13 degrees C., the average temperature of the cave habitat. The darkness was interrupted for only 15 minutes two or three times a month, when the experimenters fed the animals and examined the stage of their annual reproductive cycle. The cycle can be observed in the live animal; in the adult male the season of reproductive capability is punctuated by molting, and in the female changes in the size of the ovary and the eggs are visible through the translucent shell. Poulson and Jegla found that in the constant laboratory environment the crayfish still showed an annual cycle in reproductive readiness. The period of their circannual clock was not exactly a calendar year.

It seems highly likely that circannual clocks are widespread in the animal world. Obviously such a mechanism would be useful to migratory marine mammals and to many animals that must avoid the harshest seasonal conditions of their natural environment, as in the arctic regions and the high mountains. Circ-

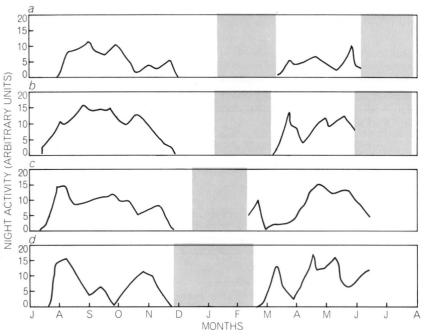

MIGRATORY BIRDS also display a circannual rhythm that is not influenced by length of day or change of temperature. The night activity of young willow warblers, an index of migratory urge, was observed in four groups of birds; in periods of molt (*bands of color*) the urge is not present. When kept at a constant temperature of 21 degrees C. (70 degrees F.) with a 12-hour period of daily illumination, one group of birds (*a*) virtually duplicated the record of night activity set by a second group (*b*) that was exposed to normal variations in day length and temperature in Munich. This was also true for birds kept at the same artificial temperature and day length (*c*) and under normal conditions (*d*) in tropical Africa.

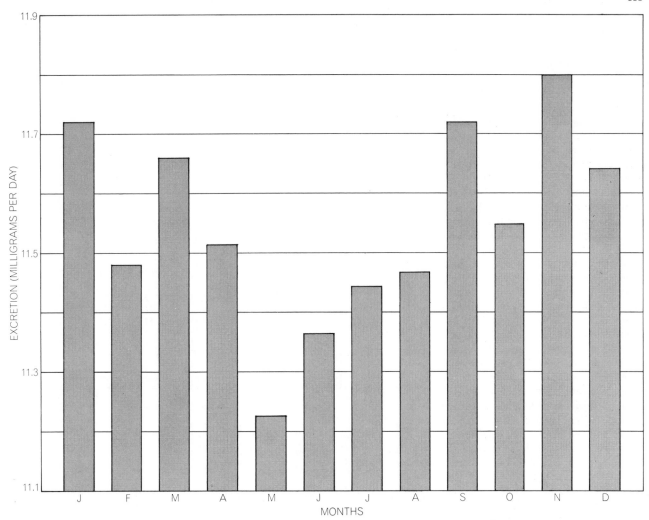

POSSIBLE CIRCANNUAL RHYTHM in man may be reflected in variations in the quantity of 17-ketosteroids found in human urine. This graph summarizes 15 years of monthly variations noted in a normal donor by Franz Halberg of the University of Minnesota.

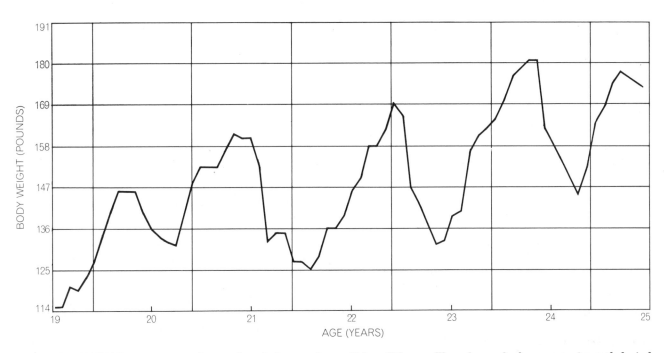

WEIGHT CHANGES in a patient diagnosed as being manic-depressive are shown for the 72-month period from the patient's 19th to 25th year. These data and others concerning pathological conditions suggest the existence of a circannual rhythm in humans.

annual clocks may, in fact, be almost as universal as circadian ones.

One may question the adaptive value of a circannual clock. Could cyclic behavior not be regulated simply by the external stimuli from seasonal changes in the environment? Actually it is easy to see that an annual endogenous clock serves several purposes. In the first place, it provides advance warning of a coming change before signs of the change appear. A hibernator, for instance, needs to begin depositing body fat before the cold weather arrives. The setting of a circannual clock gives an animal time to prepare for annual events such as reproduction, cold weather, a dry season or a period of food shortage. Secondly, it allows flexibility in a cyclic environment where events do not repeat themselves exactly each year. Furthermore, it supplies a reliable guide in situations where environmental cues are either missing or unclear. Birds wintering near the Equator, for example, get little or no signal from the environment to indicate that the time has come to fly back to their breeding place in the Temperate Zone.

Still, the circannual clock itself is not an adequate regulator. As we have seen, it is never set at exactly 365 days. If the natural environment played no part, the clock and the animal's rhythm would become increasingly out of phase with the natural seasons. Hence the animal still must depend on *Zeitgeber*, or cues from the environment, to correct the clock and thus entrain its rhythm each year. What are these cues? In the case of the circadian cycles, it is fairly well established that the daily variations in light and temperature serve such a function. No doubt these two physical parameters are also involved in regulating the annual rhythms. Probably, however, these rhythms depend as well on more subtle factors that have not yet been isolated.

Another subject that begs for investigation is the location and nature of the endogenous clock in the organism. The search for such a clock, even of the circadian variety, has barely begun. Efforts are now being made, for instance, to locate the clock in the brain of a fruit fly, and several investigators have proposed models of what an endogenous clock might be like. The living mechanism itself, however, remains quite unknown, and no doubt this problem will be very difficult to solve.

Finally, there is the intriguing question of whether or not man has a circannual clock as he most certainly has a

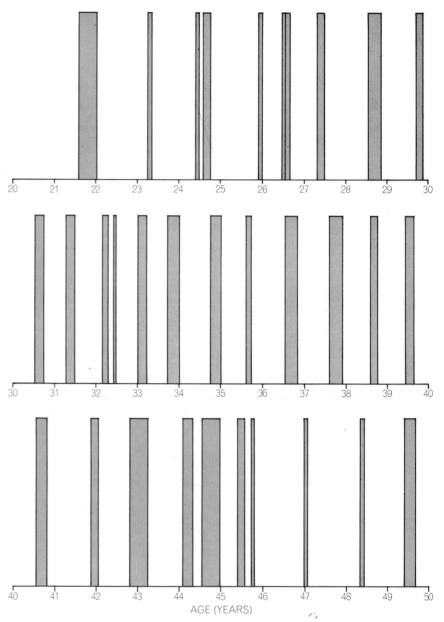

THIRTY-YEAR RECORD of manic episodes in the history of another patient is shown in this graph in three consecutive rows, each representing a 10-year period. Like the data presented at the bottom of the opposite page, the record shows a roughly circannual rhythm.

circadian one. Halberg and his associates at Minnesota recently reported some interesting data on a normal male subject whom they have been observing for about 15 years. They have measured the daily excretion of 17-ketosteroids in his urine and made a complex computer analysis of the data. The volume of these excretions shows several small oscillations, or rhythms: one with a period of about a week; another, about 20 days; another, about a month, and another, about a year. Over the 15 years of the study this subject has shown a definite annual rhythm with a low in May and highs in the fall and winter [*see top illustration on preceding page*]. The

rhythm apparently cannot be connected with any synchronous environmental signal and seems to be independent of temperature, which of course suggests that it may reflect an endogenous clock. Long-term studies of two psychotic individuals, one showing a rhythm of annual manic-depressive attacks and the other a roughly annual cycle of ups and downs in body weight, also point to a possible circannual clock.

If man does indeed possess a circannual clock and its nature becomes accessible to investigation, the implications may be as important as those of the circadian clock, to which we are all biologically bound.

THE ADAPTIVENESS
OF BEHAVIOR

II

THE ADAPTIVENESS
OF BEHAVIOR

INTRODUCTION

Every time a biologist seeks to know *why* an organism looks and acts as it does," Konrad Lorenz says in the opening article of this section, "he must resort to the comparative method." To compare species in terms of some trait, and particularly to trace their lineage on the basis of the trait, is to examine the results of an evolutionary experiment. By searching for correlates in the environment, we can sometimes establish the functions of the various forms in which the trait exists. Consider one of the more striking examples from the recent literature. The kittiwake is a gull that differs from other species in its habit of nesting on narrow ledges along cliffs and the sides of buildings instead of on the ground. As noted by Niko Tinbergen ("The Evolution of Behavior in Gulls"), the mark of this single adaptation is stamped on virtually every aspect of the kittiwake's behavior. Freed to a large extent from the necessity of defending its nests against predators, the adult bird shows little fear and has almost lost the alarm call. The territory is limited to the immediate vicinity of the nest (since so little space exists in the first place), and the birds have accordingly modified their threat behavior so that it can be performed directly on the nest site and with a minimum of movement. The chicks are unusual in the calmness of their behavior. Unlike the young of other gulls, they do not wander away from the nest and tend to stay rooted in one spot under conditions that would ordinarily cause chicks to flee. To do otherwise would be to risk a fatal plunge from the ledge. The nature and degree of these adaptations could not have been judged unless "ordinary" gull species had been used as a baseline for comparison.

A second procedure in evolutionary studies is to arrange the species in a series in order to identify the steps by which extreme forms of the trait were reached. For example, one of the most exotic forms of courtship in the animal kingdom is practiced by a few species of dance flies belonging to the family Empididae. The males gather in swarms to fly in dancing patterns. Each carries a small silken balloon, which he presents to any female of the same species that flies within reach. If only this extreme mode of courtship existed, it would have been impossible for biologists to deduce its evolutionary origin. Fortunately, however, there are many other species of empidid flies which display among themselves almost every conceivable evolutionary stage leading to the balloon ceremony. In some species the male follows the relatively primitive procedure of capturing another kind of fly and presenting it to a prospective mate. The voracious female then feeds on the gift while her suitor inseminates her. The males of other species decorate the prey with tassels of silk; others provide a complete covering of silk. As the gifts become more ornate and conspicuous, the flies tend to gather in swarms to display together. The next steps can be guessed. Some species reduce the size of the fly within the silken capsule, and others—representing the end of this particular evolutionary trend—omit the prey altogether and merely spin a balloon.

A key process in the evolution of such behavior is *ritualization*, the alteration of a previously existing behavior in a way that increases its signal value. Prey capture by dance flies, for example, has been ritualized into a courtship signal. The males of certain birds, such as the red-winged blackbird and goldeneye duck, perform ritualized flights as part of their courtship displays, sweeping the wings out in labored movements that decrease the efficiency of their flight but make the birds much more conspicuous. In the course of evolution, an impressively wide array of processes have been ritualized by one kind of animal or another: feeding and drinking; regurgitation; running, jumping, flying, and other locomotory movements; preening; respiratory movements; and even, for here is the origin of pheromones, secretion and excretion.

Having established the direction of evolutionary trends in particular cases, the biologist will sometimes then turn to the ways in which the traits are inherited. One promosing approach is to hybridize closely related species and to analyze the segregation and recombination of the traits that distinguish the species. Experiments of this kind on birds are reported in the articles by Lorenz ("The Evolution of Behavior") and William C. Dilger ("The Behavior of Lovebirds"). More recently, geneticists have begun a more concerted effort to analyze the genetic basis of behavioral variation *within* species. The goal, as explained by Seymour Benzer in "Genetic Dissection of Behavior," is to track developmental events all the way from the gene to the sense cells and neurons and thence to the behavioral patterns themselves.

The theme of evolutionary change and adaptation is pursued more extensively toward the end of Part II. The stereotyped responses that allow animals to select the correct habitat and to avoid enemies are explored in the articles by Stanley C. Wecker and by Niko Tinbergen. In "Visual Isolation in Gulls," Neal G. Smith shows how eye color is used by artic gulls to recognize members of the same species when selecting mates and hence to avoid forming hybrids with alien species. The evolution of such recognition signals is an important part of the process of the origin of new species. Finally, two of the most impressive of all achievements in behavioral evolution are presented in detail: orientation by electric fields in fish, and homing by salmon and pigeons. Both have evolved in response to special requirements and opportunities in the environment.

"The Evolution of Behavior," *by Konrad Z. Lorenz. December 1958.*
Here Lorenz states in clear terms the central themes of ethology: behavioral traits are essentially like anatomical and physiological traits; they can be isolated and their evolutionary histories inferred by the comparison of species; and their adaptive significance can be established by studying the uses of the behavior under natural conditions. Lorenz describes his own work on ducks and that of Tinbergen on gulls, placing special emphasis on the process of ritualization to form new signals in communication.

"The Evolution of Behavior in Gulls," *by Niko Tinbergen. December 1960.*
Using more detailed information on gulls, Tinbergen extends Lorenz's argument and illustrates in still greater detail the modern evolutionary approach to behavior. An important point to note in reading this article is the role of privacy in the evolution of displays at the level of the species. For alarm and threat, in which there is no need for gulls to distinguish members of their own species from those of other species, the displays are virtually identical. But in courtship, where the selection of mates belonging to the same species is a crucial act, the displays vary markedly from one kind of gull to the next.

In short, alarm and threat communication is communal at the species level, but courtship is private.

"The Behavior of Lovebirds," *by William C. Dilger. January 1962.*

In the experiments reported here, a genetic basis is demonstrated for the differences in nest-building behavior between species of lovebirds. When Dilger crossed two species in his aviary, he obtained hybrids characterized by curious mixtures of behavioral components from both parents. The final result was an incompetence in nest building. Dilger's study has a bearing on a general conceptual problem in the study of behavior. When we encounter a particular pattern of behavior, it is tempting to ask whether the pattern is "instinctive" or "learned." This question really has little meaning unless it is applied to *variation* in particular traits. Are the differences between two species (or individuals) based on differences in the genes, or are they due to differences in learning or some other environmental influence? The surest way to answer the second, more meaningful question is by breeding experiments of the kind performed by Dilger.

"Genetic Dissection of Behavior," *by Seymour Benzer. December 1973.*

The dyed-in-the-wool geneticist tends to deal with all biological problems by approximately the same approach: create genetic variation in the phenomenon, by radiation or mutagenic chemicals if necessary; track down the cause of the changes to the cellular or even molecular level; then use the information to reconstruct the various pathways in development that lead to the normal trait and its variants. The mutant pathways are like tracers; by following enough of them, one can gain a picture of which events and structures contribute to the emergence of the biological phenomenon, and which do not. This method, called genetic dissection, has enjoyed notable success in biochemistry. Here Seymour Benzer, one of the premier experimentalists of modern genetics, describes the first attempts by his research group to apply the technique to the study of innate components of behavior in the fruit fly *Drosophila.*

"Habitat Selection," *by Stanley C. Wecker. October 1964.*

Ecologists have long assumed that animals possess powers of discrimination for selecting the correct habitat, or at least a capacity to learn quickly which habitat is correct. If such abilities were lacking, many individuals would be ill-adapted, and entire species could become extinct because of this one inadequacy alone. Experiments that test the ability to select habitats are obviously difficult to design, because of the size and complexity of the environments in which most kinds of animals move. Wecker describes a successful attempt using two geographical races of field mice. One race prefers woodland and the other open fields. Wecker presents strong but not wholly conclusive evidence of the existence of a genetic component in the differences. Of equal importance is the demonstration that much of the discriminating ability can be lost by confining the mice to neutral laboratory conditions for as few as 12 to 20 generations. This is one of several cases (the others are mostly in *Drosophila*) in which the rapid evolution of behavior has been observed in laboratory stocks. The possibility of such rapid evolution is allowed by the theory of population genetics, but has seldom been put to any experimental test.

"Visual Isolation in Gulls," *by Neal Griffith Smith. October 1967.*

One of the key functions of social releasers is the identification of potential mates belonging to the same species. Failure to discriminate at this level leads to crossing between species, hybrid offspring, and genetic chaos. The adult animal that makes a mistake is not likely to contribute genes to the next generation, because hybrids are almost never as well adjusted as the pure strains of either species. To take an abstract example: where species A has been adapted during thousands of years to environment a, and species B to environment b, the hybrid A × B is specially adapted to neither a nor b, and it is likely to be outcompeted wherever it encounters one or both of the parent species. Students of evolution designate the differences that prevent species from interbreeding as "intrinsic isolating mechanisms." They have noticed that the differences are often sharpest where two closely related species are in contact, providing evidence that the populations diverged further from each other as part of the adaptation to avoid hybridization. Smith describes a series of elegant field experiments implicating eye color as the principal isolating mechanism of arctic gulls. He also cites an evolutionary change in eye color that has resulted in more efficient sexual isolation between two of the species. Smith's account exemplifies the close connections that exist between ethology and modern evolutionary theory.

"Electric Location by Fishes," *by H. W. Lissmann. March 1963.*
Everyone has heard of the electric eel, a freshwater fish that can stun its enemies with a powerful electric shock. In this fascinating article, H. W. Lissmann shows that the capacity to generate electricity has a second, more basic function in this and other similarly empowered fishes around the world. When foraging, especially at night or in cloudy water where vision is useless, the fish throws an electric field around its body. The pulses are generated from modified muscles in the tail and are monitored by sensory organs in the head region. Objects near the fish can be detected and localized because of distortions they cause in different parts of the field. The electirc fish, as Lissmann correctly puts it, lives in a world totally alien to man.

"The Homing Salmon," *by Arthur D. Hasler and James A. Larsen. August 1955.*
With a limited brain and almost no opportunity to learn by trial and error, the migratory salmon completes a lifetime journey that sometimes extends for more than a thousand miles. The fish begins life at a freshwater spawning site, swims downstream to the sea, and, after growing to maturity, swims back upstream to breed in the freshwater stream of its birth. Here Hasler and Larsen cite experiments which suggest that the entire performance is based on the memory of the odor of the home stream, imprinted on the young salmon shortly after its birth. After this article was written, Hasler and his co-workers obtained strong new evidence to support the hypothesis. Salmon exposed to an artificial chemical soon after birth were later successfully guided to streams contaminated with traces of the same substance. Thus odor training is a sufficient mechanism for long-distance homing in these fish; whether it is the only one remains to be seen.

"The Mystery of Pigeon Homing," *by William T. Keeton. December 1974.*
This article takes two very interesting but at first seemingly unrelated facts and puts them together. First, the ability of pigeons to find their way home over unknown terrain, in most weather conditions, has been a paramount mystery of animal behavior since antiquity. Second, that any kind of animal can detect magnetic fields has been doubted by some of the foremost physiologists right up to the present time. In this lucid account, William Keeton describes new experiments that indicate the presence of a magnetic sense in pigeons, a finding that may at last solve the mystery of their homing.

SUGGESTED ADDITIONAL READING for Part II

Lorenz, Konrad Z. *King Solomon's Ring.* 1952. Signet Books, New York.

Tinbergen, Niko. 1960. *The Herring Gull's World.* Harper and Row, New York. These two books are superb first-hand reports of behavioral studies done with evolutionary considerations uppermost in the observer's mind. Delightfully written, there are in many ways classic accounts.

Wickler, Wolfgang. *Mimicry in Plants and Animals.* 1968. McGraw-Hill, New York. A well-documented and superbly illustrated synthesis of a topic of long-standing evolutionary interest. The adaptive character of behavior is brought to light in the context of some of nature's own "experimental proof."

The Evolution of Behavior

by Konrad Z. Lorenz
December 1958

Beneath the varying behavior which animals learn lie unvarying motor patterns which they inherit. These behavior traits are as much a characteristic of a species as bodily structure and form

A whale's flipper, a bat's wing and a man's arm are as different from one another in outward appearance as they are in the functions they serve. But the bones of these structures reveal an essential similarity of design. The zoologist concludes that whale, bat and man evolved from a common ancestor. Even if there were no other evidence, the comparison of the skeletons of these creatures would suffice to establish that conclusion. The similarity of skeletons shows that a basic structure may persist over geologic periods in spite of a wide divergence of function.

Following the example of zoologists, who have long exploited the comparative method, students of animal behavior have now begun to ask a penetrating question. We all know how greatly the behavior of animals can vary, especially under the influence of the learning process. Psychologists have mostly observed and experimented with the behavior of individual animals; few have considered the behavior of species. But is it not possible that beneath all the variations of individual behavior there lies an inner structure of inherited behavior which characterizes all the members of a given

species, genus or larger taxonomic group —just as the skeleton of a primordial ancestor characterizes the form and structure of all mammals today?

Yes, it is possible! Let me give an example which, while seemingly trivial, has a bearing on this question. Anyone who has watched a dog scratch its jaw or a bird preen its head feathers can attest to the fact that they do so in the same way. The dog props itself on the tripod formed by its haunches and two forelegs and reaches a hindleg forward in front of its shoulder. Now the odd fact is that most birds (as well as virtu-

SCRATCHING BEHAVIOR of a dog and a European bullfinch is part of their genetic heritage and is not changed by training. The widespread habit of scratching with a hindlimb crossed over a forelimb is common to most Amniota (birds, reptiles and mammals).

DISPLAY BEHAVIOR of seagulls shows how behavior traits inherent in all gulls have adapted to the needs of an aberrant species. At top is a typical gull, the herring gull, which breeds on the shore. It is shown in the "choking" posture which advertises its nest site. In middle the herring gull is shown in the "oblique" and "long call" postures, used to defend its territory. At bottom is the aberrant kittiwake, which unlike other gulls breeds on narrow ledges and has no territory other than its nest site. The kittiwake does not use the "oblique" or "long call" postures, but employs the "choking" stance for both advertisement and defense.

ally all mammals and reptiles) scratch with precisely the same motion! A bird also scratches with a hindlimb (that is, its claw), and in doing so it lowers its wing and reaches its claw forward in front of its shoulder. One might think that it would be simpler for the bird to move its claw directly to its head without moving its wing, which lies folded out of the way on its back. I do not see how to explain this clumsy action unless we admit that it is inborn. Before the bird can scratch, it must reconstruct the old spatial relationship of the limbs of the four-legged common ancestor which it shares with mammals.

In retrospect it seems peculiar that psychologists have been so slow to pursue such clues to hereditary behavior. It is nearly 100 years since T. H. Huxley, upon making his first acquaintance with Charles Darwin's concept of natural selection, exclaimed: "How stupid of me, not to have thought of that!" Darwinian evolution quickly fired the imagination of biologists. Indeed, it swept through the scientific world with the speed characteristic of all long-overdue ideas. But somehow the new approach stopped short at the borders of psychology. The psychologists did not draw on Darwin's comparative method, or on his sense of the species as the protagonist of the evolutionary process.

Perhaps, with their heritage from philosophy, they were too engrossed in purely doctrinal dissension. For exactly opposite reasons the "behaviorists" and the "purposivists" were convinced that behavior was much too variable to permit its reduction to a set of traits characteristic of a species. The purposivist school of psychology argued for the existence of instincts; the behaviorists argued against them. The purposivists believed that instincts set the goals of animal behavior, but left to the individual animal a boundless variety of means to reach these goals. The behaviorists held that the capacity to learn endowed the individual with unlimited plasticity of behavior. The debate over instinct versus learning kept both schools from perceiving consistent, inherited patterns in behavior, and led each to preoccupation with external influences on behavior.

If any psychologist stood apart from the sterile contention of the two schools, it was Jakob von Uexküll. He sought tirelessly for the causes of animal behavior, and was not blind to structure. But he too was caught in a philosophical trap. Uexküll was a vitalist, and he denounced Darwinism as gross materialism. He believed that the regularities he observed in the behavior of species were manifestations of nature's unchanging and unchangeable "ground plan," a notion akin to the mystical "idea" of Plato.

The Phylogeny of Behavior

But even as the psychologists debated, evolutionary thought was entering the realm of behavior studies by two back doors. At Woods Hole, Mass., Charles Otis Whitman, a founder of the Marine Biological Laboratory, was working out the family tree of pigeons, which he had bred as a hobby since early childhood. Simultaneously, but unknown to Whitman, Oskar Heinroth of the Berlin Aquarium was studying the phylogeny of waterfowl. Heinroth, too, was an amateur aviculturist who had spent a lifetime observing his own pet ducks. What a queer misnomer is the word "amateur"! How unjust that a term which means the "lover" of a subject should come to connote a superficial dabbler! As a result of their "dabbling," Whitman and Heinroth acquired an incomparably detailed knowledge of pigeon and duck behavior.

As phylogenists, Whitman and Heinroth both sought to develop in detail the relationship between families and species of birds. To define a given group they had to find its "homologous" traits: the resemblances between species which bespeak a common origin. The success or failure of their detective work hinged on the number of homologous traits they could find. As practical bird-fanciers, Whitman and Heinroth came to know bird behavior as well as bird morphology, and each independently reached an important discovery: Behavior, as well as body form and structure, displays homologous traits. As Whitman phrased it just 60 years ago: "Instincts and organs are to be studied from the common viewpoint of phyletic descent."

Sometimes these traits of behavior are common to groups larger than ducks or pigeons. The scratching habit, which I have already mentioned, is an example of a behavior pattern that is shared by a very large taxonomic group, in this case the Amniota: the reptiles, birds and mammals (all of whose embryos grow within the thin membrane of the amniotic sac). This widespread motor pattern was discovered by Heinroth, who described it in a brief essay in 1930. It is noteworthy that Heinroth observed the extreme resistance of such inborn habits to changes wrought by learning. He noticed that while most bird species maintain their incongruous over-the-shoulder

"HEAD-FLAGGING" is another form of display in which the kittiwake has adapted its behavioral birthright to meet unusual needs. Most gulls—like this pair of black-faced gulls—use this stance in courtship (by averting its menacing facial and bill coloration, the bird "appeases" the aggressive instinct of its mate). Kittiwakes alone evince this posture not only in mating adults but in ledge-bound nestlings, which use it to "appease" invaders.

scratching technique, some have lost this behavior trait. Among these are the larger parrots, which feed with their claws and use the same motion—under the wing—for scratching. Parakeets, however, scratch in the unreconstructed style, reaching around the lowered wing, and do not pick up food in their claws. There are a few exceptions to this rule. The Australian broadtailed parakeet has learned to eat with its claw. When eating, it raises its claw directly to its bill. But when scratching, it still reaches its claw around its lowered wing! This oddity is evidence in itself of the obstinacy of the old scratching habit. So far no one has been able to teach a parakeet to scratch without lowering its wing or to train a parrot to scratch around a lowered wing.

Today a growing school of investigators is working in the field opened up by Whitman and Heinroth. They have set themselves the task of discovering inherited patterns of behavior and tracing them from species to species. Many of these patterns have proved to be reliable clues to the origin and relationship of large groups of animals. There is no longer any doubt that animals in general do inherit certain deep-seated behavioral traits. In the higher animals such traits tend to be masked by learned behavior, but in such creatures as fishes and birds they reveal themselves with great clarity. These patterns of behavior must somehow be rooted in the common phys-

iological inheritance of the species that display them. Whatever their physiological cause, they undoubtedly form a natural unit of heredity. The majority of them change but slowly with evolution in the species and stubbornly resist learning in the individual; they have a peculiar spontaneity and a considerable independence of immediate sensory stimuli. Because of their stability, they rank with the more slowly evolving skeletal structure of animals as ideal subjects for the comparative studies which aim to unravel the history of species.

I am quite aware that biologists today (especially young ones) tend to think of the comparative method as stuffy and old-fashioned—at best a branch of research that has already yielded its treasures, and like a spent gold mine no longer pays the working. I believe that this is untrue, and so I shall pause to say a few words in behalf of comparative morphology as such. Every time a biologist seeks to know *why* an organism looks and acts as it does, he must resort to the comparative method. Why does the ear have its peculiar conformation? Why is it mounted behind the jaw? To know the answer the investigator must compare the mammalian frame with that of other vertebrates. Then he will discover that the ear was once a gill slit. When the first air-breathing, four-legged vertebrates came out of the sea, they lost all but one pair of gill slits, each of which happened to lie conveniently near the

"INCITING" is a threatening movement used by the female duck to signal her mate to attack invaders of their territory. At left a female of the European sheldrake (*with head lowered*) incites her mate against an enemy that she sees directly before her. The female at right (*with head turned*) has seen an enemy to one side. Each female watches her enemy regardless of her own body orientation.

labyrinth of the inner ear. The water canal which opened into it became filled with air and adapted itself to conducting sound waves. Thus was born the ear.

This kind of thinking is 100 years old in zoology, but in the study of behavior it is only now coming into its own. The first studies leading to a true morphology of behavior have concentrated largely on those innate motor patterns that have the function of expression or communication within a species. It is easy to see why this should be so. Whether the mode of communication is aural, as in the case of bird songs, or visual, as in the "dis-

play" movements of courtship, many of these motor patterns have evolved under the pressure of natural selection to serve as sharply defined stimuli influencing the social behavior of fellow-members of a species. The patterns are usually striking and unambiguous. These qualities, so essential to the natural function of the behavior patterns, also catch the eye of the human observer.

Gulls, Terns and Kittiwakes

For some years N. Tinbergen of the University of Oxford has intensively

studied the innate behavior of gulls and terns: the genus *Laridae*. He has organized an international group of his students and co-workers to conduct a worldwide study of the behavior traits of gulls and terns. They are careful to observe the behavior of their subjects in the larger context of their diverse life histories and in relationship to their different environments. It is gratifying that this ambitious project has begun to meet with the success which the enthusiasm of its participants so richly deserves.

Esther Cullen, one of Tinbergen's students, has been studying an eccentric

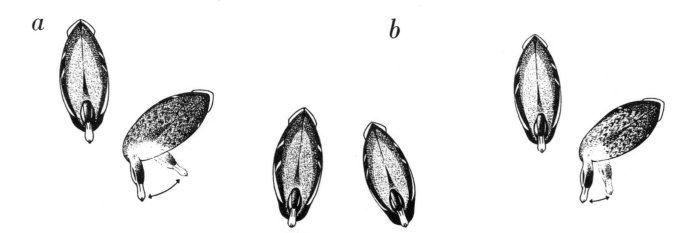

"RITUALIZED" INCITING is exhibited by mallards. In this species turning the head—as a female sheldrake does when inciting against an enemy to one side—has become an innate motor pattern. In situation *a* the female mallard turns her head toward the enemy. In *b*, with the enemy in front of her, she still turns her head even though this results in her turning it away from the enemy.

among the seagulls—the kittiwake. Most gulls are beachcombers and nest on the ground, and it is safe to assume that this was the original mode of life of the gull family. The kittiwake, however, is different. Except when it is breeding, it lives over the open sea. Its breeding ground is not a flat shore but the steepest of cliffs, where it nests on tiny ledges.

Mrs. Cullen has listed 33 points, both behavioral and anatomical, in which the kittiwake has come to differ from its sister species as a result of its atypical style of life. Just as the whale's flipper is a recognizable mammalian forelimb, so many of the kittiwake's habits are recognizably gull-like. But the kittiwake, like the whale, is a specialist; it has given its own twist to many of the behavior patterns that are the heritage of the *Laridae*.

For example, the male of most gull species stakes its claim to nesting territory by uttering the "long call" and striking the "oblique posture," its tail up and head down. To advertise its actual nesting site, it performs the "choking" movement. In the kittiwake the inherited patterns of behavior have been modified in accord with the habitat. On the kittiwake's tiny ledge, territory and nest sites are identical. So the kittiwake has lost the oblique posture and long call, and uses choking alone for display purposes.

Another example is the kittiwake gesture which Tinbergen calls "head-flagging." In other gull species a young gull which is not fully able to fly will run for cover when it is frightened by an adult bird. But its cliffside perch provides no cover for the young kittiwake. When it is frightened, the little kittiwake averts its head as a sign of appeasement. Such head-flagging does not occur in the young of other gulls, although it appears in the behavior of many adult gulls as the appeasement posture in a fight and in the rite of courtship. The kittiwake species has thus met an environmental demand by accelerating, in its young, the development of a standard motor habit of adult gulls.

Recently Wolfgang Wickler, one of my associates at the Max Planck Institute for Comparative Ethology, has found a similar case of adaptation by acceleration among the river-dwelling cichlid fishes. Most cichlids dig into the river bottom only at spawning time, when they excavate their nest pits. But there is an eccentric species (*Steatocranus*), a resident of the rapids of the Congo River, which lives from infancy in river-bottom burrows. In this cichlid the maturation of the digging urge of the mating fish is accelerated, appearing in

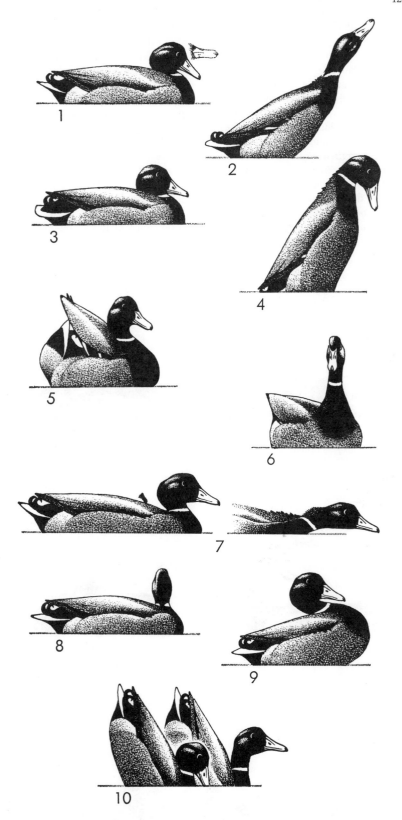

TEN COURTSHIP POSES which belong to the common genetic heritage of surface-feeding ducks are here shown as exemplified in the mallard: (1) initial bill-shake, (2) head-flick, (3) tail-shake, (4) grunt-whistle, (5) head-up—tail-up, (6) turn toward the female, (7) nod-swimming, (8) turning the back of the head, (9) bridling, (10) down-up. How the mallard and two other species form sequences of these poses is illustrated on pages 124 through 127.

the infant of the species. It is not hard to conceive how selection pressure could have led to this result.

The work of the Tinbergen school has had the important result of placing innate motor habits in their proper setting. He and his co-workers have shown that these traits are highly resistant to evolutionary change, and that they often retain their original form even when their function has diverged considerably. These findings amply justify the metaphor that describes innate patterns as the skeleton of behavior. More work of the Tinbergen kind is badly needed. There is great value in his synthetic approach, uniting the study of the physical nature and environment of animals with study of their behavior. Any such project is of course a tall order. It requires concerted field work by investigators at widely separated points on the globe.

Behavior in the Laboratory

Fortunately it is quite feasible to approach the innate motor patterns as an isolated topic for examination in the laboratory. Thanks to their stability they are not masked in the behavior of the captive animal. If only we do not forget the existence of the many other physiological mechanisms that affect behavior, including that of learning, it is legitimate for us to begin with these innate behavior traits. The least variable part of a system is always the best one to examine first; in the complex interaction of all parts, it must appear most frequently as a cause and least frequently as an effect.

Comparative study of innate motor patterns represents an important part of the research program at the Max Planck Institute for Comparative Ethology. Our

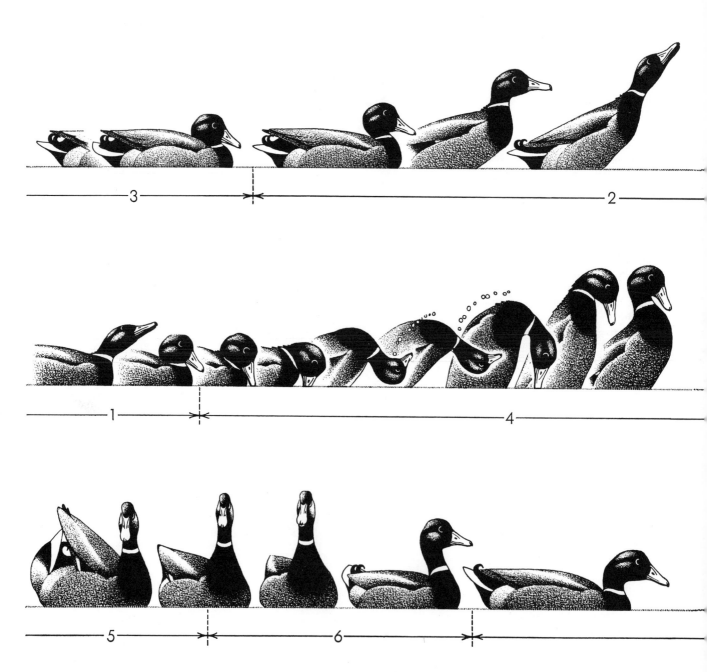

COURTSHIP SEQUENCES OF MALLARD are shown in this series of drawings, based on motion pictures made by the author at his laboratory in Seewiesen, Germany. Each sequence combines in fixed order several of the 10 innate courtship poses illustrated on

subjects are the various species of dabbling, or surface-feeding, ducks. By observing minute variations of behavior traits between species on the one hand and their hybrids on the other we hope to arrive at a phylogenetics of behavior.

Our comparative studies have developed sufficient information about the behavior traits of existing species to permit us to observe the transmission, suppression and combination of these traits in hybrid offspring. Ordinarily it is difficult to find species which differ markedly with respect to a particular characteristic and which yet will produce fertile hybrids. This is true especially with respect to behavioral traits, because these tend to be highly conservative. Species which differ sufficiently in behavior seldom produce offspring of unlimited fertility. However, closely related species which differ markedly in their patterns of sexual display are often capable of producing fertile hybrids. These motor patterns serve not only to bring about mating within a species but to prevent mating between closely allied species. Selection pressure sets in to make these patterns as different as possible as quickly as possible. As a result species will diverge markedly in sexual display behavior and yet retain the capacity to interbreed. This has turned out to be the case with dabbling ducks.

The first thing we wanted to know was how the courtship patterns of ducks become fixed. Credit is due to Sir Julian Huxley, who as long ago as 1914 had observed this process, which he called "ritualization." We see it clearly in the so-called "inciting" movement of female dabbling ducks, diving ducks, perching ducks and sheldrakes.

To see "inciting" in its original unritualized form, let us watch the female

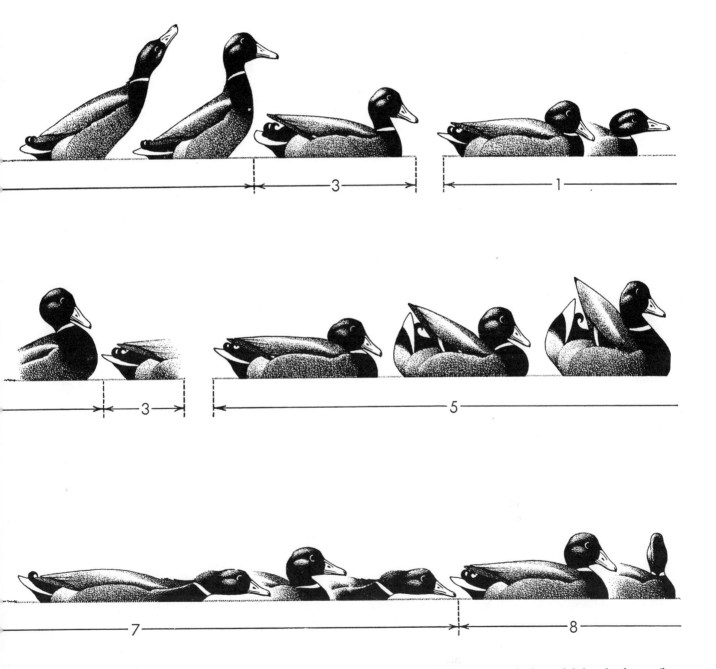

page 123. The numbers under the ducks refer to these poses. Shown here are the following obligatory sequences: tail-shake, head-flick, tail-shake; bill-shake, grunt-whistle, tail-shake; head-up tail-up, turn toward female, nod-swimming, turning back of the head.

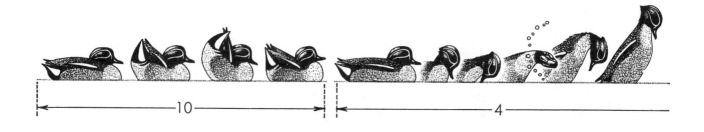

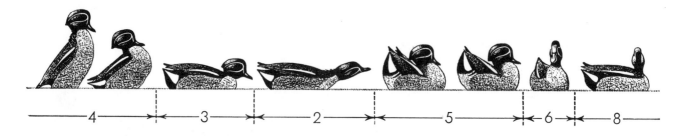

COURTSHIP OF EUROPEAN TEAL—another species of surface-feeding duck—includes tail-shake, head-flick, tail-shake (as in the mallard); down-up; grunt-whistle, tail-shake, head-flick, head-up–tail-up, turned toward the female, turning back of the head.

of the common sheldrake as she and her mate encounter another pair of sheldrakes at close quarters. Being far more excitable than her placid companion, the female attacks the "enemy" couple, that is, she adopts a threatening attitude and runs toward them at full tilt. It happens, however, that her escape reaction is quite as strong as her aggressive one. She has only to come within a certain distance of the enemy for the escape stimulus to overpower her, whereupon she turns tail and flees to the protection of her mate. When she has run a safe distance, she experiences a renewal of the aggressive impulse. Perhaps by this time she has retreated behind her mate. In that case she struts up beside him, and, as they both face the enemy, she makes threatening gestures toward them. But more likely she has not yet reached her mate when the aggressive impulse re-

turns. In that case she may stop in her tracks. With her body still oriented toward her mate, she will turn her head and threaten the enemy over her shoulder. In this stance she is said to "incite" an aggressive attitude in her partner.

Now the incitement posture of the female sheldrake does not constitute an innate behavior trait. It is the entirely plastic resultant of the pressure of two independent variables: her impulse to attack and her impulse to flee. The orientation of her head and body reflects the geometry of her position with respect to her mate and the enemy.

The same incitement posture in mallards, on the other hand, is distinctly ritualized. In striking her pose the female mallard is governed by an inherited motor pattern. She cannot help thrusting her head backward over her shoulder. She does this even if it means she must

point her bill away from the enemy! In the sheldrake this posture is the resultant of the creature's display of two conflicting impulses. In the mallard it has become a fixed motor pattern.

No doubt this motor pattern evolved fairly recently. It is interesting to note that while the female mallard is impelled to look over her shoulder when inciting, the older urge to look at the enemy is still there. Her head travels much farther backward when the enemy is behind her. If you observe closely, it is plain that her eyes are fixed on the enemy, no matter which way her head is turned.

Occasionally a female, impelled by the awkwardness of watching the enemy from the ritualized posture, will swing about and face them directly. In that case one may say that her old and new motor patterns are simultaneously active. Like the sheldrake, the mallard must once have faced the enemy during incitement. Overlying this instinct is a new one—to move her head backward over her shoulder regardless of the location of the enemy. The old orienting response survives in part. It usually displays itself at low levels of excitement. Especially at the beginning of a response, the female

mallard may stretch her neck straight forward. As her excitement mounts, however, the new motor pattern irresistibly draws her head around. This is one of many instances in which the mounting intensity of a stimulus increases the fixity of the motor coordination.

What has happened is that two independent movements have been welded together to form a new and fixed motor pattern. It is possible that all new patterns are formed by such a welding process. Sometimes two patterns remain rigidly welded. Sometimes they weld only under great excitement.

Recently we have been studying be-

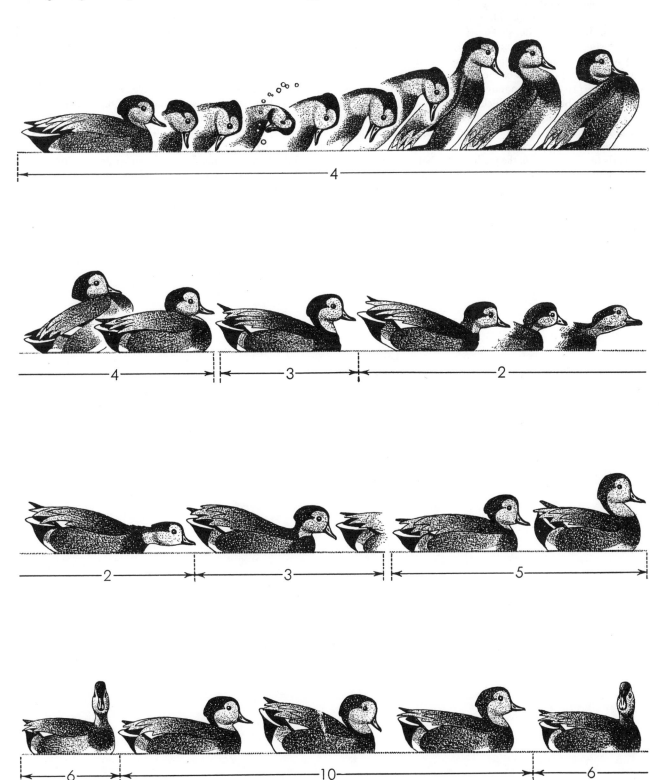

GADWALL COURTSHIP includes the grunt-whistle, always followed by the tail-shake, head-flick, tail-shake sequence also found in the other species illustrated. The head-up–tail-up (5) and the down-up (10) are always followed by a turn toward the female (6). During the most intense excitement of the courtship display, these pairs themselves become welded into the invariable sequence 5-6-10-6.

havior complexes in which more than two patterns are welded. In their courtship behavior our surface-feeding ducks display some 20 elementary innate motor patterns. We have made a special study of three species which have 10 motor patterns in common but display them welded into different combinations. As shown in the illustration on page 99, these patterns are (1) initial bill-shake, (2) head-flick, (3) tail-shake, (4) grunt-whistle, (5) head-up—tail-up, (6) turn toward the female, (7) nod-swimming, (8) turning the back of the head, (9) bridling, (10) down-up movement. Some of the combinations in which these motor patterns are displayed are shown on pages 100 through 103. In some species certain of the patterns occur independently (e.g., 1 and 10 in the mallard). Some simple combinations have wide distribution in other species as well (e.g.,

4, 3 and 5, 6 in all the species). Many combinations are more complicated, as the illustrations show.

What happens when these ducks are crossbred? By deliberate breeding we have produced new combinations of motor patterns, often combining traits of both parents, sometimes suppressing the traits of one or the other parent and sometimes exhibiting traits not apparent in either. We have even reproduced some of the behavior-pattern combinations which occur in natural species other than the parents of the hybrid. Study of our first-generation hybrids indicates that many differences in courtship patterns among our duck species may also be due to secondary loss, that is, to suppression of an inherited trait. Crosses between the Chiloe teal and the Bahama pintail regularly perform the head-up—tail-up, although neither parent is ca-

pable of this. The only possible conclusion is that one parent species is latently in possession of this behavioral trait, and that its expression in a given species is prevented by some inhibiting factor. So far our only second-generation hybrids are crosses between the Chiloe pintail and the Bahama pintail. The results look promising. The drakes of this generation differ greatly from each other and display hitherto unheard-of combinations of courtship patterns. One has even fused the down-up movement with the grunt-whistle!

Thus we have shown that the differences in innate motor patterns which distinguish species from one another can be duplicated by hybridization. This suggests that motor patterns are dependent on comparatively simple constellations of genetic factors.

The Evolution of Behavior in Gulls

by N. Tinbergen
December 1960

Gulls communicate with one another by means of calls,
postures and movements. Differences in the signaling
behavior of various species reflect the influence of
environment on gull evolution

Gulls live in flocks. They forage together the year around and nest together in the breeding season. No external force or agency compels them to this behavior; they assemble and stay together in flocks because they respond to one another. Their gregarious and often co-operative behavior is effected through communication. Each individual exhibits a considerable repertory of distinct calls, postures, movements and displays of color that elicit appropriate responses from other members of its species. Some gulls have a special food call that attracts their fellow gulls, and most have an alarm call that alerts the others. On the breeding grounds the male gull scares other males from its territory by certain calls and postures. Sex partners stimulate each other by a ritual of displays that leads to precisely timed and oriented co-operation in mating. Parent gulls attract their chicks by uttering the "mew call" or "crooning call" and lowering the beak. The chick pecks at the tip of the beak, and this stimulates the parent to regurgitate the food it has brought to the nest.

Even a nodding acquaintance with gulls suggests that their signaling behavior is just as typical of the family as their coloring and other physical conformation. Under the same circumstances the members of a given species invariably strike the same posture or act out the same ritual. Such observations suggest that signaling behavior must be largely unlearned. Investigators have found, in fact, that it is highly "environment-resistant." When a young bird is raised away from its parents or with foster parents, it does not develop a different pattern of signaling behavior but displays the repertory peculiar to its species. Moreover, gulls "understand" the meaning of various signals, apparently without the necessity of learning. The fact that many signaling movements of animals are as typical of the species as are anatomical structures and physiological mechanisms has been repeatedly stressed by Konrad Z. Lorenz of the Max Planck Institute of Comparative Ethology in Germany [see the article "The Evolution of Behavior," by Konrad Z. Lorenz, beginning on page 119].

When our group at the University of Oxford began some years ago to study the signaling behavior of gulls, we were interested primarily in finding out how the system works. We were concerned with such questions as: What is the exact function of each display? What makes a gull give a particular signal? But it was not long before another question claimed our interest. The members of our group (including Esther Cullen, Martin Moynihan and Rita and Uli Weidman) had been working at many sites around the world and observing the habits of 15 or more species of gull. We had found that the signaling systems of these species are very similar; this strengthened the conclusion, drawn from structural similarity, that gulls must have evolved from a common ancestral species. But we also found that the signaling repertories of the various species differ from one another in significant ways. Since the differences among these closely related birds are not induced by the environment, but are truly innate, it was clear that the present differences among the species must have arisen through evolutionary divergence. We decided that a comparative study of the signaling of the gulls might yield fresh insight into the evolution of their behavior.

Much as the anatomist makes comparative studies of structures in order to discover the origins and relationships of species, we have been conducting a comparative study of the signaling systems of gulls. These systems provide excellent instances for the study of behavior; precisely because of the function they serve, the signals are distinct and plain enough to be recognized even by an attentive human being. In our program the comparative method is applied in combination with our earlier methods of study. We continue to investigate the form and motivation of the displays, and this work has been facilitated by recent improvements in technique. We continue also to be concerned with the function of the displays, for this bears upon their survival value and so allows us to trace the selection pressures which must have been at work molding them. Thus the comparative study of the differences among species and the comparison of the present displays and their apparent origins make it possible to approach a description of the evolutionary changes that must have occurred as the ancestral gull family split up into the present 35 or so species of different appearances, habits and distribution.

Since there is no fossil behavior to certify our conclusions, our method of study might better be compared to that by which modern linguistics, through comparative study of languages, has worked out the family tree of the Indo-European languages, and has even reconstructed parts of the original Indo-European language [see "The Indo-European Language," by Paul Thieme; SCIENTIFIC AMERICAN, October, 1958]. The findings of such a study must always be regarded as probabilistic. On the other hand, the data of our investigation are sufficiently clear-cut. The postures and displays of each species are distinct and

FACING AWAY is an "appeasement" posture in which the gull averts its menacing beak. When a kittiwake alights on a ledge that is already occupied by a pair, he is attacked; he usually responds by facing away (*bottom bird at far right*). Females that alight

constant enough to make them useful in distinguishing and identifying the species.

The illustration on the preceding page shows eight postures and movements that occur in nearly all species of gull in more or less modified form. When they are employed for taxonomic purposes, they greatly increase the number of characteristics by which gulls may be classified. The similarities and differences

among the displays of the major subgroups correspond roughly to the classifications of the taxonomist, although studies of some of the less well-known species might force revision of their status. All of the "large gulls," among which the most familiar is the herring gull, have quite similar signaling systems. The "hooded gulls" are rather different from these, yet they are close in their habits to one another. Species that

have been placed in separate genera, such as the kittiwake and the ivory gull, have correspondingly distinct displays.

It seems clear that the signaling movements originated in more elementary behavior patterns, such as attacking, escaping, mating and nest-building. The postures and the actions themselves suggest where they came from. "Grass pulling" is a good example. In contests over territorial boundaries herring gulls and

LONG CALL of unmated males attracts females and repels other males. It is uttered in the oblique posture. This display is similar in the great skua (*far left*), the herring gull (*left center*), the kittiwake (*right center*) and the black-headed gull (*far right*).

CHOKING is stimulated by the sight of the nest. It consists of a series of down-and-up movements of the head, and in some species is accompanied by a cry. Shown here are choking displays of black-headed gull (*left*), herring gull (*center*) and kittiwake (*right*).

near a strange male also face away. Facing-away postures are similar in all species of gull, including the lesser black-backed gull (*far left*), the common gull (*second from left*), the kittiwake chick (*left center*) and the black-headed gull (*right center*).

other large gulls often peck violently at the ground, uproot plants and toss them sideways with a flick of the head [*see bottom illustration on this page*]. The pecks are indistinguishable from those aimed at rivals in actual attacks, and the pulling movements are identical with those seen when a gull seizes an opponent's wing, bill or tail. But the strange thing is that this activity is directed at the ground, not at the intruder for whose benefit the signal is displayed. The technical term for this is "redirected attack," and it may be compared to the human tendency to bang a table with the fist or kick a chair when angered. More puzzling is the sideways flick of the head that terminates this action. It is familiar to anyone who has observed gulls through more complete cycles of behavior; all gulls (and other birds as well) perform this movement when they build material into the nest. Apparently the sideways flick is stimulated by the "nest material" that the bird finds in its bill following its attack on the ground. Grass pulling may therefore be described as a redirected attack followed by a displaced nest-building movement.

"Choking" seems to have had a rather similar origin. This is the display by which the unmated male kittiwake advertises the nest site to passing females

UPRIGHT is a threat posture. The neck is stretched, bill points down and wings are raised slightly. Gulls adopt this posture when facing an intruder. Shown here are the upright postures of great skua (*left*), black-headed gull (*center*) and herring gull (*right*).

GRASS PULLING (*left*) is another threat display. The gull pecks at the ground and tears out grass in much the same way as it would peck and pull at an opponent during a fight (*right*). Bird at left is lesser black-backed gull; those at right are herring gulls.

and warns other males off; in other species it appears in boundary conflicts and is exhibited when male and female come close together, particularly at the nest site. It begins with a bending down over the nest (or any depression in the ground similar to a nest, such as a human footprint), followed by a rhythmic up-and-down movement of the head. The initial bending movement looks remarkably like the posture that the gull assumes just before it settles on the eggs, and is apparently derived from it. The rhythmic up-and-down movement appears to be a displaced nest-building, or perhaps a regurgitative movement.

But what makes a gull take up a certain posture? The controlling elements in the immediate situation in which a gull strikes one of these postures can be summed up in a single word: conflict. In a boundary dispute or in the preliminary stages of courtship the bird is in the grip of two mutually opposed impulses: to attack or to escape. The "upright," another hostile (or "agonistic") posture, is more readily recognized as a mosaic of these two behavior patterns. The gull reaches its head upward with its bill pointing down and lifts its still-folded wings from its sides. Stretching the neck and pointing the bill downward are the beginning of one form of attack, and lifting the wings is the first stage in the delivery of a blow with the folded wing. Occasionally the upright posture passes over into actual attack, but the gull usually stops short before he makes contact with his antagonist. If his opponent does not give way before his posturing, the gull will withdraw his neck, turn his bill upward and flatten his plumage; in short, he abandons the aggressive upright posture. Often he turns himself broadside to his antagonist; this maneuver appears in many postures and is a consequence of fear. It is not an uncommon sight in such a contest to see two birds walking parallel to or even around each other. Finally one gull abandons the offensive and adopts a posture of appeasement or runs away. Thus the upright is a posture that may pass over to a retreat as well as to an attack. It may be compared to the clenched-fist posture of a man whose anger is aroused, but whose action is restrained by fear or social convention. In such states of inner conflict the impulse to attack may also be redirected against inanimate objects or displaced by some irrelevant activity such as the nest-building head flick in the gull or the lighting of a cigarette in man.

Elements of conflict appear equally in the appeasement postures. The bird that yields in a contest of agonistic posturing is frequently seen to adopt the "face-away" posture: it turns its head away. The victim of an actual attack will do the same while standing his ground; this often stops the attack and always reduces it. Facing away also appears in the early phases of the mating ritual, when the members of an incipient pair are still strangers to one another. Plainly in all these situations the bird is seized simultaneously with the impulse to flee and the impulse to stay.

Somehow in the course of evolution these "involuntary" expressions of inner conflict between two incompatible behavior tendencies acquired value as signals. Other gulls responded to them, and since they facilitated adaptive social behavior, the postures or displays became incorporated in the signaling system of the species. Just how this happened in any single instance can only be imagined. On the other hand, once a given posture acquired such value, it is apparent that it tended to become quite clearly differentiated from the movement in which it originated. The original movements were primarily adapted to the functions of attack or escape, nest-building or some other activity. In becoming adapted to the signaling function they have been transformed in significant ways. Thus choking still resembles the nest-building action closely enough to be recognizable. But it also differs from that action in being much more prolonged and rhythmic. Similarly, while turning away suggests the motion that initiates fleeing, the action is usually more jerky, and the bird "freezes" in the turned-away posture. All these modifications of movement and of posture are such as to make the signal more distinct and conspicuous and so better suited to the function of providing strong stimuli to other gulls.

The same process of "ritualization" tended to make each of the postures in the repertory of a given species clearly distinguishable from one another. Thus every species has a number of agonistic displays. They are usually strikingly different, as the upright is from the choking posture. Though intermediate postures are not rare, they are much rarer than one would expect if there were a true sliding scale of displays between them. The black-headed gull, for example, has two quite distinct agonistic postures, the "oblique-*cum*-long call" and the "forward." There is good reason to believe, however, that they have a common origin. In the former the bird

tilts its body head downward and tail upward (that is, obliquely), lifts it folded wings from its sides and emits the long call. The forward looks like an extremely low oblique and, with the bill shut or half-shut, the call becomes a muffled version of the long call. When a flying stranger approaches and passes within a yard or so, the male black-headed gull exhibits first the oblique-*cum*-long call and then the forward. Though it is a smoothly changing external situation that elicits these responses, the gull does not gradually shift from one posture to the other. On the contrary, he stands in the oblique for several seconds and then abruptly (in about a fifth of a second, as shown by analysis of motion pictures) assumes the forward posture. This must mean that selection has favored distinctness of displays by suppressing intermediates among them. Since each display has a slightly different function, the elimination of ambiguity minimizes misunderstanding.

However, the evolutionary development of signaling movements has not only been controlled by the need of avoiding ambiguity between the different signals of one species, but seems also to have involved differentiation between species. The differences that distinguish the repertories of the various species are numerous. For example, the black-headed gull tilts its body head-down into an almost vertical position while choking. In this and in other postures it also lifts its folded wings in a greatly exaggerated manner compared to other gulls [see *bottom illustration on page 130*]. The great skua, which is not a gull, but belongs to the closely related jaegers, spreads its wings fully in the oblique posture when it gives the long call [see *middle illustration on page 130*]. The kittiwake opens its bill wide in all agonistic displays.

Some of these idiosyncrasies in behavior can be related readily to the characteristics of color and marking by which species are ordinarily identified. The postures have the effect of heightening the displays of these markings. Thus the kittiwake's mouth and tongue are a bright orange, and when the bird opens its bill in a posture of hostility, the color becomes visible. The great skua has bright white patches on its otherwise dark wings, which become visible as it opens its wings in the oblique. When the black-headed gull faces its opponent in the forward posture, it shows off its brown facial mask most conspicuously against the surrounding white of its body. This undoubtedly has something

REST

MEW

CHOKING

OBLIQUE
WITH LONG CALL

FORWARD

FACING AWAY

UPRIGHT

HUNCHED

PRINCIPAL DISPLAY POSTURES of herring gull are almost identical to those of other large gulls, and very similar to those of other groups in gull family. The oblique, mew, forward (and sometimes choking) are accompanied by characteristic calls.

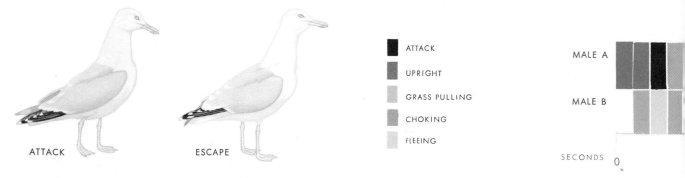

ATTACK

ESCAPE

ATTACK

UPRIGHT

GRASS PULLING

CHOKING

FLEEING

MALE A

MALE B

SECONDS   0

BEHAVIOR SEQUENCES were made from motion-picture films of a boundary clash between two gulls. Threat displays alternated with attack and defense behavior. Each block in the sequence at right represents an interval of two seconds. Legend at left center is key to the type of behavior shown by gulls during each of these intervals. Small drawings at far left show the "attack" components

to do with the fact that it employs the forward posture more frequently than other gulls do. Moreover, when this gull faces away, it turns its head around with a jerk and raises its white neck-feathers, completely concealing the mask. The hostile and the appeasement displays thus present an unmistakable contrast to the bird's antagonist. Generally speaking, however, gulls all look very much alike. Perhaps for this reason the principal interspecies differences in displays are to be found in the sharp definition of movements and postures.

The evolutionary significance of the differences among species becomes clearer when the displays are considered in terms of their functional relationship to the varying ways of life of the gulls. The most disparate of the gulls is the kittiwake. In contrast to most other species, which set up their breeding communities on dunes or grassy flats, the kittiwake is a cliff dweller. Esther Cullen, who has made an intensive study of this species, has shown convincingly that many peculiarities in its behavior are corollaries of its adaptation to this

habitat. Secure from attacks by predators, including other gulls, the kittiwake is extremely tame. It does not attack predators, and its alarm call is so rare that to elicit it one has to climb down to the narrow slanting ledges where the nests on their mud platforms cling so precariously. It is not surprising that the contrasts that distinguish the kittiwake's signaling behavior from that of other gulls emerge sharply in the displays concerned with the establishment and defense of the nesting site.

Among the grass-dwellers the agonis-

FEMALE A

MALE A

TERRITORY OF A

BOUNDARY

MALE A

MALE B

TERRITORY OF A

BOUNDARY CLASH begins with two pairs of gulls facing each other across the boundary between their territories. Display be-havior (*top row*) is a result of conflict between "anger" and "fear." Male B is choking in response to upright display of male A;

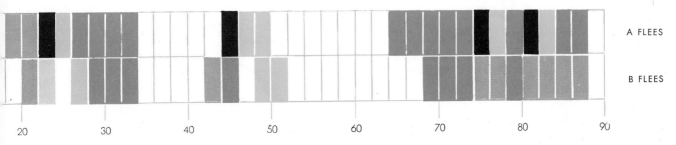

of the upright posture; gull adopts this form of upright as he runs at an intruder. If intruder fails to withdraw, or adopts the upright himself, the attacker usually stops and adopts the "escape" version of the upright (*second from left*): The bill points forward or upward, and the neck is vertical. The illustrations at the bottom of these two pages depict the behavior sequence schematically.

tic displays that maintain the territorial claims of the individual play a vital role in their adaptation to their habitat. Each male bird, having chosen a nesting site, defends an area around it, the size of which depends upon the species, the bird's own enterprise and the population pressure at the breeding ground. The spacing-out of the nests effected by this behavior prevents interference with mating; it keeps other gulls from preying on the broods, and it makes the individual nest a less inviting target for attack by predators of other species. Although the eggs and chicks of most gulls are beautifully camouflaged, this confers no absolute protection. The scattering of the nesting sites is necessary to make the camouflage effective. But the gulls are also gregarious, and this trait promotes survival by providing a warning system and social defense against predators. Thus the density of the colony appears to be a compromise between the advantages of proximity and the advantages of spacing-out.

The behavior of the male black-headed gull in defense of its territory is typi-

cal. His first response to another bird's approach in the air or on the ground is the oblique-*cum*-long call. Moving out to meet the intruder at the boundary of his territory, he may assume the upright posture. However, the farther the defender ventures from the center of his territory, the more does his aggression become attenuated. At a certain point he may find himself under counterthreat. He is then likely to adopt the choking posture, which conveys a message roughly like: "Be careful. I may not attack, but if I am attacked, I will fight back!" This

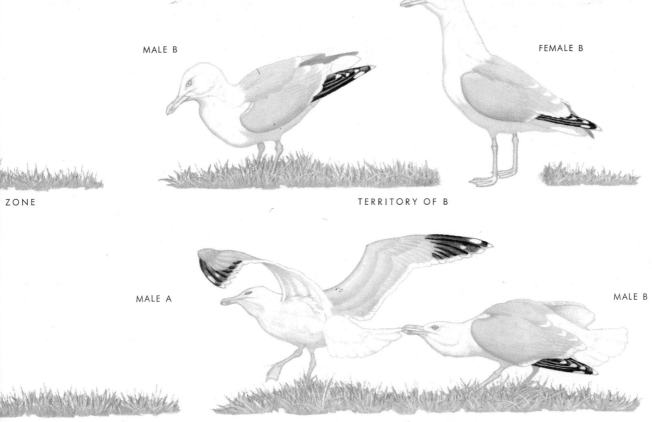

ZONE

MALE B   FEMALE B

TERRITORY OF B

MALE A   MALE B

BOUNDARY ZONE   TERRITORY OF B

female A utters mew call, while female B adopts the "escape" version of upright posture. When male B invades the territory of male A, A attacks and B flees (*bottom left*). Conversely, when male A invades the territory of B, B attacks and A flees (*bottom right*).

a

LONG CALL

FORWARD

FACING AWAY

b

LONG CALL

MEW

CHOKING

FACING AWAY

c

LONG CALL

VERTICAL

UPRIGHT WITH TILTED HEAD

d

CHOKING

LONG CALL

UPWARD CHOKING

PAIR-FORMATION involves a different series of displays in each of the four species of gulls shown here. In the black-headed gull (*a*), the herring gull (*b*) and the little gull (*c*), pair-formation begins with the long call; in kittiwake (*d*) it begins with choking.

does not cause his antagonist to withdraw, but it does reduce and often halts an attack.

The kittiwake, on his narrow ledge, acts out a quite different strategy. Most often, his first response to the approach of another kittiwake is choking. The reason for this is apparent. The kittiwake is the only species we have studied so far whose territory is so small that the bird, when it is at home at all, is standing right on the nest or the site of the nest-to-be. Now in all species of gulls choking is a posture readily elicited at the nest, and one that is probably derived from nest-building activity. The kittiwake, standing on his nest, accordingly chokes. The association of this posture with nesting activity is especially plain in the case of the kittiwake. The head-bobbing movement, so reminiscent of the action involved in putting nest material in place, is prolonged and sharply rhythmic, as befits a bird that builds with a sticky material such as mud, and must shake its head repeatedly and violently in order to put the material in place.

Another apparent effect of the nesting site upon the signaling behavior of this species is the early appearance of the face-away posture in the chick. Other gull chicks begin to roam about within a day or two after hatching, and in a couple of days think nothing of walking four feet away from the nest. When a chick is fed, its nest mates mob it in the attempt to share the meal, and the chick's natural response is to flee. But the kittiwake chick, having no place to go, has a strong tendency to stay put in the nest. Accordingly its defense against its nest mates is to face away. The effectiveness of this appeasement posture is perhaps heightened by the conspicuous black band across the back of its neck [see illustration at top of pages 130 and 131].

As might be expected, the differences among species show up most numerously and plainly in the displays associated with pair-formation and mating. Here signaling serves the obviously adaptive function of intraspecies recognition and interspecies segregation.

As contrasted with the sequence of territorial-defense postures, which induces spacing-out, the mating ritual brings the birds together. In winter most gulls lead a bachelor existence, uniting in flocks but not in pairs. Pairs form in early spring on or near the breeding site. Once gulls are paired, they know each other individually, and often re-pair without much ado each spring. Young birds mating for the first time, and widowed birds, pair up in more elaborate ways, in each case in a manner typical of their species.

The male and female of all species of gull look very much alike. Although there is evidence suggesting that the male, waiting at the nesting site he has chosen, can distinguish between males and females, he does not show it in his first response to an approaching female. In the case of the black-headed gull, the male greets the female with a display of the agonistic oblique-cum-long posture. Strange males avoid a long-calling male. But unmated females respond quite differently: they are attracted by him, and actually alight near him.

When this happens, both birds adopt a posture similar to the forward posture observed in the boundary flights of this species. The only differences are that the bill usually points up a little more than in hostile clashes, and the male and female tend to stand parallel to and not facing each other. After standing in this posture for perhaps a few seconds, first one and then the other gull suddenly jerks into the upright, another agonistic posture. But at the same time each turns its face away from the other; in other words, they moderate the hostility implied by the upright with a posture of appeasement [see "a" in illustration on page 136]. As their behavior plainly shows, the partners upon first meeting are in the thrall of three conflicting impulses—to attack, to escape and to remain near one another.

The female is usually the more timid of the two. This shows in her tendency, especially in the beginning, to stay at a safe distance from the male, and also in the extreme flattening of her plumage. Often the female, after having faced away, cautiously turns her face toward the male again, but quickly faces away each time the male moves. As a rule the female stays only a very short time, and then flies off. When the male starts calling again, she may either approach him once more, or alight near another calling male. Females can often be seen to visit a series of males in quick succession.

Sooner or later a female will become attached to one particular male. She alights near him again and again, each time going through the same performance: forward, upright and facing away. But the meeting ceremony slowly changes with repetition. Signs of aggression in the male subside, and so do signs of fear in the female. She no longer flattens her plumage so much, her neck is less stretched and she ventures nearer the male, often facing him. This proc-

ess of getting used to each other may take a short time and few visits, or it may take days and many dozens of visits. But ultimately the ceremony becomes desultory, and the birds spend a great deal of time together. The female now also begins to do the "head-tossing" movement, and then the male regurgitates and feeds her. In a short time the birds copulate.

The pair-formation ceremony is different in different species, although similar within the subgroups. All relatives of the black-headed gull follow the same ritual. But all large gulls, so far as is known, show another sequence: the oblique-cum-long call, the mew call, choking and facing away [see "b" in illustration on page 136]. The latter movement is less abrupt and less well oriented than in the black-headed gull. The little gull starts with the long call, then adopts the vertical posture (corresponding at this stage to the black-headed gull's forward) and changes to the upright, combined not with facing away but with tilting of the head, a movement also known in terms [see "c" in illustration on page 136].

All of the species show some evidence, admittedly slight, of distinctness in the character of the long call which serves the function of a mating "song" in the breeding season. The long calls are at least more distinct than the later calls and movements by which partners stimulate one another to copulate. This indicates that selection for distinctness must be much more severe for the song that first attracts the mate than for the precopulation displays that are performed after the mate has been attracted and accepted. The herring gull's long call can be distinguished even from that of its close relatives, the glaucous-winged gull and the western gull; except for this difference the mating displays of these species are practically identical.

The kittiwake follows still another routine, in keeping with the confined quarters of its nesting site. It begins by choking. This is the very display that keeps other males at their distance, but it attracts the female. When the female alights, both birds utter their version of the long call. They then return to choking once more and end by "upward choking" [see "d" in illustration]. Significantly the long call does not function as an initial advertisement in the pair-formation ceremony of this species. The kittiwake habitat probably served to give the species sufficient sexual isolation, and as a result there was no selection for a mating call with a large range of audibility.

Thus selection pressures arising in various ways from the environment brought about the differentiation in signaling behavior that distinguishes the various species of gull. The comparative method, applied to the study of behavior in much the same way as it has been applied to structure, is a powerful method for reconstructing the evolutionary history of species and for elucidating the relationships among them. Used in conjunction with motivational analysis (which shows origins) and with study of function (which shows the selection pressures), the comparative method reveals how beautifully these behavior characteristics are adapted to promote the survival of the species.

It cannot yet be said when the signaling displays that distinguish gulls from one another and from other birds first made their appearance. Some may well be much older than the gull family itself; others may have arisen later. Comparison of more species of gull and of more distantly related forms may help to solve this. Nonetheless it is clear that signaling displays are just as stable and reliable taxonomic characters as structures. As such they have much to tell about the selection process that brought the existing species of gull into being.

In our studies we have tried to follow Sherlock Holmes, by casting our nets wide to collect as many pieces of evidence as we could and taxing our brains to fit them together into a picture that makes sense. But we must confess at the end that our findings will remain probabilistic. We cannot re-enact the gull's evolutionary history, which was a unique event. But we believe that the methods I have tried to sketch allow us to reconstruct the past with some degree of probability. These methods are essentially the same as those of comparative and functional anatomy. Like its venerable senior sister, comparative ethology can provide a description of what must have happened in the past. In addition, the study of survival value or function allows us to formulate at least likely guesses about the way the evolution of a species has been controlled.

Surely it is necessary to check whether conclusions which are mainly derived from the study of structure are consistent with the facts of behavior. If we are ever to understand the evolutionary history of our own behavior, we will need the sharpest tools we can get, and there is no denying that our present tools are very blunt indeed. My main concern has been to help sharpen them a little.

The Behavior of Lovebirds

by William C. Dilger
January 1962

*The nine members of the genus Agapornis have
different rituals for such activities as nest building.
These differences shed light on the evolution of
lovebirds and on the role of heredity in behavior*

All lovebirds display the behavior that gives them their anthropomorphic common name. They pair early, and once pairs are formed they normally endure for life. The partners exhibit their mutual interest with great constancy and in a variety of beguiling activities. For the student of the evolution of animal behavior the lovebirds have special interest. The genus comprises nine forms (species or subspecies). They show a pattern of differentiation in their behavior that corresponds to their differentiation in color and morphology. By comparative study of their behavior, therefore, one can hope to reconstruct its evolution and to observe how natural selection has brought about progressive variations on the same fundamental scheme.

Together with my colleagues in the Laboratory of Ornithology at Cornell University, I have been studying both the constants and the variables in lovebird behavior for the past five years. It is not too difficult to duplicate in the laboratory the basic features of the lovebirds' natural African environment, so the birds thrive in captivity. Our work has covered all the lovebirds except Swindern's lovebird; we have not been able to obtain any specimens of this species. Our findings in two areas—sexual behavior and the defense and construction of the nest—have been particularly fruitful, because in these areas the evolutionary changes in lovebird behavior stand out in sharp relief.

Lovebirds constitute the genus *Agapornis,* and are members of the parrot family. Their closest living relatives are the hanging parakeets of Asia (the genus *Loriculus*). Three species of lovebird— the Madagascar lovebird, the Abyssinian lovebird and the red-faced lovebird—resemble the hanging parakeets and differ from all other lovebirds in two major respects. The males and females of these three species differ in color and are easily distinguishable from each other. The male and female of the other lovebirds are the same color. In these three species the primary social unit is the pair and its immature offspring. The other lovebirds are highly social and tend to nest in colonies. In these respects, then, the Madagascar lovebird, the Abyssinian lovebird and the red-faced lovebird most closely resemble the ancestral form, and the other lovebirds are more divergent.

Our study of interspecies differentiation of behavior has begun to reveal the order in which the other species arrived on the scene. Next after the three "primitive" species is Swindern's lovebird. Then comes the peach-faced lovebird and finally the four subspecies of *Agapornis personata,* commonly referred to as the white-eye-ringed forms: Fischer's lovebird, the black-masked lovebird, the Nyasaland lovebird and the black-cheeked lovebird. There are significant differences in behavior between the peach-faced lovebird and the four white-eye-ringed forms.

Perhaps the sharpest contrasts in behavior are those that distinguish the three primitive species from the species that evolved later. Even the common generic characteristic of pairing at an early age shows changes between the two groups that must be related to their contrasting patterns of life—nesting in pairs as opposed to nesting in colonies. Among the primitive species pair formation takes place when the birds are about four months old. At that time they are entirely independent of their parents and have already developed adult plumage. In the more recently evolved species, the colonial nesting pattern of which offers them access to their contemporaries virtually from the moment of their birth, pair formation takes place even earlier: the birds are about two months old and still have their juvenile plumage.

Among all the lovebird species pair formation is a rather undramatic event. Unpaired birds seek out the company of other unpaired birds and test them, as it were, by attempting to preen them and otherwise engage their interest. Couples quickly discover if they are compatible, and generally it takes no more than a few hours to establish lifelong pairs.

When the paired birds reach sexual maturity, their behavior with respect to each other becomes much more elaborate. This behavior as a whole is common to all lovebirds, and some activities are performed in the same way by all. Other activities, however, are not, and they show a gradation from the most primitive forms to the most recently evolved ones. One constant among all species is the female's frequent indifference to, and even active aggression against, the male each time he begins to woo her. Another is the essential pattern of the male's response—a combination of fear, sexual appetite, aggression and consequent frustration. Primarily motivated by both fear and sexual appetite, the male makes his first approach to his mate by sidling toward and then away from

NINE FORMS OF LOVEBIRD, as well as one hybrid (*top left*), are shown on page 143. They are arranged in their apparent order of evolution. The hybrid was bred in the laboratory for experiments on the inheritance of behavior. The letter *A.* at the beginning of each of the Latin species names stands for the genus *Agapornis.*

her while turning about on his perch. This switch-sidling, as it is called, is common to all species.

Two forms of male behavior initially associated with frustration, on the other hand, show a distinct evolutionary progression. The first of these activities is called squeak-twittering. Among the three primitive species—the Madagascar lovebird, the Abyssinian lovebird and the red-faced lovebird—the male utters a series of high-pitched vocalizations when the female thwarts him by disappearing into the nest cavity. The sounds are quite variable in pitch and purity of tone and have no recognizable rhythm. In the more recently evolved species—the peach-faced and the four white-eyeringed forms—squeak-twittering is rather different. The sound is rhythmic, purer in tone and less variable in pitch. Nor does it occur only when the female has turned her back on the male and entered the nest cavity. The male usually vocalizes even when the female is present and gives no indication whatever of thwarting him. Squeak-twittering has undergone a progressive change not only in its physical characteristics but also in the context in which it appears.

A similar evolution toward more highly ritualized behavior has occurred in another sexual activity, displacement

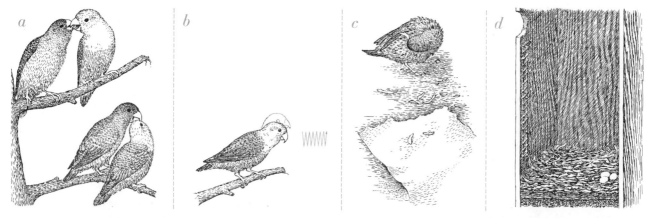

BEHAVIOR OF MADAGASCAR LOVEBIRD is outlined. Both sexes engage in courtship feeding (*a*). Accompanying head bobs are rapid and trace small arc (*b*). Nest materials, generally bark and leaves, are carried several pieces at a time and tucked among

BEHAVIOR OF PEACH-FACED LOVEBIRD suggests higher evolutionary stage. Only males perform courtship feeding; females fluff their feathers during this ritual (*a*). Slower head bobs trace wider arc (*b*). Nest materials, also bark and leaves, are

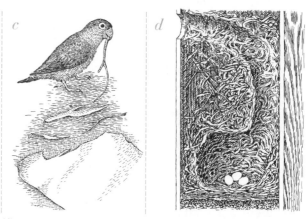

BEHAVIOR OF FISCHER'S LOVEBIRD indicates a further evolution. Courtship feeding (*a*), mobbing (*e*) and bill-fencing (*f*) are performed much as they are by the peach-faced lovebird. But other kinds of behavior are significantly different. Head bobs

scratching. This response derives from the habit, common to all species, of scratching the head with the foot when frustrated. Among the three primitive species displacement scratching is still close to its origins. Only two things distinguish it from ordinary head-scratching: its context and the fact that it is always performed with the foot nearest the female. Purely practical considerations govern this behavior: the male already has that foot raised preparatory to mounting his mate. In the more recently evolved species, displacement scratching has become primarily a form of display. Its progressive emancipation from the original motivation with which it is associated becomes more and more apparent as one observes it in the species from the peach-faced lovebird through the white-eye-ringed forms. Among all these the scratching is far more rapid and perfunctory than it is among the primitive species. Nor is it uniformly directed at the feathered portions of the head. In the peach-faced lovebird it is sometimes directed at the bill instead, and among the Nyasaland and black-cheeked lovebirds it is nearly always so directed. Moreover, these species use the far foot as well as the near one in displacement scratching; among the Nyasaland and black-cheeked lovebirds one is

all feathers of the body (c). Short strips are used to make an unshaped nest pad (d). The young join the mother in cavity-

defense display (e). In f birds show threat and appeasement display. It usually averts combat; if it fails, the birds fight furiously.

carried several at a time in back plumage (c); long strips are used to make a well-shaped nest (d). Birds join in "mobbing" to

protect nest (e). Bill-fencing (f) has a display function. It never leads to real harm; the birds bite only their opponents' toes.

(b) are still slower and trace an even wider arc. Nest materials are carried in the bill, one piece at a time (c); twigs as well

as strips of bark and leaf are used. This permits construction of an elaborate covered nest, entered through a tunnel (d).

used as often as the other. Finally, as in the case of squeak-twittering, which is often performed at the same time as displacement scratching among these species, the display occurs even when the female does not seem to be thwarting her mate.

All species engage in courtship feeding: the transfer of regurgitated food from one member of the pair to the other. In the three primitive species the female often offers food to her mate. This behavior has never been observed among the peach-faced and white-eye-ringed forms; here courtship feeding seems exclusively a male prerogative.

One can also discern an evolutionary progression in the manner in which the birds carry out the rather convulsive bobbing of the head associated with the act of regurgitation that immediately precedes courtship feeding. Among the primitive species these head-bobbings describe a small arc, are rapid and numerous and are usually followed by rather prolonged bill contacts while the food is being transferred. In the other forms the head-bobbings are slower, fewer in number and trace a wider arc; the bill contacts usually last for only a short time. Moreover, among the more recently evolved forms head-bobbing has become pure display; it is no longer accompanied by the feeding of the female. Unlike the females of the primitive lovebird species, which have no special display activity during courtship feeding, the females of the more recently evolved species play a distinctly ritualized role. They ruffle their plumage throughout the entire proceeding.

Females of all species indicate their fluctuating readiness to copulate by subtle adjustments of their plumage, particularly the feathers of the head. The more the female fluffs, the readier she is, and the more the male is encouraged. Finally she will solicit copulation by leaning forward and raising her head and tail. Females of the primitive species do not fluff their plumage during copulation; females of the more recently evolved species do. This is undoubtedly related to the morphological differences among the lovebirds. Since males and females of the more recently evolved species have the same coloring and patterning, the females must reinforce their mates' recognition of them, both in courtship and in copulation, by some behavioral means.

Although the forms of precopulatory behavior seem to be innate among all species, learning appears to play a major role in producing the changes that occur as the members of a pair become more familiar with each other. Newly formed pairs are rather awkward. The males make many mistakes and are frequently threatened and thwarted by their mates. After they have had a few broods, however, and have acquired experience, they become more expert and tend more and more to perform the right activity at the right time. As a result the female responds with aggression far less often, and the male engages more rarely in the displays that are associated with frustration and thwarting. Squeak-twittering and displacement scratching in particular become less frequent. Switch-sidling is still performed, but with a perceptibly diminished intensity. Altogether precopulatory bouts become less protracted. In spite of the male's reduced activity, the female seems to become receptive fairly quickly.

Disagreements among members of the same species are handled in quite different ways by those lovebirds that nest in pairs and those that nest colonially. Among the less social primitive species an elaborate pattern of threat and appeasement display has developed. For example, a formalized series of long, rapid strides toward an opponent signalizes aggression; a ruffling of the feathers, fear and the wish to escape. The loser in a bout of posturing may indicate submission by fleeing or by remaining quiet, turning its head away from its opponent and fluffing its plumage. By means of this code the birds can communicate rather exact items of information as to their readiness to attack or to flee. As a result actual fights seldom occur. When they do, however, the birds literally tear each other apart.

The peach-faced lovebird and the white-eye-ringed forms, which nest colonially, are thrown in contact with members of their own species much more often. This is undoubtedly related to the fact that they have developed a ritualized form of display fighting that goes far beyond a mere code of threat and appeasement and that replaces serious physical conflict. Display fighting among these more recently evolved species consists primarily of bill-fencing. The two birds parry and thrust with their bills and aim sharp nips at each other's toes.

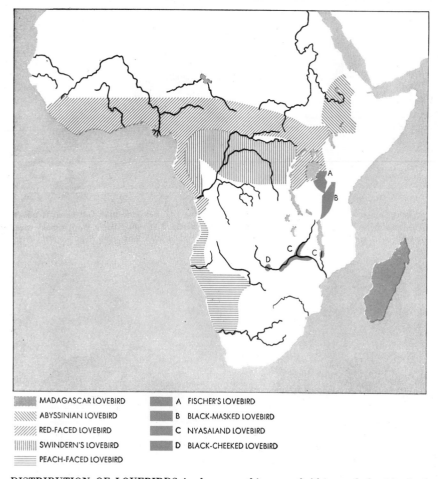

MADAGASCAR LOVEBIRD
ABYSSINIAN LOVEBIRD
RED-FACED LOVEBIRD
SWINDERN'S LOVEBIRD
PEACH-FACED LOVEBIRD

A   FISCHER'S LOVEBIRD
B   BLACK-MASKED LOVEBIRD
C   NYASALAND LOVEBIRD
D   BLACK-CHEEKED LOVEBIRD

DISTRIBUTION OF LOVEBIRDS is shown on this map of Africa and the island of Madagascar. All nine of the lovebird species and subspecies inhabit different areas.

9
BLACK-CHEEKED LOVEBIRD
(A. PERSONATA NIGRIGENIS)

8
NYASALAND LOVEBIRD
(A. PERSONATA LILIANAE)

10
PEACH-FACED LOVEBIRD × FISCHER'S LOVEBIRD
(A. ROSEICOLLIS × A. PERSONATA FISCHERI)

7
BLACK-MASKED LOVEBIRD
(A. PERSONATA PERSONATA)

5
PEACH-FACED LOVEBIRD
(A. ROSEICOLLIS)

6
FISCHER'S LOVEBIRD
(A. PERSONATA FISCHERI)

4
SWINDERN'S LOVEBIRD
(A. SWINDERNIANA)

3
RED-FACED LOVEBIRD
(A. PULLARIA)

2
ABYSSINIAN LOVEBIRD
(A. TARANTA)

1
MADAGASCAR LOVEBIRD
(A. CANA)

The toe is the only part the birds ever bite, and the inhibition against biting a member of the same species in any other place seems to be, like bill-fencing itself, an innate pattern. Though bill-fencing appears to be innate, it must be perfected by learning. The colonial nesting pattern offers young birds considerable practice with their contemporaries, and they quickly become skilled.

If lovebirds have had experience in rearing their own young, they will not rear the young of those other forms that have a natal down of a different color. On the other hand, a female that is given the egg of such a form at the time of her first egg-laying will rear the bird that emerges. Indeed, if a peach-faced lovebird has her first experience of motherhood with a newly hatched Madagascar lovebird, she will thereafter refuse to raise her own offspring. The down of the peach-faced lovebird's newly hatched young (like the down of the white-eye-ringed forms) is red, and the down of newly hatched Madagascar, Abyssinian and red-faced lovebirds is white.

Unlike most of the other members of

the parrot family, which simply lay their eggs in empty cavities, all lovebird species make nests. The red-faced lovebird constructs its nest in a hole it digs in the hard, earthy nests certain ants make in trees. All other species, however, make their nests in pre-existing cavities, which are usually reached through small entrances. The nests of the Madagascar lovebird, the Abyssinian lovebird and the red-faced lovebird are quite simple, consisting essentially of deposits of soft material on the cavity floor. These three species have developed an elaborate cavity-defense display. The moment an intruder appears, the female ruffles her feathers, partly spreads her wings and tail and utters a rapid series of harsh, buzzing sounds. If the intruder persists, she will suddenly compress her plumage, utter a piercing yip and lunge toward it. She does not bite, but she gives every indication of being about to do so. Her older offspring may join her at this time, ruffling their feathers and making grating sounds.

The effect of this performance is quite startling; it can even give pause to an experienced investigator! The Madagascar lovebird, the most primitive of all the species, is the quickest to engage in the cavity-defense display and is the only species we have seen carry the display through both stages. A stronger stimula-

tion is necessary before the Abyssinian lovebird engages in this behavior, and we have not seen the bird go any further than ruffling its body plumage and making the harsh, rasping sounds.

The white-eye-ringed lovebirds build rather elaborate nests, consisting of a roofed chamber at the end of a tunnel within the cavity. This fact and their strongly social nature combine to make their response to a threat to their nests different from the response of the primitive species. They have no cavity-defense displays at all. If a predator actually reaches the cavity, the birds within it will either cower or, if possible, flee through the entrance. But if the predator, encouraged by this show of fear, enters the cavity, it is likely to find that its troubles have just begun. It faces a journey down a narrow tunnel, defended at the end by a bird with a powerful and sharp bill. Moreover, a predator is seldom allowed to come close to the cavity. As soon as it is seen approaching, the entire colony engages in a form of behavior called mobbing: holding their bodies vertically, the birds beat their wings rapidly and utter loud, high-pitched squeaks. The sight and sound of a whole flock mobbing is quite impressive and probably serves to deter many would-be predators.

All female lovebirds prepare their nest

—— STRIPS CARRIED IN BILL
■ ■ ■ ■ ■ ■ INTENTION MOVEMENTS TO CARRY IN BILL
—— STRIPS TUCKED (NEVER CARRIED)
- - - - - INTENTION MOVEMENTS TO TUCK
    IRRELEVANT ACTIVITIES

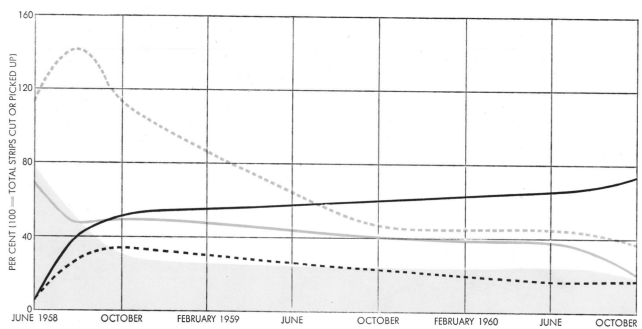

CONFLICTING PATTERNS of carrying nest-building materials are inherited by a hybrid lovebird, produced by mating the peach-faced and Fischer's lovebirds. The hybrid's behavior is charted here for a period of almost three years. As the bird progressively learns to carry nest materials as Fischer's lovebird does, the number of irrelevant movements and inappropriate activities decreases.

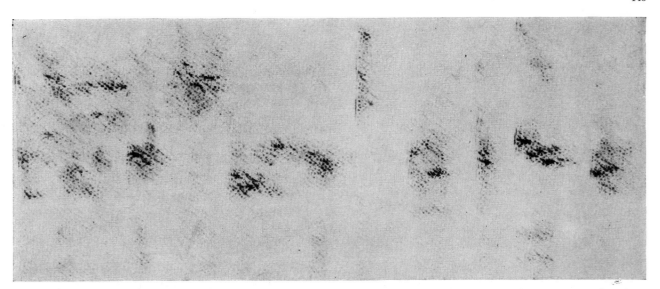

SQUEAK-TWITTERING in male Madagascar lovebird is seen on sound spectrogram. The horizontal axis represents time; the vertical axis, frequency. Uneven distribution of spots along both axes shows an arhythmic quality and a wide variation in pitch.

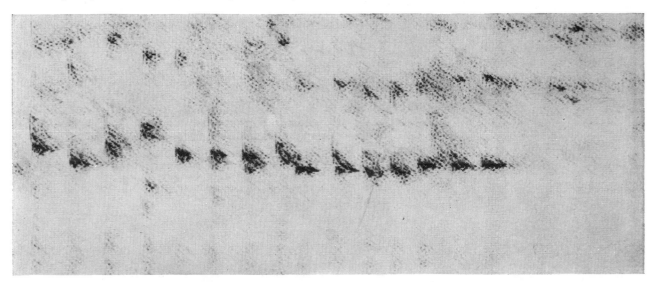

SOUND SPECTROGRAM of squeak-twittering in peach-faced lovebird shows greater rhythmicity and less variation in pitch. In Madagascar lovebird behavior is displayed only when female thwarts male. In peach-faced lovebird this is not always the case.

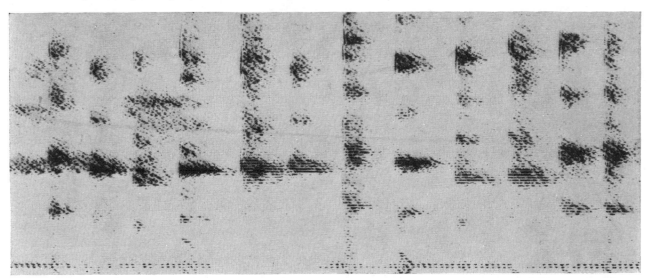

FURTHER EVOLUTION in squeak-twittering is seen in behavior of Nyasaland lovebird. Sounds are very rhythmic and show almost no variation in pitch; wide vertical distribution of spots reflects the large number of harmonics contained in the monotonous note.

materials in much the same way: by punching a series of closely spaced holes in some pliable material such as paper, bark or leaf. The material is held between the upper and lower portions of the bill, which then works like a train conductor's ticket punch. The pieces cut out in this way vary in size and shape among the various lovebirds. So do the forms of behavior that now ensue.

The three primitive species and the peach-faced lovebird tuck the pieces they have cut into the feathers of their bodies and fly off with them. The Madagascar lovebird, the Abyssinian lovebird and the red-faced lovebird use very small bits of material. (This is one of the reasons their nests are so unstructured.) The entire plumage of the bird is erected as it inserts the six to eight bits of material in place and remains erect during the whole operation. The peach-faced lovebird cuts strips that are

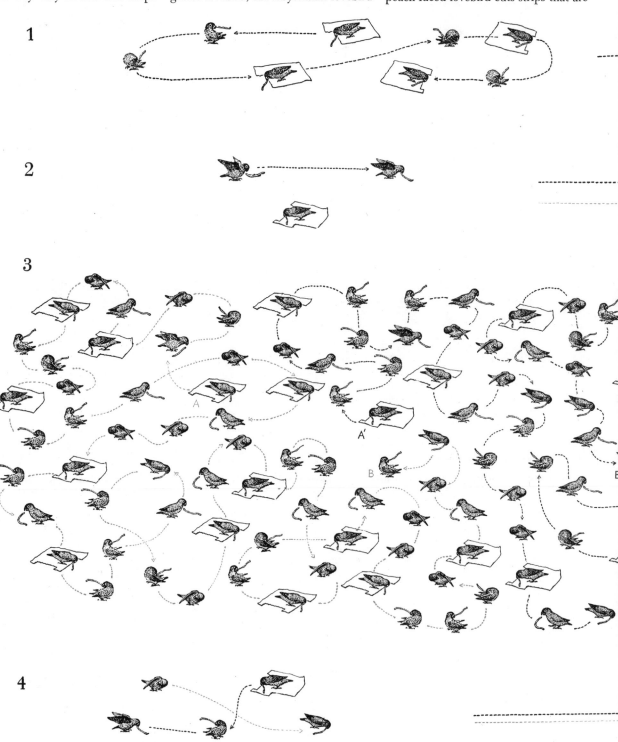

HYBRID LOVEBIRD inherits patterns for two different ways of carrying nest-building materials. From the peach-faced lovebird (1) it inherits patterns for carrying strips several at a time, in feathers. From Fischer's lovebird (2) it inherits patterns for carrying strips one at a time, in the bill. When the hybrid first begins to build a nest (3), it acts completely confused. Colored lines from A to B and black lines from A' to B' indicate the number of activities necessary for it to get two strips to the nest site, a feat achieved only when the strips are carried singly, in the bill. It takes three years before the bird perfects its bill-carrying behavior (4),

considerably longer. (This permits the more elaborate structuring of its cuplike nest.) Indeed, the strips are so long that they can be carried only in the feathers of the lower back. These are the feathers erected when the strips are tucked in, and the feathers are compressed after each strip is inserted. The peach-faced

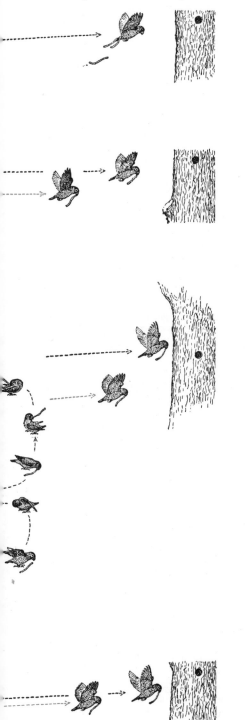

and even then it makes efforts to tuck its nest materials in its feathers. As the bird gains experience it becomes more and more proficient in this activity, which, however, never results in successful carrying.

lovebird loses about half of its cargo before it gets to its nest site; either pieces fall out while others are being cut or tucked in, or they fall out while the bird is flying. The lovebirds that use smaller bits of nest material are more successful in carrying them.

Carrying nest material in the feathers is unique to these birds and the related hanging parakeets. What is more, speculation about its origin must begin with the fact that no other parrots (with one unrelated exception) build nests at all. It is almost certain that this behavior arose from fortuitous occurrences associated with two characteristic parrot activities: chewing on bits of wood, bark and leaf to keep the bill sharp and properly worn down; and preening, which serves to keep the plumage clean and properly arranged. Some parrots that do not build nests will accidentally leave bits of the material in their feathers when they proceed directly from chewing to preening. Such oversights almost certainly initiated the evolution of the habit of carrying nest materials in the feathers.

The four white-eye-ringed forms are completely emancipated from this ancestral pattern. Fischer's lovebird, the black-masked lovebird, the Nyasaland lovebird and the black-cheeked lovebird all carry their nest materials as do most birds—in their bills. They lose little material in the process of carrying, and they pick up twigs in addition to cutting strips of pliable material. With these materials, they can build their characteristically elaborate nests.

Although the peach-faced lovebird normally carries its nest-building material in its feathers, on about 3 per cent of its trips it carries material in its bill. This peculiarity suggested an experiment. We mated the peach-faced lovebird with Fischer's lovebird (the birds hybridize readily in captivity) to see what behavior would show up in the hybrids. In confirmation of the thesis that patterns of carrying nest materials are primarily innate, the hybrid displays a conflict in behavior between the tendency to carry material in its feathers (inherited from the peach-faced lovebird) and the tendency to carry material in its bill (inherited from Fischer's lovebird).

When our hybrids first began to build their nests, they acted as though they were completely confused. They had no difficulty in cutting strips, but they could not seem to determine whether to carry them in the feathers or in the bill. They got material to the nest sites only when they carried it in the bill, and in their first effort at nest building they did carry

in their bills 6 per cent of the time. After they had cut each strip, however, they engaged in behavior associated with tucking. Even when they finally carried the material in the bill, they erected the feathers of the lower back and rump and attempted to tuck. But if they were able to press the strips into their plumage—and they were not always successful in the attempt—they could not carry it to the nest site in that fashion. Every strip dropped out.

Two months later, after they had become more experienced, the hybrids carried many more of their nest strips in their bills—41 per cent, to be exact. But they continued to make the movements associated with the intention to tuck: they erected their rump plumage and turned their heads to the rear, flying away with material in their bills only after attempting to tuck.

After two more months had passed they began to learn that strips could be picked up in the bill and carried off with a minimum of prior abortive tucking. But it took two years for them to learn to diminish actual tucking activity to any great extent, and even then they continued to perform many of the movements associated with tucking.

Today the hybrids are behaving, by and large, like Fischer's lovebird, the more recently evolved of their two parents. Only infrequently do they attempt to tuck strips into their plumage. But it has taken them three years to reach this stage—evidence of the difficulty they experience in learning to use one innate pattern at the expense of another, even though the latter is never successful. Moreover, when they do carry out the activities associated with tucking, they perform them far more efficiently than they did at first. Evidently this behavior need not achieve its normal objective in order to be improved.

So far our hybrids have proved to be sterile and therefore unable to pass on their behavior to a second generation. Even in the first generation, however, one can see the ways in which nature interweaves innate and learned elements to produce the behavior characteristic of a species. Further comparative studies can add much to our understanding not only of the behavior of lovebirds but also of the behavior of all vertebrates, including man.

15

Genetic Dissection of Behavior

by Seymour Benzer
December 1973

*By working with fruit flies that are mosaics of normal
and mutant parts it is possible to identify the genetic
components of behavior, retrace their development and
locate the sites where they operate*

When the individual organism develops from a fertilized egg, the one-dimensional information arrayed in the linear sequence of the genes on the chromosomes controls the formation of a two-dimensional cell layer that folds to give rise to a precise three-dimensional arrangement of sense organs, central nervous system and muscles. Those elements interact to produce the organism's behavior, a phenomenon whose description requires four dimensions at least. Surely the genes, which so largely determine anatomical and biochemical characteristics, must also interact with the environment to determine behavior. But how? For two decades molecular biologists were engaged in tracking down the structure and coding of the gene, a task that was pursued to ever lower levels of organization [see "The Fine Structure of the Gene," by Seymour Benzer; SCIENTIFIC AMERICAN Offprint 120]. Some of us have since turned in the opposite direction, to higher integrative levels, to explore development, the nervous system and behavior. In our laboratory at the California Institute of Technology we have been applying tools of genetic analysis in an attempt to trace the emergence of multidimensional behavior from the one-dimensional gene.

Our objectives are to discern the genetic component of a behavior, to identify it with a particular gene and then to determine the actual site at which the gene influences behavior and learn how it does so. In brief, we keep the environment constant, change the genes and see what happens to behavior. Our choice of an experimental organism was constrained by the fact that the simpler an organism is, the less likely it is to exhibit interesting behavioral patterns that are relevant to man; the more complex it is, the more difficult it may be to analyze

and the longer it takes. The fruit fly *Drosophila melanogaster* represents a compromise. In mass, in number of nerve cells, in amount of DNA and in generation time it stands roughly halfway on a logarithmic scale between the colon bacillus *Escherichia coli* (which can be regarded as having a one-neuron nervous system) and man. Although the fly's nervous system is very different from the human system, both consist of neurons and synapses and utilize transmitter molecules, and the development of both is dictated by genes. A fly has highly developed senses of sight, hearing, taste, smell, gravity and time. It cannot do everything we do, but it does some things we cannot do, such as fly and stand on the ceiling. Its visual system can detect the movement of the minute hand on a clock. One must not underestimate the little creature, which is not an evolutionary antecedent of man but is itself high up on the invertebrate branch of the phylogenetic tree. Its nervous system is a miracle of microminiaturization, and some of its independently evolved behavior patterns are not unlike our own.

Jerry Hirsch, Theodosius Dobzhansky and many others have demonstrated that if one begins with a genetically heterogeneous population of fruit flies, various behavioral characters can be enhanced by selective breeding pursued over many generations. This kind of experiment demonstrates that behavior can be genetically modified, but it depends on the reassortment of many different genes, so that it is very difficult to distinguish the effect of each one. Also, unless the selective procedure is constantly maintained, the genes may reassort, causing loss of the special behavior. For analyzing the relation of specific genes to behavior, it is more effective to begin

with a highly inbred, genetically uniform strain of flies and change the genes one at a time. This is done by inducing a mutation: an abrupt gene change that is transmitted to all subsequent generations.

A population of flies exposed to a mutagen (radiation or certain chemicals) yields some progeny with anatomical anomalies such as white eyes or forked bristles, and it also yields progeny with behavioral abnormalities. Workers in many laboratories (including ours) have compiled a long list of such mutants, each of which can be produced by the alteration of a single gene. Some mutants are perturbed in sexual behavior, which in normal *Drosophila* involves an elaborate sequence of fixed action patterns. Margaret Bastock showed years ago that some mutant males do not court with normal vigor. Kulbir Gill discovered a mutant in which the males pursue one another as persistently as they do females. The mutant *stuck*, found by Carolyn Beckman, suffers from inability to disengage after the normal 20-minute copulation period. A converse example is *coitus interruptus,* a mutant Jeffrey C. Hall has been studying in our laboratory; mutant males disengage in about half the normal time and no offspring are produced. Obviously most such mutants would not stand a chance in the competitive natural environment, but they can be maintained and studied in the laboratory.

As for general locomotor activity, some mutants are *sluggish* and others, such as one found by William D. Kaplan at the City of Hope Medical Center, are *hyperkinetic*, consuming oxygen at an exaggerated rate and dying much earlier than normal flies. Whereas normal flies show strong negative geotaxis (a tendency to move upward against the force of gravity), *nonclimbing* mutants do not.

BEHAVIOR of a normal and of a mosaic fruit fly is demonstrated in an experiment photographed by F. W. Goro. Normal flies move toward light and upward against the force of gravity. A normal fly that is placed in a glass tube with a light at the top and photographed by successive stroboscopic flashes traces a line straight up the tube (*left*). A mosaic fly, with one good eye and one blind eye, also climbs straight up if there is no light, guided by its sense of gravity. If there is a light at the top of the tube, however, the mosaic fly traces a helical path (*right*), turning its bad eye toward the light in a vain effort to balance the light input to both eyes.

"WINGS-UP" FLIES are mutants that keep their wings straight up and cannot fly. Such behavior could be the result of flaws in wing structure, in musculature or in nerve function. Mosaic experiments in the author's laboratory have traced the defect to the muscle.

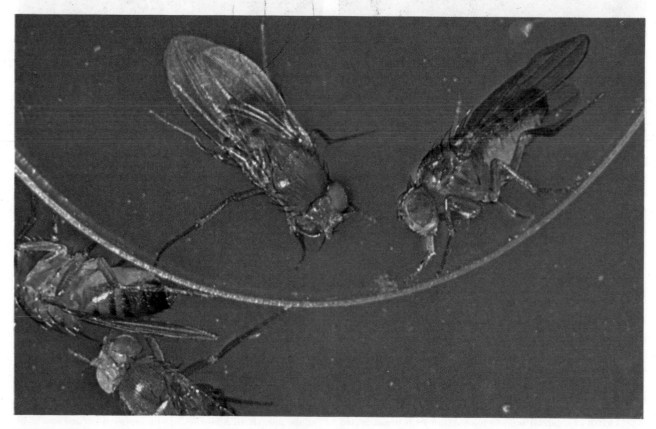

MOSAIC FLIES used for investigating behavior are gynandromorphs: partly male and partly female. The female parts are normal, the male parts mutant in one physical or behavioral trait or more. These flies have one normal red eye and one mutant white eye, and the male side of each fly also has the shorter wing that is normal for a male fly. The flies are about three millimeters long.

Flightless flies do not fly even though they may have perfectly well-developed wings and the male can raise his wing and vibrate it in approved fashion during courtship. Some individuals that appear to be quite normal may harbor hereditary idiosyncrasies that show up only under stress. Take the *easily shocked* mutants we have isolated, or the one called *tko,* found by Burke H. Judd and his collaborators at the University of Texas at Austin. When the mutant fly is subjected to a mechanical jolt, it has what looks like an epileptic seizure: it falls on its back, flails its legs and wings, coils its abdomen under and goes into a coma; after a few minutes it recovers and goes about its business as if nothing had happened. John R. Merriam and others working in our laboratory have found several different genes on the X chromosome that can produce this syndrome if they are mutated.

In many organisms mutations have been discovered that are temperature-sensitive, that is, the abnormal trait is displayed only above or below a certain temperature. David Suzuki and his associates at the University of British Columbia discovered a behavioral *Drosophila* mutant of this type called *paralyzed:* when the temperature goes above 28 degrees Celsius (82 degrees Fahrenheit), it collapses, although normal flies are unaffected; when the temperature is lowered, the mutant promptly stands up and moves about normally. We have found other mutants, involving different genes, that become similarly paralyzed at other specific temperatures. In one of these, *comatose,* recovery is not instantaneous but may take many minutes or hours, depending on how long the mutant was exposed to high temperatures. Recent experiments by Obaid Siddiqi in our laboratory have shown that action potentials in some of the motor nerves are blocked until the fly recovers.

An important feature of behavior in a wide range of organisms is an endogenous 24-hour cycle of activity. The fruit fly displays this "circadian" rhythm, and one can demonstrate the role of the genes in establishing it. A fly does well to emerge from the pupal stage around dawn, when the air is moist and cool and the creature has time to unfold its wings and harden its cuticle, or outer shell, before there is much risk of desiccation or from predators. (The name *Drosophila,* incidentally, means "lover of dew.") Eclosion from the pupa at the proper time is controlled by the circadian rhythm: most flies emerge during

a few hours around dawn and those missing that interval tend to wait until dawn on the following day or on later days. This rhythm, which has been much studied by Colin S. Pittendrigh of Stanford University, persists even in constant darkness provided that the pupae have once been exposed to light; having been set, the internal clock keeps running. The clock continues to control the activity of the individual fly after eclosion, even if the fly is kept in the dark. By monitoring the fly's movement with a photocell sensitive to infrared radiation (which is invisible to the fly)

one can observe that it begins to walk about at a certain time and does so for some 12 hours; then it becomes quiescent, as if it were asleep on its feet, for half a day. After that, at the same time as the first day's arousing or within an hour or so of it, activity begins anew. Ronald Konopka demonstrated the genetic control of this internal clock as a graduate student in our laboratory. By exposing normal flies to a mutagen he obtained mutants with abnormal rhythms or no rhythm at all. The *arrhythmic* flies may eclose at any time of day; if they are maintained in the dark

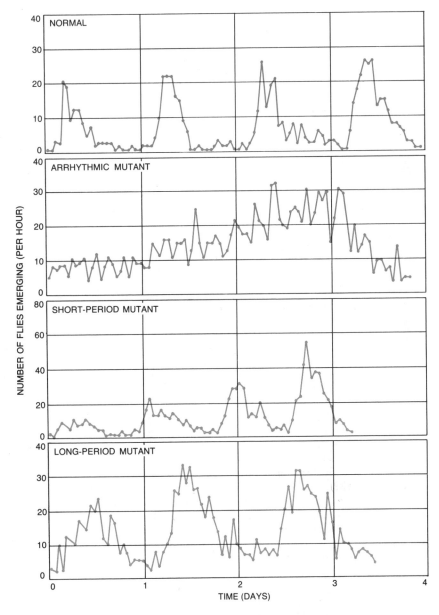

"BIOLOGICAL CLOCK" is an example of a behavioral mechanism that is genetically determined. It governs the periodicity of the time flies eclose from the pupa and also their daily cycle of activity as adults. The curves are for the eclosion of flies kept in total darkness. Normal flies emerge from the pupa at a time corresponding to dawn; those that miss dawn on one day emerge 24 hours later (*top*). Mutants include arrhythmic flies, which emerge at arbitrary times in the course of the day, and flies with 19-hour and 28-hour cycles.

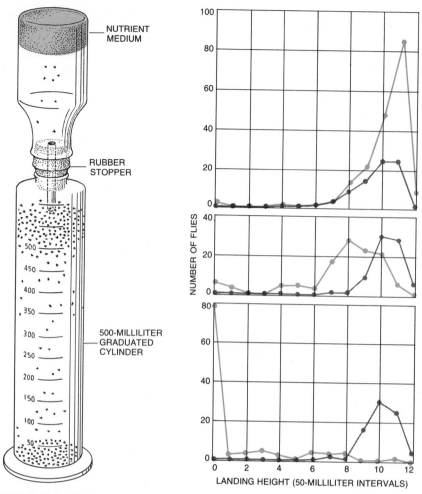

NUTRIENT MEDIUM

RUBBER STOPPER

500-MILLILITER GRADUATED CYLINDER

NUMBER OF FLIES

LANDING HEIGHT (50-MILLILITER INTERVALS)

FLIGHT TESTER is a simple device for measuring the flying ability of normal and mutant flies. It is a 500-milliliter graduated cylinder, its inside wall coated with paraffin oil. Flies are dumped in at the top. They strike out horizontally as best they can, and so the level at which they hit the wall and become stuck in the oil film reflects their flying ability. The curves compare the performance of female control flies (*gray*) with that of males (*color*) that fly normally (*top*) or poorly (*middle*) and with male *flightless* mutants (*bottom*).

after emergence, they are insomniacs, moving about during random periods throughout the day. The *short-period* mutant runs on a 19-hour cycle and the *long-period* mutant on a 28-hour cycle. (May there not be some analogy between such flies and humans who are either cheerful early birds or slow-to-awaken night owls?)

Let me now use a defect in visual behavior to illustrate in some detail how we analyze behavior. The first problem is to quantitate behavior and to detect and isolate behavioral mutants. It is possible to handle large populations of flies, treating each individual much as a molecule of behavior and fractionating the group into normal and abnormal types. We begin, using the technique devised by Edward B. Lewis at Cal Tech, by feeding male flies sugar water to which has been added the mutagen ethyl methane sulfonate, an alkylating agent that induces mutations in the chromosomes

of sperm cells. The progeny of mutagenized males are then fractionated by means of a kind of countercurrent distribution procedure [*see illustration on opposite page*], somewhat as one separates molecules into two liquid phases. Here the phases are light and darkness and the population is "chromatographed" in two dimensions on the basis of multiple trials for movement toward or away from light. Normal flies—and most of the progeny in our experiment—are phototactic, moving toward light but not away from it. Some mutants, however, do not move quickly in either direction; they are *sluggish* mutants. There are *runners*, which move vigorously both toward and away from light. A *negatively phototactic* mutant moves preferentially away from light. Finally, there are the *nonphototactic* mutants, which show a normal tendency to walk but no preference for light or darkness. They behave in light as normal flies behave in

the dark, which suggests that they are blind.

My colleague Yoshiki Hotta, who is now at the University of Tokyo, and I studied the electrical response of the nonphototactic flies' eyes. Similar mutant isolation and electrical studies have also been carried out by William L. Pak and his associates at Purdue University and by Martin Heisenberg at the Max Planck Institute for Biology at Tübingen, so that many mutants are now available, involving a series of different genes. The stimulus of a flash of light causes the photoreceptor cells of a normal fly's eye to emit a negative wave, which in turn triggers a positive spike from the next cells in the visual pathway; an electroretinogram, a record of this response, can be made rather easily with a simple wick electrode placed on the surface of the eye. In some nonphototactic mutants the photoreceptor cells respond but fail to trigger the second-order neurons; in other cases the primary receptor cells are affected so that there is no detectable signal from them even though they are anatomically largely normal. These mutants may be useful in understanding the primary transduction mechanism in the photoreceptor cells. Mutant material provides perturbations, in other words, that enable one to analyze normal function. When Hotta and I examined the eyes of some of the nonphototactic mutants, we found that the photoreceptor cells are normal in the young adult but that they degenerate with age. There are genetic conditions that produce this result in humans, and it may be that the fly's eye can provide a model system for studying certain kinds of blindness.

Now, if one knows that a certain behavior (nonphototactic, say) is produced by a single-gene mutation and that it seems to be explained by an anatomical fault (the degenerated receptors), one still cannot say with certainty what is the primary "focus" of that genetic alteration, that is, the site in the body at which the mutant gene exerts its primary effect. The site may be far from the affected organ. Certain cases of retinal degeneration in man, for example, are due not to any defect in the eye but to ineffective absorption of vitamin A from food in the intestine, as Peter Gouras of the National Institute of Neurological Diseases and Blindness has demonstrated. In order to trace the path from gene to behavior one must find the true focus at which the gene acts in the developing organism. How? A good way to troubleshoot in an electronic system—a stereophonic set with two identical channels, for example—is to interchange corresponding parts. That is in effect what we do with *Drosophila*. Rather than surgically transplanting organs from one fly to another, however, we use a genetic technique: we make mosaic flies, composite individuals in which some tissues are mutant and some have a normal genotype. Then we look to see just which part has to be mutant in order to account for the abnormal behavior.

One method of generating mosaics depends on a strain of flies in which there is an unstable ring-shaped X chromo-some. Flies, like humans, have X and Y sex chromosomes; if a fertilized egg has two X chromosomes in its nucleus, it will normally develop into a female fly; an XY egg yields a male. In *Drosophila* it is the presence of two X chromosomes that makes a fly female; if there is only one X,

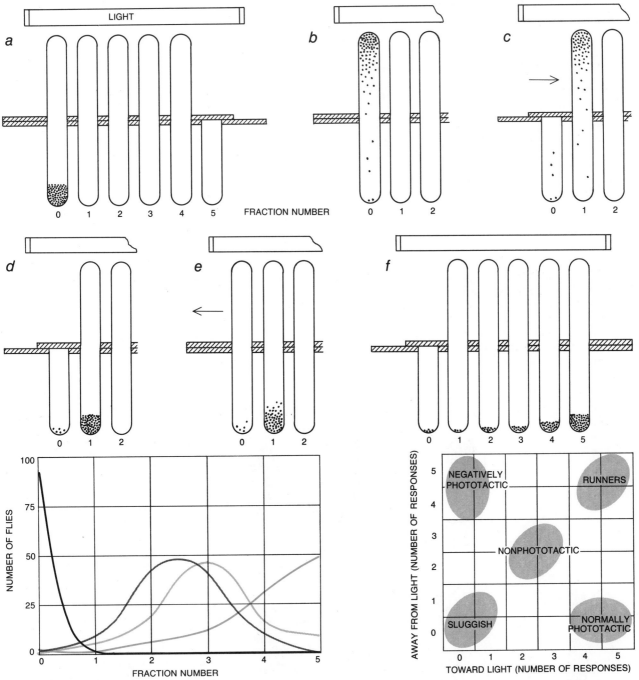

COUNTERCURRENT APPARATUS developed by the author can "fractionate" a population of flies as if they were molecules of behavior. The device consists of two sets of plastic tubes arranged in a plastic frame. Flies are put in Tube *0*; the device is held vertically and tapped to knock the flies to the bottom of the tube, and then the frame is laid flat and placed before a light at the far end of the tubes (*a*). Flies showing the phototactic response move toward the light, whereas others stay behind (*b*). After 15 seconds the top row of tubes is shifted to the right (*c*) and the responders are tapped down again (*d*), falling into Tube *1*. The upper frame is returned to the left (*e*), the frame is laid flat and again the responders move toward the light. The procedure is repeated five times in all. By then the best responders are in Tube 5, the next best in Tube 4 and so on (*f*). The curves (*bottom left*) show typical results. Phototactic flies show two very distinct peaks depending on whether the light was at the opposite end of the tubes from the starting point (*color*) or at the starting end (*black*). Nonphototactic flies, however, yield about the same curve (*light color or gray*) regardless of the position of the light. In order to distinguish variation in motor activity from phototaxis, the separation is carried out first toward light and then, processing the flies in each tube again, away from light, yielding a two-dimensional "chromatogram" (*bottom right*).

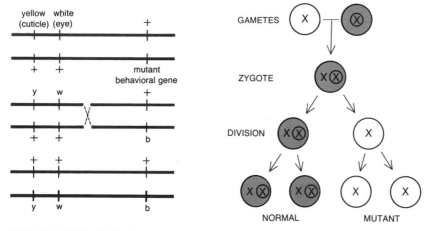

PREPARATION OF MOSAIC FLIES depends on the recombination on one X chromosome of genes for mutant "marker" traits and for a behavioral mutation. Recombination occurs through the crossing-over of segments of two homologous chromosomes, as shown at left. Males with an X chromosome carrying the desired recombination (*black*) are mated with females carrying an unstable ring-shaped X chromosome (*top row at right*). Among the resulting zygotes, or fertilized eggs, will be some carrying the mutation-loaded X and a ring X. In the course of nuclear division the ring X is sometimes lost. Tissues that stem from that nucleus are male, and mutant. In tissues that retain the ring X, however, the mutant genes are masked by the genes on the ring X, and these tissues are female and normal.

the fly will be male. The ring X chromosome has the property that it may get lost during nuclear division in the developing egg. If we start with female eggs that have one normal X and one ring X, in a certain fraction of the embryos some of the nuclei formed on divi-

sion lose the ring X and therefore have only one X chromosome left, and will therefore produce male tissues. This loss of the ring X, when it occurs, tends to happen at a very early stage in such a way as to produce about equal numbers of XX and X nuclei. The nuclei divide a

few times in a cluster and then migrate to the surface of the egg to form the early embryonic stage called a blastula: a single layer of cells surrounding the yolk [*see bottom illustration on this page*]. The nuclei tend to retain their proximity to their neighbors in the cluster, so that the female (XX) cells populate one part of the blastoderm (the surface of the blastula) and the male cells cover the rest. It is a feature of *Drosophila* that the axis of the crucial first nuclear division is oriented arbitrarily with respect to the axes of the egg. The dividing line between the XX and X cells can therefore cut the blastoderm in different ways. Once the blastoderm is formed the site occupied by a cell largely determines its fate in the developing embryo, and so the adult gynandromorph, a male-female mosaic, can have a wide variety of arrangements of male and female parts depending on how the dividing line falls in each particular embryo. The division of parts often follows the intersegmental boundaries and the longitudinal midline of the fly's exoskeleton. The reason is that the exoskeleton is an assembly of many parts, each of which was formed independently during metamorphosis from an imaginal disk in the larva that was in turn derived from a specific area of the blastoderm [*see illustrations on opposite page*].

The reader will perceive that a mosaic fly is a system in which the effects of normal and of mutant genes can be distinguished in one animal. We use this system by arranging things so that both a behavioral gene and "marker" genes that produce anatomical anomalies are combined on the same X chromosome. This is done through the random workings of the phenomenon of recombination, in which segments of two chromosomes (in this case the X) "cross over" and exchange places with each other during cell division in the formation of the egg. In this way we can, for example, produce a strain of flies that are *nonphototactic* and also have white eyes (instead of the normal red) and a yellow body color. Then we breed males of this strain with females of the ring X strain. Some of the resulting embryos will have one ring X chromosome and one mutation-loaded X chromosome. In a fraction of these embryos the ring X (carrying normal genes) will be lost at an early nuclear division. The XX body parts of the resulting adult fly will have one X chromosome with normal genes and one with mutations; because both the behavioral and the anatomical genes in question are recessive (their effect is masked by the presence of a single nor-

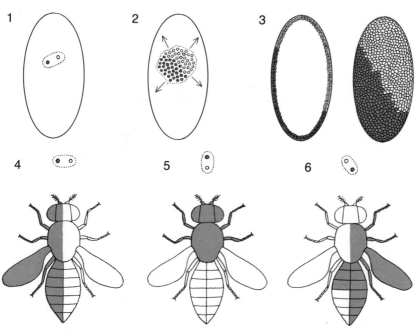

DEVELOPMENT OF A MOSAIC FLY proceeds from nuclear division (*shown in the illustration at top of page*), in which loss of the ring X occurs, producing an XX (*color*) nucleus and an X (*white*) nucleus (*1*). The nuclei divide a few times (*2*), then migrate to the surface of the egg and form a blastula: a single layer of cells, shown here in section and surface views (*3*). Note that female (XX) cells cover part of the surface and male (X) cells the other part. The arrangement of male (*white*) and female (*color*) parts in the adult fly depends on the way the boundary between the XX and X cells happened to cut the blastula, and that in turn depends on the orientation of the axis along which the original nucleus divides (*4–6*).

mal gene) the mutations will not be expressed in those parts. In the body parts having lost the ring X, however, the single X chromosome will be the one carrying the mutations. And because it is all alone the mutations will be expressed. Examination of the fly identifies the parts that have normal color and those in which the mutant genes have been uncovered. We can select from among the randomly divided gynandromorphs individuals in which the dividing line falls in various ways: a normal head on a mutant body, a mutant head on a normal body, a mutant eye and a normal one and so on. And then we can pose the question we originally had in mind: What parts must be mutant for the mutant behavior to be expressed?

When Hotta and I did that with certain visually defective mutants, for instance ones that produce no receptor potential, we found that the electroretinogram of the mutant eye was always completely abnormal, whereas the normal eye functioned properly. Even in gynandromorphs in which everything was normal except for one eye, that eye showed a defective electroretinogram. This makes it clear that the defects in those mutants are not of the vitamin A type I mentioned above; the defect must be autonomous within the eye itself.

The behavior of flies with one good eye and one bad eye is quite striking. A normal fly placed in a vertical tube in the dark climbs more or less straight up, with gravity as its cue. If there is a light at the top of the tube, the fly still climbs straight up because phototaxis (which the fly achieves by moving so as to keep the light intensities on both eyes equal) is consistent in direction with the negative geotaxis. A mosaic fly with one good eye also climbs straight up in the dark, since its sense of gravity is unimpaired. If a light is turned on at the top, however, the fly tends to trace a helical path, turning its defective eye toward the light in a futile attempt to balance input signals. If the right eye is the bad one, the fly traces a right-handed helix; if the left eye is bad, the helix is left-handed. (Sometimes it is difficult to resist the temptation, out of nostalgia for the old molecular-biology days, to put in two flies and let them generate a double helix.)

In these mutants the primary focus of the *phototactic* defect is in the affected organ itself. More frequently, however, the focus is elsewhere. A good way to see how this situation is dealt with is to consider a *hyperkinetic* mutant that

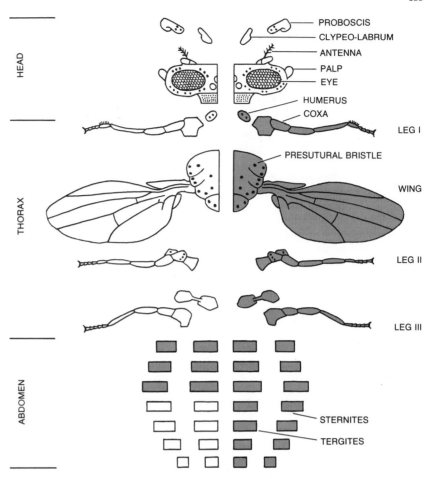

ADULT FLY is an assembly of a large number of external body parts, each of which was formed independently from a primordial group of cells of the blastula. In a mosaic fly the boundary between male and female tissues tends to follow lines of division between discrete body parts. Here the main external parts are named; black dots are the major bristles.

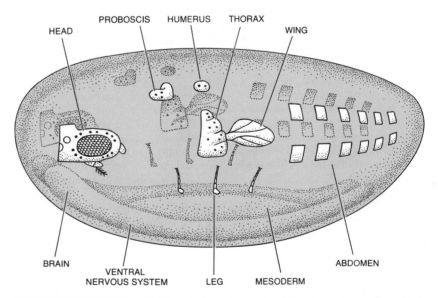

FANCIFUL DRAWING of the blastula shows how each adult body part came from a specific site on the blastula: left-side and right-side parts (or left and right halves of parts such as the head) from the left and right sides of the blastula respectively. The nervous system and the mesoderm (which gives rise to the muscles) have also been shown by embryologists to originate in specified regions of the blastula. It is clear that the probability that any two parts will have a different genotype (that is, that they will be on different sides of the mosaic boundary that cuts across the blastula) will depend on how far they are from each other on the blastoderm, the blastula surface. Conversely, the probability that two parts are of different genotypes should be a measure of their distance apart on the blastoderm.

was studied by Kaplan and Kazuo Ikeda. When such a fly is anesthetized with ether, it does not lie still but rather shakes all six of its legs vigorously. Kaplan and Ikeda found that flies that are mosaics for the gene shake some of their legs but not others and that the shaking usually correlates well with the leg's surface genotype as revealed by markers—but not always. (Suzuki and his colleagues found the same to be true of flies mosaic for the *paralyzed* mutation.) The point is that the markers are on the outside of the fly. The genotype of the surface is not necessarily the same as that of the underlying tissues, which arise from different regions of the embryo. And one might well expect that leg function would be controlled by nervous elements somewhere inside the fly's body that could have a different genotype from the leg surface. The problem is to find a way of relating internal behavioral foci to external landmarks. Hotta and I developed a method of mapping this relation by extending to behavior the idea of a "fate map," which was originally conceived by A. H. Sturtevant of Cal Tech.

Sturtevant was the genius who had earlier shown how to map the sequence of genes on the chromosome by measuring the frequency of recombination among genes. He had seen that the probability of crossing over would be greater the farther apart the genes were on the chromosome. In 1929 he proposed that one might map the blastoderm in an analogous way: the frequency with which any two parts of adult mosaics turned out to be of different genotypes could be related to the distance apart on the blastoderm of the sites that gave rise to those parts. One could look at a large group of mosaics, score structure A and structure B and record how often one was normal whereas the other was mutant, and vice versa. That frequency would represent the relative distance between their sites of origin on the blastoderm, and with enough such measurements one could in principle construct a two-dimensional map of the blastoderm. Sturtevant scored 379 mosaics of *Drosophila simulans*, put his data in a drawer and went on to something else. At Cal Tech 40 years later Merriam and Antonio Garcia-Bellido inherited those 379 yellowed sheets of paper, computed the information and found they could indeed make a self-consistent map.

When Hotta and I undertook to map behavior in *D. melanogaster*, we began by preparing our own fate map of the adult external body parts based on the scores for 703 mosaic flies [*see upper il-*lustration on these two pages]. Distances on the map are in "sturts," a unit Merriam, Hotta and I have proposed in memory of Sturtevant. One sturt is equivalent to a probability of 1 percent that the two structures will be of different genotypes.

Now back to *hyperkinetic*. We produced 300 mosaic flies and scored each for a number of surface landmarks and for the coincidence of marker mutations at those landmarks with the shaking of each leg. We confirmed the observations of Ikeda and Kaplan that the behavior of each leg (whether it shakes or not) is independent of the behavior of the other legs and that the shaking behavior and the external genotype of a leg are frequently the same—but not always. The independent behavior of the legs indicated that each had a separate focus. For each leg we calculated the distance from the shaking focus to the leg itself and to a number of other landmarks [*see lower illustration on these two pages*] and thus determined a map location for each focus. They are near the corresponding legs but below them, in the region of the blastoderm that Donald F. Poulson of Yale University years ago identified by embryological studies as the origin of the ventral nervous system. This is consistent with electrophysiological evidence, obtained by Ikeda and Kaplan, that neurons in the thoracic ganglion of the ventral nervous system behave abnormally in these mutants.

Another degree of complexity is represented by a mutant we call *drop-dead*. These flies develop, walk, fly and otherwise behave normally for a day or two after eclosion. Suddenly, however, an individual fly becomes less active, walks in an uncoordinated manner, falls on its back and dies; the transition from apparently normal behavior to death takes only a few hours. The time of onset of the syndrome among a group of flies hatched together is quite variable; after the first two days the number of survivors in the group drops exponentially, with a half-life of about two days. It is as if some random event triggers a cataclysm. The gene has been identified as a recessive one on the X chromosome. Symptoms such as these could result from malfunction almost anywhere in the body of the fly, for example from a blockage of the gut, a general biochemical disturbance or a nerve disorder. In order to localize the focus we did an analysis of 403 mosaics in which the XX parts were normal and the X body parts expressed the *drop-dead* gene and surface-marker mutations, and we scored

for *drop-dead* behavior and various landmarks.

Drop-dead behavior, unlike shaking behavior, which could be scored separately for each leg, is an all-or-none property of the entire fly. First we did a rough analysis to determine whether the behavior was most closely related to the head, thorax or abdomen, considering only flies in which the surface of each of these structures was either completely mutant or completely normal. Among mosaics in which the entire head surface was normal almost all behaved

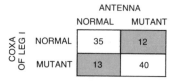

FATE MAP, a two-dimensional map of the blastoderm, is constructed by calculating the distances between the sites that gave rise to various parts. This is done by observing a large number of adult flies and recording the number of times each of two parts is mu-

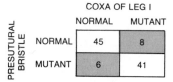

BEHAVIORAL FOCI, the sites at which a mutant gene exerts its effect on behavior, are plotted in the same way. The behavior in this example (*left*) is abnormal shaking of

normally, but six flies out of 97 died in the *drop-dead* manner; in the reciprocal class eight flies of 80 with mutant head surfaces lived. In other words, the focus was shown to be close to, but distinct from, the blastoderm site of origin of the head surface. Comparable analysis showed that the focus was substantially farther away from the thorax and farther still from the abdomen. Next we considered individuals with mosaic heads. The reader will recall that in certain visual mutants the visual defect was always observed in the eye on the mutant side of the head; flies with half-normal heads had normal vision in one eye. For *drop-dead*, on the other hand, of mosaics in which half of the head surface was mutant only about 17 percent dropped dead. All the rest survived.

Now, a given internal part should occur in normal or mutant form with equal probability, as the external parts in these mosaics did. On that reasoning, if there were a single focus inside the head of the fly, half of the bilateral-mosaic flies should have dropped dead. We formed the hypothesis, therefore, that there must be two foci, one on each side, and that they must interact. Both of them must be mutant for the syndrome to appear. In other words, a mutant focus must be "submissive" to a normal one. In that case, if an individual exhibits *drop-dead* behavior, both foci must be mutant, and if a fly survives, one focus may be normal or both of them may be.

Mapping a bilateral pair of interacting foci calls for special analysis. By considering the various ways a mosaic dividing line could fall in relation to a pair of visible external landmarks (one on

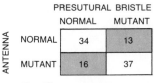

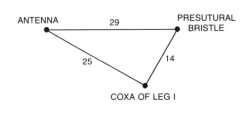

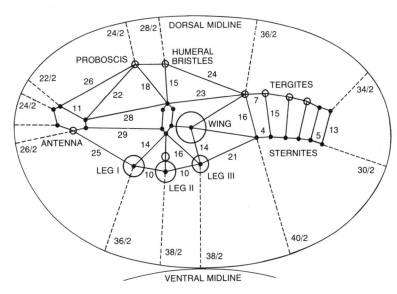

tant or normal. The numbers are entered in a matrix, as shown (*left*) for three pairs of parts. Instances in which one part is normal whereas the other part is mutant on the same fly (*colored boxes*) are totaled. That figure, divided by the total number of instances, gives the probability that the two parts are of different genotypes. And that probability is proportional to the distance between them, indicated in "sturts." Plotting the three distances triangulates the relative locations of the three sites. By thus scoring 703 mosaic flies for body parts, Yoshiki Hotta in the author's laboratory built up the fate map of external body parts (*right*). Broken lines represent distances to the blastula midlines, obtained by dividing by two the distances between homologous parts on opposite sides of the fly.

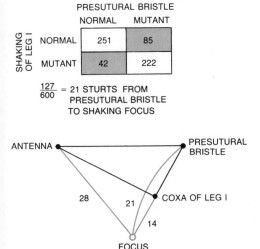

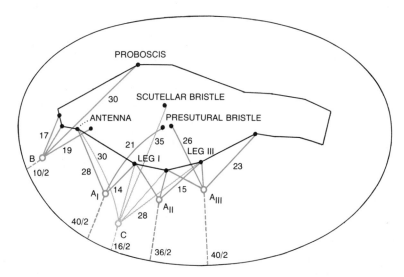

the legs under ether. The shaking is independent for each leg and here leg *I* has been scored for 300 flies—or 600 instances, since the data can be doubled to represent both sides of the fly. (The total of instances can be less than 600 because cases in which both the body part and the behavior are mosaic are eliminated.) Distances calculated (*colored lines*) triangulate the focus. In this way the foci for shaking behavior for each leg (*A*) are added to the map (*right*). Foci for *drop-dead* (*B*) and *wings-up* (*C*) behavior are also found.

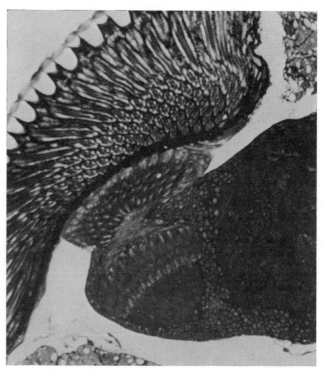

BRAINS OF DROP-DEAD MUTANTS that have reached the symptomatic stage show striking degeneration, as shown by photomicrograph (*left*) of a section of such a fly's head enlarged 300 diameters. The brain and the optic ganglia are full of holes. Sections fixed before a mutant has shown any symptoms, on the other hand, show no more degeneration than a section of a normal brain does (*right*).

each side of the body) and a symmetrical pair of internal foci, one can set up equations based on the probability of each possible configuration. Using the observed data on how many mosaic flies showed the various combinations of mutant and normal external landmarks and mutant or normal behavior, it is possible to solve these equations for the map distance from each landmark to the corresponding focus and from one focus to the homologous focus on the other side of the embryo. The *drop-dead* foci turn out to be below the head-surface area of the blastoderm, in the area embryologists have assigned to the brain. Sure enough, when we examined the brain tissue of flies that had begun to exhibit the initial stages of *drop-dead* behavior, it showed striking signs of degeneration, whereas brain tissue fixed before the onset of symptoms appeared normal. As for mosaics whose head surfaces are half-normal, those that die show degeneration of the brain on both sides; the survivors' brains show no degeneration on either side, a finding consistent with the bilateral-submissive-focus hypothesis. It appears that the normal side of the brain supplies some factor that prevents the deterioration of the side with the mutant focus.

There is another kind of bilateral focus, "domineering" rather than submissive. An example is the mutant we call *wings-up*. There are two different genes, *wup A* and *wup B*, which produce very similar overt behavior: shortly after emergence each of these flies raises its wings straight up and keeps them there. It cannot fly, but otherwise it behaves normally. Is *wings-up* the result of a defect in the wing itself, in its articulation or in the muscles or neuromuscular junctions that control the wing, or of some "psychological quirk" in the central nervous system? The study of mosaic flies shows that the behavior is more often associated with a mutant thorax than with a mutant head or abdomen. The focus cannot be in the wings themselves or anywhere on the surface of the thorax, however, because in some mosaics the wings and the thorax surface are normal and yet the wings are held up and in other mosaics the wings and thorax display all the mutant markers and yet the fly flies. These observations suggest that some structure inside the thorax could be responsible.

Once again we look at the bilateral mosaics, those with one side of the thorax carrying mutant markers and the other side appearing normal. Unlike the *drop-dead* bilateral mosaics, most of which were normal, these bilaterals are primarily mutant; well over half of them hold both wings up. Both wings seem to act together; either both are held up or both are in the normal position. This suggests two interacting foci, one on each side, with the mutant focus domineering with respect to the normal one, that is, if either of the foci is mutant, or both are, then both wings will be up. Again we can set up equations based on the probability of the various mosaic configurations and solve to find the pertinent map distances. The focus comes out to be close to the ventral midline of the blastula. That is a region known to produce the mesoderm, the part of the developing embryo from which muscle tissue is derived, which suggested that a defect in the fly's thoracic muscle tissue could be responsible for *wings-up* behavior.

The abnormality became obvious when we dissected the thorax. In the fly the raising and lowering of the wings in normal flight is accomplished by changes in the shape of the thorax, changes brought about by the alternate action of sets of vertical and horizontal muscles. Under the phase-contrast microscope these indirect flight muscles are seen to be highly abnormal in both *wings-up* mutants. Developmental studies show that in *wup A* the muscles form properly at first, then degenerate after the fly emerges. In the *wup B* mutant, on the other hand, the myoblasts that normally produce the muscles fuse properly but the muscle fibrils fail to appear. In both mutants the other muscles, such as those of the leg, appear to be quite normal,

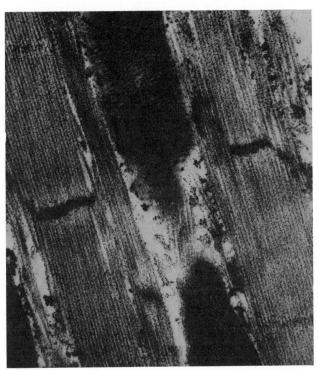

FLIGHT MUSCLE OF WINGS-UP MUTANTS shows degeneration that seems to account for their behavior. Normal flight muscle, enlarged 30,000 diameters in an electron micrograph, has bundles of filaments crossed by straight, dense bands: the Z lines (*left*). In the muscle of flies heterozygous for the gene *wup B*, which hold their wings normally but cannot fly, the Z lines are irregular (*right*).

and the flies walk and climb perfectly well. In flies that are heterozygous for *wup A* (nonmosaic flies with one mutated and one normal gene) the muscles and flight behavior are normal, that is, the gene is completely recessive. In *wup B* heterozygotes, on the other hand, the wings are held in the normal position but the flies cannot fly. Electron microscopy shows that even in these heterozygotes the indirect flight muscles are defective: the microscopic filaments that constitute each muscle fibril are arranged correctly, but the Z line, a dense region that should run straight across the fibril, is often crooked and forked. Examination of the muscles in bilateral mosaics confirmed the impression that the *wup* foci are domineering. In every mosaic that had shown *wings-up* behavior one or more muscle fibers were degenerated or missing; no fibers were seriously deficient in any flies that had displayed normal behavior. The natural shape of the thorax apparently corresponds to the *wings-up* position, and the presence of defective muscles on either side is enough to make it impossible to change the shape of the thorax, locking the wings in the vertical position.

The mutants so far mapped provide examples involving the main components of behavior: sensory receptors, the nervous system and the muscles. For some of the mutants microscopic examination has revealed a conspicuous lesion of some kind in tissue. The obvious question is whether or not fate mapping is necessary; why do we not just look directly for abnormal tissue? One answer is that for many mutants we do not know where to begin to look, and it is helpful to narrow down the relevant region. Furthermore, in many cases no lesion may be visible, even in the electron microscope. More important, and worth reiterating, is the fact that the site of a lesion is not necessarily the primary focus. For example, an anomaly of muscle tissue may result from a defect in the function of nerves supplying the muscle. This possibility has been a lively issue in the study of diseases such as muscular dystrophy. Recently, by taking nerve and muscle tissues from a dystrophic mutant of the mouse and from its normal counterpart and growing them in tissue culture in all four combinations, the British workers Belinda Gallup and V. Dubowitz were able to show that the nerves are indeed at fault.

The mosaic technique in effect does the same kind of experiment in the intact animal. In the case of the *wings-up* mutant the primary focus cannot be in the nerves, since if that were so the focus would map to the area of the blastoderm destined to produce the nervous system, not to the mesoderm, where muscle tissue is formed. The *wings-up* mutants clearly have defects that originate in the muscles themselves.

Another application of mosaics is in tagging cells with genetic labels to follow their development. The compound eye of *Drosophila* is a remarkable structure consisting of about 800 ommatidia: unit eyes containing eight receptor cells each. The arrangement of cells in an ommatidium is precise and repetitive; the eye is in effect a neurological crystal in which the unit cell contains eight neurons. Thomas E. Hanson, Donald F. Ready and I have been interested in how this structure is formed. Are the eight photoreceptor cells derived from one cell that undergoes three divisions to produce eight, or do cells come together to form the group irrespective of their lineage? This can be tested by examining the eyes of flies, mosaic for the *white* gene, in which the mosaic dividing line passes through the eye. By sectioning the eye and examining ommatidia near the border between white and red areas microscopically, it is possible to score the tiny pigment granules that are present in normal photoreceptor cells but absent in *white* mutant cells. The result is clear: A single ommatidium can contain a mixture of receptor cells of both genotypes. This proves that the eight cells cannot be derived from a single ancestral cell but have become associated in their special group of eight irrespective of lineage. The same conclusion applies to the other cells in each ommatidium, such as the normally heavily pigmented cells that surround the receptors.

Not all cells have such convenient pigment markers. It would obviously be valuable to have a way of labeling all the internal tissues as being either mutant or normal, much as yellow color labels a landmark on the surface. This can now be done for many tissues by utilizing mutants that lack a specific enzyme. If a recessive enzyme-deficient mutant gene is recombined on the X chromosome along with the *yellow, white* and behavioral genes and mosaics are produced in the usual way, the male tissues of the mosaic will lack the enzyme. By making a frozen section of the fly and staining it for enzyme activity one can identify normal and mutant cells.

In order to apply this method in the nervous system one needs to have an enzyme that is normally present there in a large enough concentration to show up in the staining procedure and a mutant that lacks the enzyme, and the lack should have a negligible effect on the behavior under study. Finally, the gene in question should be on the X chromosome. Douglas R. Kankel and Jeffrey Hall in our laboratory have developed several such mutants, including one with an acid-phosphatase-deficient gene found by Ross J. MacIntyre of Cornell University. By scoring the internal tissues they have constructed a fate map of the internal organs of the kind made earlier for surface structures. We are now adapting the staining method for electron microscopy in order to work at the level of the individual cell.

The staining procedure has demon-

strated graphically that the photoreceptor cells of the eye come from a different area of the blastoderm than do the neurons of the lamina, to which they project. In the adult fly the two groups of cells are in close apposition, but the former arise in the eye whereas the latter come from the brain. The distance between them on the fate map, determined by Kankel, is about 12 sturts, so that a considerable number of mosaic flies have a normal retina and a mutant lamina or vice versa. This makes it possible to distinguish between presynaptic and postsynaptic defects in mutants with blocks in the visual pathway. In the nonphototactic mutants Hotta and I analyzed in mosaics, the defect in the electroretinogram was always associated with the eye. In contrast, a mutant with a similar electroretinogram abnormality that was studied by Linda Hall and Suzuki

MOSAIC EYE contains a patch of cells that carry the *white* gene and therefore lack the normal red pigment. The fly's compound eye is an array of hexagonal ommatidia, each containing eight photoreceptor cells (*circles*) and two primary pigment cells (*crescents*) and surrounded by six shared secondary pigment cells (*ovals*). The fact that a single ommatidium can have *white* and normal genotypes shows its cells are not necessarily descended from a common ancestral cell. Nor is the mirror-image symmetry about the equator (*heavy line*) the result of two cell lines: mutant cells appear on both sides. Drawing is based on observations by Donald Ready.

showed, in some mosaics, a normal trace for a mutant eye—and vice versa. Fate mapping placed the focus in precisely the region corresponding to the lamina. What appeared to be similar malfunctions in two mutants were thus shown to be different, due in one case to a presynaptic block and in the other case to a postsynaptic one.

Much of what has been done so far involves relatively simple aspects of behavior chosen to establish the general methodology of mutants and mosaic analysis. Can the methodology be applied to more elaborate and interesting behavior such as circadian rhythm, sexual courtship and learning? Some beginnings have been made on all of these. By making flies that are mutant for normal and mutant rhythms, Konopka has shown that the internal clock is most closely associated with the head. Looking at flies with mosaic heads, he found that some exhibited the normal rhythm and others the mutant rhythm but that a few flies exhibited a peculiar rhythm that appears to be a sum of the two, as if each side of the brain were producing its rhythm independently and the fly responded to both of them. By applying the available cell-staining techniques it may be possible to identify the cells that control the clock.

Sexual courtship is a higher form of behavior, since it consists of a series of fixed action patterns, each step of which makes the next step more likely. The sex mosaics we have generated lend themselves beautifully to the analysis of sexual behavior. A mosaic fly can be put with normal females and its ability to perform the typical male courtship steps can be observed. Hotta, Hall and I found that the first steps (orientation toward the female and vibrating of the wings) map to the brain. This is of particular interest because the wings are vibrated by motor-nerve impulses from the thoracic ganglion; even a female ganglion will produce the vibration "song" typical of the male if directed to do so by a male brain. It would appear that the thoracic ganglion in a female must "know" the male courtship song even though she does not normally emit it. This is consistent with recent experiments by Ronald Hoy and Robert Paul at the State University of New York at Stony Brook, in which they showed that hybrid cricket females responded better to the songs of hybrid males than to males of either of the two parental species.

Sexual behavior in *Drosophila,* although complex, is a stereotyped series of instinctive actions that are performed

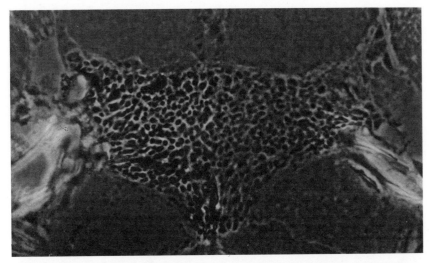

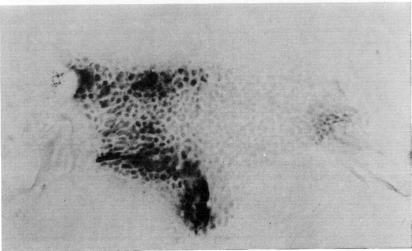

DIRECT SCORING OF NORMAL AND MUTANT CELLS within the nervous system is possible with a staining method developed by Douglas Kankel and Jeffrey Hall in the author's laboratory. Mosaics are produced in which mutant cells are deficient in the enzyme acid phosphatase. When the proper stain is applied to a section of nerve tissue, normal cells stain brown and mutant cells are unstained. Here a section of the thoracic ganglion, thus stained, is shown in phase-contrast (*top*) and bright-field (*bottom*) photomicrographs. In the bottom picture normal cells are marked by the stain, delineating the mosaic boundary.

correctly by a fly raised in isolation and without previous sexual experience. Other forms of behavior such as phototaxis also appear to be already programmed into the fly when it ecloses. Whether a fruit fly can learn has long been debated; various claims have been made and later shown to be incorrect. Recently William G. Quinn, Jr., and William A. Harris in our laboratory have shown in carefully controlled experiments that the fly can learn to avoid specific odors or colors of light that are associated with a negative reinforcement such as electric shock. This opens the door to genetic analysis of learning behavior through mutations that block it.

In tackling the complex problems of behavior the gene provides, in effect, a microsurgical tool with which to produce very specific blocks in a behavioral path-

way. With temperature-dependent mutations the blocks can be turned on and off at will. Individual cells of the nervous system can be labeled genetically and their lineage can be followed during development. Genetic mosaics offer the equivalent of exquisitely fine grafting of normal and mutant parts, with the entire structure remaining intact. What we are doing in mosaic mapping is in effect "unrolling" the fantastically complex adult fly, in which sense organs, nerve cells and muscles are completely interwoven, backward in development, back in time to the blastoderm, a stage at which the different structures have not yet come together. Filling the gaps between the one-dimensional gene, the two-dimensional blastoderm, the three-dimensional organism and its multidimensional behavior is a challenge for the future.

Habitat Selection

by Stanley C. Wecker
October 1964

*How does an animal choose its environment?
Experiments with mice that live either in fields
or in forests indicate that both heredity and
learning have played a role in the evolution
of this behavior*

*Mid pleasures and palaces
'though we may roam,
Be it ever so humble,
there's no place like home.*

If animals were capable of understanding verse, this sentiment would doubtless have as much meaning for the denizens of a rotting log as it does for the inhabitants of the most fashionable suburb. One need only visit the countryside to perceive that the plants and animals in a natural community, like their human counterparts, are not scattered haphazardly over the landscape. Each organism tends to be restricted in distribution by its behavioral and physiological responses to the environment. It follows that living things must be able to locate favorable places in which to live. Their methods of doing so are so numerous and varied, however, that it is difficult to generalize about the selection of habitat.

On the one hand, many small organisms of otherwise low mobility have evolved means for utilizing air and water currents in the dispersion of members of their species. Spores, seeds, ballooning spiders and a surprisingly large number of insects drift in the upper reaches of the atmosphere, and a wide variety of planktonic forms ride the waves of the waters below. Occasionally terrestrial organisms accidentally cross long stretches of sea on pieces of driftwood, and live fish have been transported from pond to pond by hurricanes. The end result of this passive and essentially random dissemination of individuals is that a small number of them eventually reach areas conducive to continued survival and reproduction.

For the majority of animals, on the other hand, choosing a habitat is a more active process. This does not imply that most species can make a critical evaluation of the entire constellation of factors confronting them. More probably they react automatically to certain key aspects of their surroundings. For example, a wide variety of animals, ranging from single-celled protozoans to beetles and salamanders, often select their habitat at least in part by orientation along physicochemical gradients in the environment. These include such factors as temperature, moisture, light and salinity.

Another form of behavior that results in habitat selection is the choice of egg-laying sites by insects. Among certain beetles, butterflies and wasps the gravid female instinctively selects a plant or an animal host that will satisfy the requirements of the developing larva, whether or not the needs of the larva coincide with her own. Among the birds that live in shrubbery or forest the choice of habitat has been found to be associated with the height, spacing and form of the vegetation. Even when the overall character of the vegetation is appropriate for a species, a deficiency of specific environmental cues, such as song perches and nest sites, may exclude the species from an area within its range. The British ornithologist David Lack has called this phenomenon a "psychological factor" in habitat selection. Among the higher forms of life such factors may be fully as important as stimuli more directly related to physiological tolerances.

Although many ecologists have investigated the physical and biological factors that cause mammals to occupy certain habitats and avoid others, little is known about the role of psychological factors. One genus that is particularly well suited for the study of these factors is *Peromyscus*, the deer mouse. To this hardy little mammal almost every conceivable ecological situation, ranging from desert to tropical rain forest, from barren tundra to windblown mountaintops, is home. One species, *P. maniculatus*, is among the most variable of all North American rodents. It has 66 subspecies, which are found in so many habitats that a leading ecologist has remarked that probably no environmental change short of the inundation of the entire continent would eliminate all of them! In spite of this variability, however, in the sense of ecological adaptation the species has just two principal types: the long-tailed, long-eared forest forms, and the smaller short-tailed, short-eared grassland forms.

The prairie deer mouse of the Middle Western and Plains states (*Peromyscus maniculatus bairdi*) is a strictly field-dwelling subspecies that avoids all forested areas, even those with a grassy floor. Studies comparing the food preferences and the requirements for temperature and moisture of this subspecies and a closely related woodland form, *P. m. gracilis*, have not revealed any physiological differences of sufficient magnitude to account for the difference in their choice of habitat. It has therefore been concluded that the absence of the prairie deer mouse from forested areas within its geographic range is primarily a behavioral response to its environment.

The first experimental attempt to identify the environmental cues that cause these mice to choose a place to live was undertaken in 1950 by Van T. Harris, then working with Lee R. Dice at the University of Michigan Laboratory of Vertebrate Biology. Harris presented individual prairie and woodland deer mice with a choice between a laboratory "field" and a laboratory "woods." Each type of mouse exhibited a clear preference for the artificial habitat more closely resembling its natural en-

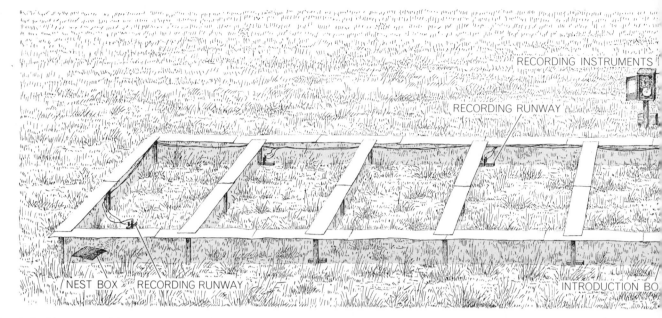

EXPERIMENTAL ENCLOSURE for testing habitat preference of prairie deer mice is 100 feet long and 16 feet wide. Five of its 10 compartments are in a field (*left*) and five are in an oak-hickory woodlot. For testing, each mouse is placed in the introduction box near the middle. It can go from there into either the field half or the woods half of the enclosure. Each partition has a run-

vironment. Since the physical conditions throughout the experimental room were uniform, Harris concluded that the mice were reacting to the character of the artificial vegetation. Moreover, laboratory-reared animals with no outdoor experience chose the "correct" artificial habitat as readily as the wild mice did. Harris therefore decided that this behavior was innate.

These experiments were not, however, designed to test the possibility that learning might also be involved. It has recently been established that early experience is of greater importance in the development of adult behavior than had once been thought [see "Early Experience and Emotional Development," by Victor H. Denenberg; SCIENTIFIC

AMERICAN Offprint 478]. Since young prairie deer mice are normally born and reared in open fields, one would expect their early experience to reinforce any innate preference for this habitat.

These considerations raise two questions: (1) Does learning actually play a role in habitat selection by *Peromyscus*? (2) Can an innate preference for field conditions be overridden by early experience in a different environment? In order to investigate these problems I constructed a 100-foot-long outdoor pen on the University of Michigan's Edwin S. George Reserve, 26 miles northwest of Ann Arbor. The project was initiated with the support of the Department of Zoology and the Muse-

um of Zoology and was carried out under the auspices of the Laboratory of Vertebrate Biology and its director, Francis C. Evans.

The long axis of the experimental pen crosses a relatively sharp boundary between an open field and an oak-hickory woodlot. The enclosure is divided into 10 compartments, five of which are in the field and five in the woods. There are two underground nest boxes, one at the end that extends farthest into the woods, the other at the end that extends farthest into the field. A third underground box in the middle of the enclosure serves as a chamber for introducing mice. Small metal runways leading from one compartment to another allow the animals to go anywhere with-

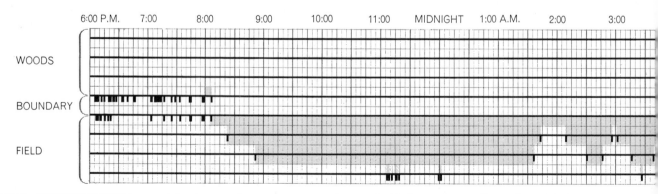

RECORD MADE BY MOUSE in two nights shows preference for field. The mouse, from laboratory stock, was in the field for 10 days when quite young, then lived in laboratory for 56 days before test. Daylight hours are omitted here because the mouse was quiet in the field nest box. The eight horizontal lines (*black*) were traced by

pens connected with the various treadles in the enclosure, ranging in order from the woods nest box at top to field nest box at bottom. Short vertical lines along the tracings are "blips" made when mouse crossed treadle. Just after 6:00 P.M. (*far left*) mouse leaves introduction box, runs back and forth across treadles at

NEST BOX

way at one end that enables the mouse to go from one compartment to the next. Two of the seven runways with recording treadles are labeled. Nest boxes, both of which have treadles, are in the last compartments at left and right. The instruments that make permanent records of movements of each mouse (*see bottom of these two pages*) are in box at top, just to left of center.

in the entire fenced area. A centrally located electric device records the time at which a mouse passes through the runways and enters the nest boxes. I place each mouse in the experimental enclosure alone and leave it there until it has nested in the same habitat for two consecutive days.

Prairie deer mice are nocturnal and are inactive during the day. I decided, therefore, that it would be most meaningful to consider the length of an animal's active and inactive periods in each environment (woods and field) as separate measures of habitat selection. Three other categories of measurement provide further data for comparing an animal's response to the woods with its response to the field. The five categories used in

this study, then, are (A) *time active,* or time spent outside the nest boxes in woods and field respectively; (B) *time inactive,* or time spent nesting in woods or field; (C) *rate of travel,* or the speed at which a mouse moves about in each of the two habitats; (D) *activity,* or the frequency with which a mouse changes compartments or enters nest boxes in woods or field; and (E) *average penetration* (in feet) into either of the two habitats each time a mouse crosses the boundary between them. In all categories except rate of travel the higher score for woods or field is taken to indicate habitat preference. In the case of rate of travel it was assumed that a mouse travels more slowly in the preferred habitat; presumably the animal

is less subject to stress in its normal environment.

In the course of the study I tested six groups of prairie deer mice, one mouse at a time, in the enclosure. Observing the 132 mice occupied the spring, summer and fall of two successive years. The two control and four experimental groups were each characterized by a different combination of two variables: hereditary background and pretest experience. The hereditary distinction was between field-caught mice (and their immediate offspring) and individuals selected from a laboratory stock. The experience was provided in the field, in the woods and in the laboratory.

The first group to be tested consisted

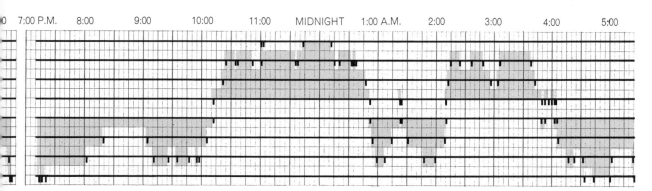

7:00 P.M.   8:00   9:00   10:00   11:00   MIDNIGHT   1:00 A.M.   2:00   3:00   4:00   5:00

habitat boundary. After brief foray into woods (7:58 P.M.) it returns to field and gradually moves all the way to the nest box (11:07 P.M.). It goes in and out, then moves back several times toward habitat boundary but never crosses it. At 5:16 A.M. it enters field nest box (*end of first night's record*) and remains throughout

the day. Record for second night shows two long periods in the woods, including two entries into the woods nest box. It was usually assumed that the mouse went at least halfway to the next treadle after crossing a treadle, as shown by colored shading. Actually the mouse could have been anywhere between the two treadles.

INTRODUCTION BOX opens into the runway (*left*) that crosses the habitat boundary. The two recording treadles for the boundary can be seen in this runway. Tiny door at end of exit tube opens outward only. The two nest boxes resemble the introduction chamber.

of individuals recently caught in old fields of the Edwin S. George Reserve, where earlier studies had clearly demonstrated the strong affinity of prairie deer mice for the field environment. My assumption was that the reactions of these adult animals would provide a basis for evaluating any unnatural effects of the enclosure itself. Accordingly I designated the eight males and four females in the group as Control Series I. At the end of the test it was obvious from all five measurements that the mice much preferred the field half of the enclosure [*see upper illustration on opposite page*]. From this I concluded that the testing situation permitted the animals to exercise their normal habitat preference.

If this preference is innate, field mice reared in the laboratory should also choose the field environment. In order to evaluate this hypothesis I tested seven males and six females from the prairie deer mouse colony of the University of Michigan Mammalian Genetics Center. These were Control Series II. The entire laboratory stock, designated *Peromyscus maniculatus bairdi* Washtenaw (for Washtenaw County), was descended from 10 pairs of animals trapped in the vicinity of Ann Arbor by Harris in 1946. According to the records the 13 individuals of Control Series II were 12 to 20 generations removed from any field experience. Their performance in the enclosure contrasted sharply with that of Control Series I [*see lower illus-*

tration on opposite page]. In three of the five categories more Control Series II individuals preferred the woods to the field! The most that can be said of the group as a whole, however, is that it did not demonstrate a well-defined preference for either habitat.

In its laboratory environment the *bairdi* Washtenaw stock has been subjected to different selective pressures from those encountered in fields. Combinations of genes that are advantageous to prairie deer mice in nature, such as those affecting response to the environment, would in the laboratory probably not be selected for and might even be selected against. One can therefore assume that the field and laboratory populations used in my experiments had genetically diverged. Since other investigators have shown that such divergence in laboratory stocks can lead to morphological changes, it seems reasonable to assume that behavioral modifications will arise also. I suggest that these contributed to the highly variable habitat response of the mice in Control Series II. It is of considerable interest that the marked preference for fields displayed by their ancestors has been lost in only 12 to 20 generations.

Thus the data from Control Series II neither support nor refute Harris' contention that the habitat preference of prairie deer mice is normally determined by heredity. The next experiment provided a more rigorous evalua-

tion. For this test I caught more wild field mice and bred them in the laboratory. The offspring, which were separated from their parents shortly after weaning, lived in laboratory cages for an average of about two months. Then I tested eight males and four females in the enclosure as Experimental Series I. None had had any previous outdoor experience.

Among these mice there was no reason to anticipate hereditary modifications of the type postulated for the laboratory stock. Thus if habitat preference is genetically determined, the behavior of the mice in Experimental Series I should approximate that of Control Series I. As the records indicate [*see top illustration on page 168*] these animals did display a pronounced affinity for the field half of the enclosure. Obviously prior experience in this environment is not a necessary prerequisite for habitat selection.

Since the animals were reared by field-caught parents, however, it is possible that some form of noninherited social interaction brought about the results. Unfortunately I have had no opportunity to evaluate this possibility, but other investigators have failed to find evidence in prairie deer mice for transfer of behavioral traits from generation to generation through learning. Litters reared by foster parents do not reveal any consistent indication of maternal, paternal or joint parental influence. It seems likely that, as Harris concluded, the habitat preference of wild populations of prairie deer mice is an expression of an innate pattern of behavior. The pattern may be elicited by certain key environmental stimuli, but it apparently does not depend on a period of habituation to the environment for its expression. This does not mean, however, that early experience has no effect on the selection of habitat by an adult animal. It seems reasonable to assume that a young deer mouse's normal association with open fields will reinforce its innate preference for this environment.

In order to ascertain the role such experience plays, I allowed pairs of laboratory animals to rear litters in a 10-by-10-foot pen constructed in the field. Located a short distance from the main enclosure, this area was divided into two compartments, each of which contained a number of nest boxes. Mice that had mated in the laboratory were moved into the nest boxes soon after they had borne litters and before the eyes of the young had opened. After an average of 31 days in the field pen 13 of the offspring were tested in the main enclosure. I labeled this group Experimental Series II.

These 13 mice—eight males and five females—displayed a well-defined preference for the field habitat. Since the laboratory stock of Control Series II had not particularly preferred the field, the highly contrasting behavior of the offspring of such stock can only be explained by their field experience. Although the laboratory animals have apparently lost the innate preference of the subspecies for fields, they have retained a capacity for learning that enables them to exercise habitat selection if they are exposed to the field environment at an early age. Whether or not early experience in a different environment would reverse normal habitat affinities, however, remained to be determined.

Accordingly field-caught prairie deer mice were allowed to raise litters in a 10-by-10-foot pen in the woods. Subsequently I tested seven of the woods-reared offspring in the large enclosure. These mice—six males and one female—constituted Experimental Series III. The two weeks of woods experience did not noticeably influence their selection of habitat. In all five categories of measurement a majority of the mice exhibited the normal field preference. It thus appears that early experience in the "wrong" environment is not enough to override the innate habitat response. Since learning assumes a more impor-

tant role in the development of a well-defined habitat preference by laboratory animals, it seemed possible that early experience in the woods might lead mice from laboratory stock to prefer the woods habitat.

In order to determine if this was the case I transferred to the woods pen litters born to *bairdi* Washtenaw females. Nine of these offspring, six males and three females, were subsequently tested in the main enclosure as Experimental Series IV. As a whole, in spite of their 24 days of woods experience, these animals did not demonstrate a pronounced tendency to select the woods half of the enclosure. On the other hand, neither did they display any special preference for the field half. One must therefore conclude that prairie deer mice can only learn to respond to environmental cues associated with the field habitat.

To summarize the six experiments, four groups (Control Series I and Experimental Series I, II and III) consistently selected the field half of the enclosure, whereas the other two (Control Series II and Experimental Series IV) did not exhibit a well-defined habitat preference. All the individuals in the four groups that preferred the field environment had either field-caught parents or field experience or both. The other two groups were offspring of laboratory animals and had had no contact with the natural field environment prior to testing in the enclosure.

The data warrant the following conclusions: (1) The choice of the field environment by *P. m. bairdi* is normally

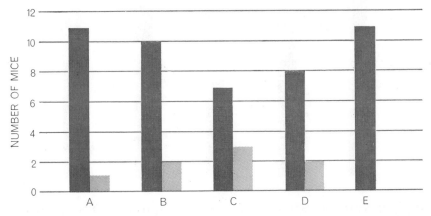

CONTROL SERIES I, consisting of 12 adult mice trapped in an open field, showed a clear preference for the field Gray bars indicate choice of field for each criterion of measurement (*A* through *E*); colored bars denote a preference for the woods. In this and the five bar graphs that follow, a pair of bars for some criteria does not add up to the total number of mice in the group. This results from a failure in the recording apparatus, or from the fact that an animal's score was the same for both woods and field, or because the mouse spent all its time in one habitat, making the comparisons in categories C, D and E impossible.

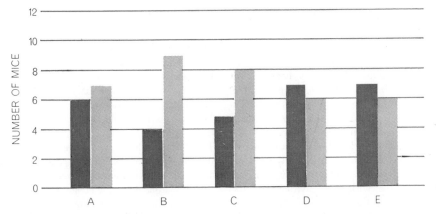

CONTROL SERIES II, consisting of 13 animals of laboratory stock, 12 to 20 generations away from the field, preferred the woods according to three categories of measurement. As a whole, however, this group cannot be said to have selected either half of the enclosure. Categories of measurement are (*A*) percent of time active in woods or field, (*B*) percent of time inactive, (*C*) rate of travel in woods or field, (*D*) activity in field or woods and (*E*) average penetration in feet by a mouse into woods or field from the habitat boundary.

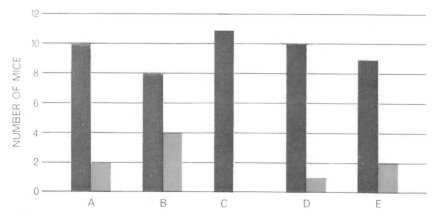

EXPERIMENTAL SERIES I, 12 mice, were first-generation offspring of field stock, reared in the laboratory. In all five measurements of habitat selection, they chose the field.

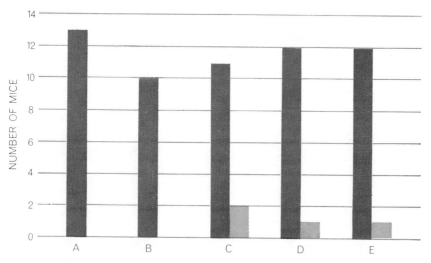

EXPERIMENTAL SERIES II, 13 animals, were laboratory stock reared in a pen in the field. By all five criteria of measurement they displayed a strong preference for the field.

determined by heredity. (2) Early field experience can reinforce this innate preference, but it is not a prerequisite for subsequent habitat selection. (3) Early experience in other environments (woods or laboratory) cannot override the normal affinity of field stock for the field habitat. (4) Confinement of the *bairdi* Washtenaw stock in the laboratory for 12 to 20 generations has apparently reduced hereditary control over the habitat response. This genetic change has markedly increased the behavioral variability of these animals when tested in the enclosure. (5) The laboratory stock did retain an innate capacity for learning from early field experience to respond positively to stimuli associated with this environment. Experience in the woods, however, did not cause them, on the whole, to select the woods habitat.

These results indicate that both heredity and experience can play a role in determining the preference of the prairie deer mouse for the field habitat,

which raises an interesting question. Since the same affinity for fields can be learned by each generation, why has natural selection produced an apparently parallel, genetically determined response?

According to the British zoologist C. H. Waddington, evolutionary changes that increase hereditary control are advantageous because they tend to limit the number of possible ways an organism can respond to a particular environmental stimulus. This is beneficial because natural selection favors only those responses conducive to survival. Therefore, as long as the environment remains relatively stable, the population as a whole will eventually become genetically adjusted to the ecological situation it is most likely to encounter and best able to exploit. The innate preference of prairie deer mice for the field environment represents such an adjustment. Why, then, does the mouse retain what appears to be an independent mechanism for habitat selection based

on learning? Furthermore, if we are dealing with two independent mechanisms, why should relaxation of natural selection under laboratory conditions remove one and not the other?

I would suggest that the innate pattern of habitat selection is not independent of the learned pattern but rather is really an extension of the learned pattern. This idea derives support from the observations on *bairdi* Washtenaw stock: the laboratory animals have lost any innate habitat preference but learn to select the "correct," or field, half of the enclosure after being reared in a pen in the field. Presumably a certain number (X) of "field-adapting" genes would give the prairie deer mouse the ability to learn to respond positively to the field environment; a larger number of such genes (X plus Y) could make this behavior innate. After 12 to 20 generations in the laboratory the mouse reverts from the X-plus-Y genotype back to the X genotype.

The behavioral evolution from learned to innate response can be explained as an example of the "Baldwin effect," originally called organic selection when postulated in 1896 by J. M. Baldwin of Princeton University. Recently George Gaylord Simpson of Harvard University has redefined the process to explain how individually acquired, nongenetic adaptations may, under the influence of natural selection, be replaced in a population by similar hereditary characteristics.

As an alternative to accepting the old Lamarckian doctrine that acquired characteristics can be directly inherited, one might apply Simpson's interpretation of the Baldwin effect to the prairie deer mouse situation as follows: As the mice became physiologically and morphologically adapted to existence in the grasslands, patterns of behavior based on some form of learning (homing, for example) tended to confine individuals to the field environment. These patterns, although not exclusively hereditary as such, were still advantageous in that they restricted the animals to the habitat best suited for their survival and reproduction. Then chance mutation created genetic factors that facilitated the development of behavior patterns whose effects resembled those acquired through learning. Finally, since natural selection favored these factors, they spread through the population.

Waddington believes, however, that the Baldwin effect, with its emphasis on chance mutation, involves an oversimplification that ignores the role of the environment in determining the manner in which particular combinations of genes will be expressed. For example, climate or some other aspect of the environment may determine what color certain animals will be, the animals themselves having a genetic potential for more than one color. Waddington maintains that natural selection operates not in favor of genes whose effects happen *by chance* to parallel acquired (nongenetic) adaptations but in favor of factors that control the capacity of an individual to respond to its surroundings. The interaction of organism and environment has the effect of reducing the number of different pathways for genetic expression, thus facilitating the production of better-adapted individuals. The more thorough this "canalization" of developmental possibilities is, the more likely it will be that favorable combinations of genes already present in the population in low frequency will find expression. Once expressed, these combinations of genes can be acted on by natural selection. Since they are favorable, the number of individuals bearing them will ultimately increase. Waddington terms this process the "genetic assimilation" of a character that is initially acquired, or nongenetic.

The results of experiments I am now conducting suggest that the *bairdi* Washtenaw stock learns to respond to the field environment very quickly and may indeed exhibit what the British zoologist W. H. Thorpe has called habitat imprinting. If imprinting is actually operating, one would expect the adult habitat response to be determined during a critical period early in the life of the animal, probably shortly after the young mouse first leaves the nest. It is significant, therefore, that young laboratory animals receiving only 10 days of early field experience still have a marked preference for that environment when they are tested in the enclosure, even after two months of confinement in laboratory cages! On the other hand, exposure of adult laboratory animals to the field environment for as long as 59 days does not cause them to develop a well-defined habitat preference.

In view of the above, it appears that one result of selection for an increased number of "field-adapting" genes has been to shift the development of the behavior patterns involved in habitat selection to earlier and earlier periods in the life of the individual. Obviously survival is enhanced by recognition of a favorable environment over successive generations through learning. It would be even more advantageous to restrict

learning capacity to include only those cues associated with the favorable environment and to reduce to an absolute minimum the time required for such learning. Finally, the necessity for learning could be eliminated altogether by selection for sets of genes that endow an individual with the capacity for making an adaptive response to the critical stimuli as soon as the stimuli are encountered. In this context a hypothetical imprinting stage may have been an important preliminary to the ultimate genetic assimilation of the habitat response of the prairie deer mouse. Indeed, the behavioral differences among the various groups of mice tested during my investigation could be taken to reflect different steps in an evolutionary sequence leading from behavior largely dependent on learning to the development of an innate pattern of control.

This sequence might have occurred as follows: (1) habitat restriction through social factors and homing, (2) recognition of the field environment through learning, (3) learning capacity reduced to cues associated with the field habitat, (4) imprinting to the field environment through exposure very early in life, (5) innate determination of the habitat response.

So far no one has identified the specific cues by which a prairie deer mouse recognizes the field environment. Fortunately the results of my investigations suggest a unique approach to this problem. Young laboratory animals do not develop a well-defined habitat preference in the absence of early field experience. It should therefore be meaningful to expose them in the laboratory to single stimuli designed to simulate different aspects of the natural field en-

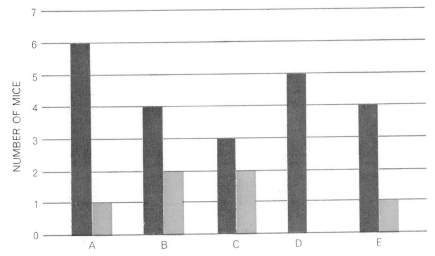

EXPERIMENTAL SERIES III, seven mice, offspring of adults trapped in the field, were conditioned by rearing in a pen in the woods. They tended to choose the field habitat.

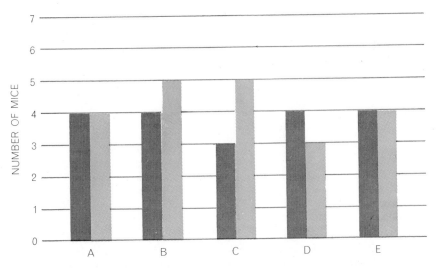

EXPERIMENTAL SERIES IV, nine animals, were laboratory stock reared in the pen in the woods. Overall they appeared to have no particular preference for either of the habitats.

vironment. These stimuli include the sight, odor and touch of field vegetation, with artificial grass as the touch stimulus. Groups of animals, each group exposed to only one such factor, could then be tested in the experimental enclosure. In fact, I am now conducting such tests, but the data are not yet extensive enough to warrant any conclusions.

Having considered the role of evolution in habitat selection, I should like to discuss briefly the part that habitat selection plays in the evolutionary process. A diversity of habitat preferences within a species favors survival by making the species more adaptable to environmental change. Such a diversity, however, might be expected to lead to genetic divergence by selective processes similar to those already described. Nevertheless, most biologists do not believe that new species can arise in this way unless some form of geographical isolation occurs. In Michigan the ranges of the prairie deer mouse and the woodland deer mouse overlap, but there is no evidence of intergradation, or interbreeding, between the two types. It is known, however, that these subspecies did not develop side by side, because they were formerly isolated geographically. Indeed, the two forms came into close contact only during the past century, when the clearing of forests by man enabled the prairie field mouse to extend its range northward.

Both Harris' experiments and mine provide evidence that the observed difference in habitat preference of these subspecies forms the basis for their continued segregation. As Ernst Mayr of Harvard University points out, ecological differences between two such overlapping forms are to be expected, since competition would otherwise prevent both from coexisting in the same area.

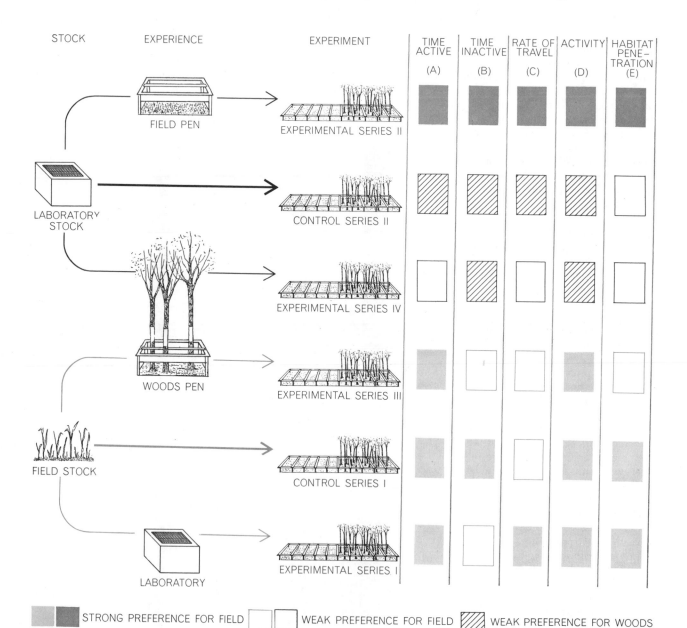

SUMMARY OF OBSERVATIONS reveals preferences of prairie deer mice from various backgrounds. Mice captured in the field were placed in the experimental pen as "Control Series I" (*thick colored arrow*). Offspring of field-caught mice were given conditioning in laboratory cages or in a pen in the woods before testing (*thin colored arrows*). Laboratory stock were tested in the experimental enclosure as "Control Series II" (*thick black arrow*), and their offspring were conditioned in woods or field pens before testing (*thin black arrows*). Results of the tests on the six groups are given in right half of the diagram. The five categories of measurement are explained in the text. Degree of preference is indicated. Results are based on mean response of each group.

Visual Isolation in Gulls

by Neal Griffith Smith

October 1967

Some species of gulls live together and look alike, yet they do not interbreed. How do the species remain isolated? Experiments in the Arctic indicate that they do so by recognizing subtle visual signals

Gulls look remarkably alike. That was the problem. Differences in appearance among the large gulls of the genus *Larus* can be subtle: a slight variation in size or a change in the color of the wing tips or of the eye and the small fleshy ring around the eye. Observing differences of this kind, an ornithologist discriminates among species of the genus. The problem arose from the fact that the gulls are equally discriminating. In some places *Larus* species that seem virtually indistinguishable nest side by side, yet they do not interbreed. How do gulls of one species avoid interbreeding with gulls of another?

The question of how species acquire and maintain their identity has received much attention in the century since the publication of Charles Darwin's *Origin of Species*. It is now well established that geographic isolation between populations is of prime importance in initiating the process by which species arise. Indeed, the gulls of the Northern Hemisphere have been cited as a classic example supporting this concept.

The common ancestor of the *Larus* gulls probably emerged in the Siberian region. As these gulls spread to the east and west, simple geographic distance began to inhibit the flow of genes between the most distant populations. By the time these populations had spread around the hemisphere and overlapped in western Europe, their respective genetic backgrounds were different enough so that hybrids between them were at some disadvantage; thus they did not interbreed. The advance and retreat of the ice during the Pleistocene epoch caused a further fracturing and recombination of these circumpolar gull populations. In some cases the differences evolved were not critical enough to confer a disadvantage on hybrids; thus the rejoined populations interbred.

It seems clear that the mechanisms by which species discriminate among one another evolved gradually during the process of species formation. In the Canadian Arctic, and probably elsewhere in the north, the ice intruded between various gull populations at different times and for different lengths of time. Accordingly the isolating mechanisms were likely to be at different stages of development in different populations. By studying populations in such an area one can uncover what these mechanisms are.

It is one thing to identify differences in the appearance of two closely related species living side by side and infer that the differences function as a barrier to interbreeding. It is quite another to demonstrate that these features are actually utilized in the isolation of species. To explain how such features work is still another step. This article is primarily concerned with the last two problems. It also considers the evolutionary history of the *Larus* gulls, because the elucidation of a feature that is utilized in species isolation can suggest what the species were like in the past and how the isolation mechanisms evolved.

The four species of *Larus* gulls I have been studying comprise the Canadian portion of the complex of gull populations around the North Pole. The fact that the four species do not interbreed has been clearly established by other workers and myself. All four gulls have a white body and a gray back and wings. The largest in body size is the glaucous gull (*Larus hyperboreus*). The tips of its wings are white, the iris of its eye is yellow and the fleshy ring around the eye is an even brighter yellow. Colonies of glaucous gulls are found throughout the polar area; in the eastern part of the Canadian Arctic they usually nest on cliffs. A more familiar species is the herring gull (*L. argentatus*), the only one of the four that breeds in the continental U.S. It is a medium-sized bird with wing tips that are partly black and partly white. Like the glaucous gull, the herring gull has a yellow iris; its eye-ring, however, is orange. In the Arctic this species usually nests on the ground in marshy areas. About the same size and coloration is Thayer's gull (*L. thayeri*), except that in this species the iris of the eye is dark brown and the eye-ring is a reddish purple. Thayer's gull nests almost exclusively on towering cliffs. The smallest of the four species (although not by very much) is Kumlien's gull (*L. glaucoides*), which also nests on cliffs. It is most like Thayer's gull: its eye-ring is reddish purple but the iris varies from clear yellow to dark brown. Its wing tips also vary in their amount of gray.

The common breeding grounds of these gulls are difficult to visit, and not much has been known about them. When I began my work, the evidence was that no one area was shared by all four species. In the course of trying to find such an area I spent three seasons (April to September) in the Canadian Arctic, during which I covered just under 2,000 miles by dogsled and canoe. During this time I studied three of the gulls (glaucous, Kumlien's and herring) I found nesting together on the south side of Baffin Island and a different trio (glaucous, Thayer's and herring) on nearby Southampton Island. Finally I discovered all four species nesting together on the east side of Baffin Island. It was never easy to find the ground-nesting herring gulls in association with the cliff-nesting species. Nesting on cliffs evolved as an adaptation against predators such as foxes; apparently competition with the other gulls for nesting sites has resulted in the herring gulls' occupying poorer sites. Nevertheless, where the surface allowed it and where the birds were safe from predators in a place such as a rocky islet, all the gulls would nest together.

There were a number of factors, for instance the habitat differences I have mentioned, that tend to reduce the possibility of mixed matings in the areas shared by different populations of gulls; here, however, I shall discuss only differences in external appearance among the species that function as major isolating mechanisms. In 1950 Finn Salomonsen, a Danish ornithologist, suggested that the color of the eye-ring might serve as a signal for differentiation between Kumlien's gull (reddish-

purple eye-ring) and the glaucous gull (yellow eye-ring). Although I tested the possible significance of all the differences in the gull's external appearance (with the exception of size), I concentrated on the color of the eye-ring.

In order to study the gulls closely it was necessary to catch them. At first I did so by stretching over a ledge a large fishnet under which food was placed. When the gulls were under the net, an Eskimo assistant and I rushed forward and dropped it, pinning the gulls to the ground. This was obviously an inefficient method, and later I used the drug tribromoethanol. Capsules of the drug were inserted into pieces of meat; after eating the meat the gulls quickly became anesthetized. In this way more than 1,800 gulls were trapped. After the gulls had been drugged they were immobilized with a surgical rubber band that pinned

their legs and wings to their bodies, and colored leg bands were put on them to make it possible to recognize individuals. Sex was determined by measuring bill, feet and wings; the males are usually larger. The determinations were confirmed by the subsequent behavior of the gulls.

One of my first thoughts had been that if markings and coloration play a role in the gulls' mating behavior, it should be possible to demonstrate it by changing these features artificially. This I now undertook to do. To change the color of the eye-ring I applied oil paint with a thin brush. The wing-tip pattern was changed with white or black ink after first wiping the feathers with alcohol so that the ink would penetrate. Judging from the behavior of the painted gulls neither of these procedures caused any physical irritation. On the

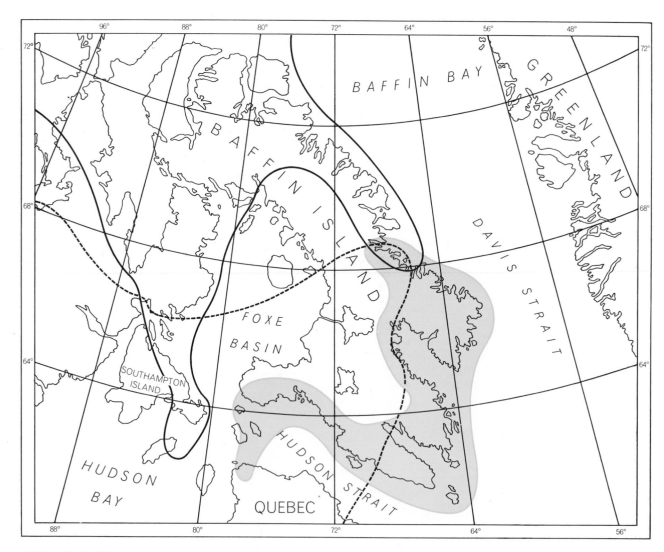

BREEDING RANGES of large *Larus* gulls lie in the eastern Canadian Arctic. Thayer's gulls (*black line*) and Kumlien's gulls (*colored area*) usually were found nesting in colonies on sea cliffs. The glaucous gull nests throughout this region; it was observed both on cliffs and on level ground. Herring gulls (*broken line*), a ground-nesting species, were found with the others only on rocky islands. Before discovering all four gulls on the east coast of Baffin Island, the author studied some species on Southampton Island.

other hand, when I attempted to change the color of a gull's back by spraying it with paint, the feathers stuck together and the gull tried repeatedly to remove the paint.

In my first season, after observing the behavior of individual pairs of glaucous, Kumlien's and herring gulls in a colony in southern Baffin Island, I captured a small group of the gulls. The eye-ring of each one was changed to the color of a different species. Over the yellow ring of the glaucous gull, for example, I painted a ring of reddish purple. All the female birds had copulated with males before the experiment but none had laid eggs. When the females returned to their nests, they were accepted by their mates. In the days that followed, however, the males would no longer mount, in spite of intense solicitation by the females. In all cases where the female's eye-ring color had been changed the pair did not remain together. Five of the males whose mates had been painted formed pairs with nonaltered females in adjacent territories. Copulation ensued, and after two weeks all the new pairs had eggs. The females I had painted left the colony.

In contrast to these findings, changing the eye-ring of a mated male gull appeared not to affect a pair's behavior. The females accepted their altered mates and the males responded to the soliciting behavior of the females. In the one case where both individuals of a mated pair were changed the results were exactly the same as they were when only the female was changed.

The results looked promising. Although the number of individuals involved was small (33 females and 30 males) and some important controls were lacking, I now had a working hypothesis, namely that in some way the eye-ring color of the females functioned as a stimulus for mounting by their mates and that this reaction was keyed to differences among the species.

The program for the next two seasons was to repeat the eye-ring experiments with the necessary controls and to explore the hypothesis in greater detail. Was it the fleshy eye-ring alone or the entire eye that functioned as a stimulus? Was the important factor color or was it contrast? In answering these questions the critical species would be Thayer's gull, with its dark eye-ring and dark iris. It was also of prime importance to test the function of eye-ring color and other physical features with unmated gulls. There was reason to be-

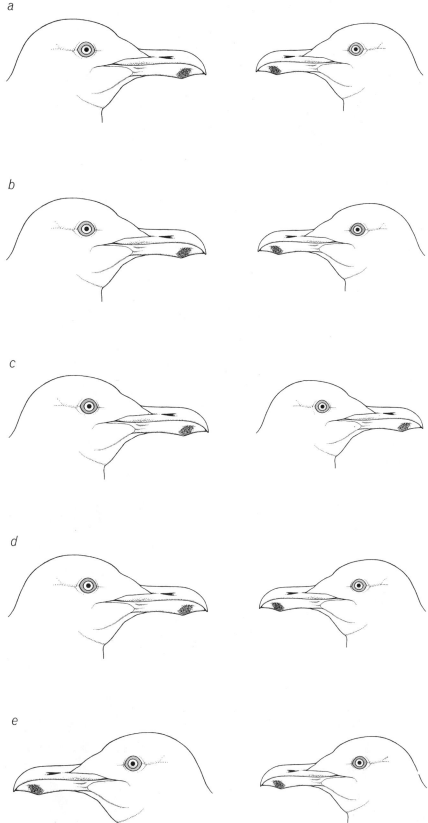

MATING BEHAVIOR OF GULLS changes when their appearance is artificially altered. Even when individuals of like species nest nearby, males and females of the same species normally mate (*a*). Such pairs still form even if an unmated female's eye-rings are painted to look like those of another species (*b*). Painted the same way, an unmated male fails to obtain a mate of his species (*c*). If the male is mated when his eye-rings are altered, his mate remains with him (*d*). When same change is made in the female, they usually separate (*e*).

lieve that the females choose the males, and it seemed unlikely that a mixed pair would form only to separate later because copulatory behavior was disrupted. There should also be an earlier isolating barrier.

In the experiments of these two seasons there were three major control groups: gulls that were drugged but not painted, gulls that were drugged and painted with their own color or pattern and gulls that were not captured but whose behavior was observed. My earlier findings with mated female gulls were confirmed in experiments that also shed light on the question of color v. contrast. A group of Kumlien's gulls was captured and their reddish-purple eye-ring was painted over either with a light color (yellow, orange or white) or a dark one (red or black). When the female gull had been painted with a light color, copulation usually stopped and the pair separated; in this regard an eye-ring painted white was the most effective. Dark colors had no significant effect. Exactly the reverse was true when the herring gull (orange eye-ring) and the glaucous (yellow) were painted in the same

way: dark colors inhibited copulation and light colors had no significant effect. Among the broken pairs were a number of female glaucous gulls whose yellow eye-ring had been changed to orange. This fitted the other results rather nicely, the orange eye-ring of the herring gull being darker than the yellow one of the glaucous gull.

When the same procedures were tried with Thayer's gull, however, there was no significant change in behavior. In this species it is not the eye-ring that stands out against the bird's white head, as it does in the other three species, but the entire orbital region—both the iris and the eye-ring are dark. This suggested that the orbital region as a whole functioned as a stimulus.

One could not paint the eye to change its color, but painting the reddish-purple eye-ring white reduces the contrast of the orbital region against the white head. After thus "erasing" the eye-ring I painted a larger one on the feathers around the eye. In making this "super-eye-ring" I used on various gulls the same assortment of colors as I had in the other experiments. One might think

HERRING GULL

EXTERNAL FEATURES vary among species. Each gull shown above differs from the

of the painted circle as the "eye-ring," the white feathers between it and the eye as the "iris" and the actual iris as the new "pupil." This may seem a bit far-fetched, and I do not mean to imply that this is what the gull sees, but the fact remains that in a significant number of cases where the female had been given a light-colored super-eye-ring copulation was inhibited and pairs separated. Apparently the stimulus to copulation was the contrast pattern of the ringed eye against the white head: dark color against white in Kumlien's and Thayer's gull and light color against white in the herring and glaucous gull.

In the course of these experiments I observed that a male occasionally mounted his altered mate but did not attempt copulation even when the female prodded his breast or rubbed her tail against his anal region. Earlier I had observed that copulation was invariably preceded by such tactile stimulation on the part of the females. I concluded that successful copulation probably involves both visual and tactile stimuli. (Auditory stimuli may also be involved, but this was not tested.) It appeared that the tactile stimuli were supplemental to other stimuli rather than independent of them; the eye-head contrast of the females played the major role.

Did the eye-head contrast play a role in the formation of pairs as well as in copulatory behavior? The eye-rings of a group of unmated female gulls were changed to determine if this made a significant difference in the species of the males with which they paired. The results showed no difference between this group and control groups. After a week or two, however, pairs in which the females had been given the "wrong" eye-head contrast separated; the males would not copulate. This of course further supported the role of eye-head

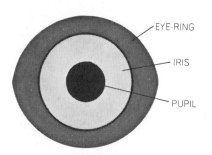

HERRING GULL

GLAUCOUS GULL

THAYER'S GULL

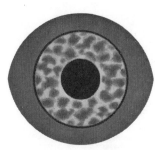

KUMLIEN'S GULL

RINGED EYES of various colors, photographed in black and white, exhibit different color values. The orange eye-ring of the herring gull shows darker than the yellow ring of the glaucous gull. Darker still is the reddish-purple ring of Thayer's and Kumlien's gulls. In Thayer's gulls brown irises enforce the contrast of the orbital region against the white head; the iris of Kumlien's is lighter. Eye-head contrast acts as an interspecies barrier among gulls.

GLAUCOUS GULL   THAYER'S GULL   KUMLIEN'S GULL

others in size, in the coloration of the orbital region, back and wings and in the pattern of the wing tips. The author's experiments suggest that the wing-tip pattern serves to supplement the signal of eye-head contrast in preventing the formation of mixed pairs.

contrast.

When the same experiment was performed with unmated male gulls, the results were quite different. In one instance 91 percent of the male Thayer's gulls that had been changed by the super-eye-ring technique to the light-eyed condition failed to obtain mates of their own species. An experiment with glaucous and herring gulls also showed that if the eye-head contrast of unmated males was "wrong," they were significantly less successful in obtaining mates of the same species than the controls were.

The results suggested that in pair formation it is indeed the females that choose the males, and that they select males with an eye-head contrast like their own. In other words, the same feature works in two ways to isolate the species: in the males it serves the purpose of pair formation and in females the purpose of copulatory behavior. What role is played by other external differences? The color of the mantle (back and wings) of the *Larus* gulls varies from one species to another. Among the four species I studied the differences in mantle coloration were not pronounced; still it seemed worthwhile to attempt an evaluation of mantle color as a possible signal. As I have indicated, however, spraying paint on a gull's back has too great an effect on the gull's behavior to make for a sound experiment, and the role played by this feature remains obscure.

Tests of the wing-tip pattern displayed at rest suggest that this feature does function as a signal in species discrimination during pair formation. There was no significant change in behavior after alteration of the wing tips of female gulls, whether mated or unmated. On the other hand, alteration of the wing tips of unmated males indicated that the wing-tip pattern functions as a stimulus to pair formation in combination with the eye-head contrast. This was shown by the fact that female gulls chose males with both "right" wing-tip pattern and eye-head contrast over males with only the "right" eye-head contrast. The wing-tip pattern alone is apparently not utilized in species discrimination during pair formation.

In several of the experiments male Thayer's gulls painted to appear light-eyed had been chosen by glaucous females. Since the females had the "wrong" eye-head contrast for the males, no copulation resulted and these mixed pairs did not remain together. After 59 Thayer's-glaucous pairs had formed I captured all but three of the glaucous females and altered them to the "right" contrast. Ten days later all 56 male Thayer's gulls had been observed to mount their altered mates, and about two weeks later 55 of the pairs had eggs. (One pair did not remain together.) Heavy ice on the rocks unfortunately forced me to abandon these colonies; I was never able to return to them. Before leaving I did collect several eggs, and they contained well-developed embryos. It may be that the mixed pairs produced hybrid offspring.

From the start of my experiments it had been clear that there was a strong correlation between the behavior that resulted from changing the eye-ring and the gonadal cycle of the gull. Two identical experiments, one performed 16 days before the first eggs were laid and the other 12 days later, yielded strikingly different results. I considered initially as a working hypothesis that the main component of the pair bond was the attachment of the individuals to each other, and that during and after the egg-laying period the main component of the bond became the attachment of the individuals to the nest and the eggs. This hypothesis could account for certain pairs of gulls that had remained together even though the males had failed to respond to the solicitations of their altered mates. It could not, however, explain instances in which males continued to mount their mates after they had been painted and before egg-laying had begun. Moreover, the hypothesis offered no answer to the crucial question of what the physiological basis for the male's behavior is.

The solution to the problem was found in the relation between the internal physiological state of the male (indicated by the weight of the testes) and the number of times a pair had copulated. All but 12 of the 168 pairs of gulls that had remained together after the female had been given a different eye-ring had copulated six or more times before she had been painted. This number of copulations could be correlated with a certain weight of the testes attained in the male's gonadal cycle. I concluded that a gull whose testes had developed to the critical weight or beyond it would respond to a mate whether or not her eye-ring had been changed. The most telling evidence was that if the female's eye-ring was changed at a time before the critical weight was reached, the testes of her mate did not increase in weight—in fact, they diminished!

It is fairly well substantiated that gonadal development in many species is stimulated by changes in the daily cycle of daylight and darkness. On arriving in the Arctic in summer gulls are subjected to periods of daylight lasting almost 24 hours. This factor alone, however, could not cause the gulls' testes to develop beyond the level attained at the end of pair formation. Certain other stimuli must interact with light, and one of them—probably the most important one

MATED FEMALES		PAIRS BROKEN	
	NUMBER PAINTED	NUMBER	PERCENT
HERRING AND GLAUCOUS GULLS			
BLACK OR RED	173	132	76.3
WHITE, YELLOW OR ORANGE	163	10	6.1
DRUG ONLY	71	5	7.0
NOT CAUGHT	93	2	2.1
THAYER'S AND KUMLIEN'S GULLS			
WHITE	389	222	57.0
PURPLE, BLACK OR RED	227	14	6.1
DRUG ONLY	134	6	4.4
NOT CAUGHT	204	12	5.8

SUMMARIZED FINDINGS document experiments with mated female gulls. Pairs separated in most cases where the female's eye contrast was changed; this feature appears to be a major stimulus to the male in copulatory behavior. Because a drug was used in capturing gulls, one control group was drugged and not painted. Another group, not captured, was observed. Some gulls were painted with eye-rings of their own color as a further control.

—is the presence of a mate with the proper eye-head contrast.

Although I have no evidence for it, it seems likely on logical grounds that a similar mechanism functions in females during pair formation. Once a pair bond is formed and a series of hormonal events is activated, inhibition of the female's gonadal development does not occur, even when the original stimulus—the eye-head contrast of the male—is removed. In the female, as in the male, the interaction of stimuli and the hormonal background at different times in the season provides a species-isolating mechanism that appears to be wholly effective.

Thus far we have been considering the two questions raised at the beginning: What are the visual factors involved in species discrimination among the gulls, and how do these factors affect reproductive behavior? At this point I should like to take up the matter of how the mechanisms that isolate species have evolved. In this regard it is instructive to examine the natural variation in iris and wing-tip color that occurs in one species: Kumlien's gull.

The eye-ring color differs little among Kumlien's gulls and Thayer's gulls but the amount of dark pigment in their irises varies considerably. Kumlien's gull is by far the more variable, ranging from individuals with completely dark irises to those with completely clear eyes of yellow. I divided this variation into six classes, Class 1 being the darkest and Class 6 the clearest. On the south coast of Baffin Island, Kumlien's gulls live together with herring gulls and glaucous gulls. In this locale Kumlien's gulls with clear irises were almost entirely absent; they occupied classes from Class 1 to Class 4 or Class 5. On the east coast of Baffin Island, where Kumlien's gulls nested with Thayer's gulls and glaucous gulls, the situation was reversed. There almost all the Kumlien's gulls fell into the last three classes, being clear-eyed or nearly so.

This pattern can be explained in terms of the natural selection of the variations that will reduce the possibility of mixed mating. According to my experiments, the contrast of the eye-ring and iris against the white head is the chief factor in species discrimination among gulls. To avoid mixed pairings, then, selection favored dark-eyed individuals where Kumlien's gulls nested with the light-eyed herring gulls and light-eyed individuals where Kumlien's gulls nested with the dark-eyed Thayer's gulls. Apparently the dark eye-ring of Kumlien's gull has been adequate for species recognition between Kumlien's gull and the yellow-eye-ring glaucous gull. The orange eye-ring of the herring gull affords a darker contrast, however, and where herring gulls and Kumlien's gulls nest together the dark iris of the latter reinforces the eye-head contrast. It is interesting to note that in Greenland, Kumlien's gulls have light eyes; there herring gulls are not found and glaucous gulls are.

The amount of dark pigment in the iris of Kumlien's gull is highly correlated with the amount of pigment in the wing tips. Individual Kumlien's gulls with light irises, as found on the east coast of Baffin Island, have white wing tips; those with dark eyes, as found on the south coast of the island, have dark blotches on their wing tips. It has been suggested that this variation in wing-tip pattern is the result of hybridization between Thayer's gulls and Kumlien's gulls, but that is not the case. The two species are most unlike each other where they nest together; they are very much like each other where they do not live together but where each is associated with glaucous gulls and herring gulls. The explanation for the variation of the wing tip is simply that it reflects the correlation between the pigment in the iris and the wing tip and the results of selection for differences in iris color in different populations.

In the course of my earlier work I had come to the conclusion that female gulls chose males that in eye-head contrast and wing-tip pattern were most like themselves. This created a problem, because it implied that the female knows what it looks like. A series of observations and one experiment on the east coast of Baffin Island provided an escape from this dilemma and also showed how responsive to very slight evolutionary pressures the visual isolating mechanisms are. The experiment was one in which I had hoped to induce mixed matings between Kumlien's gulls and Thayer's gulls by painting the eye-rings of unmated male Kumlien's gulls black to increase the contrast. The males were chosen not by Thayer's gull females, however, but by females of their own species. I concluded that other features, perhaps the wing-tip pattern, were the critical ones in discrimination between the two species.

Then further investigation of Kumlien's gulls in this area where they overlapped with Thayer's gulls revealed a curious phenomenon. Although the ma-

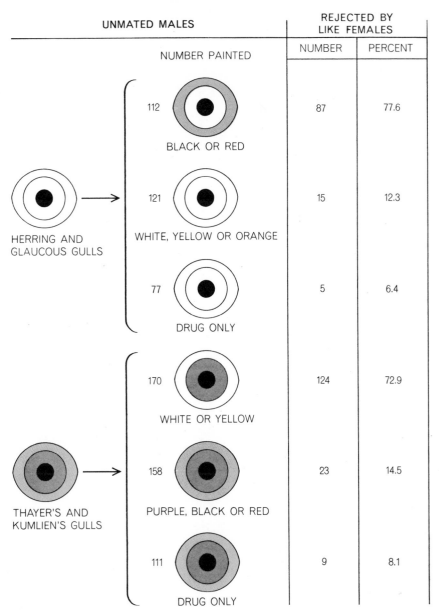

UNMATED MALES		REJECTED BY LIKE FEMALES	
NUMBER PAINTED		NUMBER	PERCENT
112 BLACK OR RED		87	77.6
121 WHITE, YELLOW OR ORANGE		15	12.3
77 DRUG ONLY		5	6.4
170 WHITE OR YELLOW		124	72.9
158 PURPLE, BLACK OR RED		23	14.5
111 DRUG ONLY		9	8.1

HERRING AND GLAUCOUS GULLS

THAYER'S AND KUMLIEN'S GULLS

MATED MALES		PAIRS BROKEN	
NUMBER PAINTED		NUMBER	PERCENT
164 DARK		4	.02
51 LIGHT		3	.06

HERRING AND GLAUCOUS GULLS

THAYER'S AND KUMLIEN'S GULLS

RESULTS of experiments in which the eye-ring of male gulls was altered indicate (*top*) that in pair formation the female chooses the male and that eye-head contrast is a factor in the choice. Changing the eye contrast of the male after a pair was formed (*bottom*) did not produce a change in mating behavior: nearly all the pairs of gulls remained together.

HEADS OF *LARUS* GULLS are almost identical except for the color of the eye and its encircling fleshy ring. At top on the opposite page are two Kumlien's gulls (*Larus glaucoides*), a species in which the iris varies from clear yellow to mottled brown. The eye-ring is reddish purple. Below appear two Thayer's gulls (*L. thayeri*). The eye-ring is the same as it is in Kumlien's gulls but the iris tends to be darker in this species. Next is a herring gull (*L. argentatus*), the only one that nests in the continental U.S. The glaucous gull (*L. hyperboreus*), shown last, has an eye-ring of yellow, which distinguishes it from the herring gull. Smallest gull is at top; largest at bottom. The painting was made by Guy Tudor.

jority of Kumlien's gulls in the area had clear yellow irises, there were many individuals (37 percent) with various amounts of iris pigmentation. If the gulls are viewed as two groups, one with iris pigmentation and one without it, a pattern emerges: in each group there is a striking preponderance of matings between individuals that look alike. Outside the overlap area mating is essentially random with respect to the presence or absence of iris pigmentation.

With this information in mind the results of the black-eye-ring experiment can be interpreted very differently. In this area dark-eyed males normally are chosen by dark-eyed females and clear-eyed males by clear-eyed females. Yet in the experiment even though 92 clear-eyed females were available for mating, they chose only four from the group of 41 clear-eyed males that had been painted to have a darker orbital region; 29 of these males were chosen by dark-eyed females. None of the 26 males with dark irises ringed in black were picked by light-eyed females. It can be seen that by increasing the eye-head contrast of unmated male Kumlien's gulls in this area, one can predict the iris coloration of the eventual mate.

Thus there was both observational and experimental evidence that the female Kumlien's gulls of this area were discerning very slight differences in the eye-head contrast of prospective mates. This mating system probably evolved as a result of two primary pressures. The presence of small numbers of herring gulls in the overlap zone probably provided pressure to maintain the delicate balance between clear-eyed and dark-eyed Kumlien's gulls. Secondly, in order to avoid mixed pairings with Thayer's gulls, selection favored individual Kumlien's gulls that perceived slight eye-head contrast differences. The mating system among Kumlien's gulls, in which like mates with like, is a by-product of such selection. This view is supported by the fact that increasing the eye-head contrast of unmated males just outside the overlap zone had no detectable effect, whereas the same alterations within the zone produced major changes in the mating system.

At first, of course, it is difficult to imagine how female gulls manage to choose mates that look like themselves. Presumably they do not actually see themselves. (Mirrors are rare in the Arctic.) The answer may nonetheless be quite simple. It is known that many birds "imprint" on their parents soon after birth, and that they choose mates that look like their parents. Possibly gulls do the same. If eye color in gulls is inherited (as seems likely, although genetic information is lacking), then female gulls choose mates that look like themselves simply because they are looking for mates that look like their parents, and in most cases they themselves look like their parents. This hypothesis suggests that the Kumlien's gull mating system may simply represent an intensification of the normal process, that is, a female chooses a male most like her parents in eye-head contrast and wing tips.

To understand the evolution of these visual signals that function as reproductive isolating mechanisms one can examine the distribution of the large *Larus* gulls throughout the Northern Hemisphere. With the exception of the glaucous gull, all the other *Larus* gulls that overlap with the herring gull and are reproductively isolated from it have the contrast pattern of a dark eye against a white head. The populations that apparently hybridize with herring gulls have dark eye-rings but light irises. As I have indicated, dark eye-rings without dark irises are insufficient as isolating mechanisms against the orange-eye-ring herring gulls. The important point here is that the darkening of the eye region (principally the eye-ring) begins to develop in an isolated population. If the population becomes genetically so different from the one from which it was separated that on coming together again the two populations remain distinct, selection will favor a further increase in the darkening of the eye region, specifically a darkening of the iris. The end result is a sealing off of gene exchange.

18

Electric Location by Fishes

H. W. LISSMANN
March 1963

It is well known that some fishes generate strong electric fields to stun their prey or discourage predators. Gymnarchus niloticus produces a weak field for the purpose of sensing its environment

Study of the ingenious adaptations displayed in the anatomy, physiology and behavior of animals leads to the familiar conclusion that each has evolved to suit life in its particular corner of the world. It is well to bear in mind, however, that each animal also inhabits a private subjective world that is not accessible to direct observation. This world is made up of information communicated to the creature from the outside in the form of messages picked up by its sense organs. No adaptation is more crucial to survival; the environment changes from place to place and from moment to moment, and the animal must respond appropriately in every place and at every moment. The sense organs transform energy of various kinds—heat and light, mechanical energy and chemical energy—into nerve impulses. Because the human organism is sensitive to the same kinds of energy, man can to some extent visualize the world as it appears to other living things. It helps in considering the behavior of a dog, for example, to realize that it can see less well than a man but can hear and smell better. There are limits to this procedure; ultimately the dog's sensory messages are projected onto its brain and are there evaluated differently.

Some animals present more serious obstacles to understanding. As I sit writing at my desk I face a large aquarium that contains an elegant fish about 20 inches long. It has no popular name but is known to science as *Gymnarchus niloticus*. This same fish has been facing me for the past 12 years, ever since I brought it from Africa. By observation and experiment I have tried to understand its behavior in response to stimuli from its environment. I am now convinced that *Gymnarchus* lives in a world totally alien to man: its most important

sense is an electric one, different from any we possess.

From time to time over the past century investigators have examined and dissected this curious animal. The literature describes its locomotive apparatus, central nervous system, skin and electric organs, its habitat and its family relation to the "elephant-trunk fishes," or mormyrids, of Africa. But the parts have not been fitted together into a functional pattern, comprehending the design of the animal as a whole and the history of its development. In this line of biological research one must resist the temptation to be deflected by details, to follow the fashion of putting the pieces too early under the electron microscope. The magnitude of a scientific revelation is not always paralleled by the degree of magnification employed. It is easier to select the points on which attention should be concentrated once the plan is understood. In the case of *Gymnarchus*, I think, this can now be attempted.

A casual observer is at once impressed by the grace with which *Gymnarchus* swims. It does not lash its tail from side to side, as most other fishes do, but keeps its spine straight. A beautiful undulating fin along its back propels its body through the water—forward or backward with equal ease. *Gymnarchus* can maintain its rigid posture even when turning, with complex wave forms running hither and thither over different regions of the dorsal fin at one and the same time.

Closer observation leaves no doubt that the movements are executed with great precision. When *Gymnarchus* darts after the small fish on which it feeds, it never bumps into the walls of its tank, and it clearly takes evasive action at some distance from obstacles placed in

its aquarium. Such maneuvers are not surprising in a fish swimming forward, but *Gymnarchus* performs them equally well swimming backward. As a matter of fact it should be handicapped even when it is moving forward: its rather degenerate eyes seem to react only to excessively bright light.

Still another unusual aspect of this fish and, it turns out, the key to all the puzzles it poses, is its tail, a slender, pointed process bare of any fin ("gymnarchus" means "naked tail"). The tail was first dissected by Michael Pius Erdl of the University of Munich in 1847. He found tissue resembling a small electric organ, consisting of four thin spindles running up each side to somewhere beyond the middle of the body. Electric organs constructed rather differently, once thought to be "pseudoelectric," are also found at the hind end of the related mormyrids.

Such small electric organs have been an enigma for a long time. Like the powerful electric organs of electric eels and some other fishes, they are derived from muscle tissue. Apparently in the course of evolution the tissue lost its power to contract and became specialized in various ways to produce electric discharges [see "Electric Fishes," by Harry Grundfest; Scientific American, October, 1960]. In the strongly electric fishes this adaptation serves to deter predators and to paralyze prey. But the powerful electric organs must have evolved from weak ones. The original swimming muscles would therefore seem to have possessed or have acquired at some stage a subsidiary electric function that had survival value. Until recently no one had found a function for weak electric organs. This was one of the questions on my mind when I began to study *Gymnarchus*.

I noticed quite early, when I placed a

ELECTRIC FISH *Gymnarchus niloticus*, from Africa, generates weak discharges that enable it to detect objects. In this sequence the fish catches a smaller fish. *Gymnarchus* takes its name, which means "naked tail," from the fact that its pointed tail has no fin.

new object in the aquarium of a well-established *Gymnarchus,* that the fish would approach it with some caution, making what appeared to be exploratory movements with the tip of its tail. It occurred to me that the supposed electric organ in the tail might be a detecting mechanism. Accordingly I put into the water a pair of electrodes, connected to an amplifier and an oscilloscope. The result was a surprise. I had expected to find sporadic discharges co-ordinated with the swimming or exploratory motions of the animal. Instead the apparatus recorded a continuous stream of electric discharges at a constant frequency of about 300 per second, waxing and waning in amplitude as the fish changed position in relation to the stationary electrodes. Even when the fish was completely motionless, the electric activity remained unchanged.

This was the first electric fish found to behave in such a manner. After a brief search I discovered two other kinds that emit an uninterrupted stream of weak discharges. One is a mormyrid relative of *Gymnarchus;* the other is a gymnotid, a small, fresh-water South American relative of the electric eel, belonging to a group of fish rather far removed from *Gymnarchus* and the mormyrids.

It had been known for some time that the electric eel generates not only strong discharges but also irregular series of weaker discharges. Various functions had been ascribed to these weak discharges of the eel. Christopher W. Coates, director of the New York Aquarium, had suggested that they might serve in navigation, postulating that the eel somehow measured the time delay between the output of a pulse and its reflection from an object. This idea was untenable on physical as well as physiological grounds. The eel does not, in the first place, produce electromagnetic waves; if it did, they would travel too fast to be timed at the close range at which such a mechanism might be useful, and in any case they would hardly penetrate water. Electric current, which the eel does produce, is not reflected from objects in the surrounding environment.

Observation of *Gymnarchus* suggested another mechanism. During each discharge the tip of its tail becomes momentarily negative with respect to the head. The electric current may thus be pictured as spreading out into the surrounding water in the pattern of lines that describes a dipole field [*see illustration on the next page*]. The exact configuration of this electric field depends on the conductivity of the water and on the distortions introduced in the field by objects with electrical conductivity different from that of the water. In a large volume of water containing no objects the field is symmetrical. When objects are present, the lines of current will converge on those that have better conductivity and diverge from the poor conductors [*see top illustration on page 184*]. Such objects alter the distribution of electric potential over the surface of the fish. If the fish could register these changes, it would have a means of detecting the objects.

Calculations showed that *Gymnarchus* would have to be much more sensitive electrically than any fish was known to be if this mechanism were to work. I had observed, however, that *Gymnarchus* was sensitive to extremely small external electrical disturbances. It responded violently when a small magnet or an electrified insulator (such as a comb that had just been drawn through a person's hair) was moved near the aquarium. The electric fields produced in the water by such objects must be very small indeed, in the range of fractions of a millionth of one volt per centimeter. This crude observation was enough to justify a series of experiments under more stringent conditions.

In the most significant of these experiments Kenneth E. Machin and I trained the fish to distinguish between objects that could be recognized only by an electric sense. These were enclosed in porous ceramic pots or tubes with thick walls. When they were soaked in water, the ceramic material alone had little effect on the shape of the electric field. The pots excluded the possibility of discrimination by vision or, because each test lasted only a short time, by a chemical sense such as taste or smell.

The fish quickly learned to choose between two pots when one contained aquarium water or tap water and the other paraffin wax (a nonconductor). After training, the fish came regularly to pick a piece of food from a thread suspended behind a pot filled with aquarium or tap water and ignored the pot filled with wax [*see bottom illustration on page 184*]. Without further conditioning it also avoided pots filled with air, with distilled water, with a close-fitting glass tube or with another nonconductor. On the other hand, when the electrical conductivity of the distilled water was matched to that of tap or aquarium water by the addition of salts or acids, the fish would go to the pot for food.

A more prolonged series of trials showed that *Gymnarchus* could distinguish mixtures in different proportions of tap water and distilled water and perform other remarkable feats of discrimination. The limits of this performance can best be illustrated by the fact that the fish could detect the presence of a glass rod two millimeters in diameter and would fail to respond to a glass rod .8 millimeter in diameter, each hidden in a

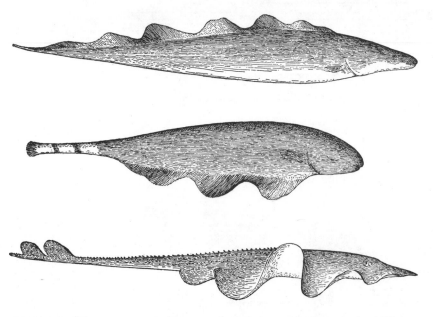

UNUSUAL FINS characterize *Gymnarchus* (*top*), a gymnotid from South America (*middle*) and sea-dwelling skate (*bottom*). All swim with spine rigid, probably in order to keep electric generating and detecting organs aligned. *Gymnarchus* is propelled by undulating dorsal fin, gymnotid by similar fin underneath and skate by lateral fins resembling wings.

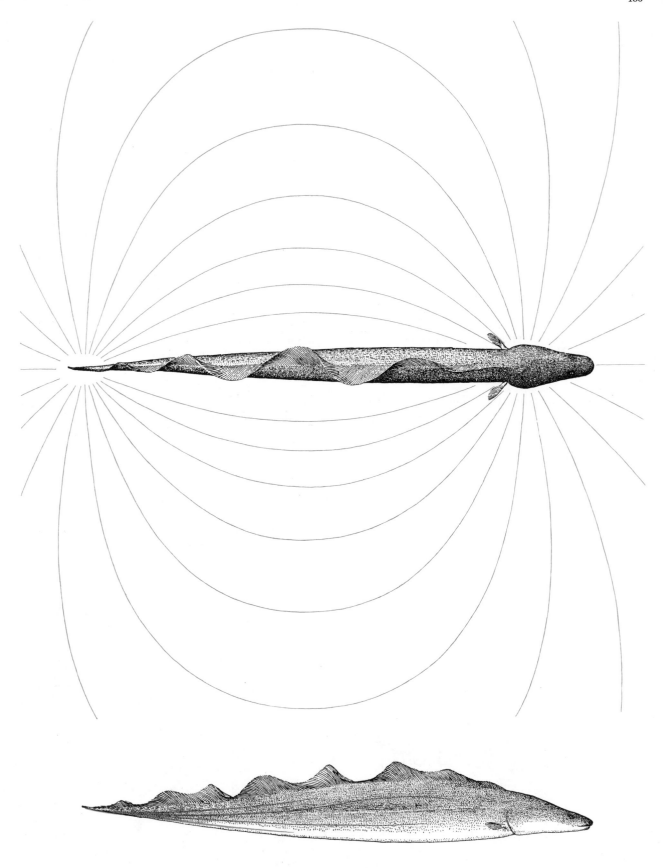

ELECTRIC FIELD of *Gymnarchus* and location of electric generating organs are diagramed. Each electric discharge from organs in rear portion of body (*color in side view*) makes tail negative with respect to head. Most of the electric sensory pores or organs are in head region. Undisturbed electric field resembles a dipole field, as shown, but is more complex. The fish responds to changes in the distribution of electric potential over the surface of its body. The conductivity of objects affects distribution of potential.

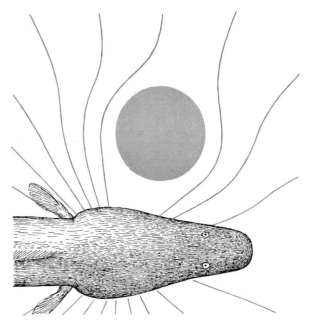
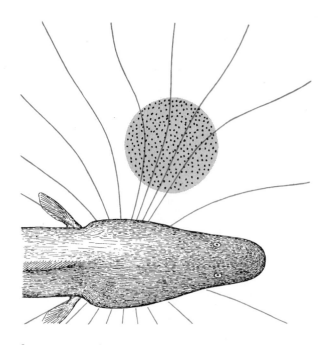

OBJECTS IN ELECTRIC FIELD of *Gymnarchus* distort the lines of current flow. The lines diverge from a poor conductor (*left*) and converge toward a good conductor (*right*). Sensory pores in the head region detect the effect and inform the fish about the object.

pot of the same dimensions. The threshold of its electric sense must lie somewhere between these two values.

These experiments seemed to establish beyond reasonable doubt that *Gymnarchus* detects objects by an electrical mechanism. The next step was to seek the possible channels through which the electrical information may reach the brain. It is generally accepted that the tissues and fluids of a fresh-water fish are relatively good electrical conductors enclosed in a skin that conducts poorly. The skin of *Gymnarchus* and of many mormyrids is exceptionally thick, with layers of platelike cells sometimes arrayed in a remarkable hexagonal pattern [*see top illustration on page 187*]. It can therefore be assumed that natural selection has provided these fishes with better-than-average exterior insulation.

In some places, particularly on and around the head, the skin is closely perforated. The pores lead into tubes often filled with a jelly-like substance or a loose aggregation of cells. If this jelly is a good electrical conductor, the arrangement would suggest that the lines of electric current from the water into the body of the fish are made to converge at these pores, as if focused by a lens. Each jelly-filled tube widens at the base into

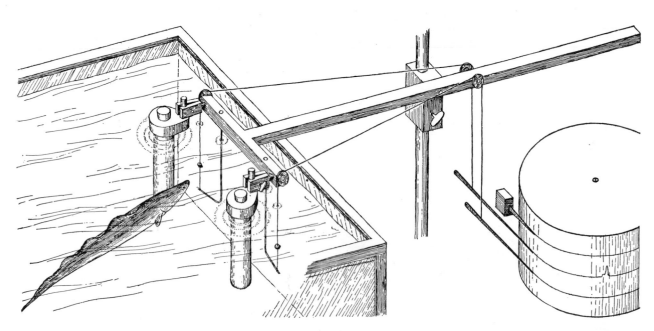

EXPERIMENTAL ARRANGEMENT for conditioned-reflex training of *Gymnarchus* includes two porous pots or tubes and recording mechanism. The fish learns to discriminate between objects of different electrical conductivity placed in the pots and to seek bait tied to string behind the pot holding the object that conducts best. *Gymnarchus* displays a remarkable ability to discriminate.

a small round capsule that contains a group of cells long known to histologists by such names as "multicellular glands," "mormyromasts" and "snout organs." These, I believe, are the electric sense organs.

The supporting evidence appears fairly strong: The structures in the capsule at the base of a tube receive sensory nerve fibers that unite to form the stoutest of all the nerves leading into the brain. Electrical recording of the impulse traffic in such nerves has shown that they lead away from organs highly sensitive to electric stimuli. The brain centers into which these nerves run are remarkably large and complex in *Gymnarchus*, and in some mormyrids they completely cover the remaining portions of the brain [*see illustration on next page*].

If this evidence for the plan as well as the existence of an electric sense does not seem sufficiently persuasive, corroboration is supplied by other weakly electric fishes. Except for the electric eel, all species of gymnotids investigated so far emit continuous electric pulses. They are also highly sensitive to electric fields. Dissection of these fishes reveals the expected histological counterparts of the structures found in the mormyrids: similar sense organs embedded in a similar skin, and the corresponding regions of the brain much enlarged.

Skates also have a weak electric organ in the tail. They are cartilaginous fishes, not bony fishes, or teleosts, as are the mormyrids and gymnotids. This means that they are far removed on the family line. Moreover, they live in the sea, which conducts electricity much better than fresh water does. It is almost too much to expect structural resemblances to the fresh-water bony fishes, or an electrical mechanism operating along similar lines. Yet skates possess sense organs, known as the ampullae of Lorenzini, that consist of long jelly-filled tubes opening to the water at one end and terminating in a sensory vesicle at the other. Recently Richard W. Murray of the University of Birmingham has found that these organs respond to very delicate electrical stimulation. Unfortunately, either skates are rather uncooperative animals or we have not mastered the trick of training them; we have been unable to repeat with them the experiments in discrimination in which *Gymnarchus* performs so well.

Gymnarchus, the gymnotids and skates all share one obvious feature: they swim in an unusual way. *Gymnarchus* swims with the aid of a fin on its back; the gymnotids have a similar fin on their

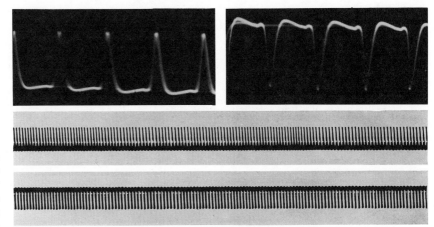

ELECTRIC DISCHARGES of *Gymnarchus* show reversal of polarity when detecting electrodes are rotated 180 degrees (*enlarged records at top*). The discharges, at rate of 300 per second, are remarkably regular even when fish is resting, as seen in lower records.

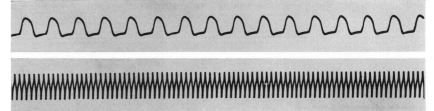

DISCHARGE RATES DIFFER in different species of gymnotids. *Sternopygus macrurus* (*upper record*) has rate of 55 per second; *Eigenmannia virescens* (*lower*), 300 per second.

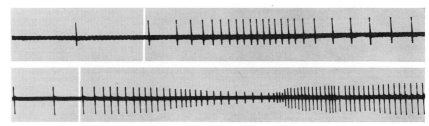

VARIABLE DISCHARGE RATE is seen in some species. Tap on tank (*white line in upper record*) caused mormyrid to increase rate. Tap on fish (*lower record*) had greater effect.

underside; skates swim with pectoral fins stuck out sideways like wings [*see illustration on page 182*]. They all keep the spine rigid as they move. It would be rash to suggest that such deviations from the basic fish plan could be attributed to an accident of nature. In biology it always seems safer to assume that any redesign has arisen for some reason, even if the reason obstinately eludes the investigator. Since few fishes swim in this way or have electric organs, and since the fishes that combine these features are not related, a mere coincidence would appear most unlikely.

A good reason for the rigid swimming posture emerged when we built a model to simulate the discharge mecha-

nism and the sensory-perception system. We placed a pair of electrodes in a large tank of water; to represent the electric organ they were made to emit repetitive electric pulses. A second pair of electrodes, representing the electric sense organ, was placed some distance away to pick up the pulses. We rotated the second pair of electrodes until they were on a line of equipotential, where they ceased to record signals from the sending electrodes. With all the electrodes clamped in this position, we showed that the introduction of either a conductor or a nonconductor into the electric field could cause sufficient distortion of the field for the signals to reappear in the detectors.

In a prolonged series of readings the

slightest displacement of either pair of electrodes would produce great variations in the received signal. These could be smoothed to some extent by recording not the change of potential but the change in the potential gradient over the "surface" of our model fish. It is probable that the real fish uses this principle, but to make it work the electrode system must be kept more or less constantly aligned. Even though a few cubic centimeters of fish brain may in some respects put many electronic computers in the shade, the fish brain might be unable to obtain any sensible information if the fish's electrodes were to be misaligned by the tail-thrashing that propels an ordinary fish. A mode of swimming that keeps the electric field symmetrical with respect to the body most of the time would therefore offer obvious advantages. It seems logical to assume that *Gymnarchus*, or its ancestors, acquired the rigid mode of swimming along with the electric sensory apparatus and subsequently lost the broad, oarlike tail fin.

Our experiments with models also showed that objects could be detected only at a relatively short distance, in spite of high amplification in the receiving system. As an object was moved farther and farther away, a point was soon reached where the signals arriving at the oscilloscope became submerged in the general "noise" inherent in every detector system. Now, it is known that minute amounts of energy can stimulate a sense organ: one quantum of light registers on a visual sense cell; vibrations of subatomic dimensions excite the ear; a single molecule in a chemical sense organ can produce a sensation, and so on. Just how such small external signals can be picked out from the general noise in and around a metabolizing cell represents one of the central questions of sensory physiology. Considered in connection with the electric sense of fishes, this question is complicated further by the high frequency of the discharges from the electric organ that excite the sensory apparatus.

In general, a stimulus from the environment acting on a sense organ produces a sequence of repetitive impulses in the sensory nerve. A decrease in the strength of the stimulus causes a lower frequency of impulses in the nerve. Conversely, as the stimulus grows stronger, the frequency of impulses rises, up to a certain limit. This limit may vary from one sense organ to another, but 500 impulses per second is a common upper limit, although 1,000 per second have been recorded over brief intervals.

In the case of the electric sense organ of a fish the stimulus energy is provided by the discharges of the animal's electric organ. *Gymnarchus* discharges at the rate of 300 pulses per second. A change in the amplitude—not the rate—of these pulses, caused by the presence of an object in the field, constitutes the effective stimulus at the sense organ. Assuming that the reception of a single discharge of small amplitude excites one impulse in a sensory nerve, a discharge of larger amplitude that excited two impulses would probably reach and exceed the upper limit at which the nerve can generate impulses, since the nerve would now be firing 600 times a second (twice the rate of discharge of the electric organ). This would leave no room

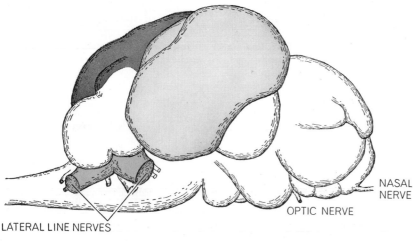

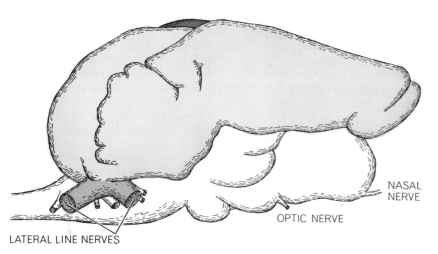

BRAIN AND NERVE ADAPTATIONS of electric fish are readily apparent. Brain of typical nonelectric fish (*top*) has prominent cerebellum (*gray*). Regions associated with electric sense (*color*) are quite large in *Gymnarchus* (*middle*) and even larger in the mormyrid (*bottom*). Lateral-line nerves of electric fishes are larger, nerves of nose and eyes smaller.

to convey information about gradual changes in the amplitude of incoming stimuli. Moreover, the electric organs of some gymnotids discharge at a much higher rate; 1,600 impulses per second have been recorded. It therefore appears unlikely that each individual discharge is communicated to the sense organs as a discrete stimulus.

We also hit on the alternative idea that the frequency of impulses from the sensory nerve might be determined by the mean value of electric current transmitted to the sense organ over a unit of time; in other words, that the significant messages from the environment are averaged out and so discriminated from the background of noise. We tested this idea on *Gymnarchus* by applying trains of rectangular electric pulses of varying voltage, duration and frequency across the aquarium. Again using the conditioned-reflex technique, we determined the threshold of perception for the different pulse trains. We found that the fish is in fact as sensitive to high-frequency pulses of short duration as it is to low-frequency pulses of identical voltage but correspondingly longer duration. For any given pulse train, reduction in voltage could be compensated either by an increase in frequency of stimulus or an increase in the duration of the pulse. Conversely, reduction in the frequency required an increase in the voltage or in the duration of the pulse to reach the threshold. The threshold would therefore appear to be determined by the product of voltage times duration times frequency.

Since the frequency and the duration of discharges are fixed by the output of the electric organ, the critical variable at the sensory organ is voltage. Threshold determinations of the fish's response to single pulses, compared with quantitative data on its response to trains of pulses, made it possible to calculate the time over which the fish averages out the necessarily blurred information carried within a single discharge of its own. This time proved to be 25 milliseconds, sufficient for the electric organ to emit seven or eight discharges.

The averaging out of information in this manner is a familiar technique for improving the signal-to-noise ratio; it has been found useful in various branches of technology for dealing with barely perceptible signals. In view of the very low signal energy that *Gymnarchus* can detect, such refinements in information processing, including the ability to average out information picked up by a large number of separate sense organs,

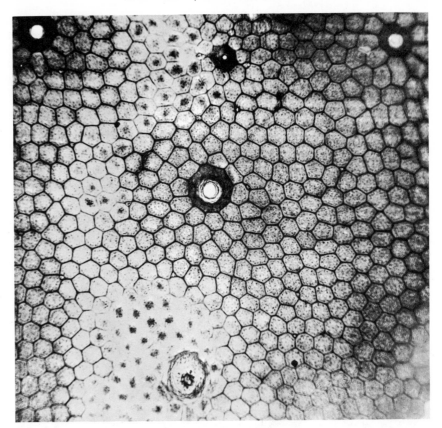

SKIN OF MORMYRID is made up of many layers of platelike cells having remarkable hexagonal structure. The pores contain tubes leading to electric sense organs. This photomicrograph by the author shows a horizontal section through the skin, enlarged 100 diameters.

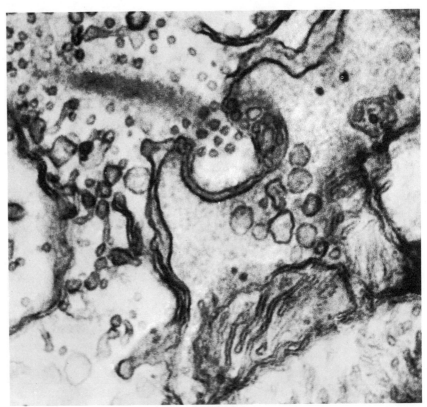

MEETING POINT of electric sensory cell (*left*) and its nerve (*right*) is enlarged 120,000 diameters in this electron micrograph by the author and Ann M. Mullinger. Bulge of sensory cell into nerve ending displays the characteristic dense streak surrounded by vesicles.

appear to be essential. We have found that *Gymnarchus* can respond to a continuous direct-current electric stimulus of about .15 microvolt per centimeter, a value that agrees reasonably well with the calculated sensitivity required to recognize a glass rod two millimeters in diameter. This means that an individual sense organ should be able to convey information about a current change as small as .003 micromicroampere. Extended over the integration time of 25 milliseconds, this tiny current corresponds to a movement of some 1,000 univalent, or singly charged, ions.

The intimate mechanism of the single sensory cell of these organs is still a complete mystery. In structure the sense organs differ somewhat from species to species and different types are also found in an individual fish. The fine structure of the sensory cells, their nerves and associated elements, which Ann M. Mullinger and I have studied with both the light microscope and the electron microscope, shows many interesting details. Along specialized areas of the boundary between the sensory cell and the nerve fiber there are sites of intimate contact where the sensory cell bulges into the fiber. A dense streak extends from the cell into this bulge, and the vesicles alongside it seem to penetrate the intercellular space. The integrating system of the sensory cell may be here.

These findings, however, apply only to *Gymnarchus* and to about half of the species of gymnotids investigated to date. The electric organs of these fishes emit pulses of constant frequency. In the other gymnotids and all the mormyrids the discharge frequency changes with the state of excitation of the fish. There is therefore no constant mean value of current transmitted in a unit of time; the integration of information in these species may perhaps be carried out in the brain. Nevertheless, it is interesting that both types of sensory system should have evolved independently in the two different families, one in Africa and one in South America.

The experiments with *Gymnarchus*, which indicate that no information is carried by the pulse nature of the discharges, leave us with a still unsolved problem. If the pulses are "smoothed out," it is difficult to see how any one fish can receive information in its own frequency range without interference from its neighbors. In this connection Akira Watanabe and Kimihisa Takeda at the University of Tokyo have made the potentially significant finding that the gymnotids respond to electric oscillations close in frequency to their own by shifting their frequency away from the applied frequency. Two fish might thus react to each other's presence.

For reasons that are perhaps associated with the evolutionary origin of their electric sense, the electric fishes are elusive subjects for study in the field. I have visited Africa and South America in order to observe them in their natural habitat. Although some respectable specimens were caught, it was only on rare occasions that I actually saw a *Gymnarchus*, a mormyrid or a gymnotid in the turbid waters in which they live. While such waters must have favored the evolution of an electric sense, it could not have been the only factor. The same waters contain a large number of

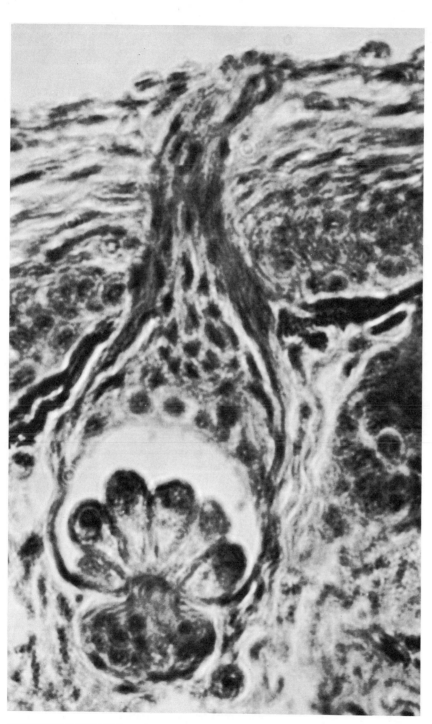

VERTICAL SECTION through skin and electric sense organ of a gymnotid shows tube containing jelly-like substance widening at base into a capsule, known as multicellular gland, that holds a group of special cells. Enlargement of this photomicrograph is 1,000 diameters.

other fishes that apparently have no electric organs.

Although electric fishes cannot be seen in their natural habitat, it is still possible to detect and follow them by picking up their discharges from the water. In South America I have found that the gymnotids are all active during the night. Darkness and the turbidity of the water offer good protection to these fishes, which rely on their eyes only for the knowledge that it is day or night. At night most of the predatory fishes, which have well-developed eyes, sleep on the bottom of rivers, ponds and lakes. Early in the morning, before the predators wake up, the gymnotids return from their nightly excursions and occupy inaccessible hiding places, where they often collect in vast numbers. In the rocks and vegetation along the shore the ticking, rattling, humming and whistling can be heard in bewildering profusion when the electrodes are connected to a loudspeaker. With a little practice one can begin to distinguish the various species by these sounds.

When one observes life in this highly competitive environment, it becomes clear what advantages the electric sense confers on these fishes and why they have evolved their curiously specialized sense organs, skin, brain, electric organs and peculiar mode of swimming. Such well-established specialists must have originated, however, from ordinary fishes in which the characteristics of the specialists are found in their primitive state: the electric organs as locomotive muscles and the sense organs as mechanoreceptors along the lateral line of the body that signal displacement of water. Somewhere there must be intermediate forms in which the contraction of a muscle, with its accompanying change in electric potential, interacts with these sense organs. For survival it may be important to be able to distinguish water movements caused by animate or inanimate objects. This may have started the evolutionary trend toward an electric sense.

Already we know some supposedly nonelectric fishes from which, nevertheless, we can pick up signals having many characteristics of the discharges of electric fishes. We know of sense organs that appear to be structurally intermediate between ordinary lateral-line receptors and electroreceptors. Furthermore, fishes that have both of these characteristics are also electrically very sensitive. We may hope one day to piece the whole evolutionary line together and express, at least in physical terms, what it is like to live in an electric world.

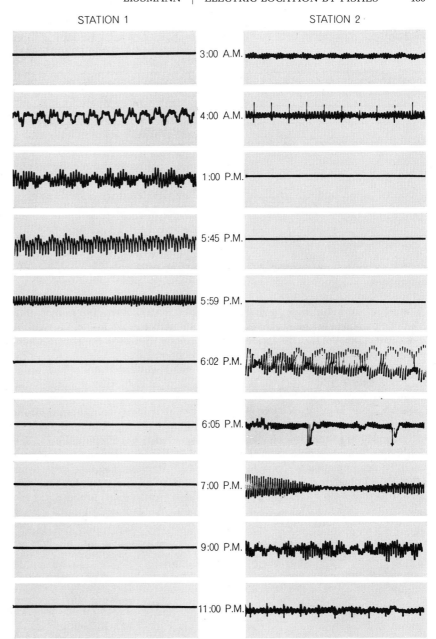

STATION 1   STATION 2

3:00 A.M.
4:00 A.M.
1:00 P.M.
5:45 P.M.
5:59 P.M.
6:02 P.M.
6:05 P.M.
7:00 P.M.
9:00 P.M.
11:00 P.M.

TRACKING ELECTRIC FISH in nature involves placing electrodes in water they inhabit. Records at left were made in South American stream near daytime hiding place of gymnotids, those at right out in main channel of stream, where they seek food at night.

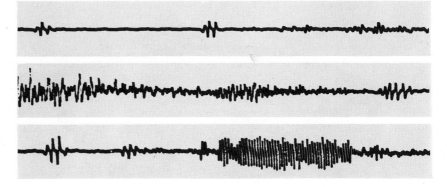

AFRICAN CATFISH, supposedly nonelectric, produced the discharges shown here. Normal action potentials of muscles are seen, along with odd regular blips and still other oscillations of higher frequency. Such fish may be evolving an electric sense or may already have one.

19

The Homing Salmon

by Arthur D. Hasler and James A. Larsen
August 1955

How do salmon find their way back to the waters of their birth? Recent experiments in the laboratory and in the field indicate that they do so by means of a remarkably refined sense of smell

A learned naturalist once remarked that among the many riddles of nature, not the least mysterious is the migration of fishes. The homing of salmon is a particularly dramatic example. The Chinook salmon of the U. S. Northwest is born in a small stream, migrates downriver to the Pacific Ocean as a young smolt and, after living in the sea for as long as five years, swims back unerringly to the stream of its birth to spawn. Its determination to return to its birthplace is legendary. No one who has seen a 100-pound Chinook salmon fling itself into the air again and again until it is exhausted in a vain effort to surmount a waterfall can fail to marvel at the strength of the instinct that draws the salmon upriver to the stream where it was born.

How do salmon remember their birthplace, and how do they find their way back, sometimes from 800 or 900 miles away? This enigma, which has fascinated naturalists for many years, is the subject of the research to be reported here. The question has an economic as well as a scientific interest, because new dams which stand in the salmon's way have cut heavily into salmon fishing along the Pacific Coast. Before long nearly every stream of any appreciable size in the West will be blocked by dams. It is true that the dams have fish lifts and ladders designed to help salmon to hurdle them. Unfortunately, and for reasons which are different for nearly every dam so far designed, salmon are lost in tremendous numbers.

There are six common species of salmon. One, called the Atlantic salmon, is of the same genus as the steelhead trout. These two fish go to sea and come back upstream to spawn year after year. The other five salmon species, all on the Pacific Coast, are the Chinook (also called the king salmon), the sockeye, the silver, the humpback and the chum. The

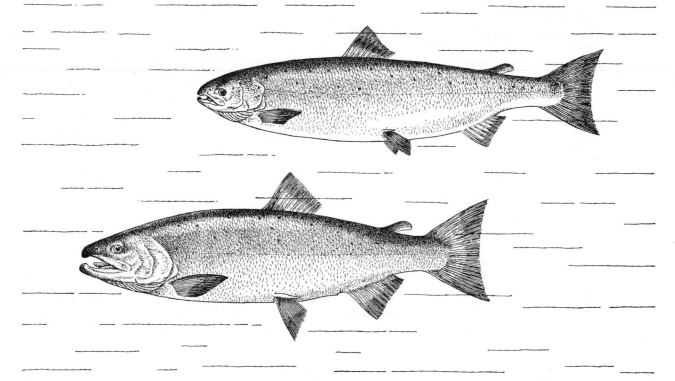

TWO COMMON SPECIES of salmon are (*top*) the Atlantic salmon (*Salmo salar*) and (*bottom*) the silver salmon (*Oncorhynchus kisutch*). The Atlantic salmon goes upstream to spawn year after year; the silver salmon, like other Pacific species, spawns only once.

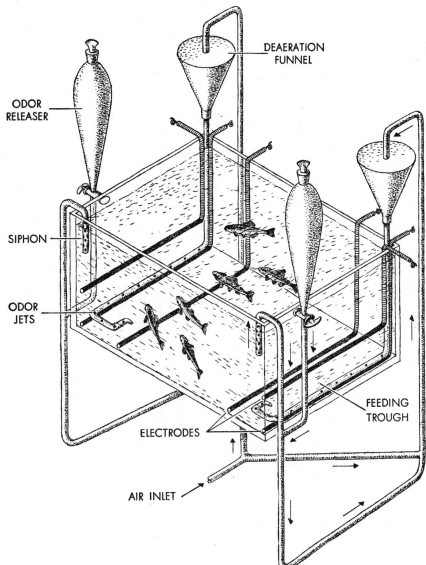

Pacific salmon home only once: after spawning they die.

A young salmon first sees the light of day when it hatches and wriggles up through the pebbles of the stream where the egg was laid and fertilized. For a few weeks the fingerling feeds on insects and small aquatic animals. Then it answers its first migratory call and swims downstream to the sea. It must survive many hazards to mature: an estimated 15 per cent of the young salmon are lost at every large dam, such as Bonneville, on the downstream trip; others die in polluted streams; many are swallowed up by bigger fish in the ocean. When, after several years in the sea, the salmon is ready to spawn, it responds to the second great migratory call. It finds the mouth of the river by which it entered the ocean and then swims steadily upstream, unerringly choosing the correct turn at each tributary fork, until it arrives at the stream where it was hatched. Generation after generation, families of salmon return to the same rivulet so consistently that populations in streams not far apart follow distinctly separate lines of evolution.

The homing behavior of the salmon has been convincingly documented by many studies since the turn of the century. One of the most elaborate was made by Andrew L. Pritchard, Wilbert A. Clemens and Russell E. Foerster in Canada. They marked 469,326 young sockeye salmon born in a tributary of the Fraser River, and they recovered nearly 11,000 of these in the same parent stream after the fishes' migration to the ocean and back. What is more, not one of the marked fish was ever found to have strayed to another stream. This remarkable demonstration of the salmon's precision in homing has presented an exciting challenge to investigators.

At the Wisconsin Lake Laboratory during the past decade we have been studying the sense of smell in fish, beginning with minnows and going on to salmon. Our findings suggest that the salmon identifies the stream of its birth by odor and literally smells its way home from the sea.

Fish have an extremely sensitive sense of smell. This has often been observed by students of fish behavior. Karl von Frisch showed that odors from the injured skin of a fish produce a fright reaction among its schoolmates. He once noticed that when a bird dropped an injured fish in the water, the school of fish from which it had been seized quickly dispersed and later avoided the area.

EXPERIMENTAL TANK was built in the Wisconsin Lake Laboratory to train fish to discriminate between two odors. In this isometric drawing the vessel at the left above the tank contains water of one odor. The vessel at the right contains water of another odor. When the valve below one of the vessels was opened, the water in it was mixed with water siphoned out of the tank. The mixed water was then pumped into the tank by air. When the fish (minnows or salmon) moved toward one of the odors, they were rewarded with food. When they moved toward the other odor, they were punished with a mild electric shock from the electrodes mounted inside the tank. Each of the fish was blinded to make sure that it would not associate reward and punishment with the movements of the experimenters.

It is well known that sharks and tuna are drawn to a vessel by the odor of bait in the water. Indeed, the time-honored custom of spitting on bait may be founded on something more than superstition; laboratory studies have proved that human saliva is quite stimulating to the taste buds of a bullhead. The sense of taste of course is closely allied to the sense of smell. The bullhead has taste buds all over the surface of its body; they are especially numerous on its whiskers. It will quickly grab for a piece of meat that touches any part of its skin. But it becomes insensitive to taste and will not respond in this way if a nerve serving the skin buds is cut.

The smelling organs of fish have evolved in a great variety of forms. In the bony fishes the nose pits have two separate openings. The fish takes water into the front opening as it swims or breathes (sometimes assisting the intake with cilia), and then the water passes out through the second opening, which may be opened and closed rhythmically by the fish's breathing. Any odorous substances in the water stimulate the nasal receptors chemically, perhaps by an effect on enzyme reactions, and the re-

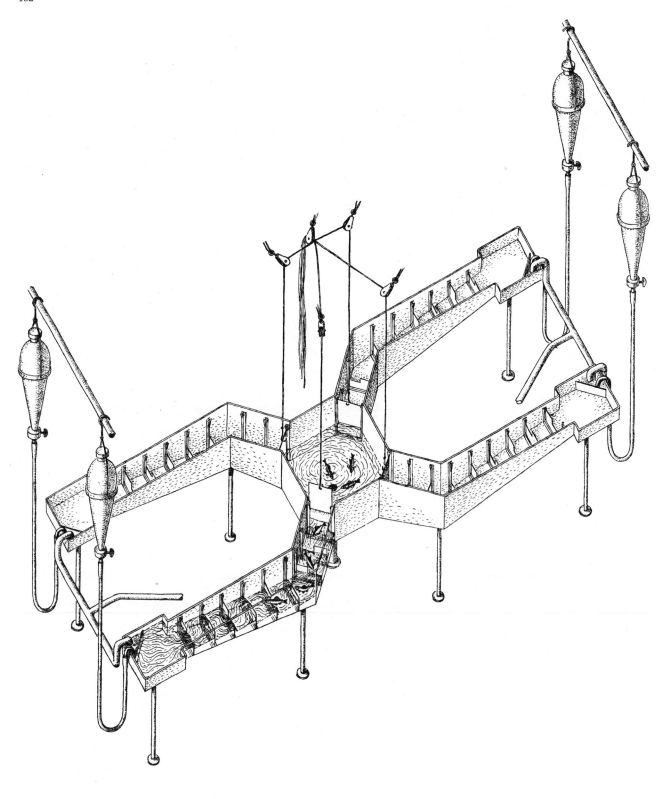

FOUR RUNWAYS are used to test the reaction of untrained salmon fingerlings to various odors. Water is introduced at the outer end of each runway and flows down a series of steps into a central compartment, where it drains. In the runway at the lower left the water cascades down to the central compartment in a series of miniature waterfalls; in the other runways the water is omitted to show the construction of the apparatus. Odors may be introduced into the apparatus from the vessels suspended above the runways. In an experiment salmon fingerlings are placed in the central compartment and an odor is introduced into one of the runways. When the four doors to the central compartment are opened, the fingerlings tend to enter the arms, proceeding upstream by jumping the waterfalls. Whether an odor attracts them, repels or has no effect is judged by the observed distribution of the fish in the runways.

sulting electrical impulses are relayed to the central nervous system by the olfactory nerve.

The human nose, and that of other land vertebrates, can smell a substance only if it is volatile and soluble in fat solvents. But in the final analysis smell is always aquatic, for a substance is not smelled until it passes into solution in the mucous film of the nasal passages. For fishes, of course, the odors are already in solution in their watery environment. Like any other animal, they can follow an odor to its source, as a hunting dog follows the scent of an animal. The quality or effect of a scent changes as the concentration changes; everyone knows that an odor may be pleasant at one concentration and unpleasant at another.

When we began our experiments, we first undertook to find out whether fish could distinguish the odors of different water plants. We used a specially developed aquarium with jets which could inject odors into the water. For responding to one odor (by moving toward the jet), the fish were rewarded with food; for responding to another odor, they were punished with a mild electric shock. After the fish were trained to make choices between odors, they were tested on dilute rinses from 14 different aquatic plants. They proved able to distinguish the odors of all these plants from one another.

Plants must play an important role in the life of many freshwater fish. Their odors may guide fish to feeding grounds when visibility is poor, as in muddy water or at night, and they may hold young fish from straying from protective cover. Odors may also warn fish away from poisons. In fact, we discovered that fish could be put to use to assay industrial pollutants: our trained minnows were able to detect phenol, a common pollutant, at concentrations far below those detectable by man.

All this suggested a clear-cut working hypothesis for investigating the mystery of the homing of salmon. We can suppose that every little stream has its own characteristic odor, which stays the same year after year; that young salmon become conditioned to this odor before they go to sea; that they remember the odor as they grow to maturity, and that they are able to find it and follow it to its source when they come back upstream to spawn.

Plainly there are quite a few ifs in this theory. The first one we tested was the question: Does each stream have its own odor? We took water from two creeks in Wisconsin and investigated whether fish

could learn to discriminate between them. Our subjects, first minnows and then salmon, were indeed able to detect a difference. If, however, we destroyed a fish's nose tissue, it was no longer able to distinguish between the two water samples.

Chemical analysis indicated that the only major difference between the two waters lay in the organic material. By testing the fish with various fractions of the water separated by distillation, we confirmed that the identifying material was some volatile organic substance.

The idea that fish are guided by odors in their migrations was further supported by a field test. From each of two different branches of the Issaquah River in the State of Washington we took a number of sexually ripe silver salmon which had come home to spawn. We then plugged with cotton the noses of half the fish in each group and placed all the salmon in the river below the fork to make the upstream run again. Most of the fish with unplugged noses swam back to the stream they had selected the first time. But the "odor-blinded" fish migrated back in random fashion, picking the wrong stream as often as the right one.

In 1949 eggs from salmon of the Horsefly River in British Columbia were hatched and reared in a hatchery in a tributary called the Little Horsefly. Then they were flown a considerable distance and released in the main Horsefly River, from which they migrated to the sea. Three years later 13 of them had returned to their rearing place in the Little Horsefly, according to the report of the Canadian experimenters.

In our own laboratory experiments we tested the memory of fish for odors and found that they retained the ability to differentiate between odors for a long period after their training. Young fish remembered odors better than the old. That animals "remember" conditioning to which they have been exposed in their youth, and act accordingly, has been demonstrated in other fields. For instance, there is a fly which normally lays its eggs on the larvae of the flour moth, where the fly larvae then hatch and develop. But if larvae of this fly are raised on another host, the beeswax moth, when the flies mature they will seek out beeswax moth larvae on which to lay their eggs, in preference to the traditional host.

With respect to the homing of salmon we have shown, then, that different streams have different odors, that salmon respond to these odors and that they remember odors to which they have been

conditioned. The next question is: Is a salmon's homeward migration guided solely by its sense of smell? If we could decoy homing salmon to a stream other than their birthplace, by means of an odor to which they were conditioned artificially, we might have not only a solution to the riddle that has puzzled scientists but also a practical means of saving the salmon—guiding them to breeding streams not obstructed by dams.

We set out to find a suitable substance to which salmon could be conditioned. A student, W. J. Wisby, and I [Arthur Hasler] designed an apparatus to test the reactions of salmon to various organic odors. It consists of a compartment from which radiate four runways, each with several steps which the fish must jump to climb the runway. Water cascades down each of the arms. An odorous substance is introduced into one of the arms, and its effect on the fish is judged by whether the odor appears to attract fish into that arm, to repel them or to be indifferent to them.

We needed a substance which initially would not be either attractive or repellent to salmon but to which they could be conditioned so that it would attract them. After testing several score organic odors, we found that dilute solutions of morpholine neither attracted nor repelled salmon but were detectable by them in extremely low concentrations—as low as one part per million. It appears that morpholine fits the requirements for the substance needed: it is soluble in water; it is detectable in extremely low concentrations; it is chemically stable under stream conditions. It is neither an attractant nor a repellent to unconditioned salmon, and would have meaning only to those conditioned to it.

Federal collaborators of ours are now conducting field tests on the Pacific Coast to learn whether salmon fry and fingerlings which have been conditioned to morpholine can be decoyed to a stream other than that of their birth when they return from the sea to spawn. Unfortunately this type of experiment may not be decisive. If the salmon are not decoyed to the new stream, it may simply mean that they cannot be drawn by a single substance but will react only to a combination of subtle odors in their parent stream. Perhaps adding morpholine to the water is like adding the whistle of a freight train to the quiet strains of a violin, cello and flute. The salmon may still seek out the subtle harmonies of an odor combination to which they have been reacting by instinct for centuries. But there is still hope that they may respond to the call of the whistle.

20

The Mystery of Pigeon Homing

by William T. Keeton
December 1974

Recent findings have upset previous explanations of how pigeons find their way home from distant locations. It appears that they have more than one compass system for determining direction

How does a homing pigeon find its way back to its home loft from hundreds of miles away? The answer does not lie in visible landmarks; pigeons taken in covered cages to areas they have never seen before have little trouble finding their way home. Nor does it lie entirely in the bird's ability to determine compass directions from the sun or the earth's magnetic field. Even when a pigeon can determine compass directions, how can it know which way home is? Although the homing prowess of the pigeon has long engaged the curiosity of man, the full story of how the bird navigates still remains a mystery. Nonetheless much has been learned about the pigeon's navigational abilities in the past two decades, particularly in the past six years.

The modern homing pigeon, a de-

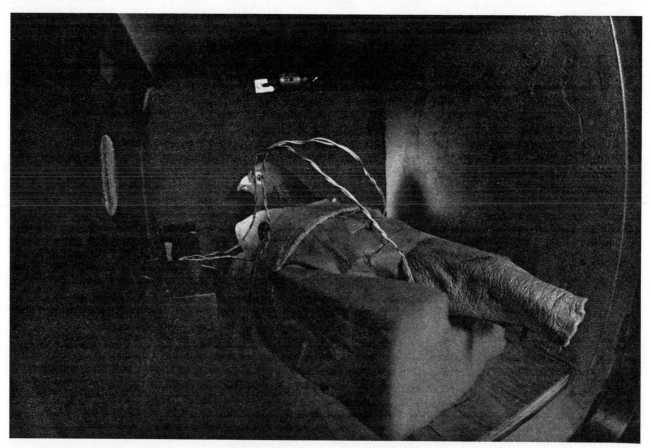

PIGEON IN ISOLATION CHAMBER has been prepared for tests of its unusual sensory capabilities. Two wires go to electrodes on the pigeon that give it a mild electric shock during the test, and two other wires are connected to electrodes that pick up the bird's heartbeat. The pigeon is restrained by a harness to keep it from moving. In a typical experiment the bird receives a shock following a specific stimulus, which might be a change in the strength of an induced magnetic field, a change of air pressure or a change in the plane of polarized light falling on the pigeon's eye. If the bird is able to sense the changes in the stimulus, it begins to anticipate the shock and its heart rate increases at the beginning of the stimulus, which is given at random time intervals. Experiments in the author's laboratory indicate that the pigeon is capable of sensing tiny fluctuations in atmospheric pressure. In addition pigeons, like honeybees, can detect changes in the plane of polarized light. The photograph was made with a camera placed inside the chamber.

scendant of several earlier breeds of pigeons, was developed in Belgium in the middle of the 19th century. Today, in addition to serving as message carriers, homing pigeons are raised for competitive racing. This sport is widespread and very popular in Europe, and has become firmly established in many parts of the U.S. as well. Often several thousand pigeons are entered in a single race. The birds are shipped to a designated point, usually between 100 and 600 miles away, and are then released simultaneously. After the owners of the birds have recorded the arrival times at home, using special devices designed for the purpose, the speed of the individual birds is calculated to determine the winners. Speeds of 50 miles per hour are common; the best pigeons can make it home from 600 miles away in a single day.

The remarkable ability of pigeons to find their way home has been known for at least as long as there has been written history. The armies of the ancient Persians, Assyrians, Egyptians and Phoenicians all sent messages by pigeon from the field. It is known that regular communication via pigeon existed in the days of Julius Caesar. During the siege of Paris in 1870 more than a million messages reached Parisians by means of pigeons that had been smuggled out of the city in balloons. Pigeons did such valuable service in both world wars that monuments in their honor were erected in Brussels and in the French city of Lille. In the U.S. some famous pigeon "heroes" were stuffed and mounted after their death; they are on display at the Army Signal Corps Museum and the National Museum.

In 1949 Gustav Kramer and his students at the Max Planck Institute for Marine Biology at Wilhelmshaven in Germany demonstrated that a pigeon in a circular cage with identical food cups at regular intervals around its periphery could easily be trained always to go to a food cup located in a particular direction, for example the northwest, even though the cage was rotated and the visual landscape around it was changed. They found that the pigeon's ability to determine a direction depended on the bird's being able to see the sun. Under a heavy overcast the bird's choice of food cups became random. If the sun's apparent position was altered by mirrors, the pigeon's choice of food cups was correspondingly altered.

It is obvious that if birds can use the sun as a compass to determine directions, they must be able to compensate for the change in the sun's apparent position during the day. In the Northern Hemi-

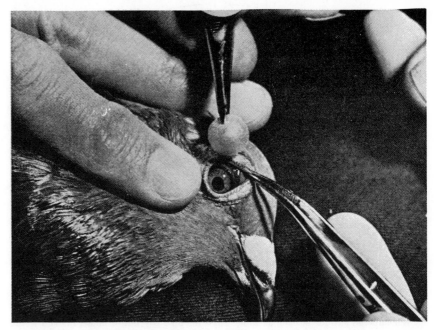

FROSTED CONTACT LENS is placed on a pigeon's eye before a test release. When both of the pigeon's eyes are covered by the lenses, the pigeon is unable to see objects that are more than a few yards away. Control pigeons have clear lenses put on their eyes and are released at the same time from the site. Experiments by Klaus Schmidt-Koenig and H. J. Schlichte of the University of Göttingen, who developed the technique, have demonstrated that pigeons wearing the frosted lenses are able to orient their flight in a homeward direction when they are released at a distant site, and that some pigeons are able to fly back to their home loft. Lenses currently in use are made of a gelatin that dissolves in a few hours.

NIGHT NAVIGATION of pigeons is being studied by Cornell workers. Pigeons with radio transmitters on their back are released and tracked by a radio receiver in the truck. In this time exposure three successive light flashes were used to illuminate the flying pigeon.

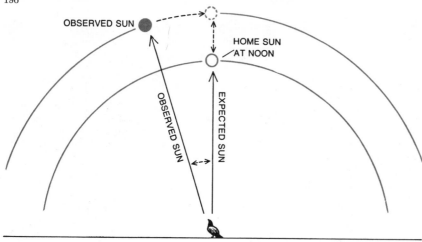

SUN-ARC HYPOTHESIS was proposed by G. V. T. Matthews of the University of Cambridge in the 1950's to explain how pigeons could obtain from the sun alone all the information required to determine their north-south and east-west displacement from home. For example, if a pigeon were released at noon at an unfamiliar site that is southwest of its home loft, the bird would observe the sun's motion and quickly extrapolate along the sun's arc across the sky to the noon position. It would then compare the sun's noon altitude with the remembered altitude at noon at home. Since the bird is south of home, the sun would be higher at the release site, and the pigeon would know that it has to fly north to make the sun appear lower. In order to determine its east-west displacement the pigeon would compare the position of the sun at the release site with the position the sun should have according to the bird's internal clock. In this instance the bird's clock would inform it that the time at home is noon. The sun at the release site, however, is at an altitude lower than that at noon, so that the bird knows it must fly east. By combining the two displacements, the bird would know that it should start flying to the northeast to get to its home loft.

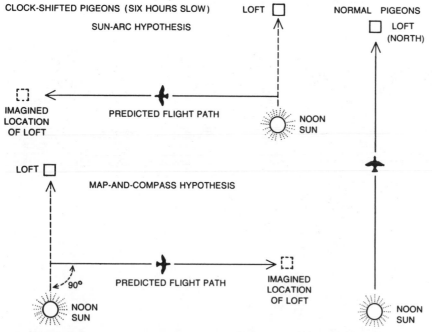

TESTS OF TWO SUN NAVIGATION HYPOTHESES were made with pigeons whose internal clocks had been shifted six hours slow by altering their day-and-night time periods in the laboratory. According to the sun-arc hypothesis, when the clock-shifted pigeons are taken south of their home loft and released at noon, their internal clock tells them that it is 6:00 A.M. at home. They observe that the sun at the release site is too far along its arc for 6:00 A.M. and that they therefore should fly west (*top left*). The alternative map-and-compass hypothesis suggests that the pigeons know where they are relative to home from some kind of map, and that they use the sun only to get compass direction. Their internal clock says it is 6:00 A.M., and they therefore assume that the sun is in the east. Since the sun is really in the south, they should begin flying east, thinking that this direction is north (*bottom left*). When the experiment was carried out, the clock-shifted pigeons flew east, thereby supporting the map-and-compass hypothesis and contradicting the sun-arc hypothesis. Normal pigeons released at the same site departed in the correct homeward direction (*right*).

sphere the sun rises in the east, moves through south at noon and sets in the west. If a pigeon is to determine a particular direction, it cannot simply select a constant angle relative to the sun. It must change the relative angle by about 15 degrees per hour, which is the average rate of change of the sun's position throughout the day. In short, the bird must have an accurate sense of time, an internal clock, and that clock must somehow be coupled with the position of the sun in the sky if an accurate determination of direction from the sun is to be possible.

In a simple but elegant fashion Kramer demonstrated that birds do indeed compensate for time when they are using the sun as a compass. He trained some birds, in this case starlings, to use the sun to go in a particular direction to get to a food cup. He then substituted a stationary light for the sun. The starlings responded to the light as though it were moving at 15 degrees per hour. Since the light was in fact stationary, the bearing taken by the birds shifted approximately 15 degrees per hour.

Klaus Hoffmann, one of Kramer's students, went an important step further in demonstrating the role of the internal clock in sun-compass orientation. He kept starlings for several days in closed rooms where the artificial lights were turned on six hours after sunrise and turned off six hours after sunset. It is known that the internal clocks of most organisms can be shifted to a new rhythm in this manner; the process is very similar to what is experienced by a human being who flies from the U.S. to Europe in a few hours and then takes several days to adjust to European time. When the starlings whose internal clocks had been shifted six hours slow were tested in a circular cage under the real sun, they selected a bearing 90 degrees to the right of the original training direction. Since their internal clocks were a quarter of a day out of phase with sun time, they made a quarter-circle error in their selection of food cups.

Although Kramer and his colleagues had clearly demonstrated that some birds, including pigeons, can use the sun as a compass, their discovery by itself cannot explain how pigeons home. As I have indicated, homing requires more than a compass. If you were taken hundreds of miles away into unfamiliar territory, given only a magnetic compass and told to start walking toward home, you would not be able to get there. Even though you could determine where north

MAGNETIC-FIELD HYPOTHESIS, proposed more than a century ago, had been rejected until recently because earlier experiments failed to show that putting a magnet on a pigeon disorients its homing. Recent tests show, however, that pigeons with bar magnets attached to them are disoriented when they are released at an unfamiliar site under a total overcast but are not disoriented when the sun is visible. Control pigeons with brass bars attached to them show little difference in their mean vanishing bearing under the sun or an overcast. The vanishing bearings of individual pigeons, as determined by an observer with binoculars, are shown by the solid circles. The broken line indicates the true home bearing. The mean vector, or directional tendency, of all the birds in a test group is shown by the arrow. The length of the mean vector is a statistical representation of the degree of agreement among the birds in selecting a direction. Perfect agreement would give a vector length equal to the circle's radius; the more scattered the departing directions, the shorter the vector.

was, you would not know where you were with respect to home, hence such compass information would be nearly useless.

In 1953 G. V. T. Matthews, who was then working at the University of Cambridge, suggested that pigeons get far more information than compass bearings from the sun. He hypothesized that the sun gives them all the information they need to carry out true bicoordinate navigation. Stated briefly, on release at a distant site a pigeon would determine its north-south displacement from home by observing the sun's motion along the arc of its path across the sky, extrapolating to the sun's noon position on that arc, measuring the sun's noon altitude and comparing it with the sun's noon altitude at home (as the bird remembered it). If the sun's noon altitude at the release site was lower than it was at home, the bird would know that it was north of home; if the sun was higher than it was at home, the bird would know that it was south of home. To calculate the east-west displacement the bird would determine local sun time by observation of the sun's position on its arc at the release site and compare the local time with home time as indicated by its internal clock. A local time ahead of home time would indicated the bird was east of home; a local time behind home time would indicated the bird was west of home. Thus, according to Matthews, the bird would determine its north-south displacement from the sun's altitude and its east-west displacement by the time difference; combining these data would indicate the homeward direction [see top illustration on page 196].

Matthews' sun-arc hypothesis was a major stimulus to further research on pigeon homing, and it formed the basis for many of the experiments conducted in the following decade. Unfortunately, however, nearly all the results of these experiments contradicted the hypothesis, and investigators actively engaged in research on pigeon homing today no longer regard it as being probable. The evidence against the hypothesis is so extensive that most of it cannot be discussed here. For the moment I shall mention only one kind of experiment to help the reader understand some of the more recent research.

Klaus Schmidt-Koenig, another of Kramer's students, showed in 1958 that when pigeons whose clocks have been artificially shifted are released at a distant site, their initial choice of direction is shifted. Their vanishing bearings (the bearings at which they vanish from the view of an observer using high-power binoculars) deviate from those of normal pigeons by 15 degrees for each hour the birds have been clock-shifted.

Let us examine a test involving clock-shifted pigeons to see whether or not the results agree with what would be predicted by the sun-arc hypothesis. Suppose we shift the birds' internal clock so that it is six hours slow and then release them at noon 100 miles south of their home loft. According to the sun-arc hypothesis, the birds would observe that it is noon at the release site, but their internal clock would tell them it is only 6:00 A.M. at home. They should therefore react as though they were thousands of miles east of home, and they should start flying almost due west. When such an experiment is actually performed, however, the birds vanish nearly due east, exactly opposite what the sun-arc hypothesis predicts [see bottom illustration on page 196].

Is there any way we can make sense of these results? The answer is yes, but to do so we must turn from Matthews' sun-arc hypothesis to an alternative model proposed by Kramer. Kramer emphasized that all the evidence supports the conclusion that pigeons get only compass information from the sun and nothing else. They appear to behave in a manner analogous to a man who uses both a map and a compass, as though they first determine from some kind of map where they are relative to home and in which direction they must fly to get home and then use the sun compass to locate that direction.

Since Kramer could never explain

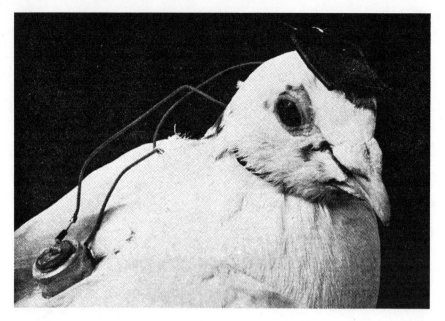

HELMHOLTZ COILS above the pigeon's head and around its neck induce a relatively uniform magnetic field through its head. The coils are powered by a small mercury battery on the bird's back. Direction of the induced field can be reversed simply by reversing the connections of the battery. The strength of the magnetic field can be varied by controlling the amount of current passing through the coils. Battery is exhausted in two or three hours.

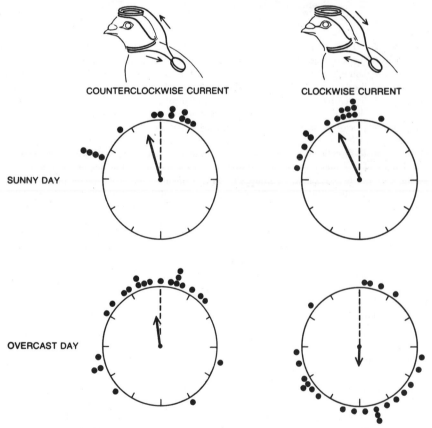

PIGEONS WITH HELMHOLTZ COILS in which the current flows counterclockwise (south-seeking pole of a compass in the induced magnetic field points up) fly almost directly homeward on both sunny and overcast days. When the current in the coils is made to flow clockwise (north-seeking pole of a compass in the induced magnetic field points up), the pigeons still fly homeward on sunny days, but on overcast days they fly almost 180 degrees away from home. These results were obtained in several experiments conducted by Charles Walcott and Robert Green of the State University of New York at Stony Brook.

what the source of the map information might be, let us for the sake of our example pretend that before we release each pigeon we whisper in its ear, "Home is due north." Now the bird must use its sun compass to locate north. Its internal clock says it is 6:00 A.M., when the sun should be in the east; hence north should be approximately 90 degrees counterclockwise from the sun. Remember, however, that the bird's clock is six hours slow; it is actually noon, when the sun is in the south. Hence the bird's choice of a bearing 90 degrees counterclockwise from the sun sends it east, not north. We can summarize by saying that no matter what combination of directions and clock-shift we use in actual experiments, the results come out consistent with the predictions of Kramer's map-and-compass model and not with Matthews' sun-arc hypothesis.

Because the pigeon's use of the sun compass in orientation was the one thing that was firmly established, there was a tendency in the 1960's for many investigators to assume that the sun is essential for homeward orientation at an unfamiliar release site. Several discrepancies, however, led me and my colleagues at Cornell University to doubt it. First, I knew of numerous instances of fast pigeon races under heavy overcast. Second, our pigeons seemed to perform well under overcast if they had first been made to fly in the vicinity of the home loft in rainy weather. Third, the published evidence that pigeons were disoriented under heavy overcast was not entirely consistent, and fourth, we and others had been able to get pigeons to home at night.

We set out to reexamine the importance of the sun in pigeon navigation. In our most important experiments we too used clock-shifted pigeons. As we expected, when pigeons whose internal clocks had been shifted six hours fast or slow were released under sunny conditions, their vanishing bearings were roughly 90 degrees to the right or left of the vanishing bearings of control pigeons whose internal clocks had not been altered. When the pigeons were released in total overcast, however, the results were quite different: both the clock-shifted birds and the control birds vanished toward home and there was no significant difference in their bearings. This was true even when the release site was completely unfamiliar to the pigeons.

These results led us to several conclusions: (1) Pigeons accustomed to flying in inclement weather are able to orient homeward under total overcast.

Since there is no difference between the bearings of control and clock-shifted birds under such conditions, it is clear they are not able to see the sun through the clouds and hence are no longer using the sun compass. (2) There must be redundancy in the pigeons' navigation system. They use the sun compass when it is available, but they can substitute information from other sources when it is not. (3) The alternative information used in lieu of the sun compass does not require time compensation. (4) The alternative system cannot be pilotage by familiar landmarks, because pigeons can correctly orient themselves homeward under overcast even in distant, unfamiliar territory.

Recognition of the fact that pigeons are able to use alternative cues, depending on the circumstances involved, meant that the results of many older experiments could no longer be accepted at face value. For example, if an experimenter altered cue A while keeping other conditions optimum, and if the pigeons continued to orient well, they may simply have used cue B as an alternative to A. Similarly, if B was altered while everything else was kept at an optimum and the birds oriented well, they may have used A as an alternative to B. In short, we would have been wrong if we had concluded from these experiments that cues A and B are not elements in the pigeons' navigation system. In fact, such experiments show only that neither A nor B alone is essential for proper orientation under the particular test conditions.

This kind of reasoning led us to conduct experiments in which we varied several possible orientational cues simultaneously, on the assumption that if we could interfere with enough of them at the same time we could hope to learn which cues are more important and how they interact with one another.

We chose first to look again at the old idea that birds might obtain directional information from the earth's magnetic field. Although this hypothesis had been known for more than a century, there was no evidence for it and much experimental evidence against it. Nonetheless, it seemed worth reexamining. And it was! When we repeated older experiments of putting bar magnets on pigeons to distort the magnetic field around them, we found, as had others before us, that the birds had no difficulty orienting on sunny days. When the test releases were conducted on totally overcast days, however, the birds carrying

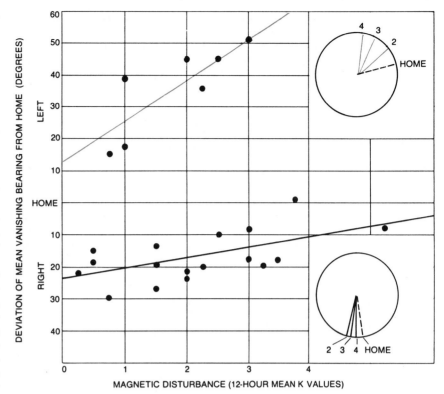

DISTURBANCES OF THE EARTH'S MAGNETIC FIELD caused by solar activity appear to affect a pigeon's initial choice of bearing when it is released at a distant site under sunny conditions. The K-index scale is used to indicate the degree of magnetic activity, ranging from quiet (less than 2) to a major magnetic storm (6 or more). In 1972 a series of releases of Cornell University pigeons from a site 45 miles north of the home lofts revealed that, as the degree of magnetic disturbance increases, the vanishing bearing of the birds steadily shifts to the left as seen by an observer facing homeward (black curve). At this release site the shift to the left brought the birds' vanishing bearings closer to the true home bearing, but success in homing was not improved. In another series of tests, pigeons from a different loft were released from a site west of their home. A similar leftward shift of vanishing bearings with increasing magnetic disturbance was found (colored curve). In this instance the shift to the left caused the vanishing bearings to recede away from the true home bearing.

magnets usually vanished randomly whereas control birds carrying brass bars of the same size and weight vanished toward home. Several other workers have since repeated these experiments, with the same results.

More recently Charles Walcott of the State University of New York at Stony Brook and his student Robert Green have gone one step further. Instead of working with bar magnets, they put a small Helmholtz coil on the pigeon's head like a cap and another coil around its neck like a collar. Power is supplied from a battery on the bird's back. This device makes it possible to induce a relatively uniform magnetic field through the bird's head. The direction of the induced magnetic field can be made to point up through the bird's head or to point down simply by hooking up the battery to make the current in the coils flow clockwise or counterclockwise. Under sunny conditions Walcott and Green found that the direction of the induced magnetic

field did not affect the pigeon's ability to orient homeward. Under total overcast, however, the direction of the induced magnetic field had a dramatic effect: when the north-seeking pole of a compass in the induced field pointed up, the pigeons flew almost directly away from home, whereas when the south-seeking pole of a compass in the induced field pointed up, the pigeons oriented toward home.

Our results, together with Walcott's, suggest that magnetic information may play a role in the pigeon navigation system. This is consistent with the recent discovery by Freidrich Merkel and Wolfgang Wiltschko of the University of Frankfurt that European robins in circular cages can use magnetic cues to orient themselves in a particular direction. William Southern of Northern Illinois University also has reported that the orientation of ring-billed gulls is influenced by magnetic activity.

Recently Martin Lindauer and Herman Martin of the University of Frankfurt have demonstrated that honeybees give orientational responses to magnetic cues several thousand times weaker than the earth's field. Only a few years ago biologists were debating whether or not any organism could detect a magnetic field as weak as the earth's (approximately half a gauss). The responses of honeybees to magnetic cues now makes us wonder if one gamma (10^{-5} gauss) will not prove to be the lower limit. Indeed, a study that my colleagues and I have recently conducted suggests that the magnetic-detection sensitivity of pigeons may rival that of honeybees. In four long series of tests over a period of three years we have found that fluctuations of less than 100 gamma (and probably less than 40 gamma) in the earth's magnetic field, caused by solar flares and sunspots, appear to have a small but significant effect on the pigeons' choice of an initial bearing at the release site.

The question of how organisms detect magnetic stimuli is unanswered. We have very little idea what a magnetic sense organ should look like, or even where in the body we should expect to find it. Since magnetic flux can pass freely through living tissue, magnetic detectors might be anywhere inside the body. The search for these detectors has already begun in our laboratory and in others throughout the world. It promises to be a challenging undertaking.

Exciting as the discovery that magnetism plays a part in avian navigation systems may be, we are in a sense back where we started. The weight of the evidence at present suggests that magnetism simply provides a second compass, not the long-sought map. Hence we must continue our search. What other sources of information might the birds have?

One possibility that comes readily to mind in this age of long-range rocketry is that the birds might be capable of inertial guidance, that they might somehow detect and record all the angular accelerations of the outward journey to the release site, then double-integrate them to determine the direction home. Intriguing as this possibility is, all the evidence is against it. Pigeons have been carried to release sites while riding on turntables or in rotating drums, yet this input of additional inertial "noise" has no effect; the birds orient homeward as accurately as control birds not so treated. Other pigeons that were carried to the release site while they were under deep anesthesia were able to determine the direction home with no difficulty. Pigeons with a variety of surgical lesions of the semicircular canals—the principal detectors of acceleration in vertebrates—orient themselves accurately, whether they are tested under sunny conditions or overcast ones.

The hypothesis that pigeons may be able to use olfactory information in navigating has been advocated by Floriano Papi and his colleagues at the University of Pisa. The probability that this is the case seems low in view of the relatively poor development of the pigeon's olfactory system. Nonetheless Papi has some interesting experimental results, and it is too early to make a judgment on his proposal. We are currently conducting experiments to test his ideas.

By now the reader may well be wondering why so little has been said about what might seem the most obvious possible cue for homing pigeons: familiar landmarks. The reason is that there is abundant evidence that landmarks play a very small role in the homing process. In the course of tracking pigeons by airplane Walcott and his colleague Martin Michener have repeatedly noted that when pigeons flying on an incorrect course encounter an area over which they have recently flown, they seldom give any indication of recognizing the familiar territory. Several other investigators, including members of our group, have found that pigeons clock-shifted six hours and released less than a mile from

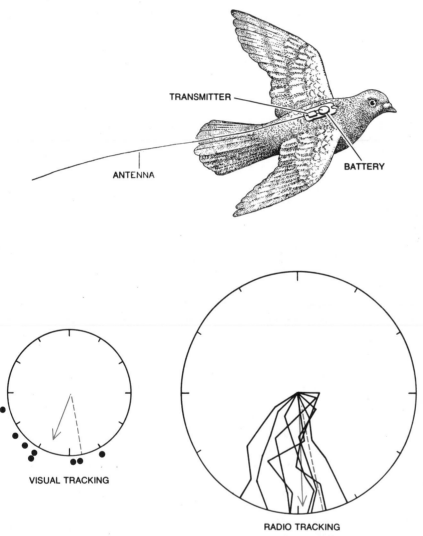

RADIO TRACKING OF PIGEONS is carried out from a receiver on the ground at the release site or sometimes by a receiver on an aircraft. An FM transmitter and a battery are glued on the pigeon's back (top). The 19-inch antenna trails behind the bird as it flies. Data from radio tracking reveal that pigeons do not continue to fly in a straight line after they leave a release site but frequently alter their course. The vanishing bearings for eight pigeons, as determined by observers with binoculars, and the bearings determined by simultaneous radio tracking from the site are compared (bottom). The scale of the two circles is arbitrary. Visual tracking extends to one or two miles, depending on the flying height of the bird. Radio tracking extends to eight miles or more. Broken line indicates home direction.

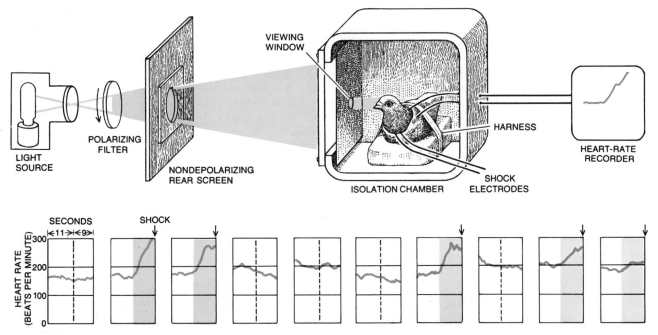

PIGEON PERCEPTION OF POLARIZED LIGHT is being tested in the author's laboratory by Melvin L. Kreithen. After electrodes for administering electric shock and for detecting the heart rate are attached, the pigeon is put in a harness and then is placed in a sealed soundproof chamber. Light is projected through a polarizing filter on a rotating mount and then through a nondepolarizing rear-projection screen. Part of the light enters the isolation chamber through a small window and falls on the pigeon's eye. The light comes on at random intervals. In some trials, which are also determined at random, the polarizing filter starts to rotate after 11 seconds; in others it does not. When the filter rotates, the pigeon receives a shock at the end of the light signal. When the filter does not rotate, no shock is given. After a number of trials the pigeon's heart rate begins to rise rapidly at the beginning of the rotation of the polarizing filter, indicating that the bird is able to sense the change in the plane of polarized light and is anticipating the shock that is to come. Recordings from a series of tests of a pigeon are shown (*bottom*). The colored block indicates the interval during which the polarizing filter rotates. In control runs during the corresponding interval (*to right of broken line*) no rotation occurs.

home, in territory over which they have flown daily during their exercise period, often vanish 90 degrees away from the home direction. Only a direct view of the loft building itself takes precedence over what their navigation system is telling them; nearby buildings or trees apparently do not serve as reference points under these conditions. In fact, even a view of the loft is not always effective, particularly at distances of a mile or more.

Perhaps most convincing of all are experiments conducted by Schmidt-Koenig and H. J. Schlichte of the University of Göttingen. They put frosted contact lenses over the eyes of pigeons, thus making it impossible for the birds to see any object that is more than a few meters away. Not only do these pigeons orient homeward when they are released as far as 80 miles away but also a surprising number of them actually get home. Schmidt-Koenig conducted some of his experiments at our Cornell lofts, and thus I had the opportunity of observing them at first hand. It was a remarkable experience. The birds arrived very high overhead and fluttered down to a landing in the fields around the loft.

Being unable to see the loft, they waited for us to pick them up and carry them the last few feet. These results suggest that the pigeon navigation system is often accurate enough to pinpoint the home location almost exactly without reliance on familiar landmarks; vision is necessary only for the final approach, frequently at a distance of less than 200 yards.

It will be apparent from all I have said that the task of uncovering the pigeon's navigation system is going to be a difficult one. The old idea that birds use a single method to determine the home direction has given way to the realization that there are probably multiple components in the system and that these components may be combined in a variety of ways, depending on such factors as weather conditions, the age of the bird and the bird's experience.

One approach that holds much promise for helping us tease apart the many elements in the system is the study of the ontogeny of navigational behavior. For example, we have found that bar magnets disrupt the initial orientation of very young pigeons released away from home for the first time in their life, even when the sun is visible. Moreover, normal first-

flight youngsters cannot orient under total overcast, even if they have previously been released for exercise in inclement weather. It seems, then, that inexperienced homing pigeons need both sun information and magnetic information. Various other manipulations that have little effect on experienced birds also disorient first-flight pigeons.

Perhaps with experience a pigeon learns to orient accurately with less information. Or perhaps experience is necessary to enable the pigeons to settle on a weighting scheme that allows them to decide what to do when they get conflicting information from different sources. Early results from some current experiments indicate that by training very young pigeons under conditions in which we severely restrict, or eliminate altogether, certain normally important environmental cues such as the sun, we may be able to induce the birds to settle on weighting schemes for evaluating directional cues that are quite different from normal. The availability of such birds would greatly facilitate the carrying out of experiments designed to clarify the roles of cues that are normally difficult to alter.

Another approach we are actively pursuing is an attempt to learn more about the sensory capabilities of pigeons. The more we learn, the more we become convinced that birds live in a sensory world very different from our own. For example, my student Melvin L. Kreithen has recently demonstrated that pigeons are remarkably sensitive to tiny changes in barometric pressure. Such a sensitivity might enable pigeons to get navigationally useful information from pressure patterns in the atmosphere. Kreithen and I have also recently obtained experimental evidence that pigeons can detect the plane of polarized light, which might

mean that pigeons, like honeybees, can continue using the sun as a compass on partially overcast days, when the sun's disk is hidden from sight but some blue sky remains visible.

It has long been known that the bearings chosen by pigeons at distant release sites, although roughly in the homeward direction, are almost never oriented directly toward home. Moreover, the mean bearings of repeated releases at any given site usually show a consistent deviation from home; there is, in effect, a relatively stable "release-site bias" that is characteristic of each location. At some

sites the bias is apparent only in the vanishing bearings obtained by visual tracking. The bias often becomes less marked when the final-contact bearings obtained by radio tracking are used, but at some sites the bias is still manifest when the birds move out of radio range (between six and 10 miles from the release site). In the hope that these biases might prove to be a key to local geophysical factors that could provide at least part of the map information for pigeons, we chose for intensive study several release sites where the biases were unusually large.

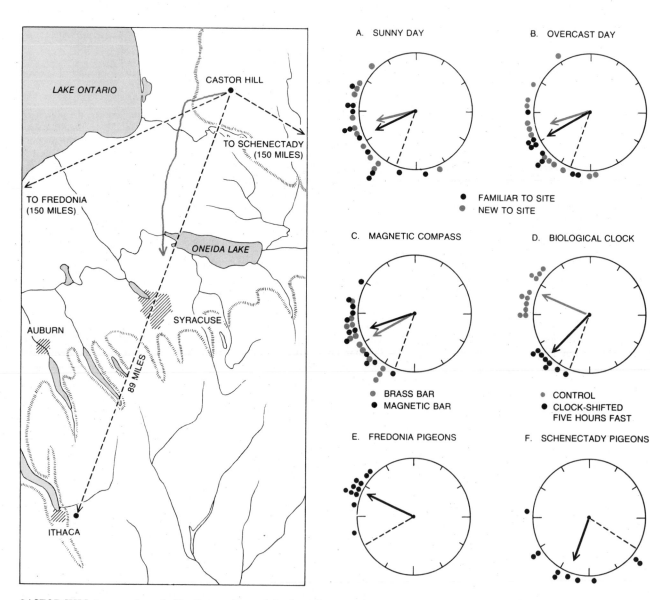

CASTOR HILL is approximately 89 miles northeast of the Cornell University pigeon lofts in Ithaca, N.Y. At Castor Hill the Cornell pigeons regularly choose an initial bearing that deviates clockwise from the true home direction. This occurs both on sunny days and on overcast days (A and B); thus the characteristic bias must not depend on the sun compass. The bias probably is not due primarily to the magnetic compass either, since under sunny conditions birds with bar magnets have about the same vanishing bearing as pigeons with brass bars (C). Pigeons whose internal

clocks have been shifted five hours fast choose a more homeward direction than normal pigeons, but the clock-shifted pigeons are less successful at getting home (D). Pigeons from another loft at Fredonia, N.Y., 150 miles west of Ithaca (E), and pigeons from Schenectady, N.Y., 150 miles east of Ithaca (F), choose bearings that also deviate clockwise from the true home bearing. It appears that the bias is a function of the site and not of the pigeons. The actual flight path of normal Cornell pigeons released at Castor Hill and tracked by an airplane is shown by the colored line on the map.

One such site is the Castor Hill Fire Tower, located 89 miles north northeast of our Cornell lofts. Here our pigeons regularly depart with a mean bearing that deviates roughly 60 degrees clockwise from the direction of home [*see illustration on opposite page*]. In a long series of experiments we have found that this clockwise bias is evident not only with experienced pigeons new to the site but also with pigeons that have been released at the site before. The same bias is found in very young pigeons on their first homing flight. It is found on both sunny and overcast days, so that it apparently has nothing to do with the sun compass, and it is found in pigeons wearing magnets, so that it probably has nothing to do with the magnetic compass either. It is even found when the pigeons are wearing frosted contact lenses; hence it must not depend on anything the pigeons see.

Wild bank swallows captured near Cornell and released at Castor Hill showed the same clockwise bias, indicating that the biasing factor, whatever it may be, affects other bird species in the same way. Pigeons borrowed from lofts 150 miles east and west of Cornell and released at Castor Hill show a similar clockwise departure bias relative to their home. Finally, pigeons clock-shifted five hours fast depart nearly straight toward home from Castor Hill but nonetheless have poorer homing success than control pigeons that depart with the usual 60-degree bias. In a joint experiment with Walcott, normal pigeons with radio transmitters attached to them were tracked by an airplane after their release from Castor Hill. We found that the birds turn onto a more homeward course when they are approximately 14 to 18 miles west of Castor Hill. It may be that the clock-shifted birds that have

poor homing success make a corresponding turn when they are a similar distance from the release site and thus become directed away from home. We hope soon to learn if this is so.

We conclude, then, that the bias in the birds' initial bearings is not a biological error, that it is due not to some peculiarity of the birds but to a peculiarity of the location. The birds are probably reading the map cues correctly but the map itself is twisted clockwise at Castor Hill. Perhaps if we can learn what geophysical factors are responsible for this distortion of the map we will finally be on the way to understanding the ancient mystery of how pigeons home.

III

THE ADAPTABILITY
OF BEHAVIOR

III   THE ADAPTABILITY OF BEHAVIOR

INTRODUCTION

When biologists distinguish adaptability from adaptiveness, they are not just playing with words. The subject of Part II was adaptiveness, the ways in which particular behaviors fit organisms to their environments. In the simplest version of the concept, most nearly realized by insects and other invertebrate animals, a behavior is viewed as merely existing; it either works in response to a given need or it does not work; in other words, it is adaptive or not. Part III introduces a new dimension to this subject. The adaptability of a behavioral pattern (as opposed to its adaptiveness) is the degree to which it can be varied to fit the requirements imposed by a changing environment. The larger the organism, the greater its potential adaptability. Its central nervous system is more extensive and complex, permitting a greater degree of learning, and a more elaborate endocrine system allows a finer adjustment of mood in accordance with needs that change from day to day. A larger animal also lives for a longer period of time. In particular, if it is a member of a species that is perennial instead of annual, it must survive the full cycle of seasons and begin over again. The longer it exists, the greater the range of contingencies it must meet, and the more complex and variable its responses must become. Finally, the larger animal forages over a wider area in search of food and shelter, a circumstance that calls for still further flexibility. For all those interlocking reasons, the behavior of vertebrates, especially that of birds and mammals, is more modifiable than that of invertebrates. Yet the difference is a matter of degree and not of kind.

All behavioral phenomena together can be thought of as a set of devices for tracking changes in the environment. Some responses track relatively infrequent, major changes that are pervasive in their effects on physiology and behavior. Examples include change from one life stage to another, and the priming of an animal for reproductive activity by its own hormones at the start of the breeding season. Other responses, such as short-term learning or the surge of epinephrine in the face of danger, track the more frequent but short-lived changes. They have a relatively localized effect on physiology and usually last no more than minutes or hours. The slow and fast responses work in concert to adjust the organism to an array of environmental changes that occur simultaneously but at different velocities.

The hierarchy of tracking devices is the theme of the *Scientific American* articles for Part III. The reader, if he follows the sequence faithfully, will first learn of the role of hormones in adjustments of reproduction and stress. He will then encounter cases of learning which occur only once in the life of an organism, creating an effect so nearly indelible that it can even be regarded as part of the development of instinct. These accounts are followed by examples of more conventional forms of learning from three very divergent groups of animals—octopi, fish, and birds. Finally, a pinnacle of behavioral evolution is described in Harry F. Harlow's article on socialization in rhesus monkeys. The fact that this author can refer unflinchingly to some of the phenomena he studied as love is indicative of his confidence, shared with many zoologists, that the evolutionary steps traced in animals do indeed lead to the very threshold of human behavior.

"The Reproductive Behavior of Ring Doves," *by Daniel S. Lehrman. November 1964.*

Hormones have long been known to turn the sexual and parental behaviors of vertebrates off and on. It was the signal contribution of the late Daniel S. Lehrman and his co-workers to trace the intricate sequence of physiological events orchestrated by whole sets of hormones. The male ring dove, when activated by his own testosterone, begins the courtship rituals by bowing and cooing to a potential mate. The mere sight of the male induces the release of pituitary hormones in the female. These in turn trigger the secretion of two additional hormones, the follicle-stimulating hormone and the luteinizing hormone, which stimulate the growth of the ovaries. From this point the reader will want to pursue the remainder of Lehrman's story of courtship, nest building, and rearing of the ring-dove nestlings. Step by step, in tight marching order, hormones lead to behavioral acts, which induce the release or shutting off of other hormones, which then elicit more behavior, and so on forward until the young are reared and the cycle completed.

"Stress and Behavior," *by Seymour Levine. January 1971.*

The "stress syndrome" of mammals, including man, is the pathological outcome of the body's complex hormonal response to physical and psychological stress. The initial response to stress is the release of ACTH from the anterior lobe of the pituitary gland. This hormone induces an outpouring of steroids from the cortex of the adrenal glands, which in turn help to reduce tissue inflammation, promote the deposition of glycogen, and in general act as a moderating influence on the body's emergency response to the stress. If the pressure is kept up, the hormones are maintained at too high a level for too long a time, causing physiological deterioration, the basis of the stress syndrome. Interest in this pathological overreaction has led to the discovery, reported here by Seymour Levine, of some of the beneficial effects of the pituitary-adrenal response in the early, more moderate stages. For example, ACTH alone has several seemingly adaptive effects on behavior. It refines the ability of laboratory rats to avoid unpleasant stimuli and prolongs their remembrance of the experience. It also slows down the speed with which the rats grow accustomed to the stimuli. In general, then, ACTH maintains the animals in a greater state of preparedness. From this example the forms of interaction between hormones and the central nervous system are seen to be quite complicated, and the subject must be judged to be in an early stage of exploration.

"Learning in the Octopus," *by Brian B. Boycott. March 1965.*

The importance of this article lies in its concern with intelligence in alien organisms. Intelligence in the octopus, such as it is, was evolved in a wholly independent fashion from that of man and the other vertebrates. It is almost as though one were examining not-very-bright visitors from another planet. The brain of the octopus, although novel in this sense, has a gross structure relatively easy to visualize. Investigators from J. Z. Young onward have found it relatively simple to excise portions of the brain and to record the subsequent changes in behavior. Brain Boycott here describes the general history of this work, stressing results that indicate the existence of separate centers for the short-term and long-term retention of memories.

"'Imprinting' in a Natural Laboratory," *by Eckhard H. Hess. August 1972.*
The stricter forms of imprinting are acts of learning that occur rather quickly
and are difficult or impossible to "unlearn." Typical imprinting is striking
in its effects, ordinarily occurs only during a limited time in the life of an
animal, and does not require a reward or punishment. In the 1930's Konrad
Lorenz provided the textbook example of the phenomenon. When he permitted the eggs of a graylag goose to hatch in an incubator, so that the new
goslings first saw Lorenz rather than the mother, the young birds followed
him afterward rather than the mother or any other goose. In the early, decisive act of learning, the goslings had been "imprinted" by the image of Lorenz. In this article Hess discusses the intricacies of the same process in
mallard ducklings. He has found that the attachment made to the mother
under natural conditions is stronger than that made in the laboratory to an
investigator or some other substitute parent. The reason turns out to be as
strange as the act of imprinting itself: under natural conditions the mother
and her young call back and forth to each other while the ducklings are still
in the eggs; this reinforcement evidently makes the bond stronger than it
would be if the ducklings learned the mother's identity only by seeing her.

"How an Instinct Is Learned," *by Jack P. Hailman. December 1969.*
An instinct is loosely defined as any stereotyped pattern of response common
to a species or at least to a local breeding population. As Hailman points out,
this concept does not imply that every instinct is entirely pre-wired in the
central nervous system; the behavior can also be shaped to some extent by
learning. The newly hatched gull chick, for example, begins with a poorly
coordinated pecking movement directed at certain shapes and movements
that are normally associated only with its parents and siblings. Through practice the chick comes to distinguish food from other objects. It improves its aim
and learns to rotate its head, enabling it not only to beg with forward pecking
movements but also turn to pick up the food regurgitated below on the nest
floor. As Hailman concludes, "It is necessary only that the learning process
be highly alike in all members of the species for a stereotyped, species-common behavioral pattern to emerge. The example of the gulls also shows clearly
that behavior cannot meaningfully be separated into unlearned and learned
components, nor can a certain percentage of the behavior be attributed to
learning."

"Love in Infant Monkeys," *by Harry F. Harlow. June 1959.*
The infant monkey, like the goslings and ducklings described earlier, is intensely attracted to its mother. Harry Harlow and his associates have been
able to identify some of the elementary stimuli by which the infant recognizes
its mother. Although psychologists in orientation, they owe their success to
essentially the same techniques used by ethologists during prior investigations
on fish and birds. Artificial "mothers" constructed of wire and cloth, and
varying in outward appearance, were presented to the infants. The models
were also provided with bottles of milk so that the young monkeys could
nurse. These experiments revealed that softness is crucial to the infants.
Models covered with soft terry cloth were accepted even when they were
supplied with large round eyes made of bicycle lamps and bizarre faces that
made them look more like gargoyles than real monkeys. Although the animals
thrived physically, they became virtually psychotic in adult life. They were
sometimes highly aggressive and sometimes autistic; in the latter state they
withdrew to rock silently back and forth. They also proved unable to mate
or to rear infants normally. The effects of social deprivation have proven
generally more severe in such mammals than in birds and other less intelligent,
complex animals. The rhesus monkey in particular is a wholly social animal,
whose survival depends on the constant exploitation of subtle, personalized

relationships with other members of the society. To remove the monkey entirely from the effects of socialization is a traumatic experience as detrimental to its welfare as malnutrition or prolonged psychological stress.

SUGGESTED ADDITIONAL READING for Part III

Deneberg, Victor H. 1972. *Readings in the Development of Behavior.* Sinauer Associates, Sunderland, Mass. An anthology of papers, mostly recent and from the psychological literature, dealing with developmental and other temporal determinants of vertebrate behavior.

Gibson, Eleanor J. 1967. *Principles of Perceptual Learning and Development.* Appleton-Century-Crofts, New York. A sophisticated, well-written summary of the psychological literature on the subject.

Hinde, R. A. 1970. *Animal Behavior: A Synthesis of Ethology and Comparative Psychology.* Second Edition. McGraw-Hill, New York. A general treatise that includes a discussion of the modifiability of behavior.

Levine, Seymour. 1972. *Hormones and Behavior.* Academic Press, New York. An advanced level book, dealing with selected aspects of the effect of hormones on vertebrate behavior.

Seligman, Martin E. P., and Joanne L. Hager. 1972. *Biological Boundaries of Learning.* Appleton-Century-Crofts, New York. A compilation of essays on "instinct and learning," for the most part excellent, that view the subject as falling within the joint province of psychology and animal behavior, two fields that are becoming increasingly integrated and interdependent.

21

The Reproductive Behavior of Ring Doves

by Daniel S. Lehrman
November 1964

*An account of experiments showing that the changes in
activity that constitute the behavior cycle are governed by
interactions of outside stimuli, the hormones and the
behavior of each mate*

In recent years the study of animal behavior has proceeded along two different lines, with two groups of investigators formulating problems in different ways and indeed approaching the problems from different points of view. The comparative psychologist traditionally tends first to ask a question and then to attack it by way of animal experimentation. The ethologist, on the other hand, usually begins by observing the normal activity of an animal and then seeks to identify and analyze specific behavior patterns characteristic of the species.

The two attitudes can be combined. The psychologist can begin, like the ethologist, by watching an animal do what it does naturally, and only then ask questions that flow from his observations. He can go on to manipulate experimental conditions in an effort to discover the psychological and biological events that give rise to the behavior under study and perhaps to that of other animals as well. At the Institute of Animal Behavior at Rutgers University we have taken this approach to study in detail the reproductive-behavior cycle of the ring dove (*Streptopelia risoria*). The highly specific changes in behavior that occur in the course of the cycle, we find, are governed by complex psycho-

REPRODUCTIVE-BEHAVIOR CYCLE begins soon after a male and a female ring dove are introduced into a cage containing nesting material (hay in this case) and an empty glass nest bowl (*1*). Courtship activity, on the first day, is characterized by the "bowing

  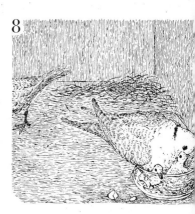

CYCLE CONTINUES as the adult birds take turns incubating the eggs (*6*), which hatch after about 14 days (*7*). The newly hatched squabs are fed "crop-milk," a liquid secreted in the gullets of the adults (*8*). The parents continue to feed them, albeit reluctantly,

biological interactions of the birds' inner and outer environments.

The ring dove, a small relative of the domestic pigeon, has a light gray back, creamy underparts and a black semicircle (the "ring") around the back of its neck. The male and female look alike and can only be distinguished by surgical exploration. If we place a male and a female ring dove with previous breeding experience in a cage containing an empty glass bowl and a supply of nesting material, the birds invariably enter on their normal behavioral cycle, which follows a predictable course and a fairly regular time schedule. During the first day the principal activity is courtship: the male struts around, bowing and cooing at the female. After several hours the birds announce their selection of a nest site (which in nature would be a concave place and in our cages is the glass bowl) by crouching in it and uttering a distinctive coo. Both birds participate in building the nest, the male usually gathering material and carrying it to the female, who stands in the bowl and constructs the nest. After a week or more of nest-building, in the course of which the birds copulate, the female be-

comes noticeably more attached to the nest and difficult to dislodge; if one attempts to lift her off the nest, she may grasp it with her claws and take it along. This behavior usually indicates that the female is about to lay her eggs. Between seven and 11 days after the beginning of the courtship she produces her first egg, usually at about five o'clock in the afternoon. The female dove sits on the egg and then lays a second one, usually at about nine o'clock in the morning two days later. Sometime that day the male takes a turn sitting; thereafter the two birds alternate, the male sitting for about six hours in the middle of each day, the female for the remaining 18 hours a day.

In about 14 days the eggs hatch and the parents begin to feed their young "crop-milk," a liquid secreted at this stage of the cycle by the lining of the adult dove's crop, a pouch in the bird's gullet. When they are 10 or 12 days old, the squabs leave the cage, but they continue to beg for and to receive food from the parents. This continues until the squabs are about two weeks old, when the parents become less and less willing to feed them as the young birds

gradually develop the ability to peck for grain on the floor of the cage. When the young are about 15 to 25 days old, the adult male begins once again to bow and coo; nest-building is resumed, a new clutch of eggs is laid and the cycle is repeated. The entire cycle lasts about six or seven weeks and—at least in our laboratory, where it is always spring because of controlled light and temperature conditions—it can continue throughout the year.

The variations in behavior that constitute the cycle are not merely casual or superficial changes in the birds' preoccupations; they represent striking changes in the overall pattern of activity and in the atmosphere of the breeding cage. At its appropriate stage each of the kinds of behavior I have described represents the predominant activity of the animals at the time. Furthermore, these changes in behavior are not just responses to changes in the external situation. The birds do not build the nest merely because the nesting material is available; even if nesting material is in the cage throughout the cycle, nest-building behavior is concentrated,

coo" of the male (2). The male and then the female utter a distinctive "nest call" to indicate their selection of a nesting site (3).

There follows a week or more of cooperation in nest-building (4), culminating in the laying of two eggs at precise times of day (5).

as the young birds learn to peck for grain themselves (9). When the squabs are between two and three weeks old, the adults ignore

them and start to court once again, and a new cycle begins (10). Physical changes during the cycle are shown on the next page.

as described, at one stage. Similarly, the birds react to the eggs and to the young only at appropriate stages in the cycle.

These cyclic changes in behavior therefore represent, at least in part, changes in the internal condition of the animals rather than merely changes in their external situation. Furthermore, the changes in behavior are associated with equally striking and equally pervasive changes in the anatomy and the physiological state of the birds. For example, when the female dove is first introduced into the cage, her oviduct weighs some 800 milligrams. Eight or nine days later, when she lays her first egg, the oviduct may weigh 4,000 milligrams. The crops of both the male and the female weigh some 900 milligrams when the birds are placed in the cage, and when they start to sit on the eggs some 10 days later they still weigh about the same. But two weeks afterward, when the eggs hatch, the parents' crops may weigh as much as 3,000 milligrams. Equally striking changes in the condition of the ovary, the weight of the testes, the length of the gut, the weight of the liver, the microscopic structure of the pituitary gland and other physiological indices are correlated with the behavioral cycle.

Now, if a male or a female dove is placed alone in a cage with nesting material, no such cycle of behavioral or anatomical changes takes place. Far from producing two eggs every six or seven weeks, a female alone in a cage lays no eggs at all. A male alone shows no interest when we offer it nesting material, eggs or young. The cycle of psychobiological changes I have described is, then, one that occurs more or less synchronously in each member of a pair of doves living together but that will not occur independently in either of the pair living alone.

In a normal breeding cycle both the male and the female sit on the eggs almost immediately after they are laid. The first question we asked ourselves was whether this is because the birds are always ready to sit on eggs or because they come into some special condition of readiness to incubate at about the time the eggs are produced.

We kept male and female doves in isolation for several weeks and then placed male-female pairs in test cages, each supplied with a nest bowl containing a normal dove nest with two eggs. The birds did not sit; they acted almost as if the eggs were not there. They courted, then built their own nest (usually on top of the planted nest and its eggs, which we had to keep fishing out to keep the stimulus situation constant!), then finally sat on the eggs—five to seven days after they had first encountered each other.

This clearly indicated that the doves are not always ready to sit on eggs; under the experimental conditions they changed from birds that did not want to incubate to birds that did want to incubate in five to seven days. What had induced this change? It could not have been merely the passage of time since their last breeding experience, because this had varied from four to six or more weeks in different pairs, whereas the variation in time spent in the test cage before sitting was only a couple of days.

Could the delay of five to seven days represent the time required for the birds to get over the stress of being handled and become accustomed to the strange cage? To test this possibility we placed pairs of doves in cages without any nest bowls or nesting material and separated each male and female by an opaque partition. After seven days we removed the partition and introduced nesting material and a formed nest with eggs. If the birds had merely needed time to recover from being handled and become acclimated to the cage, they should now have sat on the eggs immediately. They did not do so; they sat only after five to seven days, just as if they had been introduced into the cage only when the opaque partition was removed.

The next possibility we considered was that in this artificial situation stimulation from the eggs might induce the change from a nonsitting to a sitting "mood" but that this effect required five to seven days to reach a threshold value at which the behavior would change.

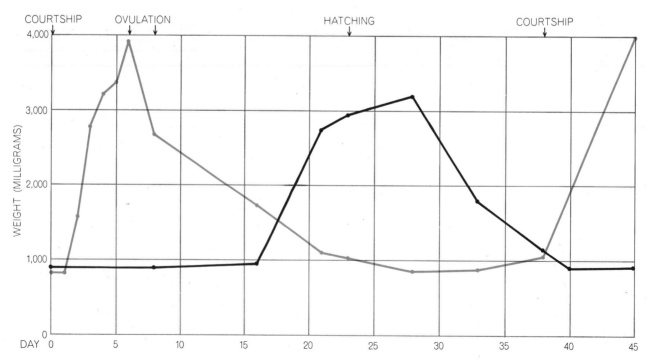

ANATOMICAL AND PHYSIOLOGICAL changes are associated with the behavioral changes of the cycle. The chart gives average weights of the crop (*black curve*) and the female oviduct (*color*) at various stages measured in days after the beginning of courtship.

We therefore placed pairs of birds in test cages with empty nest bowls and a supply of nesting material but no eggs. The birds courted and built nests. After seven days we removed the nest bowl and its nest and replaced it with a fresh bowl containing a nest and eggs. All these birds sat within two hours.

It was now apparent that some combination of influences arising from the presence of the mate and the availability of the nest bowl and nesting material induced the change from nonreadiness to incubate to readiness. In order to distinguish between these influences we put a new group of pairs of doves in test cages without any nest bowl or nesting material. When, seven days later, we offered these birds nesting material and nests with eggs, most of them did not sit immediately. Nor did they wait the full five to seven days to do so; they sat after one day, during which they engaged in intensive nest-building. A final group, placed singly in cages with nests and eggs, failed to incubate at all, even after weeks in the cages.

In summary, the doves do not build nests as soon as they are introduced into a cage containing nesting material, but they will do so immediately if the nesting material is introduced for the first time after they have spent a while together; they will not sit immediately on eggs offered after the birds have been in a bare cage together for some days, but they will do so if they were able to do some nest-building during the end of their period together. From these experiments it is apparent that there are two kinds of change induced in these birds: first, they are changed from birds primarily interested in courtship to birds primarily interested in nest-building, and this change is brought about by stimulation arising from association with a mate; second, under these conditions they are further changed from birds primarily interested in nest-building to birds interested in sitting on eggs, and this change is encouraged by participation in nest-building.

The course of development of readiness to incubate is shown graphically by the results of another experiment, which Philip N. Brody, Rochelle Wortis and I undertook shortly after the ones just described. We placed pairs of birds in test cages for varying numbers of days, in some cases with and in others without a nest bowl and nesting material. Then we introduced a nest and eggs into the cage. If neither bird sat within three hours, the test was scored as nega-

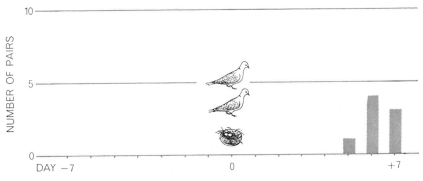

READINESS TO INCUBATE was tested with four groups of eight pairs of doves. Birds of the first group were placed in a cage containing a nest and eggs. They went through courtship and nest-building behavior before finally sitting after between five and seven days.

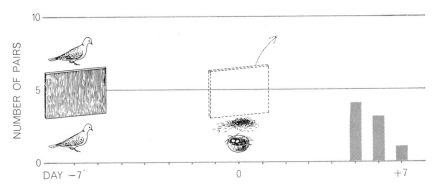

EFFECT OF HABITUATION was tested by keeping two birds separated for seven days in the cage before introducing nest and eggs. They still sat only after five to seven days.

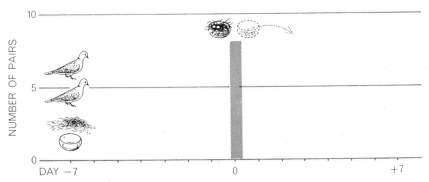

MATE AND NESTING MATERIAL had a dramatic effect on incubation-readiness. Pairs that had spent seven days in courtship and nest-building sat as soon as eggs were offered.

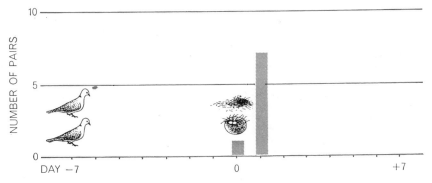

PRESENCE OF MATE without nesting activity had less effect. Birds that spent a week in cages with no nest bowls or hay took a day to sit after nests with eggs were introduced.

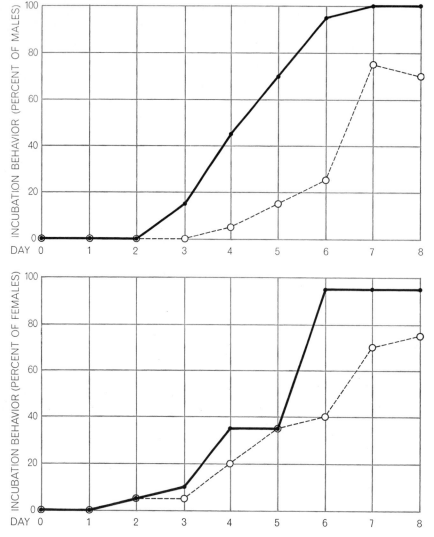

DURATION OF ASSOCIATION with mate and nesting material affects incubation behavior. The abscissas give the length of the association for different groups of birds. The plotted points show what percentage of each group sat within three hours of being offered eggs. The percentage increases for males (*top*) and females (*bottom*) as a function of time previously spent with mate (*open circles*) or with mate and nesting material (*solid dots*).

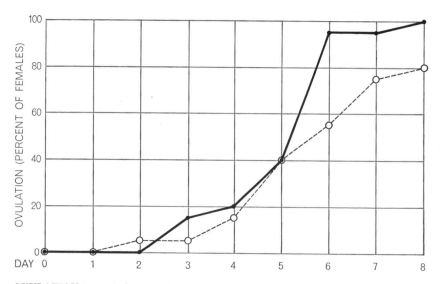

OVULATION is similarly affected. These curves, coinciding closely with those of the bottom chart above, show the occurrence of ovulation in the same birds represented there.

tive and both birds were removed for autopsy. If either bird sat within three hours, that bird was removed and the other bird was given an additional three hours to sit. The experiment therefore tested—independently for the male and the female—the development of readiness to incubate as a function of the number of days spent with the mate, with or without the opportunity to build a nest.

It is apparent [*see top illustration at left*] that association with the mate gradually brings the birds into a condition of readiness to incubate and that this effect is greatly enhanced by the presence of nesting material. Exposure to the nesting situation does not stimulate the onset of readiness to incubate in an all-or-nothing way; rather, its effect is additive with the effect of stimulation provided by the mate. Other experiments show, moreover, that the stimulation from the mate and nesting material is sustained. If either is removed, the incidence of incubation behavior decreases.

The experiments described so far made it clear that external stimuli normally associated with the breeding situation play an important role in inducing a state of readiness to incubate. We next asked what this state consists of physiologically. As a first approach to this problem we attempted to induce incubation behavior by injecting hormones into the birds instead of by manipulating the external stimulation. We treated birds just as we had in the first experiment but injected some of the birds with hormones while they were in isolation, starting one week before they were due to be placed in pairs in the test cages. When both members of the pair had been injected with the ovarian hormone progesterone, more than 90 percent of the eggs were covered by one of the birds within three hours after their introduction into the cage instead of five to seven days later. When the injected substance was another ovarian hormone—estrogen—the effect on most birds was to make them incubate after a latent period of one to three days, during which they engaged in nest-building behavior. The male hormone testosterone had no effect on incubation behavior.

During the 14 days when the doves are sitting on the eggs, their crops increase enormously in weight. Crop growth is a reliable indicator of the secretion of the hormone prolactin by the birds' pituitary glands. Since this growth coincides with the development of incubation behavior and culminates in the secretion of the crop-milk the birds feed to their young after the eggs hatch, Brody and I have recently examined the effect of injected prolactin on incubation behavior. We find that prolactin is not so effective as progesterone in inducing incubation behavior, even at dosage levels that induce full development of the crop. For example, a total prolactin dose of 400 international units induced only 40 percent of the birds to sit on eggs early, even though their average crop weight was about 3,000 milligrams, or more than three times the normal weight. Injection of 10 units of the hormone induced significant increases in crop weight (to 1,200 milligrams) but no increase in the frequency of incubation behavior. These results, together with the fact that in a normal breeding cycle the crop begins to increase in weight only after incubation begins, make it unlikely that prolactin plays an important role in the initiation of normal incubation behavior in this species. It does, however, seem to help to maintain such behavior until the eggs hatch.

Prolactin is much more effective in inducing ring doves to show regurgitation-feeding responses to squabs. When 12 adult doves with previous breeding experience were each injected with 450 units of prolactin over a seven-day period and placed, one bird at a time, in cages with squabs, 10 of the 12 fed the squabs from their engorged crops, whereas none of 12 uninjected controls did so or even made any parental approaches to the squabs.

This experiment showed that prolactin, which is normally present in considerable quantities in the parents when the eggs hatch, does contribute to the doves' ability to show parental feeding behavior. I originally interpreted it to mean that the prolactin-induced engorgement of the crop was necessary in order for any regurgitation feeding to take place, but E. Klinghammer and E. H. Hess of the University of Chicago have correctly pointed out that this was an error, that ring doves are capable of feeding young if presented with them rather early in the incubation period. They do so even though they have no crop-milk, feeding a mixture of regurgitated seeds and a liquid. We are now studying the question of how early the birds can do this and how this ability is related to the onset of prolactin secretion.

The work with gonad-stimulating hormones and prolactin demonstrates that the various hormones successively produced by the birds' glands during their reproductive cycle are capable of inducing the successive behavioral changes that characterize the cycle.

Up to this point I have described two main groups of experiments. One group demonstrates that external stimuli induce changes in behavioral status of a kind normally associated with the progress of the reproductive cycle; the second shows that these behavioral changes can also be induced by hormone administration, provided that the choice of hormones is guided by knowledge of the succession of hormone secretions during a normal reproductive cycle. An obvious—and challenging—implication of these results is that external stimuli may induce changes in hormone secretion, and that environment-induced hormone secretion may constitute an integral part of the mechanism of the reproductive behavior cycle. We have attacked the problem of the environmental stimulation of hormone secretion in a series of experiments in which, in addition to examining the effects of external stimuli on the birds' behavioral status, we have examined their effects on well-established anatomical indicators of the presence of various hormones.

Background for this work was provided by two classic experiments with the domestic pigeon, published during the 1930's, which we have verified in the ring dove. At the London Zoo, L. H. Matthews found that a female pigeon would lay eggs as a result of being placed in a cage with a male from whom she was separated by a glass plate. This was an unequivocal demonstration that visual and/or auditory stimulation provided by the male induces ovarian development in the female. (Birds are quite insensitive to olfactory stimulation.) And M. D. Patel of the University of Wisconsin found that the crops of breeding pigeons, which develop strikingly during the incubation period, would regress to their resting state if the incubating birds were removed from their nests and would fail to develop at all if the birds were removed before crop growth had begun. If, however, a male pigeon, after being removed from his nest, was placed in an adjacent cage from which he could see his mate still sitting on the eggs, his crop would develop just as if he were himself incubating! Clearly stimuli arising from participation in incubation, including visual stimuli, cause the doves' pituitary glands to secrete prolactin.

Our autopsies showed that the incidence of ovulation in females that had associated with males for various periods coincided closely with the incidence of incubation behavior [*see bottom illustration on opposite page*]; statistical analysis reveals a very high degree of association. The process by which the dove's ovary develops to the point of ovulation includes a period of estrogen secretion followed by one of progesterone secretion, both induced by appropriate ovary-stimulating hormones from the pituitary gland. We therefore conclude that stimuli provided by the male, augmented by the presence of the nest bowl and nesting material, induce the secretion of gonad-stimulating hormones by the female's pituitary, and that the onset of readiness to incubate is a result of this process.

As I have indicated, ovarian development, culminating in ovulation and egg-laying, can be induced in a female dove merely as a result of her seeing a male through a glass plate. Is this the result of the mere presence of another bird or of something the male does because he is a male? Carl Erickson and I have begun to deal with this question. We placed 40 female doves in separate cages, each separated from a male by a glass plate. Twenty of the stimulus animals were normal, intact males, whereas the remaining 20 had been castrated several weeks before. The intact males all exhibited vigorous bow-cooing immediately on being placed in the cage, whereas none of the castrates did so. Thirteen of the 20 females with intact males ovulated during the next seven days, whereas only two of those with the castrates did so. Clearly ovarian development in the female is not induced merely by seeing another bird but by seeing or hearing it act like a male as the result of the effects of its own male hormone on its nervous system.

Although crop growth, which begins early in the incubation period, is apparently stimulated by participation in incubation, the crop continues to be large and actively secreting for quite some time after the hatching of the eggs. This suggests that stimuli provided by the squabs may also stimulate prolactin secretion. In our laboratory Ernst Hansen substituted three-day-old squabs for eggs in various stages of incubation and after four days compared the adults' crop weights with those of birds that had continued to sit on their eggs dur-

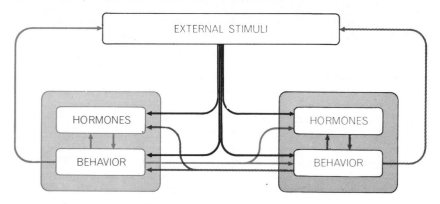

INTERACTIONS that appear to govern the reproductive-behavior cycle are suggested here. Hormones regulate behavior and are themselves affected by behavioral and other stimuli. And the behavior of each bird affects the hormones and the behavior of its mate.

ing the four days. He found that the crops grow even faster when squabs are in the nest than when the adults are under the influence of the eggs; the presence of squabs can stimulate a dove's pituitary glands to secrete more prolactin even before the stage in the cycle when the squabs normally appear.

This does not mean, however, that any of the stimuli we have used can induce hormone secretion at *any* time, regardless of the bird's physiological condition. If we place a pair of ring doves in a cage and allow them to go through the normal cycle until they have been sitting on eggs for, say, six days and we then place a glass partition in the cage to separate the male from the female and the nest, the female will continue to sit on the eggs and the male's crop will continue to develop just as if he were himself incubating. This is a simple replication of one of Patel's experiments. Miriam Friedman and I have found, however, that if the male and female are separated from the beginning, so that the female must build the nest by herself and sit alone from the beginning, the crop of the male does

not grow. By inserting the glass plate at various times during the cycle in different groups of birds, we have found that the crop of the male develops fully only if he is not separated from the female until 72 hours or more after the second egg is laid. This means that the sight of the female incubating induces prolactin secretion in the male only if he is in the physiological condition to which participation in nest-building brings him. External stimuli associated with the breeding situation do indeed induce changes in hormone secretion.

The experiments summarized here point to the conclusion that changes in the activity of the endocrine system are induced or facilitated by stimuli coming from various aspects of the environment at different stages of the breeding cycle, and that these changes in hormone secretion induce changes in behavior that may themselves be a source of further stimulation.

The regulation of the reproductive cycle of the ring dove appears to depend, at least in part, on a double set of reciprocal interrelations. First, there

is an interaction of the effects of hormones on behavior and the effects of external stimuli—including those that arise from the behavior of the animal and its mate—on the secretion of hormones. Second, there is a complicated reciprocal relation between the effects of the presence and behavior of one mate on the endocrine system of the other and the effects of the presence and behavior of the second bird (including those aspects of its behavior induced by these endocrine effects) back on the endocrine system of the first. The occurrence in each member of the pair of a cycle found in neither bird in isolation, and the synchronization of the cycles in the two mates, can now readily be understood as consequences of this interaction of the inner and outer environments.

The physiological explanation of these phenomena lies partly in the fact that the activity of the pituitary gland, which secretes prolactin and the gonad-stimulating hormones, is largely controlled by the nervous system through the hypothalamus. The precise neural mechanisms for any complex response are still deeply mysterious, but physiological knowledge of the brain-pituitary link is sufficiently detailed and definite so that the occurrence of a specific hormonal response to a specific external stimulus is at least no more mysterious than any other stimulus-response relation. We are currently exploring these responses in more detail, seeking to learn, among other things, the precise sites at which the various hormones act. And we have begun to investigate another aspect of the problem: the effect of previous experience on a bird's reproductive behavior and the interactions between these experiential influences and the hormonal effects.

Stress and Behavior

by Seymour Levine
January 1971

*The chain of pituitary and adrenal hormones that
regulates responses to stress plays a major role in
learning and other behaviors. It may be that effective
behavior depends on some optimum level of stress*

Hans Selye's concept of the general "stress syndrome" has surely been one of the fruitful ideas of this era in biological and medical research. He showed that in response to stress the body of a mammal mobilizes a system of defensive reactions involving the pituitary and adrenal glands. The discovery illuminated the causes and symptoms of a number of diseases and disorders. More than that, it has opened a new outlook on the functions of the pituitary-adrenal system. One can readily understand how the hormones of this system may defend the body against physiological insult, for example by suppressing inflammation and thus preventing tissue damage. It is a striking fact, however, that the system's activity can be evoked by all kinds of stresses, not only by severe somatic stresses such as disease, burns, bone fractures, temperature extremes, surgery and drugs but also by a wide range of psychological conditions: fear, apprehension, anxiety, a loud noise, crowding, even mere exposure to a novel environment. Indeed, most of the situations that activate the pituitary-adrenal system do not involve tissue damage. It appears, therefore, that these hormones in animals, including man, may have many functions in addition to the defense of tissue integrity, and as a psychologist I have been investigating possible roles of the pituitary-adrenal system in the regulation of behavior.

The essentials of the system's operation in response to stress are as follows. Information concerning the stress (coming either from external sources through the sensory system or from internal sources such as a change in body temperature or in the blood's composition) is received and integrated by the central nervous system and is presumably delivered to the hypothalamus, the basal area of the brain. The hypothalamus secretes a substance called the corticotropin-releasing factor (CRF), which stimulates the pituitary to secrete the hormone ACTH. This in turn stimulates the cortex of the adrenal gland to step up its synthesis and secretion of hormones, particularly those known as glucocorticoids. In man the glucocorticoid is predominantly hydrocortisone; in many lower animals such as the rat it is corticosterone.

The entire mechanism is exquisitely controlled by a feedback system. When the glucocorticoid level in the circulating blood is elevated, the central nervous system, receiving the message, shuts off the process that leads to secretion of the stimulating hormone ACTH. Two experimental demonstrations have most clearly verified the existence of this feedback process. If the adrenal gland is removed from an animal, the pituitary puts out abnormal amounts of ACTH, presumably because the absence of the adrenal hormone frees it from restriction of this secretion. On the other hand, if crystals of glucocorticoid are implanted in the hypothalamus, the animal's secretion of ACTH stops almost completely, just as if the adrenal cortex were releasing large quantities of the glucocorticoid.

Now, it is well known that a high level of either of these hormones (ACTH or glucocorticoid) in the circulating blood can have dramatic effects on the brain. Patients who have received glucocorticoids for treatment of an illness have on occasion suffered severe mental changes, sometimes leading to psychosis. And patients with a diseased condition of the adrenal gland that caused it to secrete an abnormal amount of cortical hormone have also shown effects on the brain, including changes in the pattern of electrical activity and convulsions.

Two long-term studies of my own, previously reported in SCIENTIFIC AMERICAN [see "Stimulation in Infancy," Offprint 436, and "Sex Differences in the Brain," Offprint 498], strongly indicated that hormones play an important part in the development of behavior. One study showed that rats subjected to shocks and other stresses in early life developed normally and were able to cope well with stresses later, whereas animals that received no stimulation in infancy grew up to be timid and deviant in behavior. At the adult stage the two groups differed sharply in the response of the pituitary-adrenal system to stress: the animals that had been stimulated in infancy showed a prompt and effective hormonal response; those that had not been stimulated responded slowly and ineffectively. The other study, based on the administration or deprivation of sex hormones at a critical early stage of development in male and female rats, indicated that these treatments markedly affected the animals' later behavior, nonsexual as well as sexual. It is noteworthy that the sex hormones are steroids rather similar to those produced by the adrenal cortex.

Direct evidence of the involvement of the pituitary-adrenal system in overt be-

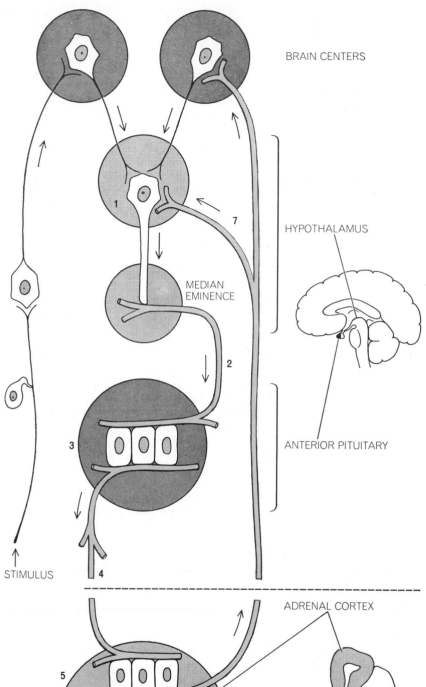

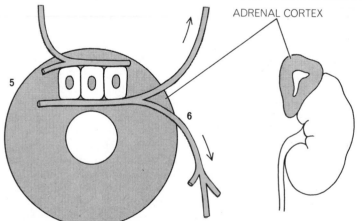

PITUITARY-ADRENAL SYSTEM involves nerve cells and hormones in a feedback loop.
A stress stimulus reaching neurosecretory cells of the hypothalamus in the base of the brain
(*1*) stimulates them to release corticotropin-releasing factor (CRF), which moves through
short blood vessels (*2*) to the anterior lobe of the pituitary gland (*3*). Pituitary cells thereupon release adrenocorticotrophic hormone (ACTH) into the circulation (*4*). The ACTH
stimulates cells of the adrenal cortex (*5*) to secrete glucocorticoid hormones (primarily
hydrocortisone in man) into the circulation (*6*). When glucocorticoids reach neurosecretory
cells or other brain cells (it is not clear which), they modulate CRF production (*7*).

havior was reported by two groups of
experimenters some 15 years ago. Morti-
mer H. Appley, now at the University of
Massachusetts, and his co-workers were
investigating the learning of an avoidance response in rats. The animals were
placed in a "shuttle box" divided into
two compartments by a barrier. An electric shock was applied, and if the animals
crossed the barrier, they could avoid or
terminate the shock. The avoidance response consisted in making the move
across the barrier when a conditioned
stimulus, a buzzer signaling the onset of
the shock, was sounded. Appley found
that when the pituitary gland was removed surgically from rats, their learning of the avoidance response was severely retarded. It turned out that an injection of ACTH in pituitary-deprived
rats could restore the learning ability to
normal. At about the same time Robert
E. Miller and Robert Murphy of the University of Pittsburgh reported experiments showing that ACTH could affect
extinction of the avoidance response.
Normally if the shocks are discontinued,
so that the animal receives no shock
when it fails to react to the conditioned
stimulus (the buzzer in this case), the
avoidance response to the buzzer is gradually extinguished. Miller and Murphy
found that when they injected ACTH in
animals during the learning period, the
animals continued to make the avoidance response anyway, long after it was
extinguished in animals that had not received the ACTH injection. In short,
ACTH inhibited the extinction process.

These findings were not immediately
followed up, perhaps mainly because
little was known at the time about the
details of the pituitary-adrenal system
and only rudimentary techniques were
available for studying it. Since then purified preparations of the hormones involved and new techniques for accurate
measurement of these substances in the
circulating blood have been developed,
and the system is now under intensive
study. Most of the experimental investigation is being conducted at three centers: in the Institute of Pharmacology
at the University of Utrecht under David
de Wied, in the Institute of Physiology
at the University of Pecs in Hungary under Elemér Endroczi and in our own
laboratories in the department of psy-
chiatry at Stanford University.

The new explorations of the pituitary-
adrenal system began where the ground
had already been broken: in studies of
the learning and extinction of the avoidance response, primarily by use of the

shuttle box. De Wied verified the role of ACTH both in avoidance learning and in inhibiting extinction of the response. He did this in physiological terms by means of several experiments. He verified the fact that removal of the pituitary gland severely retards the learning of a conditioned avoidance response. He also removed the adrenal gland from rats and found that the response was then not extinguished, presumably because adrenal hormones were no longer present to restrict the pituitary's output of ACTH. When he excised the pituitary, thus eliminating the secretion of ACTH, the animals returned to near-normal behavior in the extinction of the avoidance response.

In further experiments De Wied injected glucocorticoids, including corticosterone, the principal steroid hormone of the rat's adrenal cortex, into animals that had had the adrenal gland, but not the pituitary, removed; as expected, this had the effect of speeding up the extinction of the avoidance response. Similarly, the administration to such animals of dexamethasone, a synthetic glucocorticoid that is known to be a potent inhibitor of ACTH, resulted in rapid extinction of the avoidance response; the larger the

dose, the more rapid the extinction. Curiously, De Wied found that corticosterone and dexamethasone promoted extinction even in animals that lacked the pituitary gland, the source of ACTH. This indicated that the glucocorticoid can produce its effect not only through suppression of ACTH but also, in some way, by acting directly on the central nervous system. It has recently been found, on the other hand, that there may be secretions from the pituitary other than ACTH that can affect learning and inhibit extinction of the avoidance response. The inhibition can be produced, for example, by a truncated portion of the ACTH molecule consisting of the first 10 amino acids in the sequence of 39 in the rat's ACTH—a molecular fragment that has no influence on the adrenal cortex. The same fragment, along with other smaller peptides recently isolated by De Wied, can also overcome the deficit in avoidance learning that is produced by ablation of the pituitary.

With an apparatus somewhat different from the shuttle box we obtained further light in our laboratory on ACTH's effects on behavior. We first train the animals to press a bar to obtain water. After this learning has been established

the animal is given an electric shock on pressing the bar. This causes the animal to avoid approaching the bar (called "passive avoidance") for a time, but after several days the animal will usually return to it in the effort to get water and then will quickly lose its fear of the bar if it is not shocked. We found, however, that if the animal was given doses of ACTH after the shock, it generally failed to return to the bar at all, even though it was very thirsty. That is to say, ACTH suppressed the bar-pressing response, or, to put it another way, it strengthened the passive-avoidance response. In animals with the pituitary gland removed, injections of ACTH suppressed a return to bar-pressing after a shock but injections of hydrocortisone did not have this effect.

The experiments I have described so far have involved behavior under the stress of fear and anxiety. Our investigations with the bar-pressing device go on to reveal that the pituitary-adrenal system also comes into play in the regulation of behavior based on "appetitive" responses (as opposed to avoidance responses). Suppose we eliminate the electric shock factor and simply arrange that

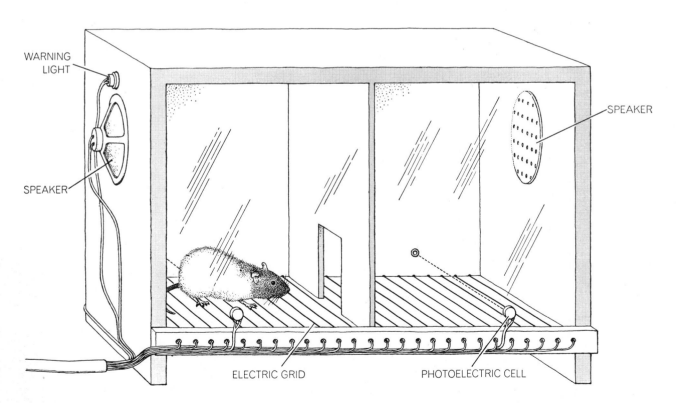

"SHUTTLE BOX" used for studying avoidance behavior is a two-compartment cage. The floor can be electrically charged. A shock is delivered on the side occupied by the rat (detected by the photocell). The rat can avoid the shock by learning to respond to the conditioned stimulus: a light and noise delivered briefly before the shock. The avoidance response, once learned, is slowly "extinguished" if the conditioned stimulus is no longer accompanied by a shock. Injections of ACTH inhibited the extinction process.

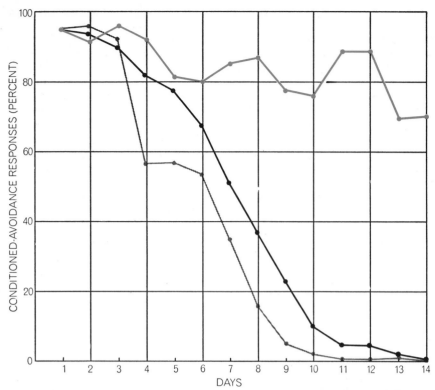

EXTINCTION of the avoidance response was studied by David de Wied of the University of Utrecht. Removal of the adrenal gland inhibited extinction (*color*); the rats responded to the conditioned stimulus in the absence of shock, presumably because adrenal hormones were not available to restrict ACTH output. When the pituitary was removed, the rate of extinction (*gray*) was about the same as in rats given only a sham operation (*black*).

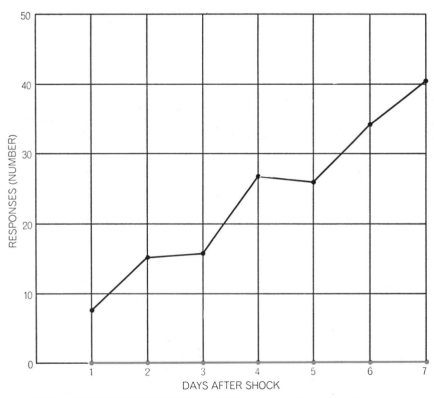

PASSIVE AVOIDANCE BEHAVIOR is studied by observing how rats, trained to press a bar for water, avoid the bar after they get a shock on pressing it. Before being shocked rats pressed the bar about 75 times a day. After the shock the control animals returned to the bar and, finding they were not shocked, gradually increased their responses (*black curve*). Rats injected with ACTH stayed away (*color*): ACTH strengthens the avoidance response.

after the animal has learned to press the bar for water it fails to obtain water on later trials. Normally the animal's bar-pressing behavior is then quickly extinguished. We found, however, that when we injected ACTH in the animals in these circumstances, the extinction of bar-pressing was delayed; the rats went on pressing the bar for some time although they received no water as reinforcement. Following up this finding, we measured the corticosterone levels in the blood of normal, untreated rats both when they were reinforced and when they were not reinforced on pressing the bar. The animals that received no water reinforcement, with the result of rapid extinction of bar-pressing, showed a marked rise in activity of the pituitary-adrenal system during this period, whereas in animals that received water each time they pressed the bar there was no change in the hormonal output. In short, the extinction of appetitive behavior in this case clearly involved the pituitary-adrenal system.

Further investigations have now shown that the system affects a much wider range of behavior than learning and extinction. One of the areas that has been studied is habituation: the gradual subsidence of reactions that had appeared on first exposure to a novel stimulus when the stimulus is repeated. An organism presented with an unexpected stimulus usually exhibits what Ivan Pavlov called an orientation reflex, which includes increased electrical activity in the brain, a reduction of blood flow to the extremities, changes in the electrical resistance of the skin, a rise in the level of adrenal-steroid hormones in the blood and some overt motor activity of the body.

If the stimulus is repeated frequently, these reactions eventually disappear; the organism is then said to be habituated to the stimulus. Endroczi and his co-workers recently examined the influence of ACTH on habituation of one of the reactions in human subjects—the increase of electrical activity in the brain, as indicated by electroencephalography. The electrical activity evoked in the human brain by a novel sound or a flickering light generally subsides, after repetition of the stimulus, into a pattern known as electroencephalogram (EEG) synchronization, which is taken to be a sign of habituation. Endroczi's group found that treatment of their subjects with ACTH or the 10-amino-acid fragment of ACTH produced a marked delay in the appearance of the synchronization pattern, indicating that the hormone

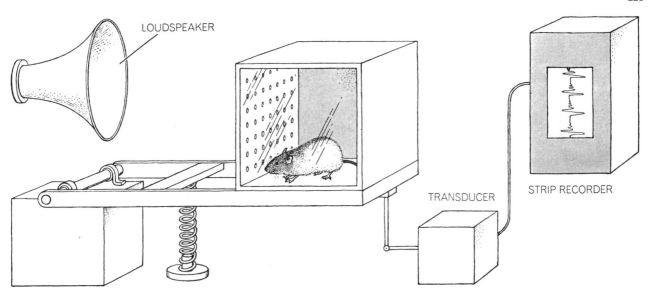

"STARTLE" RESPONSE is measured by placing a rat in a cage with a movable floor and exposing it to a sudden, loud noise. The rat tenses or jumps, and the resulting movement of the floor is transduced into movement of a pen on recording paper. After a number of repetitions of the noise the rat becomes habituated to it and the magnitude of the animal's startle response diminishes.

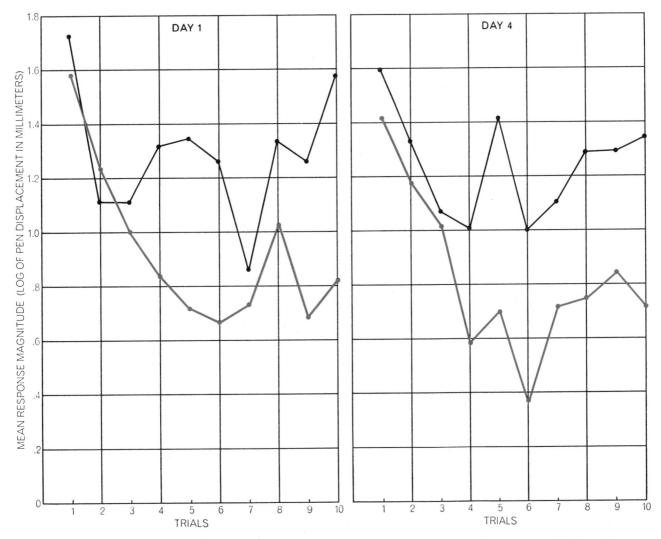

HABITUATION is affected by the pituitary-adrenal system. If a crystal of the adrenal hormone hydrocortisone is implanted in a rat's hypothalamus, preventing ACTH secretion, habituation is speeded up, as shown here. The mean startle response (shown as the logarithm of the recording pen's movement) falls away more rapidly in implanted rats (*color*) than in control animals (*black*).

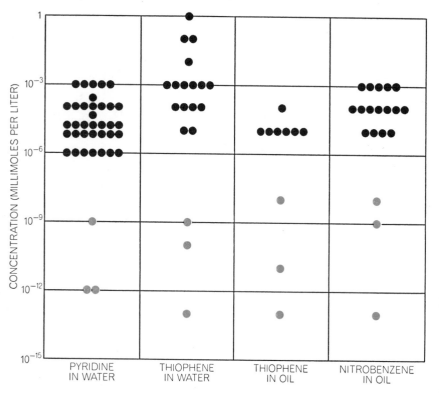

SENSORY FUNCTION is also affected by adrenocortical hormones. Robert I. Henkin of the National Heart Institute found that patients whose adrenal-hormone function is poor are much more sensitive to odor. Placing various chemicals in solution, he measured the detection threshold: the concentration at which an odor could be detected in the vapor. The threshold was much lower in the patients (*color*) than in normal volunteers (*black*).

inhibits the process of habituation.

Experiments with animals in our laboratory support that finding. The stimulus we used was a sudden sound that produces a "startle" response in rats, which is evidenced by vigorous body movements. After a number of repetitions of the sound stimulus the startle response fades. It turned out that rats deprived of the adrenal gland (and consequently with a high level of ACTH in their circulation) took significantly longer than intact animals to habituate to the sound stimulus. An implant of the adrenal hormone hydrocortisone in the hypothalamus, on the other hand, speeded up habituation.

A series of studies by Robert I. Henkin of the National Heart Institute has demonstrated that hormones of the adrenal cortex play a crucial role in the sensory functions in man. Patients whose adrenal gland has been removed surgically or is functioning poorly show a marked increase in the ability to detect sensory signals, particularly in the senses of taste, smell, hearing and proprioception (sensing of internal signals). On the other hand, patients with Cushing's syndrome, marked by excessive secretion

from the adrenal cortex, suffer a considerable dulling of the senses. Henkin showed that sensory detection and the integration of sensory signals are regulated by a complex feedback system involving interactions of the endocrine system and the nervous system. Although patients with a deficiency of adrenal cortex hormones are extraordinarily sensitive in the detection of sensory signals, they have difficulty integrating the signals, so that they cannot evaluate variations in properties such as loudness and tonal qualities and have some difficulty understanding speech. Proper treatment with steroid hormones of the adrenal gland can restore normal sensory detection and perception in such patients.

Henkin has been able to detect the effects of the adrenal corticosteroids on sensory perception even in normal subjects. There is a daily cycle of secretion of these steroid hormones by the adrenal cortex. Henkin finds that when adrenocortical secretion is at its highest level, taste detection and recognition is at its lowest, and vice versa.

In our laboratory we have found that the adrenal's steroid hormones can have a truly remarkable effect on the ability

of animals to judge the passage of time. Some years ago Murray Sidman of the Harvard Medical School devised an experiment to test this capability. The animal is placed in an experimental chamber and every 20 seconds an electric shock is applied. By pressing a bar in the chamber the animal can prevent the shock from occurring, because the bar resets the triggering clock to postpone the shock for another 20 seconds. Thus the animal can avoid the shock altogether by appropriate timing of its presses on the bar. Adopting this device, we found that rats learned to press the bar at intervals averaging between 12 and 15 seconds. This prevented a majority of the shocks. We then gave the animals glucocorticoids and found that they became significantly more efficient! They lengthened the interval between bar presses and took fewer shocks. Evidently under the influence of the hormones the rats were able to make finer discriminations concerning the passage of time. Monkeys also showed improvement in timing performance in response to treatment with ACTH.

The mechanism by which the pituitary-adrenal hormones act to regulate or influence behavior is still almost completely unknown. Obviously they must do so by acting on the brain. It is well known that hormones in general are targeted to specific sites and that the body tissues have a remarkable selectivity for them. The uterus, for instance, picks up and responds selectively to estrogen and progesterone among all the hormones circulating in the blood, and the seminal vesicles and prostate gland of the male select testosterone. There is now much evidence that organs of the brain may be similarly selective. Bruce Sherman McEwen of Rockefeller University has recently reported that the hippocampus, just below the cerebral cortex, appears to be a specific receptor site for hormones of the adrenal cortex, and other studies indicate that the lateral portion of the hypothalamus may be a receptor site for gonadal hormones. We have the inviting prospect, therefore, that exploration of the brain to locate the receptor sites for the hormones of the pituitary-adrenal system, and studies of the hormones' action on the cells of these sites, may yield important information on how the system regulates behavior. Bela Bohun in Hungary has already demonstrated that implantation of small quantities of glucocorticoids in the reticular formation in the brain stem facilitates extinction of an avoidance response.

Since this system plays a key role in learning, habituation to novel stimuli, sensing and perception, it obviously has a high adaptive significance for mammals, including man. Its reactions to moderate stress may contribute greatly to the behavioral effectiveness and stability of the organism. Just as the studies of young animals showed, contrary to expectations, that some degree of stress in infancy is necessary for the development of normal, adaptive behavior, so the information we now have on the operations of the pituitary-adrenal system indicates that in many situations effective behavior in adult life may depend on exposure to some optimum level of stress.

23

Learning in the Octopus

by Brian B. Boycott
March 1965

The animal cooperates readily in laboratory experiments. Tests of its capacities before and after brain surgery lend support to the idea that there are two kinds of memory: long-term and short-term

In recent years a number of British students of animal behavior, of whom I am one, have done much of their experimental work at the Stazione Zoologica in Naples. The reason why these investigations have been pursued in Naples rather than in Britain is that our chosen experimental animal—*Octopus vulgaris,* or the common European octopus—is found in considerable numbers along the shores of the Mediterranean. *Octopus vulgaris* is a cooperative experimental subject. If it is provided with a shelter of bricks at one end of a tank of running seawater, it takes up residence in the shelter. When a crab or some other food object is placed at the other end of the tank, the octopus swims or walks the length of the tank, catches the prey with its arms and carries it home to be poisoned and eaten. Since it responds so consistently to the presence of prey, the animal is readily trained. It is also tolerant of surgery and survives the removal of the greater part of its brain. This makes the octopus an ideal animal with which to test directly the relation between the various parts of the brain and the various kinds of perception and learning.

There are many unanswered questions about such relations. We now know a great deal about conduction in nerve fibers, transmission from nerve fiber to nerve fiber at the synapses and the integrative action of nerve fibers in such aggregations of nerve cells as the spinal cord; we are almost wholly ignorant, however, of the levels of neural integration involved in such long-term activities as memory. We can still quote with sympathy the remark of the late Karl S. Lashley of Harvard University: "I sometimes feel, in reviewing the evidence on the localization of the memory trace, that the necessary conclusion is that learning is just not possible!"

It was J. Z. Young, then at the University of Oxford, who first began to exploit the possibility of using for memory studies various marine mollusks of the class Cephalopoda. Shortly before World War II he undertook to work with the cuttlefish *Sepia officinalis.* In a simple experiment he and F. K. Sanders removed from a cuttlefish that part of the brain known as the vertical lobe. They found that a cuttlefish so deprived would respond normally—that is, attack—when it was shown a prawn. If the prawn was pulled out of sight around a corner after the attack began, however, the cuttlefish could not pursue it. The animal might advance to where the prawn had first been presented, but it was apparently unable to make whatever associations were necessary to follow the prawn around the corner. One might say it could not remember to hunt when the prey was no longer in sight. Young and Sanders found that surgical lesions in certain other parts of the cuttlefish's brain did not affect this hunting behavior.

In 1947 I had the privilege of joining Young in his studies. Financed by the Nuffield Foundation, we began

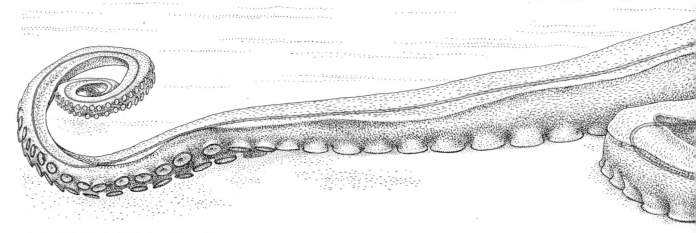

COMMON EUROPEAN OCTOPUS (*Octopus vulgaris*) is the experimental animal the author and his fellow-workers in Naples use for their investigations of perception and learning. The animal's brain (*in color between the eyes*) is about two cubic

work at the Stazione Zoologica, where both seawater aquariums and *Octopus vulgaris* were in abundant supply. The octopus was chosen in preference to the other common laboratory cephalopods—cuttlefishes and squids—because they do not survive so well in tanks and are less tolerant of surgery. At Naples today, in addition to Young's associates from University College London, there are investigators from the University of Oxford led by Stuart Sutherland and from the University of Cambridge led by Martin J. Wells, all going their various ways toward using the brains of octopuses for the analysis of perception and memory. At present most of the work is financed by the Office of Aerospace Research of the U.S. Air Force.

In our early experiments we attempted to train octopuses to do a variety of things, such as taking crabs out of one kind of pot but not out of another, to run a maze and so on. Our most successful experiment was to put a crab in the tank together with some kind of geometric figure—say a Plexiglas square five centimeters on a side—and give the octopus an electric shock when it made the normal attacking response. With this simple method we found that octopuses could learn not to attack a crab shown with a square but to go on attacking a crab shown without one [*see bottom illustrations on pages 228 and 229*]. Or we could train the animals to stop taking crabs but to go on eating sardines or vice versa. The purpose of these experiments was to elucidate the anatomy and connections of the animal's brain and relate them to its learning behavior.

Like the brains of most other invertebrates, the brain of the octopus surrounds its esophagus [*see illustrations on next page*]. The lobes of the brain under the esophagus contain nerve fibers that stimulate peripheral nerve centers, for example the ganglia in the arms and the mantle. These peripheral ganglia contain the nerve cells whose fibers in turn stimulate the muscles and other effectors of the body; through

them local reflexes can occur. When all of the brain except the lobes under the esophagus is removed, the octopus remains alive but lies at the bottom of the tank; it breathes regularly but maintains no definite posture. If it is sufficiently stimulated, it responds with stereotyped behavior.

A greater variety of behavior can be obtained if some of the brain lobes above the esophagus are left intact. For instance, the upper brain's median basal lobe and anterior basal lobe send their fibers down to the lower lobes and through them evoke the patterns of nerve activity involved in walking and swimming. Above these two lobes are the vertical lobe, the superior frontal lobe and the inferior frontal lobe; their surgical removal does not result in any defects of behavior that are immediately obvious.

It is with these three lobes and the two optic lobes—which lie on each side of the central mass of the brain—that this article is mostly concerned. Using the electric-shock method of training

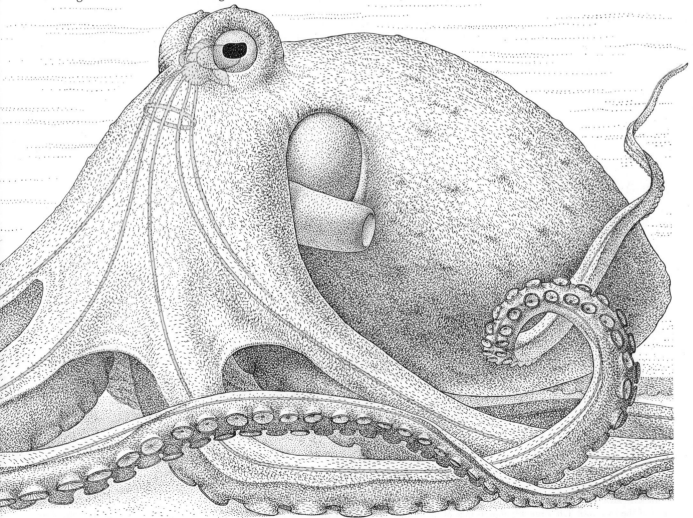

centimeters in size; the basket-like structure below it is composed of the eight major nerves of the arms, some of which are also outlined in color. The octopus adapts readily to life in a tank of seawater and can be trained easily through reward and punishment.

226

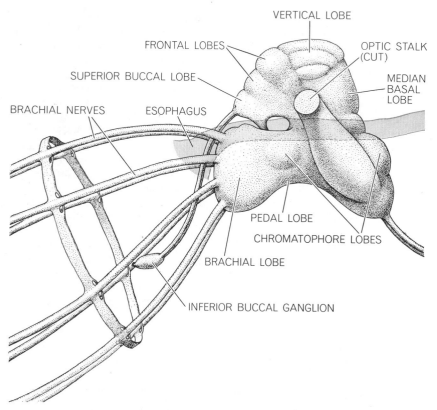

OCTOPUS BRAIN is shown in a side view with the left optic lobe removed (*see top view of brain below*). The labels identify external anatomical features of the brain and its nerve connections. As is the case with many other invertebrates, the brain of the octopus completely surrounds the animal's esophagus. Excision of the entire upper part of the brain is not fatal, but the octopus's behavior then exhibits neither learning nor memory.

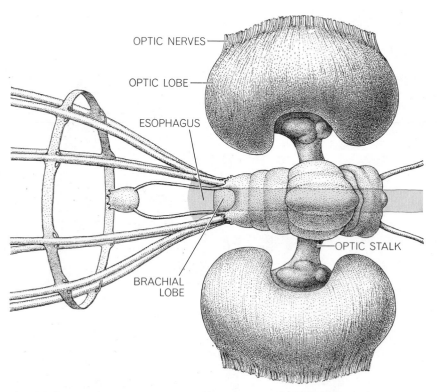

TOP VIEW OF BRAIN relates the two large optic lobes and their stalks to the central brain structure situated above and below the octopus's esophagus (*color*). Combined, the mass of the two optic lobes roughly equals that of the brain's central structure; the fringe of nerves at each lobe's outer edge connects to the retinal structures of the octopus's eyes.

we soon found that, as far as visual learning goes, removing either the vertical lobe, the superior frontal lobe or both, or cutting the nerve tracts between these two lobes, left the octopus unable to learn the required discriminations (or, if they had already been learned, unable to retain them). Since operations on other parts of the brain—performed on control animals—had no effect either on learning or on previously learned behavior, we seemed to have demonstrated that the vertical lobe and superior frontal lobe of the octopus brain are memory centers. In a sense they are, but this is an unduly simple view; in a recent summary of findings Young has listed no fewer than six different effects caused by the removal of or damage to the vertical-lobe system alone.

Karl Lashley, who studied the cerebral cortex of mammals, concluded that, in the organization of a memory, the involvement of specific groups of nerve cells is not as important as the total number of nerve cells available for organization. A similar situation appears to hold true in the functioning of the vertical lobe of the octopus brain; there is a definite relation between the amount of vertical lobe left intact and the accuracy with which a learned response is performed [*see top illustration on page 230*]. This seems to suggest that, at least in the octopus's vertical lobe and the mammalian cerebral cortex, memory is both everywhere and nowhere in particular.

Some of the difficulties such a conclusion presents may be due to a failure to distinguish experimentally between the two constituents of a memory. Whatever its nature, a memory must consist not only of a representation, in neural terms, of the learned situation but also of a mechanism that enables that representation to persist. A distinction must be made between the topology of what persists (the coding and spatial relations involved in the memory of a particular animal) and the mechanisms of persistence (the neural change that is presumably the same in the memory of any animal). Indeed, it may be that some of the theoretical confusion in the study of memory arises from the fact that experiments showing a quantitative relation between memory and nerve tissue tell us something about how the neural representation of memory is organized but nothing about how the representation is kept going.

In our experiments demonstrating that an octopus deprived of its vertical lobe could not be trained to discriminate between a crab alone (that is, reward) and a crab accompanied by a geometric figure and a shock (that is, punishment) our groups of trials were separated by intervals of approximately two hours. When we spaced the trials so that they were only five minutes apart, however, we found that such animals were capable of learning [see bottom illustration on page 230]. Using the number of trials required as a criterion of learning, we found that these animals attained a level of performance as good as that of normal animals trained with longer time intervals between trials.

One significant difference remained: a normal octopus has a learning-retention period of two weeks or longer, but animals without a vertical lobe had retention periods of only 30 minutes to two hours. These observations suggest that the establishment of a memory involves two mechanisms. There is first a short-term, or transitory, memory that, by its continuing activity between intervals of training, leads to a long-term change in the brain. If there were no reinforcement, the short-term memory

would wane; with reinforcement it keeps going and so induces the long-term—and by implication slower—changes that enable a brain to retain memories for long periods.

In 1957 Eliot Stellar of the University of Pennsylvania School of Medicine pointed out the parallels between our results with invertebrates and the unexpected discovery of a similar effect in man by Wilder Penfield, Brenda Milner and W. B. Scoville of the Montreal Neurological Institute. Epileptic patients who have been treated by surgical removal of the temporal lobes of the brain score as well in I.Q. tests after the operation as they do in tests before the onset of epilepsy. They remember their past, their profession and their relatives. They cannot, however, retain new information for more than short periods. Articles can be read and understood, but they are not remembered once they are finished and another topic is taken up. A relative may die but his death goes unremembered after an hour or so. This surgery involves the hippocampal system of the

human brain; its effects seem to suggest that, although man's cerebral cortex incorporates a long-term memory system, the hippocampal system is essential to the establishment of new long-term memories.

Today a considerable body of behavioral and psychological evidence favors the separation of memory into short-term and long-term systems. At the neurological level this distinction has brought about a reaffirmation of the role in memory of what are called self-reexciting chains. A few years ago the concept of such chains had gone out of fashion because it had been found that neither convulsive shocks nor cooling the brain to a temperature so low that all activity ceased would abolish learned responses. It is now known that if such treatments are given during the early stages of learning—that is, before a memory is fully established—they have an effect; supposedly this is because they have interfered with the more active part of the process. As the surgical operations for epilepsy indicate, a long-term memory system is intact after removal of the temporal lobes. A short-

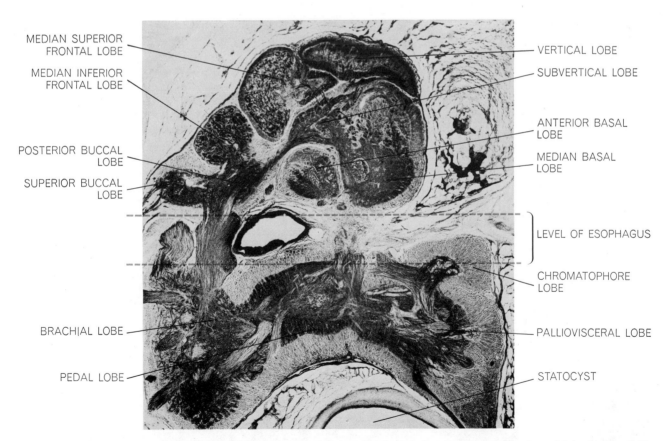

SAGITTAL SECTION stained with silver reveals some of the structures in the octopus brain. Broken lines (color) show the route of the esophagus, the boundary between the upper and lower parts of the brain. Labels identify eight lobes in the upper brain and four in the lower; experiments before and after surgical removal show that the vertical lobe (top right) plays a role in visual learning and that the inferior frontal lobe (top left) is one of two involved in tactile learning. The statocyst (bottom right) is not a part of the brain; it is one of the twin organs responsible for the octopus's sense of balance. Magnification is 15 diameters.

UNSCHOOLED OCTOPUS leaves the shelter at one end of its tank (*first photograph*) and walks toward the bait at the opposite end. The advancing animal uses only one of its eyes to guide it. When the bait, a crab, is in range, the octopus throws its leading

term memory system must also remain, however, because the patients can remember new information for short periods, particularly when they use mnemonic devices. On the basis of this interpretation it would appear that the hippocampal system may have the role of linking the two memory mechanisms—whatever that may mean.

For octopuses in our training situation it seems at first that when the vertical lobe of the brain is removed, the long-term memory system of the animal is completely abolished, leaving only the short-term system. We obtain a different result, however, if instead of showing such an animal a crab with or without a geometric figure we present it with figures only, rewarding it with a crab for an attack on one figure and punishing it with a shock for an attack on another. Under such conditions an octopus without its vertical lobe can learn the required discriminations and retain them. At least two conclusions can be drawn from this kind of result. The first is that the vertical lobe is essential to the memory system if the learned response involves a change in what might be termed innate behavior toward an object as familiar to an octopus as a crab. The second is that a long-term memory system for some responses

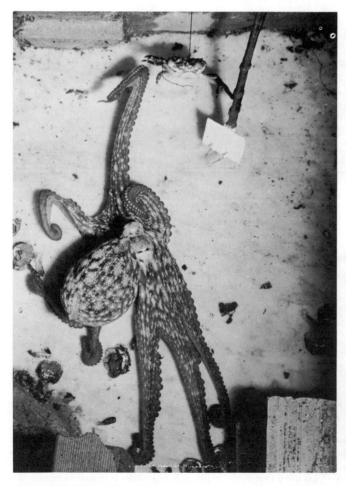

TRAINED OCTOPUS is cautious in its approach when a crab and a geometric figure are presented together (*first photograph*). If the animal seizes the crab, it receives an electric shock (*note darkened region at the base of the arm in second photograph*). As

arms forward to seize it (*second photograph*). Next it tucks the crab up toward its mouth (*third photograph*). The octopus then returns to its shelter (*fourth photograph*), where it kills the crab with a poisonous secretion from its salivary glands and eats it.

can be maintained in the absence of the vertical lobe.

Since we do not know (and probably never will know because it is so difficult to rear *Octopus vulgaris* from its larval stage) whether the octopus's response to a crab is learned or innate, our studies over the past eight years have involved experiments in which reward or punishment is given only after the animal has responded to an artificial situation, that is, the presentation of a figure of a given size, shape or color. It has been shown that animals without a vertical lobe can learn to attack unfamiliar figures for a reward, although they do so more slowly than normal animals. Once they have learned to attack such figures these octopuses retain their response for as long a time as normal animals do. If octopuses without a vertical lobe are required to reverse a learned visual response, however, they find it particularly difficult. When a shock is received for attacking a figure that formerly brought a reward, the animals can still learn to discriminate, but they make between four and five times as many mistakes as normal animals; moreover, their period of retention is shorter.

In addition to its large visual system

 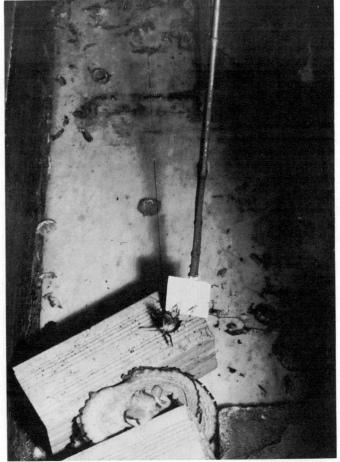

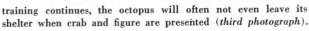

training continues, the octopus will often not even leave its shelter when crab and figure are presented (*third photograph*).

If crab and figure are brought near a fully trained animal, it pales and pumps a jet of water at them (*fourth photograph*).

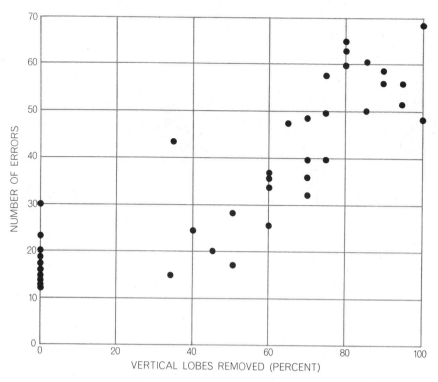

CONTRAST IN PERFORMANCE of normal (*far left*) and surgically altered octopuses shows that the number of errors increased more than threefold as larger and larger portions of the brain's vertical lobe were excised. This finding supports the conclusion that the organization of memory depends primarily on the number of brain cells available.

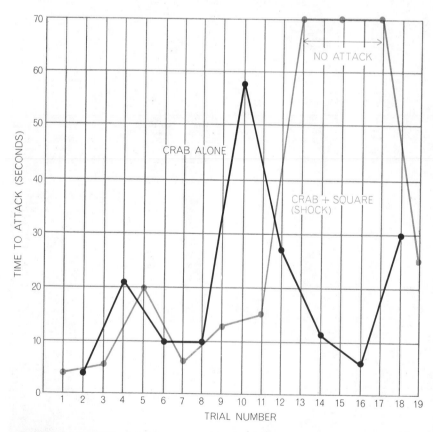

ABILITY TO LEARN can be demonstrated by an octopus deprived of the vertical lobes of its brain, provided that the trials are only a few minutes apart. In the example illustrated, little learning was apparent during 12 alternating exposures to negative and positive stimuli. Thereafter three successive negative stimuli were avoided by the octopus.

the octopus has a complex chemo-tactile sensory system. Most of the investigation of this system has been done by Martin Wells and his wife Joyce. By applying methods similar to those used for training the animals to make visual discriminations, they have been able to show that tactile learning in the octopus is about as rapid as visual learning. Octopuses have been trained to discriminate between a live bivalve and a counterfeit one consisting of shells of the same species that have been cleaned and filled with wax. They can discriminate between a bivalve with a ribbed shell and another species of comparable size but with a smooth shell. Just recently Wells has found that octopuses can detect hydrochloric acid, sucrose or quinine dissolved in sea-water at concentrations 100 times less than those the human tongue can detect in distilled water. Presented with artificial objects, they can distinguish between grooved cylinders and smooth ones, although they cannot distinguish between two grooved objects that differ only in the direction in which the grooves run [*see illustration on opposite page*]. After intensive training they can discriminate a cube from a sphere about 75 percent of the time.

Through each arm of an octopus, which is studded with two rows of suckers, runs a cord of nerve fibers and ganglia. In these ganglia occur local reflexes along the arm and between the rows of suckers. It is supposed that, when the octopus makes a tactile discrimination, the state of excitation in the ganglia above each sucker is determined by the proportion of sense organs excited, and the degree to which these sense organs are stimulated determines the frequency with which nerve impulses are discharged in the fibers running from the ganglia to the brain. Learning in the isolated arm ganglia is probably not possible. Wells has found that for tactile learning to occur the upper brain's median inferior frontal lobe and subfrontal lobe are necessary. Damage to these regions of the brain does not affect visual learning, and for that reason the two lobes have often been used as the sites for control lesions in the investigation of visual learning.

The role of the median inferior frontal lobe seems to be to interrelate the information received from each of the octopus's eight arms; if the lobe is removed and one arm is trained to reject an object, then the other arms continue to accept the object. Without the subfrontal lobe the animals cannot even learn to reject objects by touch. As in the case of the vertical lobe in visual learning, the retention of small portions of the subfrontal lobe allows adequate learned performance. Wells believes that as few as 13,000 of the five million subfrontal-lobe cells may be sufficient for some learning to occur. The subfrontal lobe is structurally very similar to the vertical lobe; it must be considered the vertical lobe's counterpart in the chemotactile system. Removal of the vertical lobe nonetheless has an effect on chemotactile discrimination, mainly in the direction of slowing the rate at which learning occurs.

This account has discussed the main lines of work on memory systems that have been carried out with octopuses as experimental animals, together with some comparisons with human memory. Recently Young has summarized all the work on the cephalopod brain of the past 17 years and has devised a scheme of how such brains may work in the formation, storage and translation of memory into effective action.

Young proposes that in the course of evolution chemotactile and visual centers developed out of a primitive taste-and-bite reflex mechanism. As these "distance receptor" systems evolved, providing information as to where food might be obtained other than that received from direct contact, there came to be a more indirect relation between a change in the environment and the responses that such a change produced in the animal. As this happened, signal systems of greater duration than are provided by simple reflex mechanisms also had to evolve; learning had to become possible so that the animal could assess the significance of each distant environmental change.

Suppose, for example, a crab appears at a distance in the visual field of an octopus; as a result of what can be called "cue signals" there arises in the octopus brain a system for producing "graduated commands to attack." This command system will be weak at first but will grow stronger with reinforcement. The actual strengthening process will vary according to the reward or punishment met at each attack, because the outcome of each attack gives rise to a "result" signal. Such signals condition the distance-receptor systems that initially cued the attack—in the present example, the visual-receptor system. These result signals

become distributed throughout the nervous tissue that carries a record of a particular event.

There is, of course, a delay between the moment the cue signals are received in the brain and the moment the result signals arrive. If the result signals are to produce the appropriate conditioning of memory elements, the address of these elements, so to speak, has to be held to allow correct delivery of the information of, say, taste or pain. In the brain of the octopus each optic lobe contains "classifying" cells, among them vertically and horizontally oriented sets of nerve fibers that are presumably related to the vertical and horizontal arrangement of elements in the retina of the octopus's eye. These classifying cells form synapses with "memory" cells in the optic lobes that in their turn activate the cells that signal either attack or retreat. According to Young's hypothesis, each of the memory cells at first has a pair of alternative pathways; the actual neural change during learning consists in closing one of the two pathways. This closing may be accomplished by small cells that are

abundant in these learning centers and that can perhaps be switched on so as to produce a substance that inhibits transmission.

Suppose an attack has been evoked by means of this system; the memory cells activate not only an attack circuit but also a circuit reaching the vertical lobe of the upper brain. The signals indicating the results of the attack, such as taste or pain, arrive back and further reinforce the memory cells in the optic lobes, which have been under the influence of the appropriate pathways set up in the vertical lobe during the time interval between the cue signal and the result signal [see illustration on next page].

The hypothesis that the actual change represented by memory is produced by the small cells agrees with the fact that these cells are also present in the part of the brain that was shown by the Wellses to be the minimum necessary for tactile memory. Young suggests that the small cells were originally part of the primitive taste-and-bite reflex system, serving the function of

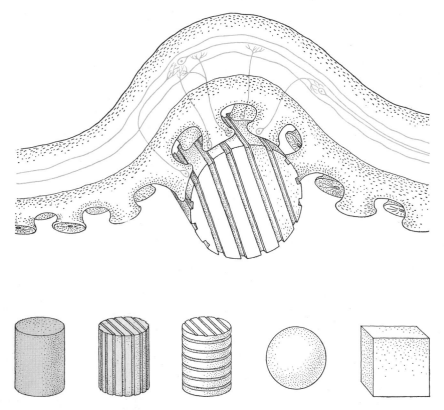

LEARNING BY TOUCH in the octopus was investigated by presenting objects with a variety of shapes and textures. In the case of a grooved cylinder (top) only the sense organs in contact with the surface are excited; those resting over the grooves remain inactive. Thus the octopus can learn to discriminate between a smooth cylinder (gray) and a grooved one (color), and even between a cube and a sphere; it cannot, however, discriminate between two cylinders that differ only in the orientation of the grooves.

temporary inhibition. The evolution of the memory consisted in making the inhibition last longer. The sets of auxiliary lobes associated with the memory system arose to allow for various combinations of inputs to be set up, to be combined with the signals that report

the results of actions and finally to be "delivered to the correct address" in the memory.

There is much that is speculative about this description, but the fact remains that both the visual and the tactile memory systems of the octopus

embrace sets of brain lobes arranged in similar circuits. This organization provides opportunities for study of the memory process that are made more challenging by Young's conviction that comparable circuits exist in the brains of mammals, including man.

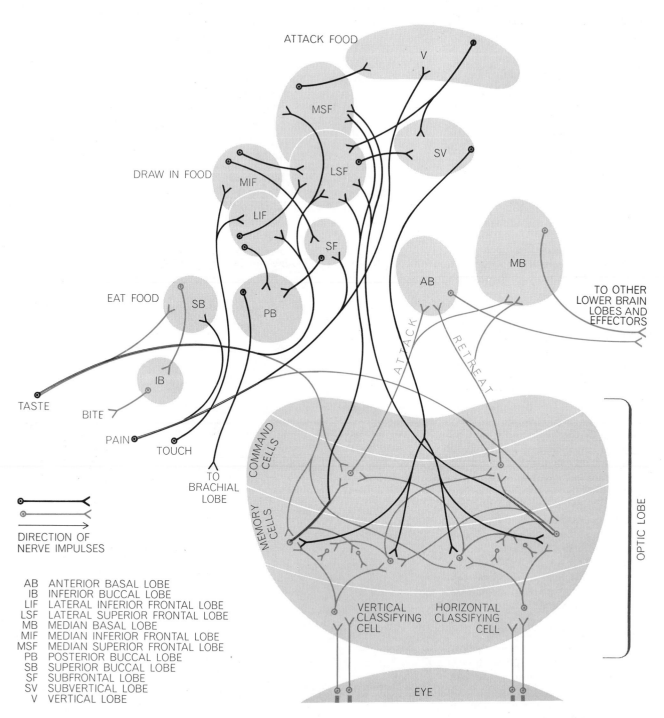

DUAL NATURE OF MEMORY can be traced out on an exploded view of the octopus brain. Circuits leading from the optic lobe (*color*) are the first to be activated on receipt of a visual cue by the lobe's classifying cells. The cue is then recorded in the memory cells and relayed to the command cells; the latter induce the octopus to attack or retreat. If an attack is rewarded, the returning "result" signal will reinforce a memory that registers the initiating cue favorably. If, instead, the attack brings pain,

the reinforced memory will register the cue unfavorably and any similar cues encountered in the future will be channeled to the circuits governing retreat rather than attack. Additional circuits (*black*) connect the memory and command regions of the optic lobe to various lobes of the upper brain; thus each event and its outcome are also recorded and reinforced in these nervous tissues. In due course what appear to be the long-term components of the memory system become localized in individual upper brain lobes.

"Imprinting" in a Natural Laboratory

by Eckhard H. Hess
August 1972

*A synthesis of laboratory and field techniques has
led to some interesting discoveries about imprinting,
the process by which newly hatched birds rapidly
form a permanent bond to the parent*

In a marsh on the Eastern Shore of Maryland, a few hundred feet from my laboratory building, a female wild mallard sits on a dozen infertile eggs. She has been incubating the eggs for almost four weeks. Periodically she hears the faint peeping sounds that are emitted by hatching mallard eggs, and she clucks softly in response. Since these eggs are infertile, however, they are not about to hatch and they do not emit peeping sounds. The sounds come from a small loudspeaker hidden in the nest under the eggs. The loudspeaker is connected to a microphone next to some hatching mallard eggs inside an incubator in my laboratory. The female mallard can hear any sounds coming from the laboratory eggs, and a microphone beside her relays the sounds she makes to a loudspeaker next to those eggs.

The reason for complicating the life of an expectant duck in such a way is to further our understanding of the phenomenon known as imprinting. It was through the work of the Austrian zoologist Konrad Z. Lorenz that imprinting became widely known. In the 1930's Lorenz observed that newly hatched goslings would follow him rather than their mother if the goslings saw him before they saw her. Since naturally reared geese show a strong attachment for their parent, Lorenz concluded that some animals have the capacity to learn rapidly and permanently at a very early age, and in particular to learn the characteristics of the parent. He called this process of acquiring an attachment to the parent *Prägung*, which in German means "stamping" or "coinage" but in English has been rendered as "imprinting." Lorenz regarded the phenomenon as being different from the usual kind of learning because of its rapidity and apparent permanence. In fact, he was hesitant at first to regard imprinting as a form of learn-

ing at all. Some child psychologists and some psychiatrists nevertheless perceived a similarity between the evidence of imprinting in animals and the early behavior of the human infant, and it is not surprising that interest in imprinting spread quickly.

From about the beginning of the 1950's many investigators have intensively studied imprinting in the laboratory. Unlike Lorenz, the majority of them have regarded imprinting as a form

of learning and have used methods much the same as those followed in the study of associative learning processes. In every case efforts were made to manipulate or stringently control the imprinting process. Usually the subjects are incubator-hatched birds that are reared in the laboratory. The birds are typically kept isolated until the time of the laboratory imprinting experience to prevent interaction of early social experience and the imprinting experience. Various objects

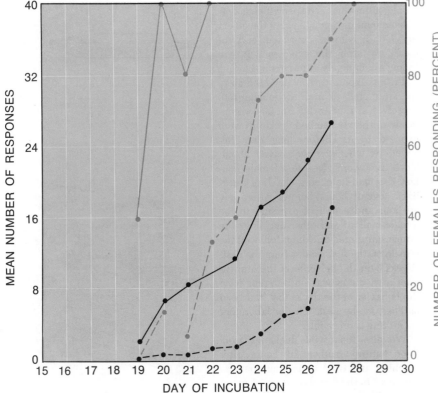

VOCAL RESPONSES to hatching-duckling sounds of 15 female wild mallards (*broken curves*) and five human-imprinted mallards (*solid curves*), which were later released to the wild, followed the same pattern, although the human-imprinted mallards began responding sooner and more frequently. A tape recording of the sounds of a hatching duckling was played daily throughout the incubation period to each female mallard while she was on her nest. Responses began on the 19th day of incubation and rose steadily until hatching.

have been used as artificial parents: duck decoys, stuffed hens, dolls, milk bottles, toilet floats, boxes, balls, flashing lights and rotating disks. Several investigators have constructed an automatic imprinting apparatus into which the newly hatched bird can be put. In this kind of work the investigator does not observe the young bird directly; all the bird's movements with respect to the imprinting object are recorded automatically.

Much of my own research during the past two decades has not differed substantially from this approach. The birds I have used for laboratory imprinting studies have all been incubated, hatched and reared without the normal social and environmental conditions and have then been tested in an artificial situation. It is therefore possible that the behavior observed under such conditions is not relevant to what actually happens in nature.

It is perhaps not surprising that studies of "unnatural" imprinting have produced conflicting results. Lorenz' original statements on the permanence of natural imprinting have been disputed. In many instances laboratory imprinting experiences do not produce permanent and exclusive attachment to the object selected as an artificial parent. For example, a duckling can spend a considerable amount of time following the object to which it is to be imprinted, and immediately after the experience it will follow a completely different object.

In one experiment in our laboratory we attempted to imprint ducklings to ourselves, as Lorenz did. For 20 continuous hours newly hatched ducklings were exposed to us. Before long they followed us whenever we moved about. Then they were given to a female mallard that had hatched a clutch of ducklings several hours before. After only an hour and a half of exposure to the female mallard and other ducklings the human-imprinted ducklings followed the female on the first exodus from the nest. Weeks later the behavior of the human-imprinted ducks was no different from the behavior of the ducks that had been hatched in the nest. Clearly laboratory imprinting is reversible.

We also took wild ducklings from their natural mother 16 hours after hatching and tried to imprint them to humans. On the first day we spent many hours with the ducklings, and during the next two months we made lengthy attempts every day to overcome the ducklings' fear of us. We finally gave up. From the beginning to the end the ducks

remained wild and afraid. They were released, and when they had matured, they were observed to be as wary of humans as normal wild ducks are. This result suggests that natural imprinting, unlike artificial laboratory imprinting, is permanent and irreversible. I have had to conclude that the usual laboratory imprinting has only a limited resemblance to natural imprinting.

It seems obvious that if the effects of natural imprinting are to be understood, the phenomenon must be studied as it operates in nature. The value of such studies was stressed as long ago as 1914 by the pioneer American psychologist John B. Watson. He emphasized that field observations must always be made to test whether or not conclusions drawn from laboratory studies conform to what actually happens in nature. The disparity between laboratory results and what happens in nature often arises from the failure of the investigator to really look at the animal's behavior. For years I have cautioned my students against shutting their experimental animals in "black boxes" with automatic recording

devices and never directly observing how the animals behave.

This does not mean that objective laboratory methods for studying the behavior of animals must be abandoned. With laboratory investigations large strides have been made in the development of instruments for the recording of behavior. In the study of imprinting it is not necessary to revert to imprecise naturalistic observations in the field. We can now go far beyond the limitations of traditional field studies. It is possible to set up modern laboratory equipment in actual field conditions and in ways that do not disturb or interact with the behavior being studied, in other words, to achieve a synthesis of laboratory and field techniques.

The first step in the field-laboratory method is to observe and record the undisturbed natural behavior of the animal in the situation being studied. In our work on imprinting we photographed the behavior of the female mallard during incubation and hatching. We photographed the behavior of the ducklings during and after hatching. We recorded

CLUCKS emitted by a female wild mallard in the fourth week of incubating eggs are shown in the sound spectrogram (upper illustration). Each cluck lasts for about 150 milliseconds

all sounds from the nest before and after hatching. Other factors, such as air temperature and nest temperature, were also recorded.

A detailed inventory of the actual events in natural imprinting is essential for providing a reference point in the assessment of experimental manipulations of the imprinting process. That is, the undisturbed natural imprinting events form the control situation for assessing the effects of the experimental manipulations. This is quite different from the "controlled" laboratory setting, in which the ducklings are reared in isolation and then tested in unnatural conditions. The controlled laboratory study not only introduces new variables (environmental and social deprivation) into the imprinting situation but also it can prevent the investigator from observing factors that are relevant in wild conditions.

My Maryland research station is well suited for the study of natural imprinting in ducks. The station, near a national game refuge, has 250 acres of marsh

and forest on a peninsula on which there are many wild and semiwild mallards. Through the sharp eyes of my technical assistant Elihu Abbott, a native of the Eastern Shore, I have learned to see much I might otherwise have missed. Initially we looked at and listened to the undisturbed parent-offspring interaction of female mallards that hatched their own eggs both in nests on the ground and in specially constructed nest boxes. From our records we noticed that the incubation time required for different clutches of eggs decreased progressively between March and June. Both the average air temperature and the number of daylight hours increase during those months; both are correlated with the incubation time of mallard eggs. It is likely, however, that temperature rather than photoperiod directly influences the duration of incubation. In one experiment mallard eggs from an incubator were slowly cooled for two hours a day in a room with a temperature of seven degrees Celsius, and another set of eggs was cooled in a room at 27 degrees C. These temperatures re-

spectively correspond to the mean noon temperatures at the research station in March and in June. The eggs that were placed in the cooler room took longer to hatch, indicating that temperature affects the incubation time directly. Factors such as humidity and barometric pressure may also play a role.

We noticed that all the eggs in a wild nest usually hatch between three and eight hours of one another. As a result all the ducklings in the same clutch are approximately the same age in terms of the number of hours since hatching. Yet when mallard eggs are placed in a mechanical incubator, they will hatch over a two- or three-day period even when precautions are taken to ensure that all the eggs begin developing simultaneously. The synchronous hatching observed in nature obviously has some survival value. At the time of the exodus from the nest, which usually takes place between 16 and 32 hours after hatching, all the ducklings would be of a similar age and thus would have equal motor capabilities and similar social experiences.

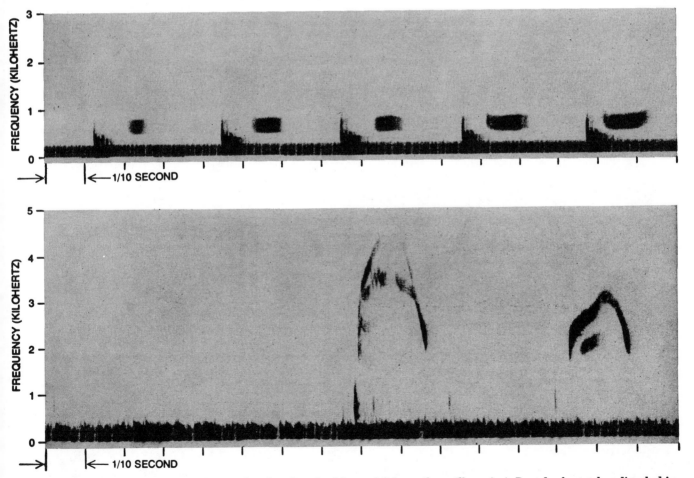

and is low in pitch: about one kilohertz or less. Sounds emitted by ducklings inside the eggs are high-pitched, rising to about four kilohertz (*lower illustration*). Records of natural, undisturbed imprinting events in the nest provide a control for later experiments.

Over the years our laboratory studies and actual observations of how a female mallard interacts with her offspring have pointed to the conclusion that imprinting is related to the age after hatching rather than the age from the beginning of incubation. Many other workers, however, have accepted the claim that age from the beginning of incubation determines the critical period for maximum effectiveness of imprinting. They base their belief on the findings of Gilbert Gottlieb of the Dorothea Dix Hospital in Raleigh, N.C., who in a 1961 paper described experiments that apparently showed that maximum imprinting in ducklings occurs in the period between 27 and 27½ days after the beginning of incubation. To make sure that all the eggs he was working with started incubation at the same time he first chilled the eggs so that any partially developed embryos would be killed. Yet the 27th day after the beginning of incubation can hardly be the period of maximum imprinting for wild ducklings that hatch in March under natural conditions, because such ducklings take on the average 28 days to hatch. Moreover, if the age of a duckling is measured from the beginning of incubation, it is hard to explain why eggs laid at different times in a hot month in the same nest will hatch within six to eight hours of one another under natural conditions.

Periodic cooling of the eggs seems to affect the synchronization of hatching. The mallard eggs from an incubator that were placed in a room at seven degrees C. hatched over a period of a day and a half, whereas eggs placed in the room at 27 degrees hatched over a period of two

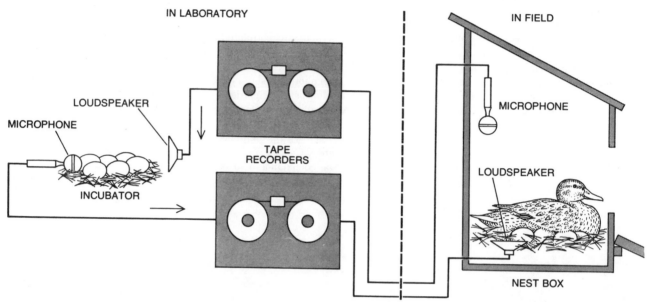

FEMALE MALLARD sitting on infertile eggs hears sounds transmitted from mallard eggs in a laboratory incubator. Any sounds she makes are transmitted to a loudspeaker beside the eggs in the laboratory. Such a combination of field and laboratory techniques permits recording of events without disturbing the nesting mallard and provides the hatching eggs with nearly natural conditions.

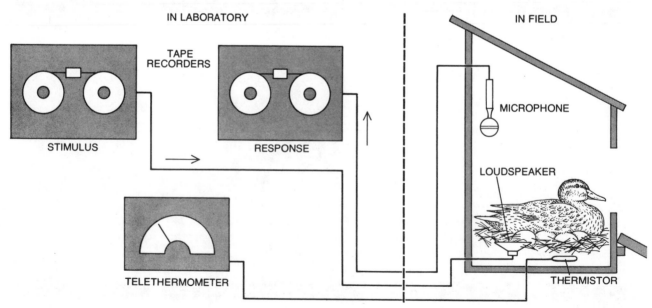

REMOTE MANIPULATION of prehatching sounds is accomplished by placing a sensitive microphone and a loudspeaker in the nest of a female wild mallard who is sitting on her own eggs. Prerecorded hatching-duckling sounds are played at specified times through the loudspeaker and the female mallard's responses to this stimulus are recorded. A thermistor probe transmits the temperature in the nest to a telethermometer and chart recorder. The thermistor records provide data about when females are on nest.

and a half days (which is about normal for artificially incubated eggs). Cooling cannot, however, play a major role. In June the temperature in the outdoor nest boxes averages close to the normal brooding temperature while the female mallard is absent. Therefore an egg laid on June 1 has a head start in incubation over those laid a week later. Yet we have observed that all the eggs in clutches laid in June hatch in a period lasting between six and eight hours.

We found another clue to how the synchronization of hatching may be achieved in the vocalization pattern of the brooding female mallard. As many others have noted, the female mallard vocalizes regularly as she sits on her eggs during the latter part of the incubation period. It seemed possible that she was vocalizing to the eggs, perhaps in response to sounds from the eggs themselves. Other workers had observed that ducklings make sounds before they hatch, and the prehatching behavior of ducklings in response to maternal calls has been extensively reported by Gottlieb.

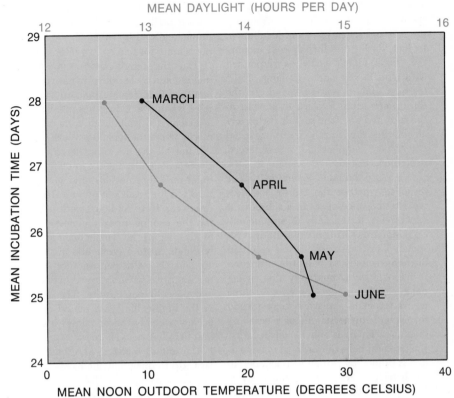

INCUBATION TIME of mallard eggs hatched naturally in a feral setting at Lake Cove, Md., decreased steadily from March to June. The incubation period correlated with both the outdoor temperature (*black curve*) and the daily photoperiod (*colored curve*).

We placed a highly sensitive microphone next to some mallard eggs that were nearly ready to hatch. We found that the ducklings indeed make sounds while they are still inside the egg. We made a one-minute tape recording of the sounds emitted by a duckling that had pipped its shell and was going to hatch within the next few hours. Then we made a seven-minute recording that would enable us to play the duckling sounds three times for one minute interspersed with one-minute silences. We played the recording once each to 37 female mallards at various stages of

NEST EXODUS takes place about 16 to 32 hours after hatching. The female mallard begins to make about 40 to 65 calls per minute and continues while the ducklings leave the nest to follow her. The ducklings are capable of walking and swimming from hatching.

incubation. There were no positive responses from the female mallards during the first and second week of incubation. In fact, during the first days of incubation some female mallards responded with threat behavior: a fluffing of the feathers and a panting sound. In the third week some females responded to the recorded duckling sounds with a few clucks. In the fourth week maternal clucks were frequent and were observed in all ducks tested.

We found the same general pattern of response whether the female mallards were tested once or, as in a subsequent experiment, tested daily during incubation. Mallards sitting on infertile eggs responded just as much to the recorded duckling sounds as mallards sitting on fertile eggs did. Apparently after sitting on a clutch of eggs for two or three weeks a female mallard becomes ready to respond to the sounds of a hatching duckling. There is some evidence that the parental behavior of the female mallard is primed by certain neuroendocrine mechanisms. We have begun a study of the neuroendocrine changes that might accompany imprinting and filial behavior in mallards.

To what extent do unhatched ducklings respond to the vocalization of the female mallard? In order to find out we played a recording of a female mallard's vocalizations to ducklings in eggs that had just been pipped and were scheduled to hatch within the next 24 hours. As before, the sounds were interspersed with periods of silence. We then recorded all the sounds made by the ducklings during the recorded female mallard vocalizations and also during the silent periods on the tape. Twenty-four hours before the scheduled hatching the ducklings emitted 34 percent of their sounds during the silent periods, which suggests that at this stage they initiate most of the auditory interaction. As hatching time approaches the ducklings emit fewer and fewer sounds during the silent periods. The total number of sounds they make, however, increases steadily. At the time of hatching only 9 percent of the sounds they make are emitted during the silent periods. One hour after hatching, in response to the same type of recording, the ducklings gave 37 percent of their vocalizations during the silent periods, a level similar to the level at 24 hours before hatching.

During the hatching period, which lasts about an hour, the female mallard generally vocalizes at the rate of from zero to four calls per one-minute interval. Occasionally there is an interval in which she emits as many as 10 calls.

When the duckling actually hatches, the female mallard's vocalization increases dramatically to between 45 and 68 calls per minute for one or two minutes.

Thus the sounds made by the female mallard and by her offspring are complementary. The female mallard vocalizes most when a duckling has just hatched. A hatching duckling emits its cries primarily when the female is vocalizing.

After all the ducklings have hatched the female mallard tends to be relatively quiet for long intervals, giving between zero and four calls per minute. This continues for 16 to 32 hours until it is time for the exodus from the nest. As the exodus begins the female mallard quickly builds up to a crescendo of between 40 and 65 calls per minute; on rare occasions we have observed between 70 and 95 calls per minute. The duration of the high-calling-rate period depends on how quickly the ducklings leave the nest to follow her. There is now a change in the sounds made by the female mallard. Up to this point she has been making clucking sounds. By the time the exodus from the nest takes place some of her sounds are more like quacks.

The auditory interaction of the female

mallard and the duckling can begin well before the hatching period. As I have indicated, the female mallard responds to unhatched-duckling sounds during the third and fourth week of incubation. Normally ducklings penetrate a membrane to reach an air space inside the eggshell two days before hatching. We have not found any female mallard that vocalized to her clutch before the duckling in the egg reached the air space. We have found that as soon as the duckling penetrates the air space the female begins to cluck at a rate of between zero and four times per minute. Typically she continues to vocalize at this rate until the ducklings begin to pip their eggs (which is about 24 hours after they have entered the air space). As the eggs are being pipped the female clucks at the rate of between 10 ·and 15 times per minute. When the pipping is completed, she drops back to between zero and four calls per minute. In the next 24 hours there is a great deal of auditory interaction between the female and her unhatched offspring; this intense interaction may facilitate the rapid formation of the filial bond after hatching, although it is quite possible that synchrony

SOUND SPECTROGRAM of the calls of newly hatched ducklings in the nest and the mother's responses is shown at right. The high-pitched peeps of the ducklings are in the

DISTRESS CALLS of ducklings in the nest evoke a quacklike response from the female mallard. The cessation of the distress calls and the onset of normal duckling peeping sounds

of hatching is the main effect. Already we have found that a combination of cooling the eggs daily, placing them together so that they touch one another and transmitting parent-young vocal responses through the microphone-loudspeaker hookup between the female's nest and the laboratory incubator causes the eggs in the incubator to hatch as synchronously as eggs in nature do. In fact, the two times we did this we found that all the eggs in the clutches hatched within four hours of one another. It has been shown in many studies of imprinting, including laboratory studies, that auditory stimuli have an important effect on the development of filial attachment. Auditory stimulation, before and after hatching, together with tactile stimulation in the nest after hatching results in ducklings that are thoroughly imprinted to the female mallard that is present.

Furthermore, it appears that auditory interaction before hatching may play an important role in promoting the synchronization of hatching. As our experiments showed, not only does the female mallard respond to sounds from her eggs but also the ducklings respond to her clucks.

Perhaps the daily cooling of the eggs when the female mallard leaves the nest to feed serves to broadly synchronize embryonic and behavioral development, whereas the auditory interaction of the mother with the ducklings and of one duckling with another serves to provide finer synchronization. Margaret Vince of the University of Cambridge has shown that the synchronization of hatching in quail is promoted by the mutual auditory interaction of the young birds in the eggs.

Listening to the female mallards vocalize to their eggs or to their newly hatched offspring, we were struck by the fact that we could tell which mallard was vocalizing, even when we could not see her. Some female mallards regularly emit single clucks at one-second intervals, some cluck in triple or quadruple clusters and others cluck in clusters of different lengths. The individual differences in the vocalization styles of female mallards may enable young ducklings to identify their mother. We can also speculate that the characteristics of a female mallard's voice are learned by her female offspring, which may then adopt a simi-

lar style when they are hatching eggs of their own.

The female mallards not only differ from one another in vocalization styles but also emit different calls in different situations. We have recorded variations in pitch and duration from the same mallard in various nesting situations. It seems likely that such variations in the female mallard call are an important factor in the imprinting process.

Studies of imprinting in the laboratory have shown that the more effort a duckling has to expend in following the imprinting object, the more strongly it prefers that object in later testing. At first it would seem that this is not the case in natural imprinting; young ducklings raised by their mother have little difficulty following her during the exodus from the nest. Closer observation of many nests over several seasons showed, however, that ducklings make a considerable effort to be near their parent. They may suffer for such efforts, since they can be accidentally stepped on, squeezed or scratched by the female adult. The combination of effort and punishment may actually strengthen imprinting. Work in my laboratory showed

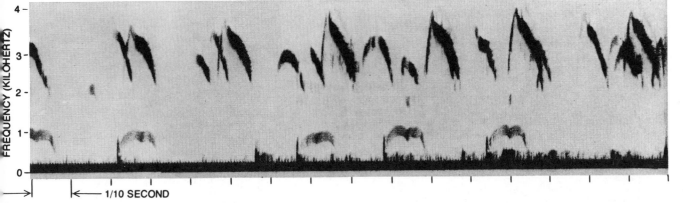

two-to-four-kilohertz range. They normally have the shape of an inverted *V*. The female mallard's clucks are about one kilohertz and last about 130 milliseconds. After the eggs hatch the vocalization of the female changes both in quantity and in quality of sound.

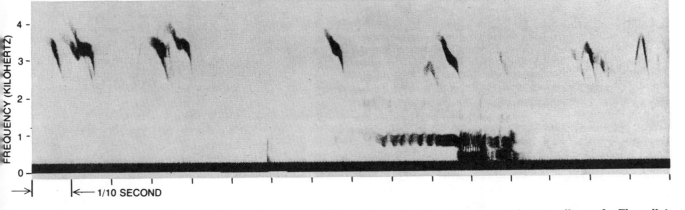

is almost immediate, as can be seen in this sound spectrogram. The female mallard's quacklike call is about one kilohertz in pitch and has a duration of approximately 450 milliseconds. The call is emitted about once every two seconds in response to distress cries.

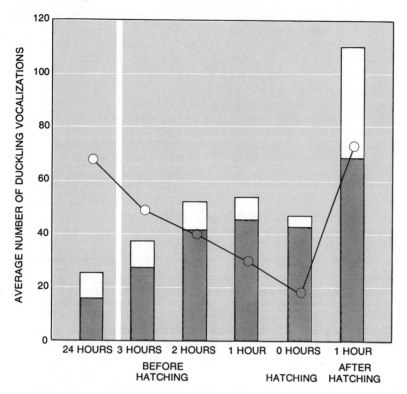

NUMBER OF SOUNDS from ducklings before and after hatching are shown. The duck-lings heard a recording consisting of five one-minute segments of a female mallard's cluck-ing sounds interspersed with five one-minute segments of silence. The recording was played to six mallard eggs and the number of vocal responses by the ducklings to the clucking segments (*gray bars*) and to the silent segments (*white bars*) were counted. Twenty-four hours before hatching 34 percent of the duckling sounds were made during the silent interval, indicating the ducklings initiated a substantial portion of the early audi-tory interaction. As hatching time approached the ducklings initiated fewer and fewer of the sounds and at hatching vocalized most in response to the clucks of the female mallard.

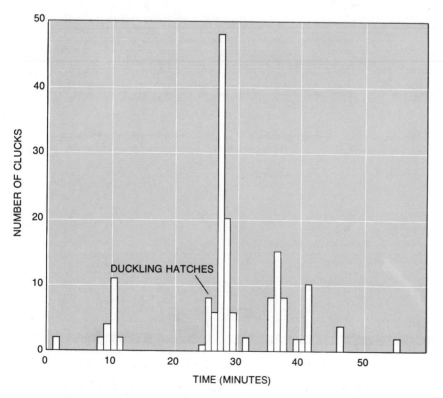

CLUCKING RATE of a wild, ground-nesting female mallard rose dramatically for about two minutes while a duckling hatched and then slowly declined to the prehatching rate. Each bar depicts the number of clucks emitted by the female during a one-minute period.

that chicks given an electric shock while they were following the imprinting ob-ject later showed stronger attachment to the object than unshocked chicks did. It is reasonable to expect similar results with ducklings.

Slobodan Petrovich of the University of Maryland (Baltimore County) and I have begun a study to determine the rel-ative contributions of prehatching and posthatching auditory experience on im-printing and filial attachment. The audi-tory stimuli consist of either natural mal-lard maternal clucks or a human voice saying "Come, come, come." Our results indicate that prehatching stimulation by natural maternal clucks may to a degree facilitate the later recognition of the characteristic call of the mallard. Duck-lings lacking any experience with a ma-ternal call imprint as well to a duck de-coy that utters "Come, come, come" as to a decoy that emits normal mallard clucks. Ducklings that had been exposed to a maternal call before hatching im-printed better to decoys that emitted the mallard clucks. We found, however, that the immediate posthatching experiences, in this case with a female mallard on the nest, can highly determine the degree of filial attachment and make imprinting to a human sound virtually impossible.

It is important to recognize that almost all laboratory imprinting experiments, including my own, have been depriva-tion experiments. The justification for such experiments has been the ostensible need for controlling the variables of the phenomenon, but the deprivation may have interfered with the normal behav-ioral development of the young duck-lings. Whatever imprinting experiences the experimenter allows therefore do not produce the maximum effect.

Although our findings are far from complete, we have already determined enough to demonstrate the great value of studying imprinting under natural con-ditions. The natural laboratory can be profitably used to study questions about imprinting that have been raised but not answered by traditional laboratory ex-periments. We must move away from the in vitro, or test-tube, approach to the study of behavior and move toward the in vivo method that allows interac-tion with normal environmental factors. Some of the questions are: What is the optimal age for imprinting? How long must the imprinting experience last for it to have the maximum effect? Which has the greater effect on behavior: first experience or the most recent experi-ence? Whatever kind of behavior is being studied, the most fruitful approach may well be to study the behavior in its natural context.

How an Instinct is Learned

25

by Jack P. Hailman
December 1969

*A study of the feeding behavior of sea gull chicks
indicates that an instinct is not fully developed at
birth. Its normal development is strongly affected
by the chick's experience*

The term "instinct," as it is often applied to animal and human behavior, refers to a fairly complex, stereotyped pattern of activity that is common to the species and is inherited and unlearned. Yet braking an automobile and swinging a baseball bat are complex, stereotyped behavioral patterns that can be observed in many members of the human species, and these patterns certainly cannot be acquired without experience. Perhaps stereotyped behavior patterns of animals also require subtle forms of experience for development. In other words, perhaps instincts are at least partly learned.

In order to investigate this possibility, I chose a typical animal instinct for study: the feeding behavior of sea gull chicks. My colleagues and I have observed the animals in their natural environment and in the laboratory, where we have conducted a number of experiments designed to elucidate the development of the feeding behavior. Our conclusion is that this particular pattern of behavior requires a considerable amount of experience if it is to develop normally. Moreover, the study strongly suggests that other instincts involve a component of learning.

Sitting quietly in a blind near the nest of a common laughing gull, which breeds on coastal marsh islands in eastern North America, one can watch the feeding of the chicks. The parent lowers its head and points its beak downward in front of a week-old chick. If some time has passed since the last feeding, the chick will aim a complexly coordinated pecking motion at the bill of the parent, grasping the bill and stroking it downward. After repeated pecking the parent regurgitates partly digested food. The pecking motion of the chick is thus seen to be a form of begging for food. If one watches further, one sees the chick peck at the food, tearing pieces away and swallowing them. Pecking is therefore also a feeding action. When the chick and the one or two other chicks in the nest have had their fill, the parent picks up the remaining food and swallows it.

Further observation reveals several intricacies in the interaction of the parent and the chicks. If the parent fails to elicit pecking by merely pointing its bill downward in front of the chicks, it may swing its beak gently from side to side. Such a motion usually stimulates pecking. After the parent has regurgitated food onto the floor of the nest it waits for the chicks to feed. If they do not, the parent lowers its beak again and appears to point at the food. This action is likely to stimulate pecking. If it fails, the parent picks up the food in its mandibles and holds it in front of a chick. If this action elicits pecking, the parent drops the food again so that the chick can eat it readily.

We find in this apparently simple pecking behavior a number of questions concerning the possible role of experience in the development of begging. How does the chick come to stroke the parent's bill with its begging peck and to tear at the food with its feeding peck when the two movements are basically so similar? Why does the chick rotate its head sideways in the begging peck but not in the feeding peck? Does the chick require practice to perfect its aim and coordination? How does the chick come to peck when it is hungry and not peck when it is sated? Why does the chick not peck at the parent's red legs or other objects in its environment? How does the chick recognize food?

In order to answer these and many other questions our group studied chicks experimentally from the time of hatching through the first week of life. By that time the feeding behavior is well established. Moreover, by restricting the study to a short period after hatching we could be sure of controlling several of the elements of development in order to assess their contribution to the behavior. As is often the case, the study raised more questions than it answered, but it also provided a good deal of information.

Let us consider first the accuracy of the pecking aim. In order to investigate this matter we painted diagrammatic pictures of parent gulls on small cards [*see top illustration on page 242*]. The card in use was mounted on a pivoting rod that could be moved horizontally back and forth in front of a chick. We collected eggs in the field and hatched them in a dark incubator so that the chicks would not have received any visual stimuli before the test. Each chick was confronted with the two-dimensional model of the parent during the day of hatching and was allowed to make about a dozen pecks at the moving model. Each peck was marked on the card with a penciled dot.

Having made sure that the chick could be identified later, we put it in a nest in the field in exchange for a pipping egg (one that was almost ready to hatch). The chick thus began to experience normal rearing by its foster parents. On the first, third and fifth day after hatching we went to the nesting area and gathered up half of the chicks for further tests, and on the second, fourth and sixth day we tested the others. On each gathering day a chick was tested again on the model and then put back in the nest.

The tests showed that on the average only a third of the pecks by a newly hatched chick strike the model. On the first day after hatching more than half of the pecks are accurate, and by two days after hatching the accuracy reaches a steady level of more than 75 percent.

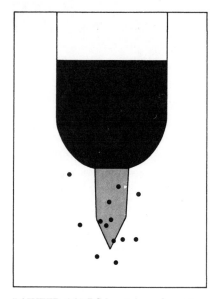 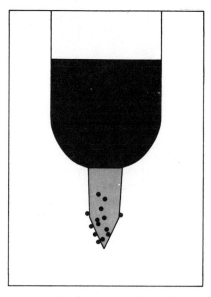

PAINTED CARDS bearing a schematic representation of the head of an adult gull were presented to chicks to test their pecking accuracy. Pecks are identified by dots. At left is the erratic record of a newborn chick, at right what the same chick did two days later.

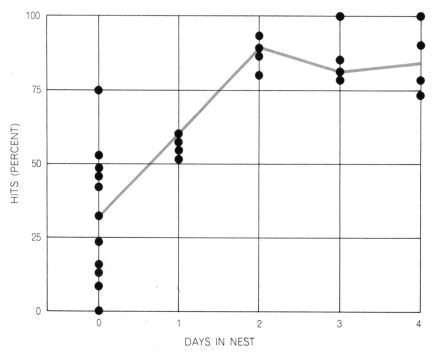

IMPROVING ACCURACY of pecking by an experimental group of chicks is charted. Each circle represents the record of one chick. After testing on cards at the time of hatching, the chicks were put into nests and fed by adult gulls. Half of the group was tested on the cards again on the first and third day and the other half on the second and fourth day. Fewer chicks are represented on those days than on the day of birth because it was not always possible to find each chick. The colored line shows the median accuracy for the chicks tested.

The record of a typical chick shows that the strokes become much more closely grouped and that in particular the horizontal error is greatly reduced.

How does this rapid increase in accuracy come about? In order to find out we designed a more extensive experiment involving the use as controls of two groups of chicks reared in the wild. Three experimental groups were reared in dark brooders so that they would not have any visually coordinated experience in pecking. One experimental group was force-fed in the brooder. A second group received no food for two days; the chicks lived on their ample reserves of yolk. The members of the third group did not hatch normally; instead the experimenter broke open the egg as soon as pipping started, took the chick out and placed it in an incubator. The reason for this procedure was to see if the movements of normal hatching had any effect on the accuracy of pecking.

On various days after hatching the chicks of all groups were photographed pecking at the stuffed head of an adult laughing gull. From the films we could ascertain the percentage of accurate pecks. Chicks of all five groups demonstrated an increasing accuracy with age, but only in the two control groups did the figure reach the normal level of more than 75 percent hits. The denial of the hatching experience had no effect on accuracy, but the denial of visual experience after hatching had a strong effect.

The most conservative interpretation of these results is that visual experience is necessary for the development of full accuracy in pecking but that a certain amount of improvement in accuracy is achieved without experience. Perhaps this amount results from improved steadiness of stance. Here again an element of experience may enter in, since the improvement in stance can most plausibly be attributed to the chick's practice in standing in the dark incubator.

How does the chick come to position itself at the correct distance to strike the bill or the model accurately? Our observations of newly hatched chicks suggest that a self-regulating form of behavior based on depth perception is at work. If an inexperienced chick is too close to the target at first, its pecking thrust against the bill or model is so strong that the chick is thrown backward as much as an inch. If the chick starts out too far from the target, the pecking thrust misses and the chick falls forward as much as two inches. Older chicks rarely make such gross errors, suggesting that the experience of overshots and undershots has helped the chick learn to adjust its distance.

It has often been implied that hunger is a learned motivation. Our experiments suggest that hunger has at least an unlearned basis from which to develop further. Several experiments showed that, as one might expect, if chicks were fed to satiation and tested with models at various times after feeding, the pecking rate increased with time since feeding. The same pattern appeared, however, with chicks we gave no opportunity to "learn" hunger. Chicks hatched and reared in dark incubators were force-fed to satiation when they were between 24 and 48 hours old and then were tested in light on models. At one hour after feeding they pecked at a mean rate of 6.2 pecks per two minutes and at two hours at 10.2 pecks, which is a statistically reliable difference.

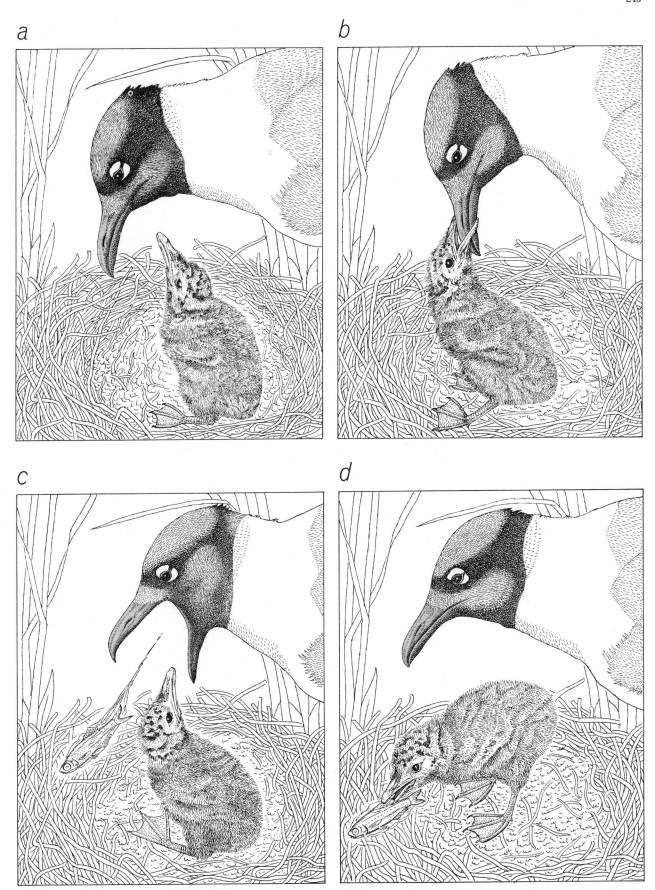

NORMAL FEEDING BEHAVIOR of a laughing gull chick includes two separate but related types of pecking. A chick about three days old, which has largely perfected the feeding pattern, is portrayed. As the parent lowers its head (*a*) the chick aims a high- ly coordinated begging peck at the parent's beak, grasping the beak (*b*) and stroking it downward. Parent then regurgitates partly digested food onto the floor of the nest (*c*) and the chick begins to eat it (*d*) with a pecking action that is called the feeding peck.

By means of high-speed motion pictures we analyzed in detail the motor pattern of the begging peck. In a chick several days old the pattern, by then well developed, includes four major components: (1) the opening and subsequent closing of the bill; (2) the motion of the head up and forward toward the beak of the parent and then down and back toward the chick's body; (3) the rotation of the head to the side in anticipation of grasping the parent's vertical beak and then the return rotation to the vertical position, and (4) a slight push upward and forward with the legs [see illustration on page 243]. A frame-by-frame analysis of the motion pictures revealed considerable variation in the synchronization of components from peck to peck of individual chicks and from chick to chick. The variation decreases somewhat with age. Presumably the decrease reflects increasing coordination, although we have not investigated the phenomenon in detail.

Among the interesting points that emerged from the films was the observation that with chicks reared in the wild the anticipatory rotation of the head became more frequent with age. To see what effect visually guided pecking experience had on this change, we analyzed the films of the five groups of chicks used in the experiments on pecking accuracy. The results indicated that chicks reared without pecking experience seldom showed any development of the head-rotation component of pecking. A chick reared in the wild does not show the rotation on the day of hatching but then acquires it and improves it rapidly.

We do not know how experience brings about this development, but the films provide a suggestion. Sometimes when a naïve chick is striking forward with its mandibles spread apart and the parental bill or the model is not exactly vertical, the chick's upper mandible goes to one side of the target and the lower mandible goes to the other side. The thrust of the forward head movement then forcibly rotates the chick's head to the side. Perhaps this is how the rotary movement in anticipation of grasping the parental bill is learned.

One of the most interesting questions about pecking is how the chick recognizes its parent. Observations from a blind show that chicks do peck at objects other than the bill of a parent, including other parts of the parent's body, but that most of the pecks are aimed at the parental bill—increasingly so with age. These observations suggest that the newly hatched chick has only a vague mental picture of the parent and that the picture becomes sharper with age and experience. We investigated the question by making a number of models of heads and beaks [see illustration on next page]. By systematically eliminating or changing parts of the models we could discover the most effective stimulus for pecking. Usually a model was mounted on a rod that could be moved on a pivot in time to a metronome, so that the speed of movement would be known and could be controlled. In each experiment a chick was presented with five models, which were offered in a random order, and the number of pecks (usually per 30 seconds) was recorded.

The first problem was to find the most

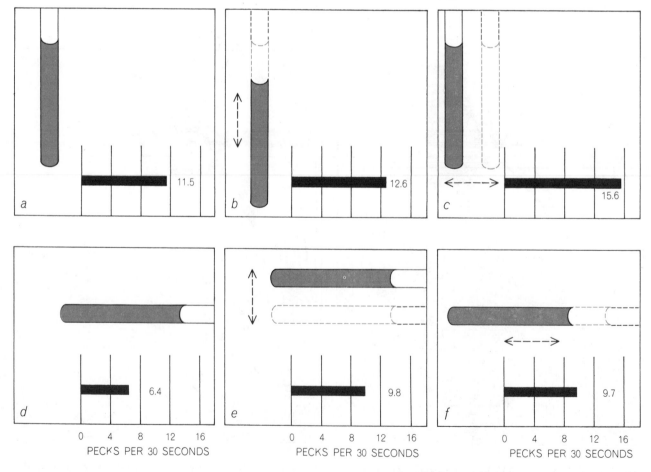

TESTS OF STIMULI were made with wooden dowels approximately the width of an adult gull's beak. The mean number of pecks in 30 seconds by a group of 25 chicks was recorded for (a) a vertical rod held stationary; (b) the same rod moved vertically; (c) the rod moved horizontally; (d) a stationary horizontal rod; (e) same rod moved vertically, and (f) the rod moved horizontally.

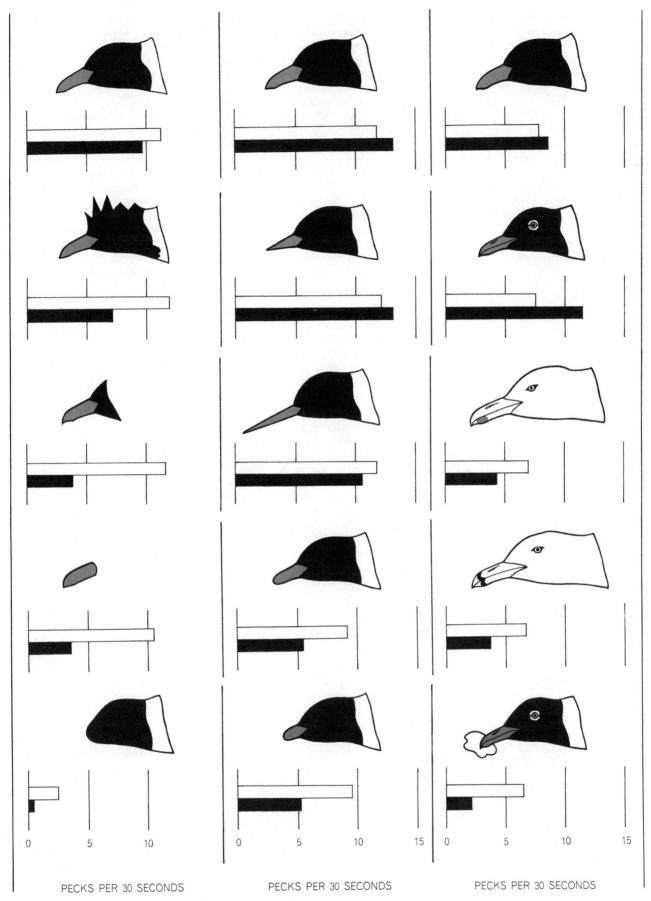

PECKS PER 30 SECONDS PECKS PER 30 SECONDS PECKS PER 30 SECONDS

FEATURES OF HEAD that elicit the most pecks were tested with models presented in groups of five to newly hatched chicks (*white bars*) and older chicks (*black bars*). All the chicks were laughing gulls. The group of models in the column at right included models of the herring gull (*third from top*) and the ring-billed gull (*fourth from top*). A model holding food in its bill is also depicted.

effective stimulus for naïve chicks, which had been kept in dark incubators until being tested about 24 hours after hatching. In the first three experiments the naïve chicks responded equally to all five models presented except for the model lacking a bill. These experiments established that newly hatched chicks are responding primarily to features of the parent's bill rather than to features of the head or even to the presence of the head.

One of the first three experiments revealed an unexpected finding: laughing gull chicks do not discriminate between a model of their own parent and a model of an adult herring gull. The adults of these two species are strikingly different in appearance. The laughing gull has a black head with a red bill; the herring gull has a white head and a yellow bill with a red spot on the lower mandible. The laughing gull chicks responded to this red spot on the model.

If laughing gull chicks fail to distinguish their parent from the herring gull, what would herring gull chicks do? In order to answer this question and others we went to a large colony of herring gulls on an island of the Grand Manan archipelago in the Bay of Fundy and tested herring gull chicks with models of both species. The result was the same: newly hatched chicks failed to discriminate between the two species in their pecking. This result suggested that the chicks of both species were responding to simple features of form and movement provided by the red bill of the laughing gull or the red spot of the herring gull.

Even though the optimum stimulus for eliciting pecks is evidently a simple one, it apparently has features that enable the chick to distinguish it from other simple forms in its environment, such as the red leg of a parent or a blade of grass, because pecks are rarely aimed at such targets. We investigated the matter with simple dowel rods made of wood and painted red. A rod was presented to a chick both vertically and horizontally, and in each orientation it was held stationary, moved vertically and moved horizontally. Every vertical stimulus received higher peck rates than every horizontal one, but the most effective vertical rod was the one moved horizontally. This result accords well with the natural situation in which the parent's vertical bill is likely to be moved horizontally in front of the chick.

Further analysis of the results showed that the vertically moved vertical rod elicited no more pecks than the stationary vertical rod. In addition, both horizontal and vertical movement of the horizontal rod were equally effective and curiously were more effective than the stationary horizontal rod. The most cautious interpretation of these results is that two kinds of movement are instrumental in eliciting pecks from the newly hatched laughing gull chick. The first is horizontal movement, and the second is movement across the long axis of the rod. This interpretation explains why a vertical rod moved vertically is no more effective than a stationary vertical rod, since both stimuli lack both horizontal movement and movement across an axis. With a horizontal rod, however, vertical movement is across the long axis and therefore is as effective as horizontal movement (along the axis), and each kind of movement is better than no movement.

The next step was to try various speeds of movement of a vertical rod. We used five different speeds with three diameters of rod. The results showed that a width of about eight millimeters was preferred, independent of the rate at which the rod was moved, and that a speed of movement of 12 centimeters per second was chosen over higher and lower speeds regardless of the width. These results demonstrate well the exactness of the match that evolution has brought about between stimulus and responsiveness. The parent's beak measures 10.6 millimeters from front to back and 3.1 millimeters laterally; the mean width of beak seen by the chick is thus about eight millimeters. Furthermore, the horizontal soliciting movement of the parent's beak was calculated from high-speed motion pictures made at the nest and was found to average 14.5 centimeters per second.

A recent experiment has added another item to this picture of a chick's ideal stimulus. Vertical rods projecting from above the chick's eye level are much preferred to those that come from below. The preference is also shown for oblique objects. Such a choice would reduce the chick's responsiveness to the parent's legs, which of course project from the ground and join the parent's body at just about the chick's eye level.

We now understood, at least roughly, how the newly hatched chick discriminates between its parent's beak and other objects in its environment. The question we addressed next was whether or not this perception changes during the first few days of life. We presented the same three series of five simple cardboard models each to week-old chicks and found a large difference between those results and the results with newly hatched chicks. Older chicks were sensitive to small differences in shape and detail of the head and beak. Moreover, the older chicks discriminated readily between their laughing gull parent and the model of the herring gull. To see if older herring gull chicks also came to prefer their own parent, we marked individual chicks and tested them on about the fourth day of life and about the seventh. The longer they had been in the nest, the greater their response to the model of their own parent and the less to the model of the laughing gull.

Was this change in perception due to conditioning experience with the parent that feeds the chick? To find out we worked with herring gull chicks that had been reared in an incubator. We divided them into three groups. If a chick was in the first group, it was fed a small amount of food when it delivered several pecks to a model of a laughing gull; if it was in the second group, it was fed for pecking at a model of its own species. The third group, which served as a control, was fed without first pecking at any model. At the end of two days of training each group was responding more, in discrimination tests without a reward of food, to the model on which it had been trained. Although the experiment is preliminary, it suggests that conditioning with a reward of food could account for the changes seen in wild chicks.

In sum, our findings indicate that the newly hatched chick responds best to a very simple stimulus situation. Although the experimenter can construct a model that is even more effective than the parent, the characteristics of the parent match the chick's ideal more closely than any other object in the environment. As a chick is fed by its parents, however, it develops a much more specific mental picture of the parent. Chicks a week old peck only at models that closely resemble the parent.

Our results did not appear consistent with certain earlier findings of the ethologists Nikolaas Tinbergen and A. C. Perdeck, who studied the herring gull. They found that if the red spot on the beak of the parent gull was moved to the forehead of a model, the model received few pecks. Since all the stimulus elements were thought to be the same in the two models, merely arranged differently, this classic experiment has been interpreted as showing the highly configural nature of the newly hatched herring gull's perception of the parent.

We thought the question of whether or not all the stimulus elements are in

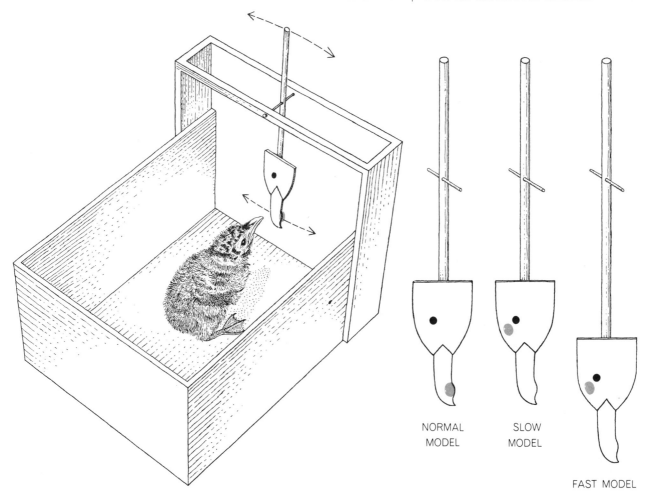

NORMAL
MODEL

SLOW
MODEL

FAST MODEL

LABORATORY APPARATUS was designed to test the effect of three models of the herring gull on the pecking rate of chicks. The normal model had the gull's red spot in its natural position on the beak. The "slow" model had the spot on the forehead; the spot moved more slowly than the one on the "fast" model when it was swung back and forth because it was closer to the pivot point.

fact the same needed investigation. The Tinbergen-Perdeck models were hand-held, so that when the model was moved in a pendulum-like manner with the hand as the pivot point, the spot on the forehead moved more slowly than the spot on the beak and through a shorter arc. Moreover, a chick had to stretch higher to peck at the forehead spot.

For these reasons we repeated the Tinbergen-Perdeck experiment, adding a third model. It was a forehead-spot model mounted on a rod in such a way that the forehead spot was the same distance from the pivot point as the bill spot was in the other model [see illustration above]. In addition our apparatus had a floor of adjustable height so that the chick's eye could always be positioned level with the red spot, whether the spot was on the bill or on the forehead. We called our third model the "fast" forehead-spot model, because the spot on it moved faster than the one on the "slow" Tinbergen-Perdeck model. If our hypothesis about movement were correct, the new, fast forehead-spot model should

be as effective as the bill-spot model.

The results were unequivocal. Newly hatched chicks responded as readily to the fast forehead-spot model as to the conventional model with the spot on the bill. Now we tested the same chicks after they had had three days in the nest and then seven days. This test showed that, as we had now come to expect, the bill-spot model improved steadily in relation to both the old and the new version of the forehead-spot model.

The classical interpretation of the Tinbergen-Perdeck experiment was that it demonstrated the existence of an innate releasing mechanism, which was conceived to be an unlearned perceptual mechanism that is activated by highly configural stimuli. Our experiment shows the gull chick's perception to be activated by a less configural simple shape when it is unlearned and by a highly configural shape when it is learned. The experiment suggests the need for reinvestigation of other results thought to be examples of innate releasing mechanisms.

The results also suggest the need for reinterpretation of another widely held concept in the behavioral sciences: classical conditioning. In the familiar example represented by the experiments of Ivan Pavlov an animal is presented with a new conditioning stimulus before it receives or just as it receives the usual stimulus that elicits the response of interest. After a number of these paired presentations the animal comes to respond to the conditioned stimulus alone. Pavlov's classic experiment involved ringing a bell before a dog was exposed to the smell of food; in time the dog would salivate merely in response to the bell.

Psychologists have long wondered how useful this cross-modal conditioning is to animals in their ordinary activities. Why should such a learning capability be evolved when it seems to be so little used under normal conditions? Our results suggest an answer worthy of further testing. As the chick develops its perceptual preference it is responding to simple features of the parent (the unconditioned stimuli) but is being presented simultaneously with all the complexities of the parental head (the conditioned stimulus). As a result of feeding, the chick comes to demand the more subtle features of the stimulus before it will peck.

This developmental process, which I have termed "perceptual sharpening," can be distinguished from the classical conditioning of laboratory experiments by the fact that the conditioned and unconditioned stimuli are physically identical. Perhaps the capability for classical conditioning has evolved primarily as a mechanism for perceptual sharpening, and the traditional experiments involving classical conditioning are in fact dealing with what is essentially an artifact of perceptual sharpening. At this stage my argument is no more than a hypothesis.

Although we have studied many more aspects of pecking that cannot be related briefly here, one should be mentioned: the recognition of food. Do newly hatched chicks recognize food when they encounter it? To find out, we placed food in dishes in the four corners of a small box and watched incubator-reared herring gull chicks find their first meal. The number of seconds taken to find food was inversely related to the pecking rate. This is the result to be expected if chicks are finding the food solely by trial and error.

If the chicks are allowed to feed until satiated and are then removed from the box until they are hungry again, they find food in the box much more quickly. The time required reaches a minimum by the third trial. This change cannot be attributed to an increase in the pecking rate, since the pecking rates in the second and subsequent trials are only slightly higher than in the first trial. The experiment shows that chicks can learn rapidly to identify food, or at least its location.

If the newly hatched chick does not initially recognize food, must it rely on trial-and-error searching to get its first bite and thereby initiate the rapid learning? Observation and experiment show that several mechanisms exist to help accelerate the first discovery of food. Recall that if a chick does not peck at food, the parent picks the food up in its beak. Quite often, in first feedings I have seen from a blind, the chick continues to peck at the parent's beak after the parent has regurgitated food onto the floor of the nest. Eventually the parent picks the food up and the chick strikes it during a peck at the bill. The observation suggests that the poor pecking accuracy of newly hatched chicks may be adaptive, ensur-

ing that the chick at least occasionally misses the parent's bill and strikes the food instead.

Another mechanism to assist the rapid learning of what food is involves the siblings in the nest. A chick will often peck at the white bill tip of another chick. In an ordinary clutch of three eggs the chicks hatch at intervals of about 12 hours, so that the older chicks have already been fed by the time a younger one appears. If a younger chick pecks at the bill of an older one while the older one is pecking at food, which it now recognizes, the younger chick's peck will probably also strike the food. Observa-

tion from a blind has shown that the first bite of food does come about in this way at times.

We tested the recognition of food with three groups of incubator-reared herring gull chicks. In one control group each chick was put alone into the small box with food; each chick in the two experimental groups had a companion. In one group the companion was equally naïve about food, and in the other group the companion had eaten earlier in the box. The results showed that the solitary chicks took the longest time to find food. The chicks with equally naïve companions took the next-longest time, and the

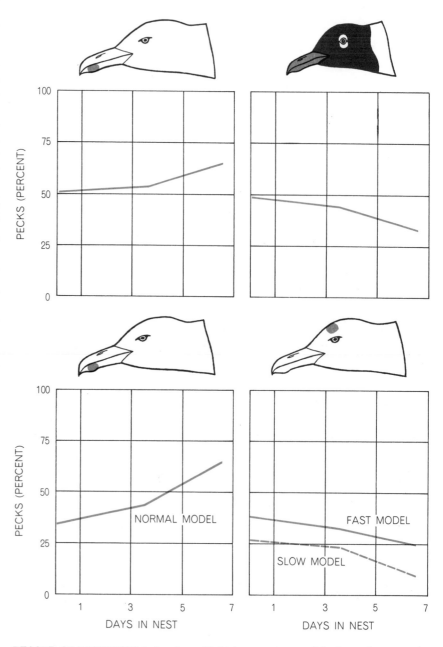

RECORD OF RESPONSES by herring gull chicks to various models shows changes as the chicks grow older. The model at top left is of a herring gull's head; the one at top right is of a laughing gull's. At bottom are two models of the herring gull; the one at left has the red spot in its normal place, and in the one at right it is on the forehead of the model.

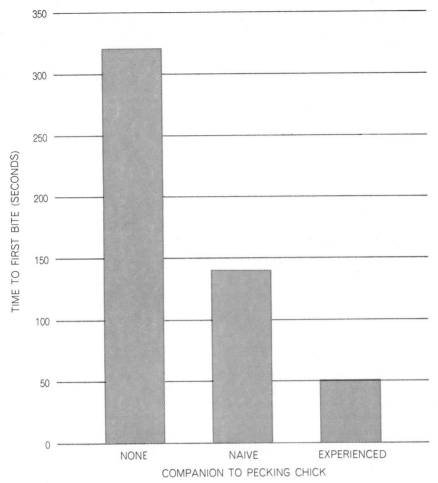

NEWLY HATCHED CHICKS were tested for speed in finding their first bite of food with and without another chick. The time was longest when the chick had no companion, shorter when the companion was equally naïve and shortest with a companion experienced in finding food. The test demonstrated that social interaction helps chicks learn to recognize food.

therefore to find food more quickly than the solitary chicks did. The experienced companion must have had a further effect, inasmuch as each naïve chick in this group first discovered food when it was pecking at the bill of its companion while the companion was pecking at food.

We can now summarize the developmental picture yielded by our investigation. The newly hatched gull chick begins life with a clumsily coordinated, poorly aimed peck motivated by hunger and elicited by simple stimulus properties of shape and movement provided only by a parent or sibling. The chick cannot recognize food, but by aiming at the bills of its relatives and missing it strikes food and rapidly learns to recognize it. As a result of the reward embodied in the food, the chick comes to learn the visual characteristics of the parent. Through practice in pecking its aim and depth perception improve steadily. The chick also learns to rotate its head when begging from the parent, and thus its begging peck and feeding peck become differentiated.

The picture strongly suggests that the normal development of other instincts entails a component of learning. It is necessary only that the learning process be highly alike in all members of the species for a stereotyped, species-common behavioral pattern to emerge. The example of the gulls also shows clearly that behavior cannot meaningfully be separated into unlearned and learned components, nor can a certain percentage of the behavior be attributed to learning. Behavioral development is a mosaic created by continuing interaction of the developing organism and its environment.

chicks with experienced companions took the least time [*see illustration above*].

These differences cannot be accounted for merely on the basis of increased exploratory trial-and-error pecking by the newly hatched chicks with companions, because the companion's bill actually diverted pecks from exploration. The naïve companion did cause the newly hatched chick to move about more in the box and

26

Love in Infant Monkeys

by Harry F. Harlow
June 1959

Affection in infants was long thought to be generated by the satisfactions of feeding. Studies of young rhesus monkeys now indicate that love derives mainly from close bodily contact

The first love of the human infant is for his mother. The tender intimacy of this attachment is such that it is sometimes regarded as a sacred or mystical force, an instinct incapable of analysis. No doubt such compunctions, along with the obvious obstacles in the way of objective study, have hampered experimental observation of the bonds between child and mother.

Though the data are thin, the theoretical literature on the subject is rich. Psychologists, sociologists and anthropologists commonly hold that the infant's love is learned through the association of the mother's face, body and other physical characteristics with the alleviation of internal biological tensions, particularly hunger and thirst. Traditional psychoanalysts have tended to emphasize the role of attaining and sucking at the breast as the basis for affectional development. Recently a number of child psychiatrists have questioned such simple explanations. Some argue that affectionate handling in the act of nursing is a variable of importance, whereas a few workers suggest that the composite activities of nursing, contact, clinging and even seeing and hearing work together to elicit the infant's love for his mother.

Now it is difficult, if not impossible, to use human infants as subjects for the studies necessary to break through the present speculative impasse. At birth the infant is so immature that he has little or no control over any motor system other than that involved in sucking. Furthermore, his physical maturation is so slow that by the time he can achieve precise, coordinated, measurable responses of his head, hands, feet and body, the nature and sequence of development have been hopelessly confounded and obscured. Clearly research into

the infant-mother relationship has need of a more suitable laboratory animal. We believe we have found it in the infant monkey. For the past several years our group at the Primate Laboratory of the University of Wisconsin has been employing baby rhesus monkeys in a study that we believe has begun to yield significant insights into the origin of the infant's love for his mother.

Baby monkeys are far better coordinated at birth than human infants. Their responses can be observed and evaluated with confidence at an age of 10 days or even earlier. Though they mature much more rapidly than their human contemporaries, infants of both species follow much the same general pattern of development.

Our interest in infant-monkey love grew out of a research program that involved the separation of monkeys from their mothers a few hours after birth. Employing techniques developed by Gertrude van Wagenen of Yale University, we had been rearing infant monkeys on the bottle with a mortality far less than that among monkeys nursed by their mothers. We were particularly careful to provide the infant monkeys with a folded gauze diaper on the floor of their cages, in accord with Dr. van Wagenen's observation that they would tend to maintain intimate contact with such soft, pliant surfaces, especially during nursing. We were impressed by the deep personal attachments that the monkeys formed for these diaper pads, and by the distress that they exhibited when the pads were briefly removed once a day for purposes of sanitation. The behavior of the infant monkeys was reminiscent of the human infant's attachment to its blankets, pillows, rag dolls or cuddly teddy bears.

These observations suggested the series of experiments in which we have sought to compare the importance of nursing and all associated activities with that of simple bodily contact in engendering the infant monkey's attachment to its mother. For this purpose we contrived two surrogate mother monkeys. One is a bare welded-wire cylindrical form surmounted by a wooden head with a crude face. In the other the welded wire is cushioned by a sheathing of terry cloth. We placed eight newborn monkeys in individual cages, each with equal access to a cloth and a wire mother [see *illustration on opposite page*]. Four of the infants received their milk from one mother and four from the other, the milk being furnished in each case by a nursing bottle, with its nipple protruding from the mother's "breast."

The two mothers quickly proved to be physiologically equivalent. The monkeys in the two groups drank the same amount of milk and gained weight at the same rate. But the two mothers proved to be by no means psychologically equivalent. Records made automatically showed that both groups of infants spent far more time climbing and clinging on their cloth-covered mothers than they did on their wire mothers. During the infants' first 14 days of life the floors of the cages were warmed by an electric heating pad, but most of the infants left the pad as soon as they could climb on the unheated cloth mother. Moreover, as the monkeys grew older, they tended to spend an increasing amount of time clinging and cuddling on her pliant terry-cloth surface. Those that secured their nourishment from the wire mother showed no tendency to spend more time on her than feeding required, contradicting the idea that affection is a response that is learned or derived in asso-

CLOTH AND WIRE MOTHER-SURROGATES were used to test the preferences of infant monkeys. The infants spent most of their time clinging to the soft cloth "mother," (*foreground*) even when nursing bottles were attached to the wire mother (*background*).

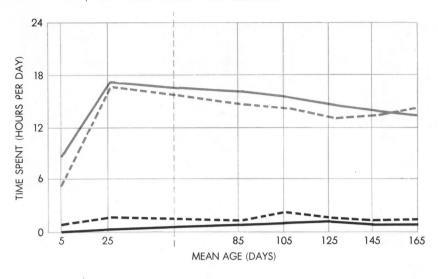

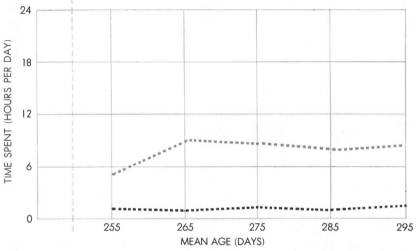

STRONG PREFERENCE FOR CLOTH MOTHER was shown by all infant monkeys.
Infants reared with access to both mothers from birth (*top chart*) spent far more time on
the cloth mother (*colored curves*) than on the wire mother (*black curves*). This was true
regardless of whether they had been fed on the cloth (*solid lines*) or on the wire mother
(*broken lines*). Infants that had known no mother during their first eight months (*bottom
chart*) soon came to prefer cloth mother, but spent less time on her than the other infants.

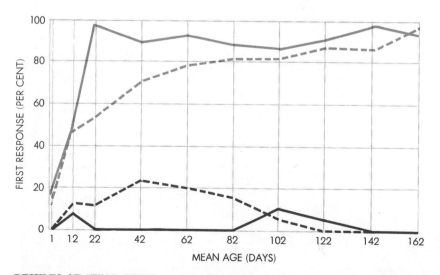

RESULTS OF "FEAR TEST" (*see photographs on opposite page*) showed that infants
confronted by a strange object quickly learned to seek reassurance from the cloth mother
(*colored curves*) rather than from the wire mother (*black curves*). Again infants fed on
the wire mother (*broken lines*) behaved much like those fed on cloth mother (*solid lines*)

ciation with the reduction of hunger or
thirst.

These results attest the importance—
possibly the overwhelming importance—
of bodily contact and the immediate
comfort it supplies in forming the infant's attachment for its mother. All
our experience, in fact, indicates that
our cloth-covered mother surrogate is an
eminently satisfactory mother. She is
available 24 hours a day to satisfy her
infant's overwhelming compulsion to
seek bodily contact; she possesses infinite patience, never scolding her baby
or biting it in anger. In these respects we
regard her as superior to a living monkey mother, though monkey fathers
would probably not endorse this opinion.

Of course this does not mean that
nursing has no psychological importance. No act so effectively guarantees
intimate bodily contact between mother
and child. Furthermore, the mother who
finds nursing a pleasant experience will
probably be temperamentally inclined
to give her infant plenty of handling and
fondling. The real-life attachment of the
infant to its mother is doubtless influenced by subtle multiple variables, contributed in part by the mother and in
part by the child. We make no claim to
having unraveled these in only two years
of investigation. But no matter what
evidence the future may disclose, our
first experiments have shown that contact comfort is a decisive variable in this
relationship.

Such generalization is powerfully supported by the results of the next
phase of our investigation. The time
that the infant monkeys spent cuddling
on their surrogate mothers was a strong
but perhaps not conclusive index of
emotional attachment. Would they also
seek the inanimate mother for comfort
and security when they were subjected
to emotional stress? With this question
in mind we exposed our monkey infants
to the stress of fear by presenting them
with strange objects, for example a
mechanical teddy bear which moved forward, beating a drum. Whether the infants had nursed from the wire or the
cloth mother, they overwhelmingly
sought succor from the cloth one; this
differential in behavior was enhanced
with the passage of time and the accrual
of experience. Early in this series of experiments the terrified infant might rush
blindly to the wire mother, but even if it
did so it would soon abandon her for the
cloth mother. The infant would cling to
its cloth mother, rubbing its body against
hers. Then, with its fears assuaged
through intimate contact with the moth-

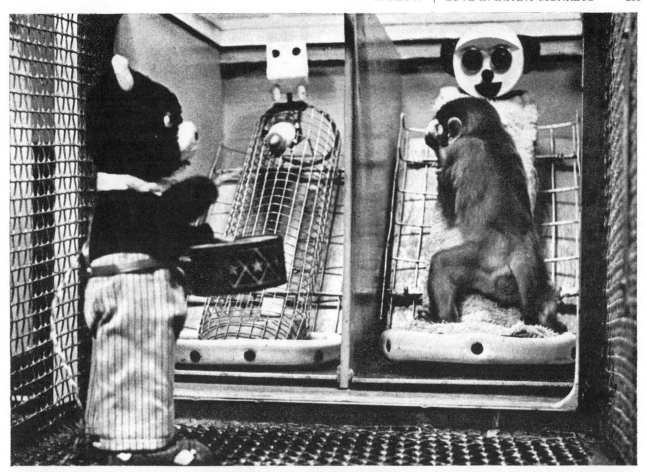

FRIGHTENING OBJECTS such as a mechanical teddy bear caused almost all infant monkeys to flee blindly to the cloth mother, as in the top photograph. Once reassured by pressing and rubbing against her, they would then look at the strange object (*bottom*).

"OPEN FIELD TEST" involved placing a monkey in a room far larger than its accustomed cage; unfamiliar objects added an additional disturbing element. If no mother was present, the infant would typically huddle in a corner (*left*). The wire mother did

er, it would turn to look at the previously terrifying bear without the slightest sign of alarm. Indeed, the infant would sometimes even leave the protection of the mother and approach the object that a few minutes before had reduced it to abject terror.

The analogy with the behavior of human infants requires no elaboration. We found that the analogy extends even to less obviously stressful situations. When a child is taken to a strange place, he usually remains composed and happy so long as his mother is nearby. If the mother gets out of sight, however, the child is often seized with fear and distress. We developed the same response in our infant monkeys when we exposed them to a room that was far larger than the cages to which they were accustomed. In the room we had placed a number of unfamiliar objects such as a small artificial tree, a crumpled piece of paper, a folded gauze diaper, a wooden block and a doorknob [*a similar experiment is depicted in the illustrations on these two pages*]. If the cloth mother was in the room, the infant would rush wildly to her, climb upon her, rub against her and cling to her tightly. As in the previous experiment, its fear then sharply diminished or vanished. The infant would begin to climb over the mother's body and to explore and manipulate her face. Soon it would leave the mother to investigate the new world, and the unfamiliar objects would become playthings. In a typical behavior sequence, the infant might manipulate the tree, return to the mother, crumple the wad of paper, bring it to the mother, explore the block, ex-

plore the doorknob, play with the paper and return to the mother. So long as the mother provided a psychological "base of operations" the infants were unafraid and their behavior remained positive, exploratory and playful.

If the cloth mother was absent, however, the infants would rush across the test room and throw themselves facedown on the floor, clutching their heads and bodies and screaming their distress. Records kept by two independent observers—scoring for such "fear indices" as crying, crouching, rocking and thumb- and toe-sucking—showed that the emotionality scores of the infants nearly tripled. But no quantitative measurement can convey the contrast between the positive, outgoing activities in the presence of the cloth mother and the stereotyped withdrawn and disturbed behavior in the motherless situation.

The bare wire mother provided no more reassurance in this "open field" test than no mother at all. Control tests on monkeys that from birth had known only the wire mother revealed that even these infants showed no affection for her and obtained no comfort from her presence. Indeed, this group of animals exhibited the highest emotionality scores of all. Typically they would run to some wall or corner of the room, clasp their heads and bodies and rock convulsively back and forth. Such activities closely resemble the autistic behavior seen frequently among neglected children in and out of institutions.

In a final comparison of the cloth and wire mothers, we adapted an experiment originally devised by Robert A. Butler

at the Primate Laboratory. Butler had found that monkeys enclosed in a dimly lighted box would press a lever to open and reopen a window for hours on end for no reward other than the chance to look out. The rate of lever-pressing depended on what the monkeys saw through the opened window; the sight of another monkey elicited far more activity than that of a bowl of fruit or an empty room [see "Curiosity in Monkeys," by Robert A. Butler; SCIENTIFIC AMERICAN Offprint 426]. We now know that this "curiosity response" is innate. Three-day-old monkeys, barely able to walk, will crawl across the floor of the box to reach a lever which briefly opens the window; some press the lever hundreds of times within a few hours.

When we tested our monkey infants in the "Butler box," we found that those reared with both cloth and wire mothers showed as high a response to the cloth mother as to another monkey, but displayed no more interest in the wire mother than in an empty room. In this test, as in all the others, the monkeys fed on the wire mother behaved the same as those fed on the cloth mother. A control group raised with no mothers at all found the cloth mother no more interesting than the wire mother and neither as interesting as another monkey.

Thus all the objective tests we have been able to devise agree in showing that the infant monkey's relationship to its surrogate mother is a full one. Comparison with the behavior of infant monkeys raised by their real mothers confirms this view. Like our experimental monkeys, these infants spend many

 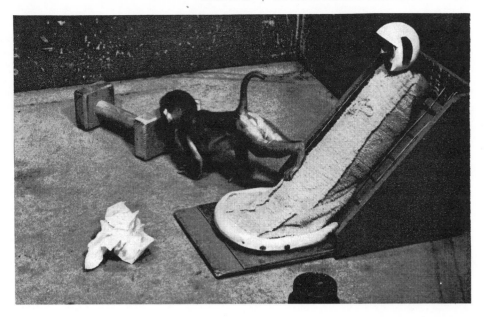

not alter this pattern of fearful behavior, but the cloth mother provided quick reassurance. The infant would first cling to her (*center*) and then set out to explore the room and play with the objects (*right*), returning from time to time for more reassurance.

hours a day clinging to their mothers, and run to them for comfort or reassurance when they are frightened. The deep and abiding bond between mother and child appears to be essentially the same, whether the mother is real or a cloth surrogate.

While bodily contact clearly plays the prime role in developing infantile affection, other types of stimulation presumably supplement its effects. We have therefore embarked on a search for these other factors. The activity of a live monkey mother, for example, provides her infant with frequent motion stimulation. In many human cultures mothers bind their babies to them when they go about their daily chores; in our own culture parents know very well that rocking a baby or walking with him somehow promotes his psychological and physiological well-being. Accordingly we compared the responsiveness of infant monkeys to two cloth mothers, one stationary and one rocking. All of them preferred the rocking mother, though the degree of preference varied considerably from day to day and from monkey to monkey. An experiment with a rocking crib and a stationary one gave similar results. Motion does appear to enhance affection, albeit far less significantly than simple contact.

The act of clinging, in itself, also seems to have a role in promoting psychological and physiological well-being. Even before we began our studies of affection, we noticed that a newborn monkey raised in a bare wire cage survived with difficulty unless we provided it with a cone to which it could cling. Re-

cently we have raised two groups of monkeys, one with a padded crib instead of a mother and the other with a cloth mother as well as a crib. Infants in the latter group actually spend more time on the crib than on the mother, probably because the steep incline of the mother's cloth surface makes her a less satisfactory sleeping platform. In the open-field test, the infants raised with a crib but no mother clearly derived some emotional support from the presence of the crib. But those raised with both showed an unequivocal preference for the mother they could cling to, and they evidenced the benefit of the superior emotional succor they gained from her.

Still other elements in the relationship remain to be investigated systematically. Common sense would suggest that the warmth of the mother's body plays its part in strengthening the infant's ties to her. Our own observations have not yet confirmed this hypothesis. Heating a cloth mother does not seem to increase her attractiveness to the infant monkey, and infants readily abandon a heating pad for an unheated mother surrogate. However, our laboratory is kept comfortably warm at all times; experiments in a chilly environment might well yield quite different results.

Visual stimulation may forge an additional link. When they are about three months old, the monkeys begin to observe and manipulate the head, face and eyes of their mother surrogates; human infants show the same sort of delayed responsiveness to visual stimuli. Such stimuli are known to have marked ef-

fects on the behavior of many young animals. The Austrian zoologist Konrad Lorenz has demonstrated a process called "imprinting"; he has shown that the young of some species of birds become attached to the first moving object they perceive, normally their mothers [see "'Imprinting' in Animals," by Eckhard H. Hess; SCIENTIFIC AMERICAN Offprint 416]. It is also possible that particular sounds and even odors may play some role in the normal development of responses or attention.

The depth and persistence of attachment to the mother depend not only on the kind of stimuli that the young animal receives but also on when it receives them. Experiments with ducks show that imprinting is most effective during a critical period soon after hatching; beyond a certain age it cannot take place at all. Clinical experience with human beings indicates that people who have been deprived of affection in infancy may have difficulty forming affectional ties in later life. From preliminary experiments with our monkeys we have found that their affectional responses develop, or fail to develop, according to a similar pattern.

Early in our investigation we had segregated four infant monkeys as a general control group, denying them physical contact either with a mother surrogate or with other monkeys. After about eight months we placed them in cages with access to both cloth and wire mothers. At first they were afraid of both surrogates, but within a few days they began to respond in much the same way as the other infants. Soon they were

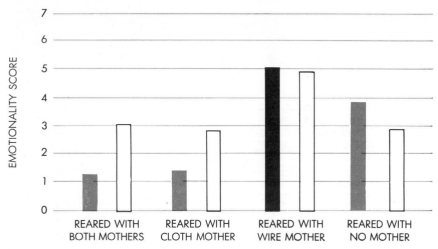

SCORES IN OPEN FIELD TEST show that all infant monkeys familiar with the cloth mother were much less disturbed when she was present (*color*) than when no mother was present (*white*); scores under 2 indicate unfrightened behavior. Infants that had known only the wire mother were greatly disturbed whether she was present (*black*) or not (*white*).

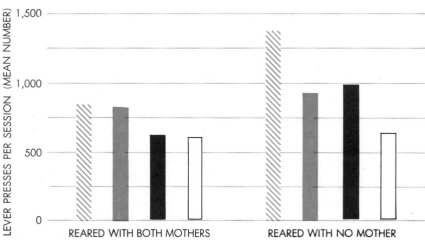

"CURIOSITY TEST" SHOWED THAT monkeys reared with both mothers displayed as much interest in the cloth mother (*solid color*) as in another monkey (*hatched color*); the wire mother (*black*) was no more interesting than an empty chamber (*white*). Monkeys reared with no mother found cloth and wire mother less interesting than another monkey.

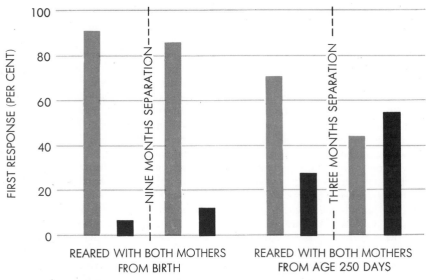

EARLY "MOTHERING" produced a strong and unchanging preference for the cloth mother (*color*) over the wire mother (*black*). Monkeys deprived of early mothering showed less marked preferences before separation and no significant preference subsequently.

spending less than an hour a day with the wire mother and eight to 10 hours with the cloth mother. Significantly, however, they spent little more than half as much time with the cloth mother as did infants raised with her from birth.

In the open-field test these "orphan" monkeys derived far less reassurance from the cloth mothers than did the other infants. The deprivation of physical contact during their first eight months had plainly affected the capacity of these infants to develop the full and normal pattern of affection. We found a further indication of the psychological damage wrought by early lack of mothering when we tested the degree to which infant monkeys retained their attachments to their mothers. Infants raised with a cloth mother from birth and separated from her at about five and a half months showed little or no loss of responsiveness even after 18 months of separation. In some cases it seemed that absence had made the heart grow fonder. The monkeys that had known a mother surrogate only after the age of eight months, however, rapidly lost whatever responsiveness they had acquired. The long period of maternal deprivation had evidently left them incapable of forming a lasting affectional tie.

The effects of maternal separation and deprivation in the human infant have scarcely been investigated, in spite of their implications concerning child-rearing practices. The long period of infant-maternal dependency in the monkey provides a real opportunity for investigating persisting disturbances produced by inconsistent or punishing mother surrogates.

Above and beyond demonstration of the surprising importance of contact comfort as a prime requisite in the formation of an infant's love for its mother —and the discovery of the unimportant or nonexistent role of the breast and act of nursing—our investigations have established a secure experimental approach to this realm of dramatic and subtle emotional relationships. The further exploitation of the broad field of research that now opens up depends merely upon the availability of infant monkeys. We expect to extend our researches by undertaking the study of the mother's (and even the father's!) love for the infant, using real monkey infants or infant surrogates. Finally, with such techniques established, there appears to be no reason why we cannot at some future time investigate the fundamental neurophysiological and biochemical variables underlying affection and love.

IV

SOCIAL BEHAVIOR

IV SOCIAL BEHAVIOR

INTRODUCTION

Social behavior permits a quantum jump in biological organization. There is a strict limit to what any single ant, wolf, or even human being is able to accomplish. In contrast, the society can evolve to much greater complexity by the dual processes of differentiation and integration, even without any increase in the powers of its individual members. At the same time, the society becomes increasingly efficient, larger, and more spatially structured, its members become more specialized into roles or castes (Figure IV.1), and their relationships become more precisely defined by means of better communication. Whole new ways of life—the practice of agriculture, the storage of information, travel over greater distances, and more—await the social group that can correctly engineer the division of labor of its members. Some ants and termites practice true agriculture, cultivating edible fungi on beds constructed of bits of leaves, dead wood, or other decaying materials. In the northern hemisphere there exist parasitic ants that capture the young workers of other kinds of ants and use them as slaves to perform the domestic chores of the colony. The nests of some tropical termites are elaborate domes six meters or more in height, their internal structure a complex of shafts and living quarters arranged in a way that provides a steady flow of fresh air. These social insects are a source of wonder to human beings, yet the members of the colonies are individually simpler and probably less intelligent than wasps, cockroaches, and other solitary insects. Their achievement is based entirely on division of labor and the specialized forms of communication that bind the castes together.

A society is defined as a group of individuals belonging to the same species and organized in a cooperative manner. A pair of animals engaged in basic courtship and sexual activity is not included in this definition, nor are individuals who simply defend their territories from one another. Sexual behavior and territoriality are important properties of societies and are correctly referred to as social behavior, but having them is not sufficient to qualify a group as a society. Bird flocks, wolf packs, and locust swarms are good examples of true elementary societies. So are parents and offspring if they communicate reciprocally. Although this last example may seem a bit extreme at first, parent-offspring interactions are in fact often very complex, and in many kinds of organisms, from social insects to the mammals, the most advanced societies appear to have evolved directly from family units.

This final section of *Scientific American* articles begins with essays that analyze the general properties of communication in all kinds of societies. By keeping these properties in mind, the reader will more easily make sense out of the great heterogeneity of societies described in the succeeding articles. The social species are presented in the order of their position in the over-all

Figure IV.1. Division of labor in insect societies is sometimes accompanied by spectacular specialization. In the termite *Rhynchotermes peramatus* (above), there is a special soldier caste, the members of which are little more than ambulatory spray guns. Through the long nozzle at the front of their heads they spray a sticky irritant fluid that protects the termites against ants and other enemies. A group of soldiers is here seen guarding the flank of a foraging column of workers in the Panamanian rain forest. In ants of the genus *Myrmecocystus* (below), the bodies of certain individuals are enormously distended from fluid food reserves stored in their crops. These individuals are permanently confined to the nests, where they act as "living honey casks," which other members of the colony tap for food by inducing them to regurgitate.

taxonomic scale: first the aggregating amoebae of the cellular slime molds; then the social insects; and finally the vertebrates, represented here by prairie dogs and baboons. A surprising reverse correlation is encountered in this series: the more advanced an organism is in its general anatomical and physiological traits, the less well-integrated is the society to which it belongs. The slime-mold amoebae actually join one another physically to create a slug-like multicellular creature, the pseudoplasmodium. A colony of ants or termites is not unified physically, but its members are differentiated into castes and operate with such a high degree of coordination and altruism that the colony is often referred to with some justice as a "superorganism." In contrast, a vertebrate society is little more than a loose confederation of families and individuals. Even when they exist as subordinate members of societies, vertebrates remain relatively selfish and aggressive. The single outstanding exception to this trend is man himself, who has retained the basic vertebrate traits, but has managed to balance them with coalitions, contracts, vastly improved communication, and long-range planning that includes premeditated acts of altruism.

Why should the lowliest creatures have the most nearly "perfect" societies? The answer is partly a matter of kinship. Because the amoebae multiply by simple cellular fission, many of the individuals are genetically identical to one another. They can afford to be extremely cooperative and even altruistic, because to advance the interests of a fellow amoeba is virtually the same as promoting one's own genes. Recent studies have shown that in most of the social insects, in particular among the members of the order Hymenoptera (ants, wasps, and bees), sisters are more closely related to one another than they are among most other kinds of animals. This fact alone might explain the extremes to which some of the hymenopterans have been able to take their social evolution. The vertebrates, in contrast, always breed sexually and exhibit no such extraordinarily close genetic ties between sisters or other close relatives.

Sociobiology promises to develop into one of the most important disciplines in all of science, because it can build a substantial bridge between biology and the social sciences. One of the principal goals of science is to provide firm underpinnings for the analysis of human nature. The Animal Kingdom contains literally thousands of other social species that can usefully be compared with man. They display among themselves an amazing diversity of communication systems and group structures, each representing a social solution to a particular set of environmental conditions. An understanding of these solutions can be the substance of a rigorous new discipline of sociobiology, which will then reinterpret much of human behavior as a special case of general, deeply understood principles.

"Animal Communication," *by Edward O. Wilson. September 1972.*
This article was originally written as a general introduction to communication in all kinds of organisms, and it can serve the same function here.

"The Language of Birds," *by W. H. Thorpe. October 1956.*
Thanks to the pioneering research of W. H. Thorpe, Peter Marler, and other zoologists during the past twenty years, rich new meaning has been attached to the many previously baffling qualities of bird song. Here Thorpe describes the physical properties that make certain calls ideal as alarm-and-assembly signals, others as signals that alarm without disclosing the location of the caller, and still others that permit the recognition of individual birds.

"The Fighting Behavior of Animals," *by Irenäus Eibl-Eibesfeldt. December 1961.*
One of the most intriguing discoveries of ethology has been that aggression

within species of animals is almost entirely ritualized. Although the stakes are high—sole possession of a territory or high rank within a dominance order—hostility does not ordinarily escalate into all-out fighting. Instead, animals perform elaborate displays in which each participant attempts to appear as large and formidable as possible. Physical contact, if it occurs at all, typically consists of a shoving or wrestling match, in which potentially lethal weapons are kept out of play. No example is more instructive than the combat dance of male rattlesnakes. The contestants neck-wrestle with one another until a clear winner emerges; they rarely if ever use the deadly fangs that could decide the matter within seconds. Why do such animals "fight clean"? The probable answer is that escalation can damage the winner as well as the loser, with the result that no contest at all would have been the better choice. The ritualized forms of aggression allow the contestants to evaluate their own chances in the event of an all-out struggle. When all other conditions are at least approximately equal, it is usually the smaller individual who submits and leaves the field. It is as though the winner thinks, "I would win in the event of a real fight, but I might be injured; so I will limit myself to displays until my opponent understands its relative weakness and leaves." The loser acts as though it were reflecting in turn, "If I stay here and pursue the matter, we might have a battle in which I would be destroyed; I will leave."

"Hormones in Social Amoebae and Mammals," *by John Tyler Bonner. June 1969.*

Organisms with outwardly simple structure sometimes possess surprisingly complex behavior. This disparity is carried to an extreme in cellular slime molds, such as *Dictyostelium*. The essence of the life cycle of these soil-dwelling organisms can be briefly summarized as follows. At one stage there are only single-celled amoebae that feed separately and multiply by simple fission. The amoebae constitute the principal vegetative form; as single-celled organisms, they are able to grow and to multiply faster than their multicellular counterparts. When the environment deteriorates, however, the amoebae swarm together and coalesce into a single, slug-like organism called a pseudoplasmodium. After traveling about for a while, this creature undergoes a complete internal reorganization. A stalk sprouts up, forming a sphere at the apex from which a large number of spores are shed. When the spores land in a favorable spot, they germinate as amoebae to start the vegetative portion of the cycle over again. For years the nature of the signal that calls the amoebae together has remained an intriguing mystery. It was known to be a specific chemical and even given the name acrasin. In this article John T. Bonner describes the experiments that finally identified acrasin as cyclic AMP. This substance is by no means an exotic pheromone peculiar to slime molds. It also occurs in mammalian cells, where it serves as a messenger substance between hormones (such as epinephrine) arriving at the outside of the cell and the enzymes that control the activity of the cell. As Bonner points out, cyclic AMP is also produced by bacteria, and the amoebae track it in order to find these organisms, on which they feed. Why has the very same substance appeared in such widely divergent functions? The answer must await further biochemical investigation.

"Communication Between Ants and Their Guests," *by Bert Hölldobler. March 1971.*

Because insect societies are organized by use of only a few releasers, mostly chemical in nature (see the introduction to Part I), they are susceptible to invasion by social parasites. Any insect that can mimic the correct combination of odors can get itself accepted into a colony, at least long enough to rob food stores, to eat some of the members, or to exploit the colony in other ways. Here Bert Holldöbler describes some of the methods by which the parasites have "broken the code" of their ant hosts. The study of these bizarre creatures offers a bonus; by analyzing their simplified techniques, we may be able to list with greater certainty the minimal set of signals needed by the ants themselves to organize their colonies.

"Dialects in the Language of the Bees," *by Karl von Frisch. August 1962.* Here the celebrated waggle dance of the honeybee is explained by the scientist who first decoded it in 1945. Karl von Frisch also describes "dialects" in the dance found in different geographic races, the "dialects" here being variations in the precise scale used to translate information on the distance of the target.

"The Social Behavior of Prairie Dogs," *by John A. King. October 1959.* The prairie dogs of the American plains have the most elaborate social systems of any of the known species of rodents around the world. They form dense local populations, within which social groups called "coteries" defend small territories. The members of each coterie share their burrows and know one another on an individual basis. Two prairie dogs "kiss" in order to check identification, and if they prove to belong to the same group, they often proceed to lie down and to groom one another. The most notable single feature of the social life of these animals is the fact that the territorial limits of the coterie are passed on by tradition. The population of each coterie constantly changes within a few months or years, by birth, death, and emigration. But the boundary of each group remains about the same, being learned by each prairie dog born within it. The communication system of the prairie dogs, which consists to a large extent of vocalizations and odors, is also exceptionally rich. In sum, these rodents have evolved social behavior which in many respects rivals that of the monkeys and apes.

"The Social Life of Baboons," *by S. L. Washburn and Irven DeVore. June 1961.* Impelled by the dream of providing a baseline for the study of human behavior, anthropologists and zoologists have in recent years enormously expanded the study of the social behavior of monkeys and apes. Prior to 1950 no more than 50 man-months had been devoted to field studies of these animals. By 1966 the cumulative field time had reached 1500 man-months, involved hundreds of investigators, and was accelerating. The amount of research conducted in the four years from 1962 through 1965 alone exceeded that of all the research before it. This article by Washburn and DeVore describes the results of one of the earliest and best studies of this genre. The authors show that baboon troops are organized in a manner very different from that previously conceived: sexual attraction is not the primary bond of the society. The core of the group is an association of females and their young. They are ruled over and protected by a coalition of males, who possess different ranks in a dominance order. Baboons are of particular interest because they are among the very few species of monkeys and apes that live primarily in the open grassland, the habitat of early man. The social adaptations they show to this special environment can hopefully shed some light on the intermediate steps that led to human social organization. For example, the tendency to feed and to travel in tight groups when crossing open terrain appears to be an adaptation to protect the baboons from the attacks of predators. The aggressive, domineering behavior of the males may well have originated as part of the same defensive strategy. When the troop is threatened, the adult males approach the predator as a group, forming a protective screen between it and the rest of the troop. Once this behavioral trait became established in evolution, it was easier for aggressiveness to permeate the relationships of the males to each other and to the females.

SUGGESTED ADDITIONAL READING FOR PART IV

Crook, J. H., ed. 1970. *Social Behaviour in Birds and Mammals: Essays on the Social Ethology of Animals and Man.* Academic Press, New York. This is an interesting collection of both general and specialized articles that reflect some of the current work on social systems in vertebrates.

Eisenberg, J. F., and W. Dillon, eds. 1971. *Man and Beast: Comparative Social Behavior.* Smithsonian Institution Press, Washington, D.C. The aim of this book, authored by experts on social behavior from around the world, was to examine in depth the relevance of the new studies of animal behavior to the understanding of human nature.

Kummer, H. 1971. *Primate Societies: Group Techniques of Ecological Adaptation.* Aldine-Atherton, Chicago. A brief but perceptive review of the basic qualities of social life in monkeys and apes, with special emphasis on baboons, which the author has studied for many years.

Wilson, E. O. 1971. *The Insect Societies.* Belknap Press of Harvard University Press, Cambridge, Mass. Written for both the scholar and the general reader, this monograph presents a comprehensive review of the biology of all of the social insects.

Wilson, E. O. 1975. *Sociobiology: The New Synthesis.* Belknap Press of Harvard University Press, Cambridge, Mass. (In press.) This book reviews social phenomena in all organisms, from bacteria and slime molds to man, reinterpreting them within the framework of modern physiology and population biology.

Animal Communication

by Edward O. Wilson

*Animals ranging from insects to mammals
communicate by means of chemicals, movements, and
sounds. Man also uses these modes of communication,
but he adds his own unique kind of language.*

The most instructive way to view the communication systems of animals is to compare these systems first with human language. With our own unique verbal system as a standard of reference we can define the limits of animal communication in terms of the properties it rarely—or never—displays. Consider the way I address you now. Each word I use has been assigned a specific meaning by a particular culture and transmitted to us down through generations by learning. What is truly unique is the very large number of such words and the potential for creating new ones to denote any number of additional objects and concepts. This potential is quite literally infinite. To take an example from mathematics, we can coin a nonsense word for any number we choose (as in the case of the googol, which designates a 1 followed by 100 zeros). Human beings utter their words sequentially in phrases and sentences that generate, according to complex rules also determined at least partly by the culture, a vastly larger array of messages than is provided by the mere summed meanings of the words themselves. With these messages it is possible to talk about the language itself, an achievement we are utilizing here. It is also possible to project an endless number of unreal images: fiction or lies, speculation or fraud, idealism or demagogy, the definition depending on whether or not the communicator informs the listener of his intention to speak falsely.

Now contrast this with one of the most sophisticated of all animal communication systems, the celebrated waggle dance of the honeybee (*Apis mellifera*), first decoded in 1945 by the German biologist Karl von Frisch. When a foraging worker bee returns from the field after discovering a food source (or, in the course of swarming, a desirable new nest site) at some distance from the hive, she indicates the location of this target to her fellow workers by performing the waggle dance. The pattern of her movement is a figure eight repeated over and over again in the midst of crowds of sister workers. The most distinctive and informative element of the dance is the straight run (the middle of the figure eight), which is given a particular emphasis by a rapid lateral vibration of the body (the waggle) that is greatest at the tip of the abdomen and least marked at the head.

The complete back-and-forth shake of the body is performed 13 to 15 times per second. At the same time the bee emits an audible buzzing sound by vibrating its wings. The straight run represents, quite simply, a miniaturized version of the flight from the hive to the target. It points directly at the target if the bee is dancing outside the hive on a horizontal surface. (The position of the sun with respect to the straight run provides the required orientation.) If the bee is on a vertical surface inside the darkened hive, the straight run points at the appropriate angle away from the vertical,

so that gravity temporarily replaces the sun as the orientation cue.

The straight run also provides information on the distance of the target from the hive, by means of the following additional parameter: the farther away the goal lies, the longer the straight run lasts. In the Carniolan race of the honeybee a straight run lasting a second indicates a target about 500 meters away, and a run lasting two seconds indicates a target two kilometers away. During the dance the follower bees extend their antennae and touch the dancer repeatedly. Within minutes some begin to leave the nest and fly to the target. Their searching is respectably accurate: the great majority come down to search close to the ground within 20 percent of the correct distance.

Superficially the waggle dance of the honeybee may seem to possess some of the more advanced properties of human language. Symbolism occurs in the form of the ritualized straight run, and the communicator can generate new messages at will by means of the symbolism. Furthermore, the target is "spoken of" abstractly: it is an object removed in time and space. Nevertheless, the waggle dance, like all other forms of nonhuman communication studied so far, is severely limited in comparison with the verbal language of human beings. The straight run is after all just a reenactment of the flight the bees will take, complete with wing-buzzing to represent the actual motor activity required. The separate messages are not devised arbitrarily. The rules they follow are genetically fixed and always designate, with a one-to-one correspondence, a certain direction and distance.

In other words, the messages cannot be manipulated to provide new classes of information. Moreover, within this

COURTSHIP RITUAL of grebes is climaxed by the "penguin dance" shown on the opposite page. In this ritual the male and the female present each other with a beakful of the waterweed that is used as a nest-building material. A pair-bonding display, the dance may have originated as "displacement" behavior, in this instance a pantomime of nest-building triggered by the conflict within each partner between hostility and sexual attraction. The penguin dance was first analyzed in 1914 by Julian Huxley, who observed the ritual among great crested grebes in Europe. Shown here are western grebes in southern Saskatchewan.

rigid context the messages are far from being infinitely divisible. Because of errors both in the dance and in the subsequent searches by the followers, only about three bits of information are transmitted with respect to distance and four bits with respect to direction. This is the equivalent of a human communication system in which distance would be gauged on a scale with eight divisions and direction would be defined in terms of a compass with 16 points. Northeast could be distinguished from north by northeast, or west from west by southwest, but no more refined indication of direction would be possible.

The waggle dance, in particular the duration of the straight run denoting distance, illustrates a simple principle that operates through much of animal communication: the greater the magnitude to be communicated, the more intense and prolonged the signal given. This graduated (or analogue) form of communication is perhaps most strikingly developed in aggressive displays among animals. In the rhesus monkey, for example, a low-intensity aggressive display is a simple stare. The hard look a human receives when he approaches a caged rhesus is not so much a sign of curiosity as it is a cautious display of hostility.

Rhesus monkeys in the wild frequent-ly threaten one another not only with stares but also with additional displays on an ascending scale of intensity. To the human observer these displays are increasingly obvious in their meaning. The new components are added one by one or in combination: the mouth opens, the head bobs up and down, characteristic sounds are uttered and the hands slap the ground. By the time the monkey combines all these components, and perhaps begins to make little forward lunges as well, it is likely to carry through with an actual attack. Its opponent responds either by retreating or by escalating its own displays. These hostile exchanges play a key role in maintaining dominance relationships in the rhesus society.

Birds often indicate hostility by ruffling their feathers or spreading their wings, which creates the temporary illusion that they are larger than they really are. Many fishes achieve the same deception by spreading their fins or extending their gill covers. Lizards raise their crest, lower their dewlaps or flatten the sides of their body to give an impression of greater depth. In short, the more hostile the animal, the more likely it is to attack and the bigger it seems to become. Such exhibitions are often accompanied by graded changes both in color and in vocalization, and even by the release of characteristic odors.

The communication systems of insects, of other invertebrates and of the lower vertebrates (such as fishes and amphibians) are characteristically stereotyped. This means that for each signal there is only one response or very few responses, that each response can be evoked by only a very limited number of signals and that the signaling behavior and the responses are nearly constant throughout entire populations of the same species. An extreme example of this rule is seen in the phenomenon of chemical sex attraction in moths. The female silkworm moth draws males to her by emitting minute quantities of a complex alcohol from glands at the tip of her abdomen. The secretion is called bombykol (from the name of the moth, *Bombyx mori*), and its chemical structure is *trans*-10-*cis*-12-hexadecadienol.

Bombykol is a remarkably powerful biological agent. According to estimates made by Dietrich Schneider and his co-workers at the Max Planck Institute for Comparative Physiology at Seewiesen in Germany, the male silkworm moths start searching for the females when they are immersed in as few as 14,000 molecules of bombykol per cubic centimeter of air. The male catches the molecules on some 10,000 distinctive sensory hairs on each of its two feathery antennae. Each hair is innervated by one or two receptor cells that lead inward to the main antennal nerve and ultimately through connecting nerve cells to centers in the brain. The extraordinary fact that emerged from the study by the Seewiesen group is that only a single molecule of bombykol is required to activate a receptor cell. Furthermore, the cell will respond to virtually no stimulus other than molecules of bombykol. When about 200 cells in each antenna are activated, the male moth starts its motor response. Tightly bound by this extreme signal specificity, the male performs as little more than a sexual guided missile, programmed to home on an increasing gradient of bombykol centered on the tip of the female's abdomen—the principal goal of the male's adult life.

Such highly stereotyped communication systems are particularly important in evolutionary theory because of the possible role the systems play in the origin of new species. Conceivably one small change in the sex-attractant molecule induced by a genetic mutation, together with a corresponding change in the antennal receptor cell, could result in the creation of a population of individuals that would be reproductively isolated from the parental stock. Persuasive

WAGGLE DANCE of the honeybee, first decoded by Karl von Frisch in 1945, is performed by a foraging worker bee on its return to the hive after the discovery of a food source. The pattern of the dance is a repeated figure eight. During the straight run in the middle of the figure the forager waggles its abdomen rapidly and vibrates its wings. As is shown in the illustrations on the opposite page, the direction of the straight run indicates the line of flight to the food source. The duration of the straight run shows workers how far to fly.

evidence for the feasibility of such a mutational change has recently been adduced by Wendell L. Roelofs and Andre Comeau of Cornell University. They found two closely related species of moths (members of the genus *Bryotopha* in the family Gelechiidae) whose females' sex attractants differ only by the configuration of a single carbon atom adjacent to a double bond. In other words, the attractants are simply different geometric isomers. Field tests showed not only that a *Bryotopha* male responds solely to the isomer of its own species but also that its response is inhibited if some of the other species' isomer is also present.

A qualitatively different kind of specificity is encountered among birds and mammals. Unlike the insects, many of these higher vertebrates are able to distinguish one another as individuals on the basis of idiosyncrasies in the way they deliver signals. Indigo buntings and certain other songbirds learn to discriminate the territorial calls of their neighbors from those of strangers that occupy territories farther away. When a recording of the song of a neighbor is played near them, they show no unusual reactions, but a recording of a stranger's song elicits an agitated aggressive response.

Families of seabirds depend on a similar capacity for recognition to keep together as a unit in the large, clamorous colonies where they nest. Beat Tschanz of the University of Bern has demonstrated that the young of the common murre (*Uria aalge*), a large auk, learn to react selectively to the call of their parents in the first few days of their life and that the parents also quickly learn to distinguish their own young. There is some evidence that the young murres can even learn certain aspects of the adult calls while they are still in the egg. An equally striking phenomenon is the intercommunication between African shrikes (of the genus *Laniarius*) recently analyzed by W. H. Thorpe of the University of Cambridge. Mated pairs of these birds keep in contact by calling antiphonally back and forth, the first bird vocalizing one or more notes and its mate instantly responding with a variation of the first call. So fast is the exchange, sometimes taking no more than a fraction of a second, that unless an observer stands between the two birds he does not realize that more than one bird is singing. In at least one of the species, the boubou shrike (*Laniarius aethiopicus*), the members of the pair learn to sing duets with each other. They work out combinations of phrases that are sufficiently individual

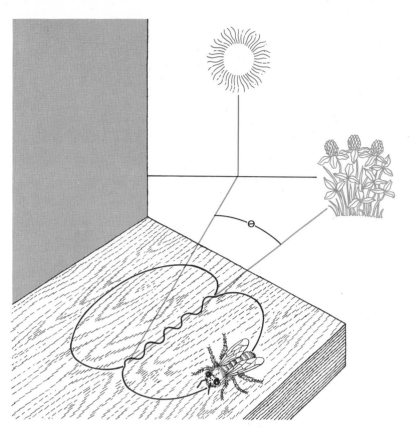

DANCING OUTSIDE THE HIVE on a horizontal surface, the forager makes the straight run of its waggle dance point directly at the source of food. In this illustration the food is located some 20 degrees to the right of the sun. The forager's fellow workers maintain the same orientation with respect to the sun as they leave for the reported source of food.

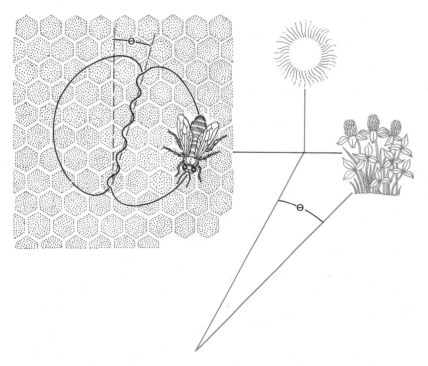

DANCING INSIDE THE HIVE on the vertical face of the honeycomb, the forager uses gravity for orientation. The straight line of the waggle dance that shows the line of flight to the source of food is oriented some 20 degrees away from the vertical. On leaving the hive, the bee's fellow workers relate the indicated orientation angle to the position of the sun.

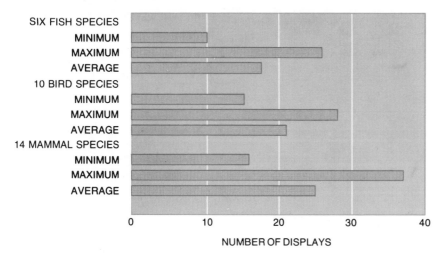

SIX FISH SPECIES
MINIMUM
MAXIMUM
AVERAGE
10 BIRD SPECIES
MINIMUM
MAXIMUM
AVERAGE
14 MAMMAL SPECIES
MINIMUM
MAXIMUM
AVERAGE

0   10   20   30   40

NUMBER OF DISPLAYS

COMMUNICATIVE DISPLAYS used by 30 species of vertebrate animals whose "languages" have been studied vary widely within each of the classes of animals represented: fishes, birds and mammals. The average differences between the classes, however, are comparatively small. The largest and smallest number of displays within each class and the average for each class are shown in this graph. Six of the fish species that have been studied use an average of some 17 displays, compared with an average of 21 displays used by 10 species of birds and an average of 25 displays among 14 species of mammals. Martin H. Moynihan of the Smithsonian Institution compiled the display data. The 30 vertebrates and the number of displays that each uses are illustrated on the opposite page and on page 270.

to enable them to recognize each other even though both are invisible in the dense vegetation the species normally inhabits.

Mammals are at least equally adept at discriminating among individuals of their own kind. A wide range of cues are employed by different species to distinguish mates, offspring and in the case of social mammals the subordinate or dominant rank of the peers ranged around them. In some species special secretions are employed to impart a personal odor signature to part of the environment or to other members in the social group. As all dog owners know, their pet urinates at regular locations within its territory at a rate that seems to exceed physiological needs. What is less well appreciated is the communicative function this compulsive behavior serves: a scent included in the urine identifies the animal and announces its presence to potential intruders of the same species.

Males of the sugar glider (*Petaurus breviceps*), a New Guinea marsupial with a striking but superficial resemblance to the flying squirrel, go even further. They mark their mate with a secretion from a gland on the front of their head. Other secretions originating in glands on the male's feet, on its chest and near its arms, together with its saliva, are used to mark its territory. In both instances the odors are distinctive enough for the male to distinguish them from those of other sugar gliders.

As a rule we find that the more highly social the mammal is, the more complex the communication codes are and the more the codes are utilized in establishing and maintaining individual relationships. It is no doubt significant that one of the rare examples of persistent individual recognition among the lower animals is the colony odor of the social insects: ants and termites and certain social bees and wasps. Even here, however, it is the colony as a whole that is recognized. The separate members of the colony respond automatically to certain caste distinctions, but they do not ordinarily learn to discriminate among their nestmates as individuals.

By human standards the number of signals employed by each species of animal is severely limited. One of the most curious facts revealed by recent field studies is that even the most highly social vertebrates rarely have more than 30 or 35 separate displays in their entire repertory. Data compiled by Martin H. Moynihan of the Smithsonian Institution indicate that among most vertebrates the number of displays varies by a factor of only three or four from species to species. The number ranges from a minimum of 10 in certain fishes to a maximum of 37 in the rhesus monkey, one of the primates closest to man in the complexity of their social organization. The full significance of this rule of relative inflexibility is not yet clear. It may be that the maximum number of messages any animal needs in order to be fully adaptive in any ordinary environment, even a social one, is no more than

30 or 40. Or it may be, as Moynihan has suggested, that each number represents the largest amount of signal diversity the particular animal's brain can handle efficiently in quickly changing social interactions.

In the extent of their signal diversity the vertebrates are closely approached by social insects, particularly honeybees and ants. Analyses by Charles G. Butler at the Rothamsted Experimental Station in England, by me at Harvard University and by others have brought the number of individual known signal categories within single species of these insects to between 10 and 20. The honeybee has been the most thoroughly studied of all the social insects. Apart from the waggle dance its known communicative acts are mediated primarily by pheromones: chemical compounds transmitted to other members of the same species as signals. The glandular sources of these and other socially important substances are now largely established. Other honeybee signals include the distinctive colony odor mentioned above, tactile cues involved in food exchange and several dances that are different in form and function from the waggle dance.

Of the known honeybee pheromones the "queen substances" are outstanding in the complexity and pervasiveness of their role in social organization. They include *trans*-9-keto-2-decenoic acid, which is released from the queen's mandibular glands and evokes at least three separate effects according to the context of its presentation. The pheromone is spread through the colony when workers lick the queen's body and regurgitate the material back and forth to one another. For the substance to be effective in the colony as a whole the queen must dispense enough for each worker bee to receive approximately a tenth of a microgram per day.

The first effect of the ketodecenoic acid is to keep workers from rearing larvae in a way that would result in their becoming new queens, thus preventing the creation of potential rivals to the mother queen. The second effect is that when the worker bees eat the substance, their own ovaries fail to develop; they cannot lay eggs and as a result they too are eliminated as potential rivals. Indirect evidence indicates that ingestion of the substance affects the corpora allata, the endocrine glands that partly control the development of the ovaries, but the exact chain of events remains to be worked out. The third effect of the pheromone is that it acts as a sex attractant. When a virgin queen flies from

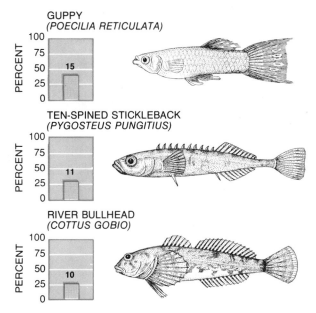

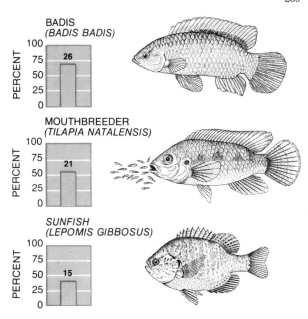

DISPLAYS BY FISHES range from a minimum of 10, used by the river bullhead (*bottom left*), to a maximum of 26, used by the badis (*top right*). The badis repertory is thus more extensive than those of eight of the 10 birds and nine of the 14 mammals studied. The bar beside each fish expressed the number of its displays in percent; 37 displays, the maximum in the study, equal 100 percent.

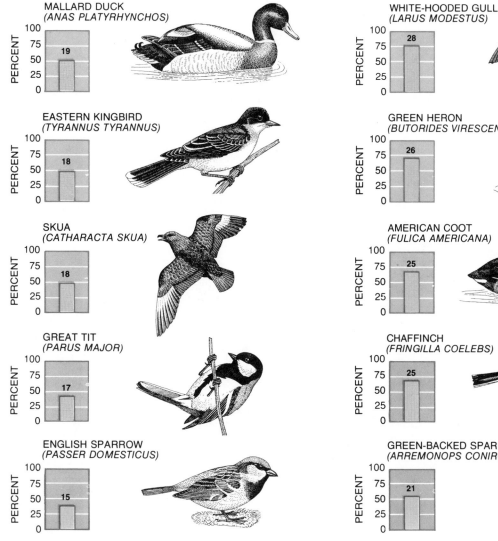

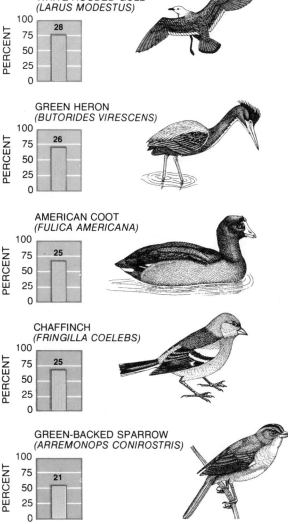

DISPLAYS BY BIRDS range from a minimum of 15, used by the English sparrow (*bottom left*), to a maximum of 28, used by the white-headed gull (*top right*). The maximum repertory among birds thus proves to be little greater than the fishes' maximum.

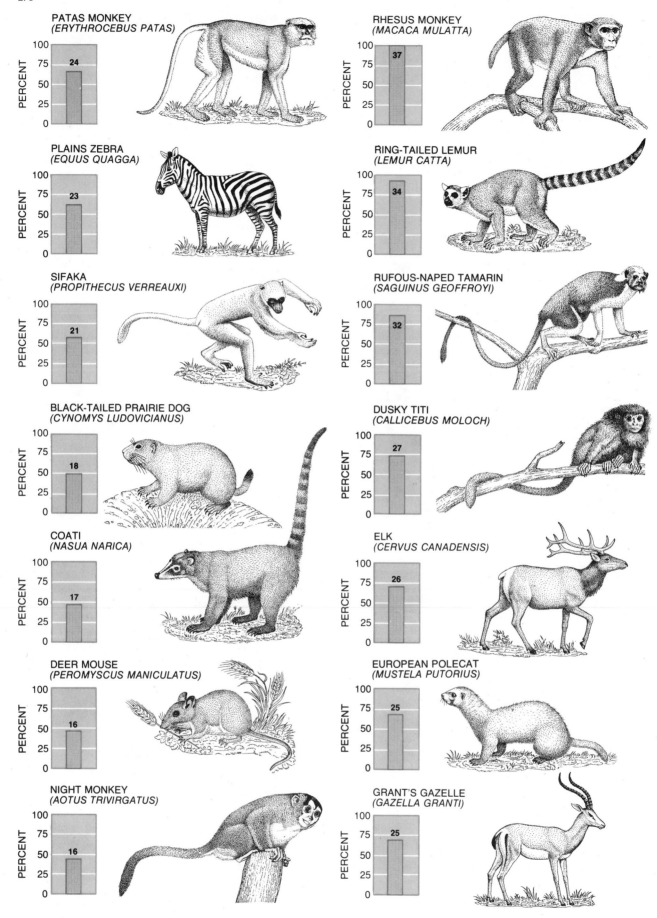

PATAS MONKEY
(ERYTHROCEBUS PATAS)

RHESUS MONKEY
(MACACA MULATTA)

PLAINS ZEBRA
(EQUUS QUAGGA)

RING-TAILED LEMUR
(LEMUR CATTA)

SIFAKA
(PROPITHECUS VERREAUXI)

RUFOUS-NAPED TAMARIN
(SAGUINUS GEOFFROYI)

BLACK-TAILED PRAIRIE DOG
(CYNOMYS LUDOVICIANUS)

DUSKY TITI
(CALLICEBUS MOLOCH)

COATI
(NASUA NARICA)

ELK
(CERVUS CANADENSIS)

DEER MOUSE
(PEROMYSCUS MANICULATUS)

EUROPEAN POLECAT
(MUSTELA PUTORIUS)

NIGHT MONKEY
(AOTUS TRIVIRGATUS)

GRANT'S GAZELLE
(GAZELLA GRANTI)

DISPLAYS BY MAMMALS range from a minimum of 16, used both by the deer mouse and by the night monkey (*left, bottom and* *next to bottom*), to a maximum of 37, used by the rhesus monkey (*top right*). Two other primates rank next in number of displays.

the hive on her nuptial flight, she releases a vapor trail of the ketodecenoic acid in the air. The smell of the substance not only attracts drones to the queen but also induces them to copulate with her.

Where do such communication codes come from in the first place? By comparing the signaling behavior of closely related species zoologists are often able to piece together the sequence of evolutionary steps that leads to even the most bizarre communication systems. The evolutionary process by which a behavior pattern becomes increasingly effective as a signal is called "ritualization." Commonly, and perhaps invariably, the process begins when some movement, some anatomical feature or some physiological trait that is functional in quite another context acquires a secondary value as a signal. For example, one can begin by recognizing an open mouth as a threat or by interpreting the turning away of the body in the midst of conflict as an intention to flee. During ritualization such movements are altered in some way that makes their communicative function still more effective. In extreme cases the new behavior pattern may be so modified from its ancestral state that its evolutionary history is all but impossible to imagine. Like the epaulets, shako plumes and piping that garnish military dress uniforms, the practical functions that originally existed have long since been obliterated in order to maximize efficiency in display.

The ritualization of vertebrate behavior commonly begins in circumstances of conflict, particularly when an animal is undecided whether or not to complete an act. Hesitation in behavior communicates the animal's state of mind—or, to be more precise, its probable future course of action—to onlooking members of the same species. The advertisement may begin its evolution as a simple intention movement. Birds intending to fly, for example, typically crouch, raise their tail and spread their wings slightly just before taking off. Many species have ritualized these movements into effective signals. In some species white rump feathers produce a conspicuous flash when the tail is raised. In other species the wing tips are flicked repeatedly downward, uncovering conspicuous areas on the primary feathers of the wings. The signals serve to coordinate the movement of flock members, and also may warn of approaching predators.

Signals also evolve from the ambivalence created by the conflict between two or more behavioral tendencies.

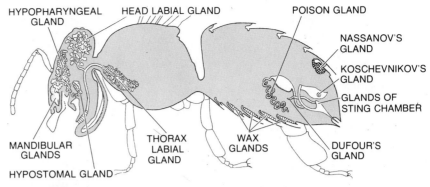

PHEROMONES OF THE HONEYBEE are produced by the glands shown in this cutaway figure of a worker. The glands perform different functions in different castes. In workers, for example, the secretion of the mandibular glands serves as an alarm signal. In a queen, however, the mandibular secretion that is spread through the colony as a result of grooming inhibits workers from raising new queens and also prevents workers from becoming egg-layers. It is also released as a vaporous sex attractant when the new queen leaves the hive on her nuptial flight. The "royal jelly" secreted by the hypopharyngeal gland serves as a food and also acts as a caste determinant. The labial glands of head and thorax secrete a substance utilized for grooming, cleaning and dissolving. Action of the hypostomal-gland secretion is unknown, as is the action of Dufour's gland. The wax glands yield nest-building material, the poison gland is for defense and the sting-chamber glands provide an alarm signal. The secretion of Nassanov's gland assists in assembling workers in conjunction with the waggle dance; that of Koschevnikov's gland renders queens attractive to workers.

When a male faces an opponent, unable to decide whether to attack or to flee, or approaches a potential mate with strong tendencies both to threaten and to court, he may at first make neither choice. Instead he performs a third, seemingly irrelevant act. The aggression is redirected at a meaningless object nearby, such as a pebble, a blade of grass or a bystander that serves as a scapegoat. Or the animal may abruptly commence a "displacement" activity: a behavior pattern with no relevance whatever to the circumstance in which the animal finds itself. The animal may preen, start ineffectual nest-building movements or pantomime feeding or drinking.

Such redirected and displacement activities have often been ritualized into strikingly clear signals. Two classic examples involve the formation of a pair bond between courting grebes. They were among the first such signals to be recognized; Julian Huxley, the originator of the concept of ritualization, analyzed the behavior among European great crested grebes in 1914. The first ritual is "mutual headshaking." It is apparently derived from more elementary movements, aimed at reducing hostility, wherein each bird simply directs its bill away from its partner. The second ritual, called by Huxley the "penguin dance," includes headshaking, diving and the mutual presentation by each partner to its mate of the waterweeds that serve as nesting material. The collection and presentation of the waterweeds may have evolved from displacement nesting behavior initially produced by the conflict between hostility and sexuality.

A perhaps even more instructive example of how ritualization proceeds is provided by the courtship behavior of the "dance flies." These insects include a large number of carnivorous species of dipterans that entomologists classify together as the family Empididae. Many of the species engage in a kind of courtship that consists in a simple approach by the male; this approach is followed by copulation. Among other species the male first captures an insect of the kind that normally falls prey to empids and presents it to the female before copulation. This act appears to reduce the chances of the male himself becoming a victim of the predatory impulses of the female. In other species the male fastens threads or globules of silk to the freshly captured offering, rendering it more distinctive in appearance, a clear step in the direction of ritualization.

Increasing degrees of ritualization can be observed among still other species of dance flies. In one of these species the male totally encloses the dead prey in a sheet of silk. In another the size of the offered prey is smaller but its silken covering remains as large as before: it is now a partly empty "balloon." The male of another species does not bother to capture any prey object but simply offers the female an empty balloon. The last display is so far removed from the original behavior pattern that its evolutionary origin in this empid species might have remained a permanent mystery if biolo-

AGGRESSIVE DISPLAYS by a rhesus monkey (*top*) and a green heron (*bottom*) illustrate a major principle of animal communication: the greater the magnitude to be communicated, the more prolonged and intense the signal is. In the rhesus what begins as a display of low intensity, a hard stare (*left*), is gradually escalated as the monkey rises to a standing position (*middle*) and then, with an open mouth, bobs its head up and down (*right*) and slaps the ground with its hands. If the opponent has not retreated by now, the monkey may actually attack. A similarly graduated aggressive display is characteristic of the green heron. At first (*middle*) the heron raises the feathers that form its crest and twitches the feathers of its tail. If the opponent does not retreat, the heron opens its beak, erects its crest fully, ruffles all its plumage to give the illusion of increased size and violently twitches its tail (*right*). Thus in both animals the likelier the attack, the more intense the aggressive display. Andrew J. Meyerriecks of the University of South Florida conducted the study of heron display and Stuart A. Altmann of the University of Chicago conducted the rhesus display study.

gists had not discovered what appears to be the full story of its development preserved step by step in the behavior of related species.

One of the most important and most difficult questions raised by behavioral biology can be phrased in the evolutionary terms just introduced as follows: Can we hope to trace the origin of human language back through intermediate steps in our fellow higher primates—our closest living relatives, the apes and monkeys—in the same way that entomologists have deduced the origin of the empty-balloon display among the dance flies? The answer would seem to be a very limited and qualified yes. The most probable links to investigate exist within human paralinguistics: the extensive array of facial expressions, body postures, hand signals and vocal tones and emphases that we use to supplement verbal speech. It might be possible to match some of these auxiliary signals with the more basic displays in apes and monkeys. J. A. R. A. M. van Hooff of the State University of Utrecht, for example,

has argued persuasively that laughter originated from the primitive "relaxed open-mouth display" used by the higher primates to indicate their intention to participate in mock aggression or play (as distinct from the hostile open-mouth posture described earlier as a low-intensity threat display in the rhesus monkey). Smiling, on the other hand, van Hooff derives from the primitive "silent bared-teeth display," which denotes submission or at least nonhostility.

What about verbal speech? Chimpanzees taught from infancy by human trainers are reported to be able to master the use of human words. The words are represented in some instances by sign language and in others by metal-backed plastic symbols that are pushed about on a magnetized board. The chimpanzees are also capable of learning rudimentary rules of syntax and even of inventing short questions and statements of their own. Sarah, a chimpanzee trained with plastic symbols by David Premack at the University of California at Santa Barbara, acquired a vocabulary of 128 "words," including a different "name"

for each of eight individuals, both human and chimpanzee, and other signs representative of 12 verbs, six colors, 21 foods and a rich variety of miscellaneous objects, concepts, adjectives and adverbs. Although Sarah's achievement is truly remarkable, an enormous gulf still separates this most intelligent of the anthropoid apes from man. Sarah's words are given to her, and she must use them in a rigid and artificial context. No chimpanzee has demonstrated anything close to the capacity and drive to experiment with language that is possessed by a normal human child.

The difference may be quantitative rather than qualitative, but at the very least our own species must still be ranked as unique in its capacity to concatenate a large vocabulary into sentences that touch on virtually every experience and thought. Future studies of animal communication should continue to prove useful in helping us to understand the steps that led man across such a vast linguistic chasm in what was surely the central event in the evolution of the human mind.

The Language of Birds

by W. H. Thorpe
October 1956

*The songs and other calls of birds are not merely
joyous outbursts. They make up a complex
communications system, with the various sounds
suited to various types of message*

If birds were suddenly eliminated from the world, the absence of their song would at once change the whole aspect of nature out of doors. The reasons why we appreciate bird song so keenly are several. We are attracted by its beauty, by its association with spring and all the promise which that time of the year suggests. Perhaps strongest of all is the feeling expressed by W. H. Davies in the lines:

*I could not sleep again, for such
    wild cries,
And went out early into their green
    world:
And then I saw what set their little
    tongues
To scream for joy—they saw the
    East in gold.*

The idea that bird song is often an expression of irrepressible joy actually has some scientific justification. But we are apt to overlook the fact that bird songs and calls are not merely a spontaneous emotional outlet, that they are in fact the language in which the chirpers communicate with one another.

The sounds that birds make have two main functions: to arouse an emotional state (by way of warning, wooing, etc.) and to convey precise information. The sounds themselves can conveniently be divided into two categories: call notes and true song. In general the characteristic call notes of a species are inherited, whereas the true songs may be either entirely inherited, partly inherited and partly learned or almost entirely learned. We have been making a detailed study of bird songs and calls for several years at the Madingley Ornithological Field Station of the University of Cambridge. In this investigation we have enlisted the help of tape recorders and

electronic equipment for analyzing sound. The chief bird partner in our study has been the common English chaffinch (*Fringilla coelebs*). The calls and the songs of the chaffinch illustrate very beautifully some of the chief generalizations we are now able to draw

concerning the musical language of birds.

A bird's call note, in contrast to its song, is a brief sound with a relatively simple acoustic structure. Its main function, in the case of small birds, is to sound a warning of the presence of a dangerous enemy, such as a hawk or

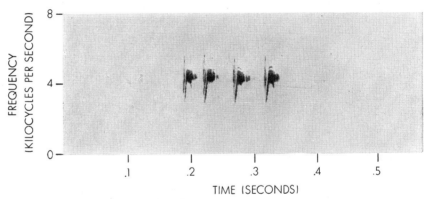

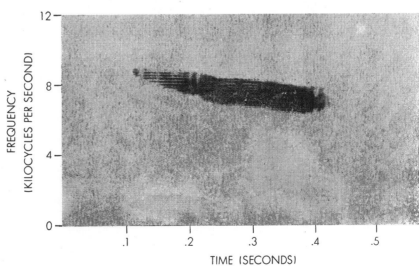

WARNING CALLS differ according to the circumstances of their use (*see drawing opposite*). "Chink" (*top oscillogram*) is easy to locate because of its long wavelength, brief duration and abrupt beginning and end. "Seeet" (*bottom oscillogram*) has short wavelength, long duration and begins and ends imperceptibly, making its direction difficult to trace.

DEFENSE ACTION of small birds against a predator takes two forms, which require different types of audible message. When the enemy is perched in a tree (*upper drawing*) the small birds may practice "mobbing." They swarm around the larger bird, uttering sharp, easily located "chinks" to advertise its presence. Against a hawk on the wing (*lower drawing*) the little birds take cover under foliage, and their warning cry is now the high-pitched, drawn-out "seeet" which is difficult to trace (*see diagrams on preceding page*).

owl. If the bird of prey is perched conspicuously in a tree, the small birds will often make themselves conspicuous too by behavior known as "mobbing." They set up a chorus of cries which points out the predator, so to speak, to all and sundry. If the bird of prey is in flight, on the other hand, the small birds race to the nearest bush or other cover, and utter their warning cries from that haven. Now the calls are very different in the two cases. Chaffinches mobbing a perched predator give forth relatively low-pitched sounds known as "chink" calls. But when they have fled to cover, the males give a high, thin note designated as the "seeet" call, the effect of which is to cause other chaffinches also to dash for the nearest shelter, and to peer cautiously upward looking for the hawk in the sky.

The significant difference between the two calls is that the "chink" note is easy to locate whereas the "seeet" is extremely difficult. The low frequencies of the "chink" sound are of a wavelength which allows the two ears of a hawk (or a man) to detect phase differences; the call also gives clues to its direction in the form of intensity and time of arrival of the sound at the two ears—the latter because the sound comes as clicklike pulses. On the other hand, the "seeet" call is composed of certain high frequencies which probably permit no clues to its location by phase or intensity difference, and it probably also fails to give a time clue, because it begins and ends imperceptibly, instead of coming as a sharp click [see spectrograms on page 273]. So it seems that the "seeet" call is admirably adjusted to avoid giving positional clues of any kind to predators. Yet it is just as effective as any other sound would be in warning neighboring chaffinches.

The full song of the male chaffinch performs the function of keeping other males from its territory and attracting unmated females. It is not too inadequately described by the mnemonic jingle *chip-chip-chip; tell-tell-tell; cherry-erry-erry; tissy-chee-wee-oo.* As sound spectrograms show [see opposite page], the chaffinch song is sufficiently complex not only to identify the species but also to allow wide individual variation, so that individual birds are recognizable, even by human beings, by their personal signature tune.

Study of these songs is extremely interesting from the standpoint of heredity and learning. We have found evidence that even complex songs may depend primarily on the inherited make-up of a bird. For example, the spectro-

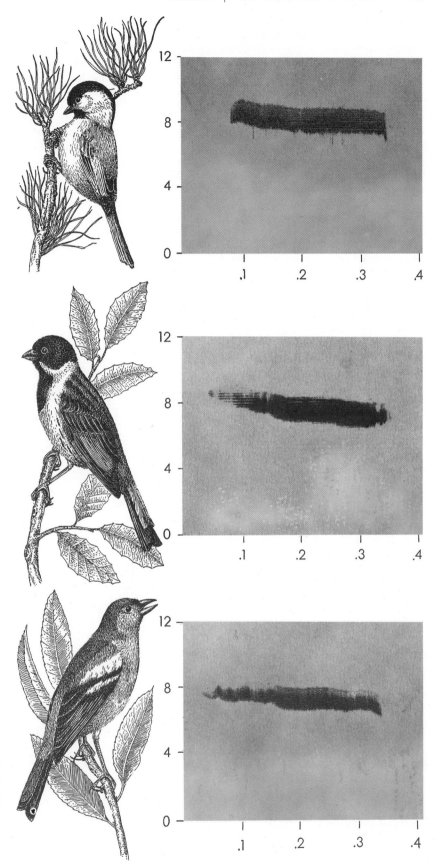

CALLS OF DIFFERENT SPECIES are sometimes remarkably alike. Above are sketches of three small birds (the blue tit, a close relative of the American chickadee, top, the bunting, center, and the chaffinch, bottom) together with oscillograms of their "seeet" calls. The oscillograms show time in seconds horizontally and kilocycles per second on the vertical axis.

grams of the songs of the European wren and of the American winter wren are remarkably alike. It appears that the song pattern of some birds must rest upon an inborn pattern of activity in the central nervous system.

On the other hand, everyone who has listened to bird songs attentively and with a trained ear has detected individual differences. We have lately proved, by experiments with chaffinches, that these individual differences are not the expression of genetic differences but develop by learning during the early life of the bird. This problem fascinated field ornithologists in England and Germany as far back as the early 18th century, but its precise study has become possible only within the last few years, when apparatus for the accurate analysis of sounds became available.

When a young chaffinch is taken from the nest and reared separately out of hearing of all chaffinch song, its song development is greatly restricted. The bird eventually produces a song of about the normal length (two to three seconds), but it fails to divide the first part of the song into phrases, as a normally reared chaffinch does, or to end its song with the elaborate flourish—the *tissy-chee-wee-oo*—which is one of the most striking characteristics of the chaffinch song in the wild.

The simple, restricted song of the isolated bird can be taken to represent the inherited basis of a chaffinch's performance. Now if after babyhood two or more such birds are put together in a room, but still with no opportunity to hear experienced chaffinches, they develop more complex songs. The attempt to sing in company provides mutual stimulation which encourages the production of complexity. The group of birds will, by mutual imitation, build up a distinctive community pattern. The members of the group conform so closely to this pattern that it is barely possible to distinguish the songs one from another, even by the most minute electronic analysis. Their song may be quite as complex as that of a normal wild chaffinch. But it bears practically no resemblance to the characteristic chaffinch song!

There is a short and critical six-week period, at about the 11th month of life, during which the chaffinch develops its song pattern. Once this critical learning period is over, the song is fixed for life: no matter how much a bird is exposed to other chaffinch songs afterward, it continues year after year to sing only the song or songs it worked out as a youngster.

Such experiments make clear what is happening in the wild. In nature young chaffinches must certainly learn some details of song from their parents or from other adults in the first few weeks of life. At this stage a young bird absorbs the general pattern of division of the song into two or three phrases with a flourish at the end. But not until the critical learning period, during the following spring, does the bird develop the finer details of its song. This is the time when the young wild chaffinch first sings in a territory in competition with neighboring birds of the same species, and there is good evidence that it learns details of song from these neighbors. It may learn two or three different songs, sometimes even more, from neighbors on different sides of its territory. Many field naturalists have observed local dialects of the song of a given species.

So the full chaffinch song is a simple integration of inborn and learned song patterns, the former constituting the basis for the latter. While isolated chaffinches can, as we have seen, work out for themselves very strange songs, those in the wild are circumscribed by the general pattern of the chaffinch song characteristic of the locality, though they may develop individual variations in detail.

The chaffinch, like many other small birds, has a subsong consisting of an irregular sequence of chirps and rattles. This subsong, in contrast to the full song, is usually uttered from dense cover and appears to have little or no communicatory function. Heard in the early spring, it seems to be associated in some degree with rising production of sex hormones by the gonads. The subsong

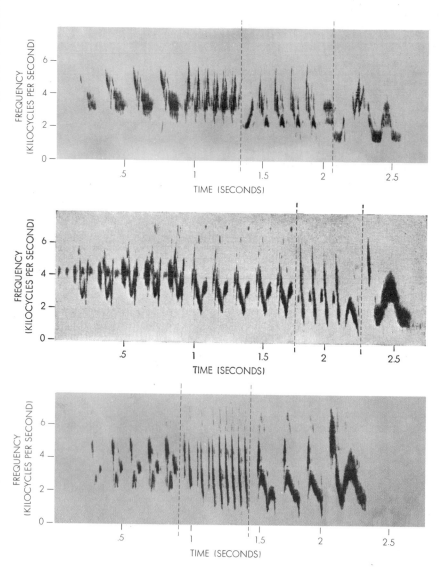

SONGS OF DIFFERENT BIRDS of the same species have similar over-all patterns but may show striking individual differences. Above are oscillograms of the songs of three different chaffinches. The phrases into which the songs are divided are marked off by dotted lines.

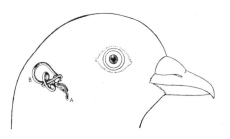

INNER EAR of a bird contains the cochlea (A), or organ of hearing, and the highly developed semicircular canals (B) which provide acute sense of balance needed for flight.

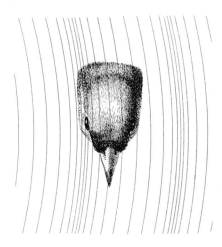

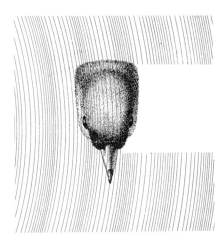

DIRECTION from which a sound is coming may be judged in various ways. Wavelengths comparable to the width of the bird's head give phase differences at its two ears (top). Shorter waves may be blocked by the head so that the more distant ear is in a "sound shadow" and receives little energy (bottom).

in a sense constitutes the raw material out of which the true song is constructed. It contains a much bigger range of frequencies than does the full song, and the bird drops some of these frequencies in the full song. This suggests a similarity with the way in which human children learn to speak. A baby will produce at random almost every conceivable sound its vocal mechanisms are capable of making, but as it grows it ceases to produce those sounds which it does not hear uttered by its parents and others around it.

Everyone knows that many birds are good mimics: besides the proverbial parrot there are the mockingbird in the U. S. and the starling and other species in Europe. To its own innate song such a bird adds notes and phrases from a wide variety of other species, and sometimes even sounds of inanimate origin. (The biological function of this imitation is still obscure: if the song of the starling, for instance, is a territorial proclamation like that of the chaffinch, to imitate many other species would seem merely to lead to confusion.) We have recently found that a chaffinch caged with, say, canaries may introduce a very good imitation of a phrase of canary song into its subsong, but never into its full song. How this restriction is maintained, we do not know. It is being studied, however, and we hope that its investigation will in due course throw light on the relationship between subsongs and full songs.

We have just commenced a very interesting study of the songs of hybrid birds—born of two different species. A curious and puzzling first result is that the hybrid is sometimes more imitative than either parent bird. For example, the song of a hybrid offspring from a goldfinch and a greenfinch is a virtually per-

fect imitation of a chaffinch song which the bird heard in its aviary.

It was said at the beginning that most people tend to think of bird song as a sort of emotional release rather than the process of communication that it usually is. Actually the possibility of birds singing for pleasure is by no means ruled out. The evidence is far from negligible that the songs of some thrushes, the warblers and the nightingale exhibit elaborate esthetic improvement far beyond what strict biological necessity requires. To be sure, the purity of tone which characterizes the best singers is potentially advantageous to them, for it helps to provide an additional dimension for distinctiveness. On the other hand, we have seen that in birds' imitation of song there seems to be no immediate or obvious biological reward, and that often the matching of sound patterns must itself constitute the reward. Much of it, of course, may be merely part of the trial-and-error process of learning, but in those cases of vocal imitation where new phrases are produced only after long delay and apparently without specific practice, another influence must be at work.

29

The Fighting Behavior
of Animals

by Irenäus Eibl-Eibesfeldt

Combat between members of a species serves useful functions. Death or serious injury to a contender is avoided by formal tournaments, the behavior patterns for which appear to be innate in the species

Fighting between members of the same species is almost universal among vertebrates, from fish to man. Casual observation suggests the reason: Animals of the same kind, occupying the same niche in nature, must compete for the same food, the same nesting sites and the same building materials. Fighting among animals of the same species therefore serves the important function of "spacing out" the individuals or groups in the area they occupy. It thereby secures for each the minimum territory required to support its existence, prevents overcrowding and promotes the distribution of the species. Fighting also arises from competition for mates, and thus serves to select the stronger and fitter individuals for propagation of the species. It is no wonder, then, that herbivores seem to fight each other as readily as do carnivores, and that nearly all groups of vertebrates, except perhaps some amphibians, display aggressive behavior.

A complete investigation of fighting behavior must take account, however, of another general observation: Fights between individuals of the same species almost never end in death and rarely result in serious injury to either combatant. Such fights, in fact, are often highly ritualized and more nearly resemble a tournament than a mortal struggle. If this were not the case—if the loser were killed or seriously injured—fighting would have grave disadvantages for the species. The animal that loses a fight is not necessarily less healthy or less viable; it may simply be an immature animal that cannot withstand the attack of a mature one.

In view of the disadvantages of serious injury to a member of the species, evolution might be expected to have exerted a strong selective pressure against aggressive behavior. But spacing out through combat was apparently too important to permit a weakening of aggressive tendencies; in fact, aggressiveness seems to have been favored by natural selection. It is in order to allow spacing out—rather than death or injury—to result from fighting that the ceremonial combat routines have evolved.

Investigators of aggressive behavior, often strongly motivated by concern about aggressive impulses in man, have usually been satisfied to find its origin in the life experience of the individual animal or of the social group. Aggressiveness is said to be learned and so to be preventable by teaching or conditioning. A growing body of evidence from observations in the field and experiments in the laboratory, however, points to the conclusion that this vital mode of behavior is not learned by the individual but is innate in the species, like the organs specially evolved for such combat in many animals. The ceremonial fighting routines that have developed in the course of evolution are highly characteristic for each species; they are faithfully followed in fights between members of the species and are almost never violated.

All-out fights between animals of the same species do occur, but usually in species having no weapons that can inflict mortal injury. Biting animals that can kill or seriously injure one another are usually also capable of quick flight. They may engage in damaging fights, but these end when the loser makes a fast getaway. They may also "surrender," by assuming a submissive posture that the winner respects. Konrad Z. Lorenz of the Max Planck Institute for the Physiology of Behavior in Germany has described such behavior in wolves and dogs. The fight begins with an exchange of bites; as soon as one contestant begins to lose, however, it exposes its vulnerable throat to its opponent by turning its head away. This act of submission immediately inhibits further attack by its rival. A young dog often submits by throwing itself on its back, exposing its belly: a pet dog may assume this posture if its master so much as raises his voice. Analogous behavior is common in birds: a young rail attacked by an adult turns the back of its head —the most sensitive part of its body— toward the aggressor, which immediately stops pecking. Lorenz has pointed out that acts of submission play a similar role in fights between men. When a victim throws himself defenseless at his enemy's feet, the normal human being is strongly inhibited from further aggression. This mechanism may now have lost its adaptive value in human affairs, because modern weapons can kill so quickly and from such long distances that the attacked individual has little opportunity to appeal to his opponent's feelings.

Most animals depend neither on flight nor on surrender to avoid damaging fights. Instead they engage in a ceremonial struggle, in the course of which the contestants measure their strength in bodily contact without harming each other seriously. Often these contests begin with a duel of threats—posturings,

CICHLID FISH (*Aequidens portalegrensis*) perform a ritual fight that begins with a threat and proceeds to bodily contact without damage to either. After a formal display (*a*) the fish fan their tails to propel currents of water at each other (*b*). Then the rivals grasp each other with their thick-lipped mouths and push and pull (*c*) until one gives up and swims away (*d*).

a

b

c

MALE MARINE IGUANA (*Amblyrhynchus cristatus*) of the Galápagos Islands defends his territory against intruding males. As the rival approaches (*a*), the territory owner struts and nods his head. Then the defender lunges at the intruder and they clash head on (*b*), each seeking to push the other back. When one iguana (*left at "c"*) realizes he cannot win, he drops to his belly in submission.

movements and noises—designed to cow the opponent without any physical contact at all. Sometimes this competition in bravado brings about a decision; usually it is preliminary to the remainder of the tournament.

On the lava cliffs of the Galápagos Islands a few years ago I observed such contests between marine iguanas (*Amblyrhynchus cristatus*), large algae-eating lizards that swarm by the hundreds over the rocks close to shore. During the breeding season each male establishes a small territory by defending a few square yards of rock on which he lives with several females. If another male approaches the territorial border, the local iguana responds with a "display." He opens his mouth and nods his head, presents his flank to his opponent and parades, stiff-legged, back and forth, his apparent size enlarged by the erection of his dorsal crest. If this performance does not drive the rival off, the resident of the territory attacks, rushing at the intruder with his head lowered. The interloper lowers his head in turn and the two clash, the tops of their heads striking together. Each tries to push the other backward. If neither gives way, they pause, back off, nod at each other and try again. (In an apparent adaptation to this mode of combat the head of the marine iguana is covered with horn-like scales.) The struggle ends when one of the iguanas assumes the posture of submission, crouching on its belly. The winner thereupon stops charging and waits in the threatening, stiff-legged stance until the loser retreats [*see illustration on opposite page*]. A damaging fight is triggered only when an invader does not perform the ceremonies that signal a tournament; when, for example, the animal is suddenly placed in occupied territory by a man, or crosses another animal's territory in precipitous flight from an earlier contest. On these occasions the territory owner attacks by biting the intruder in the nape of the neck. Female iguanas, on the other hand, regularly engage in damaging fights for the scarce egg-laying sites, biting and shaking each other vigorously.

The lava lizard (*Tropidurus albemarlensis*) of the larger Galápagos Islands engages in a similar ceremonial fight that begins with the rivals facing each other, nodding their heads. Suddenly one of them rushes forward, stands alongside his opponent and lashes him with his tail once or several times, so hard that the blows can be heard several yards away. The opponent may reply with a

RATTLESNAKES (*Crotalus ruber*) perform the combat dance shown in these drawings based on a study by Charles E. Shaw of the San Diego Zoo. The rivals move together (*a*) and then "Indian wrestle" head to head (*b*). Sometimes they face each other, weaving and rubbing their ventral scales (*c*). Finally one lashes out and throws (*d*) and pins (*e*) the other.

tail-beating of his own. Then the attacker turns and retreats to his original position. The entire procedure is repeated until one of the lizards gives up and flees.

According to Gertraud Kitzler of the University of Vienna, fights between lizards of the central European species *Lacerta agilis* may terminate in a curious manner. After an introductory display one lizard grasps the other's neck in his jaws. The attacked lizard waits quietly for the grip to loosen, then takes his turn at biting. The exchange continues until one lizard runs away. Often, however, it is the biter, not the bitten, that does the fleeing. The loser apparently recognizes the superiority of the

winner not only by the strength of the latter's bite but also by his unyielding resistance to being bitten.

Beatrice Oehlert-Lorenz of the Max Planck Institute for the Physiology of Behavior has described a highly ritualized contest between male cichlid fish (*Aequidens portalegrensis*). The rivals first perform a display, presenting themselves head on and side on, with dorsal fins erected. Then they beat at each other with their tails, making gusty currents of water strike the other's side. If this does not bring about a decision, the rivals grasp each other jaw to jaw and pull and push with great force until the loser folds his fins and

swims away [*see illustration on page 279*]. John Burchard of the same institution raised members of another cichlid species (*Cichlasoma biocellatum*) in isolation from the egg stage and found that they fought each other in the manner peculiar to their species.

The ritualization of fighting behavior assumes critical importance in contests between animals that are endowed with deadly weapons. Rattlesnakes, for example, can kill each other with a single bite. When male rattlesnakes fight, however, they never bite. Charles E. Shaw of the San Diego Zoo has described the mode of combat in one species (*Crotalus ruber*) in detail. The two snakes glide along, side by side, each with the forward third of its length raised in the air. In this posture they push head against head, each trying to force the other sideways and to the ground, in accordance with strict rules reminiscent of those that govern "Indian wrestling." The successful snake pins the loser for a moment with the weight of his body and then lets the loser escape [*see illustration on preceding page*]. Many other poisonous snakes fight in a similar fashion.

Among mammals, the fallow deer (*Dama dama*) engages in a particularly impressive ceremonial fight. The rival stags march along side by side, heads raised, watching each other out of the corners of their eyes. Suddenly they halt, turn face to face, lower their heads and charge. Their antlers clash and they wrestle for a while. If this does not lead to a decision, they resume their march. Fighting and marching thus alternate until one wins. What is notable about this struggle is that the stags attack only when they are facing each other. A motion picture made by Horst Siewert of the Research Station for German Wildlife records an occasion on which one deer turned by chance and momentarily exposed his posterior to his opponent. The latter did not take advantage of this opportunity but waited for the other to turn around before he attacked. Because of such careful observance of the rules, accidents are comparatively rare.

Mountain sheep, wild goats and antelopes fight similar duels with their horns and foreheads, the various species using their horns in highly specific ways. From observation of clashes between rapier-horned oryx antelope (*Oryx gazella beisa*) and other African antelope, Fritz Walther of the Opel Open-Air Zoo for Animal Research concludes that the function of the horns is to lock the heads of the animals together as they engage in a pushing match. In one instance a

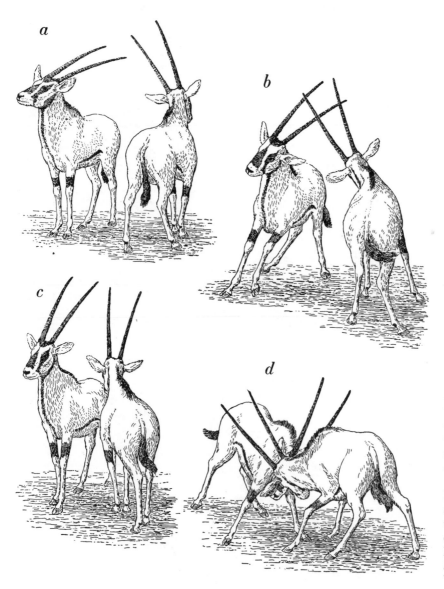

ORYX ANTELOPE (*Oryx gazella beisa*) has rapier-shaped horns but does not gore his rival. Two bulls begin combat with a display (*a*), then fence with the upper portion of their horns (*b*). After a pause (*c*) the rivals clash forehead to forehead (*d*) and push each other, using their horns to maintain contact. Drawings are based on observations by Fritz Walther.

duel between two oryx antelope in the wilderness was observed to begin with a display in which the two bulls stood flank to flank, heads held high. Then they came together in a first clash, only the upper third of their horns making contact. After a pause the animals charged again, this time forehead to forehead. They maintained contact by touching and beating their horns together [*see illustration on preceding page*]. Oryx antelope never use their horns as daggers in intraspecies fights. One hornless bull observed by Walther carried out the full ritual of combat as if he still had horns. He struck at his opponents' horns and missed by the precise distance at which his nonexistent horns would have made contact. Equally remarkable, his opponents acted as though his horns were in place and responded to his im-

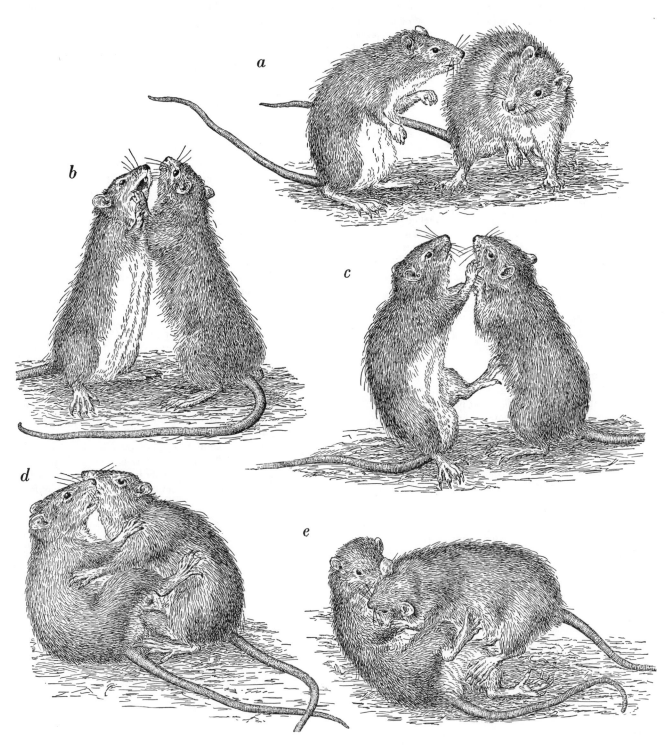

NORWAY RATS (*Rattus norvegicus*) fight in a species-specific manner whether they are raised in isolation or with a group of other rats. The aggressor approaches, displaying his flank and arching his back (*a*). Then, standing on their hind legs, the two rats wrestle. They push with their forelegs (*b*) and sometimes kick with their hind legs (*c*). If one rat is forced to his back as they tussle (*d*), he sometimes gives up; otherwise the tournament phase ends and the real fight begins with a serious exchange of bites (*e*).

aginary blows.

Until field observations of this kind had accumulated in support of the innateness of fighting behavior, laboratory experiments had made a strong case for the notion that such behavior is learned. Experiments by J. P. Scott of the Roscoe B. Jackson Memorial Laboratory in Bar Harbor, Me., had indicated, for example, that a rat or a mouse reacts aggressively toward another rat or mouse primarily because of pain inflicted by a nestmate early in life. Scott suggests that aggressiveness should therefore be controllable by a change in environment; in other words, rats that have had no early experience of pain inflicted by another rat should be completely unaggressive.

To test this conclusion I raised male Norway rats in isolation from their 17th day of life, an age at which they do not show any aggressive behavior. When each was between five and six months old, I put another male rat in the cage with him. At first the hitherto isolated rat approached the stranger, sniffed at him and sometimes made social overtures. But this never lasted long. The completely inexperienced rat soon performed the species-specific combat display—arching his back, gnashing his teeth, presenting his flank and uttering ultrasonic cries. Then the two rats pushed, kicked and wrestled, standing on their hind legs or falling together to the ground. Sometimes the fights ended at this point, the rat that landed on his back giving up and moving away. But usually the rats went on to exchange damaging bites. The patterns of display, tussling and biting were essentially the same in the case of the inexperienced rats as in the case of those who had

been brought up with other rats and were faced by an outsider [see illustration on page 283]. The steps in the ritual are apparently innate and fixed behavior patterns; many of the movements seem to be available to each rat like tools in a toolkit.

Raising rats in groups, where there was an opportunity for young rats to undergo early painful experiences, provided another check on the Scott theory. The members of a group displayed almost no aggressive behavior toward each other. The few fights that took place rarely included biting; for the most part the animals merely pushed each other with their paws. But when a stranger was introduced into the group, he was attacked viciously and was hurt. This agrees with observations of wild Norway rats; they live peacefully together in large packs but attack any rat not a member of their group. Because the attacked animal is able to escape, the species has not developed a tournament substitute for biting. In the laboratory a strange rat introduced into a colony from which it cannot escape is likely to be killed. In sum, the experiments demonstrate that aggressiveness is aroused in adult male rats whenever a stranger enters the territory, even when the defender has had no painful experience with members of its species. Similar experiments on polecats (*Putorius putorius*) have shown the same results.

The view that aggressiveness is a basic biological phenomenon is supported by physiological studies of the underlying neural and hormonal processes. Some investigators have actually elicited fighting behavior in birds and mammals by stimulating specific areas of the brain

with electrical currents. The mind of a newborn animal is not a blank slate to be written on by experience. Aggressive behavior is an adaptive mechanism by which species members are spaced out and the fittest selected for propagation. Learning is no prerequisite for such behavior, although it probably has an influence on the intensity and detailed expression of aggressiveness.

In the human species, it seems likely, aggressive behavior evolved in the service of the same functions as it did in the case of lower animals. Undoubtedly it was useful and adaptive thousands of years ago, when men lived in small groups. With the growth of supersocieties, however, such behavior has become maladaptive. It will have to be controlled—and the first step in the direction of control is the realization that aggressiveness is deeply rooted in the history of the species and in the physiology and behavioral organization of each individual.

In this connection, it should be emphasized that aggressiveness is not the only motive governing the interaction of members of the same species. In gregarious animals there are equally innate patterns of behavior leading to mutual help and support, and one may assert that altruism is no less deeply rooted than aggressiveness. Man can be as basically good as he can be bad, but he is good primarily toward his family and friends. He has had to learn in the course of history that his family has grown, coming to encompass first his clan, then his tribe and his nation. Perhaps man will eventually be wise enough to learn that his family now includes all mankind.

Hormones in Social Amoebae and Mammals

by John Tyler Bonner
June 1969

The substance that attracts social amoebae to one another to form a sluglike mass has recently been identified. It turns out that the same substance also acts as a "messenger" in mammalian cells

One of the pleasures of science is to see two distant and apparently unrelated pieces of information suddenly come together. In a flash what one knows doubles or triples in size. It is like working on two large but separate sections of a jigsaw puzzle and, almost without realizing it until the moment it happens, finding that they fit into one. I recently had this pleasure, although my own work directly concerned only one section of the puzzle. The assembling of the other section began some years ago, when a substance in the family of adenosine phosphates was found to be a "chemical messenger" intimately involved in the action of many mammalian hormones, including man's. The substance is cyclic AMP, or cyclic-3′,5′-adenosine monophosphate. ("Cyclic" refers to the fact that the atoms of the phosphate group form a ring.) The relatives of cyclic AMP are, among others, the more familiar, noncyclic ADP (adenosine diphosphate) and ATP (adenosine triphosphate), substances that play key roles in plant and animal metabolism.

Cyclic AMP was discovered by Earl W. Sutherland, Jr., now at the Vanderbilt University School of Medicine, and Theodore W. Rall of the Western Reserve University School of Medicine in 1958. They and their collaborators have since learned a number of remarkable facts about the substance, the bare bones of which I shall outline here. To deal with the basic biochemistry first, cyclic AMP is formed from ATP in a one-step reaction; the enzyme responsible for the conversion is adenyl cyclase. The substance is subject to further modification, being converted into 5′ AMP (5′-adenosine monophosphate) in another one-step reaction. The enzyme that carries out this step is a phosphodiesterase specific for the reaction.

The hormone that Sutherland and Rall first showed to be involved with cyclic AMP was adrenalin, or epinephrine, as it is now usually called in the U.S. One is taught in elementary biology that the stress of anger, pain or fear produces a great surge of epinephrine in the blood, with a resulting quick rise in the amount of blood sugar available as energy for emergency action. Inevitably one develops the simpleminded notion that the epinephrine directly splits glycogen, or animal starch, into its subunits of glucose, or blood sugar. This is far from being the case. As Sutherland and Rall discovered, cyclic AMP is implicated in a much more complex process. I shall go into a small amount of detail here because the example of epinephrine illustrates the way cyclic AMP works with hormones generally.

Skipping the step that puts the epinephrine into the blood, we can start with the liver, which is one of the places where the hormone does its work. Cells in the liver are suddenly bathed in epinephrine, brought to them by neighboring blood vessels. Adenyl cyclase, the enzyme that converts ATP into cyclic AMP, is attached in some way to the surface membrane of liver cells. When the epinephrine reaches the cell surface, it specifically stimulates the enzyme, the conversion takes place and cyclic AMP is formed inside the cell.

The cyclic AMP inside the cell now stimulates a second enzyme, changing it from an inactive form to an active one. Once activated, the second enzyme acts on a third to activate still another enzyme: phosphorylase. This fourth enzyme cleaves glycogen to produce glucose-1-phosphate. Before the sugar can escape from the cell, however, it must go through two more enzyme steps, one that converts it to glucose-6-phosphate and one that converts glucose-6-phos-phate to glucose, the sugar that enters the blood [*see illustration on page 287*]. The entire sequence is remarkable for many reasons, not least of which is the extraordinary number of chemical steps involved in what originally was thought to be a simple process.

At the moment more than a dozen mammalian hormones are known to use cyclic AMP as this kind of "second messenger," to use the name Sutherland and his co-workers have given it. I shall not list them all here but prominent among them are glucagon and insulin, both of which are involved in blood-sugar levels, three hormones of the anterior pituitary and one of the posterior pituitary. In addition large amounts of the substance are present in brain tissue. It has been known for some time that the synaptic gaps between nerves are bridged by neurohormones such as acetylcholine. Bruce McL. Breckenridge of the Rutgers University Medical School has recently put forward the hypothesis that cyclic AMP is somehow involved in the bridging sequence.

One of the many unanswered questions about cyclic AMP is how the single enzyme adenyl cyclase can respond to so many different hormones and do so in such a specific way. There is some evidence that the enzyme, which always appears to be associated with cell membranes, differs in different cell systems: one type can respond only to one hormone, another type only to another hormone, and so on. This raises further questions of deep interest. For example, how could such a chemical system have evolved? It would almost seem as though the fundamental part of the hormone system is the cyclic AMP, and that the variety of methods for turning on the system had evolved subsequently. I shall return to some evolutionary considerations presently, but the main basis for such a

a

ADENOSINE
TRIPHOSPHATE

ADENYL
CYCLASE

b

CYCLIC-3',5'-
ADENOSINE
MONOPHOSPHATE

PHOSPHO-
DIESTERASE

c

5'-ADENOSINE
MONOPHOSPHATE

CYCLIC AMP (*b*) **is formed from the more familiar adenosine triphosphate, or ATP** (*a*), **in a one-step reaction catalyzed by the enzyme adenyl cyclase. The word "cyclic" refers to the ring shape of the molecule's phosphate group** (*color*). **The substance is converted into 5′ AMP** (*c*) **by a phosphodiesterase that is specific for the reaction.**

contention is that cyclic AMP appears to be a common component of many cells and tissues and is even found in bacteria.

I should now like to ask the reader to forget momentarily what has gone before while we look at the other piece of the puzzle. This piece involves an entire series of experiments, including some of my own, in a quite specialized area of biology. Not until the very last step, however, was there the faintest suspicion that cyclic AMP played any part in the work.

The social amoebae known as the cellular slime molds were first described by Oskar Brefeld of Germany exactly a century ago. A few years later their peculiar life history was elucidated, and today a score or more species are known. Here I am going to concentrate mainly on the species *Dictyostelium discoideum,* which was discovered in the 1930's by Kenneth B. Raper of the University of Wisconsin. *Dictyostelium* amoebae live in the soil; they play a key role in the food chain because they are the immediate predators of the soil bacteria.

In appearance and size the amoebae are similar to human leukocytes, the white blood cells, and like leukocytes they engulf bacteria by surrounding them with protoplasm. After they clear an area of food and enter a period of starvation, the individual amoebae stream together into central collection points. In a reasonably well-populated area an aggregate may number 100,000 or more cells, although if food is scarce, the number will be fewer.

The assembled amoebae gather into a cartridge-shaped mass, one or two millimeters long, which has a distinct front and hind end. The mass crawls through the cavities in the soil like a slug and shows an ability to orient toward light and heat, an ability that is completely lacking in the separate amoebae before they aggregate. As I have argued elsewhere, spore dispersal must be of key importance in the natural selection of soil organisms, and the formation of a slug and its migration to "a better place in the sun" as it wanders through the soil or among rotting leaves in the humus is therefore an important adaptation for more effective spore dispersal.

After a period of migration the slug upends itself until it points at a right angle to the surface. Some of its cells will now form a stalk, while others will go to make up a spherical mass of spores. The stalk cells swell, become hollow like the pith of a plant and die. The result is a slender, tapering hollow rod that rises a few millimeters into the air and is filled with dead cells that serve as internal

trusses. As the stalk grows the spore mass is carried aloft with it so that ultimately the delicate stalk bears a sphere of spores at its tip. Any one of the spores is capable of starting a new generation.

Because this organism is a relatively simple system of cells that differentiate (that is, take up specialized functions) it has been an object of interest to students of developmental biology for some years. One of the prime targets of study has been the social amoeba's mechanism of aggregation. How are the individuals oriented so that they stream into the central collecting points? The action is a rather pure example of morphogenesis, or formative movement, a process that is known to be most important in the initial stages of the development of all animal embryos.

At the turn of the century it was suggested that the aggregation might be due to the attraction of the individual amoebae by some kind of chemical stimulus, but there was no evidence to support this idea. Then in 1942 Ernest H. Runyon of Agnes Scott College showed that if he put an aggregation center on one side of a semipermeable membrane (such as a sausage casing) and put amoebae that were ready to aggregate on the other side, the amoebae would react. In spite of the membrane barrier they would swarm to a point on their side of the surface exactly opposite the aggregation center on the other side. Since it was known that large molecules, such as proteins, cannot readily penetrate such membranes, Runyon suggested that the chemical attractant might be a small molecule.

More evidence in support of the hypothesis of a chemical attractant was needed; it was clear that other forces, for example electric forces, could also penetrate a membrane. Studying the attractant problem further, I found in 1947 that if the amoebae began to aggregate underwater, and if one moved the water slowly past them (as though they were at the bottom of a brook), then the amoebae upstream became completely disoriented but the ones downstream oriented beautifully toward the aggregation center, even over large distances. The obvious conclusion was that aggregation was mediated by a free-diffusing agent; it might be heat or it might be a chemical substance, as Runyon had suggested. For various reasons I favored the idea of a chemical, and I gave the unidentified attractant the name acrasin because the proper name of the cellular slime molds is Acrasiales, and also because in Edmund Spenser's

Faerie Queene there is a witch named Acrasia who attracted men and transformed them into beasts.

The final proof that acrasin really existed came from a remarkable set of experiments performed by Brian M. Shaffer of the University of Cambridge. At Princeton we had been unable to find any trace of the attractant at aggregation centers we had killed. Yet we felt that they must contain acrasin, because a few seconds earlier, when they were still alive, they had been actively attracting amoebae. Unknown to us, Shaffer was having the same difficulty; like us he suspected that for some reason the acrasin was rapidly disappearing.

To test this suspicion Shaffer sandwiched some active amoebae between a glass slide and a small block of agar [*see top illustration on page 291*]. To

the edges of the sandwich he added drops of water that had been near an aggregation center (and therefore contained acrasin) every few seconds. In a few minutes all the amoebae under the block streamed toward the wet edges. If he let a minute or more pass between drops, there was no attraction; acrasin evidently disappeared quickly and, in order to attract, it had to be applied at short successive intervals. When Shaffer froze water that had been near aggregation centers in capillary tubes and three months later used the contents of the capillaries at short intervals, the amoebae were attracted; acrasin was apparently stable at low temperatures. In any case, there was no longer any doubt about the attractant's existence. As Shaffer wrote to me after these experiments, "I have bottled it."

The next problem was to find out why acrasin was so unstable. Shaffer showed conclusively that the reason was that a companion substance, an enzyme, rapidly destroyed it. He did this while visiting Princeton, so that I had a close view. His principal method was to plunge water containing acrasin into methanol, which denatured the accompanying enzyme; the remaining acrasin was stable. This was an economical experiment, but I always had special admiration for what might be called his Rube Goldberg demonstration. He suspended a long cylinder of sausage casing in a moist chamber; then he painted the outside of the casing with amoebae that were about to aggregate and dripped water along the inside of the casing. The water collected in a beaker at the bottom of the casing. The acrasin it contained was stable at room

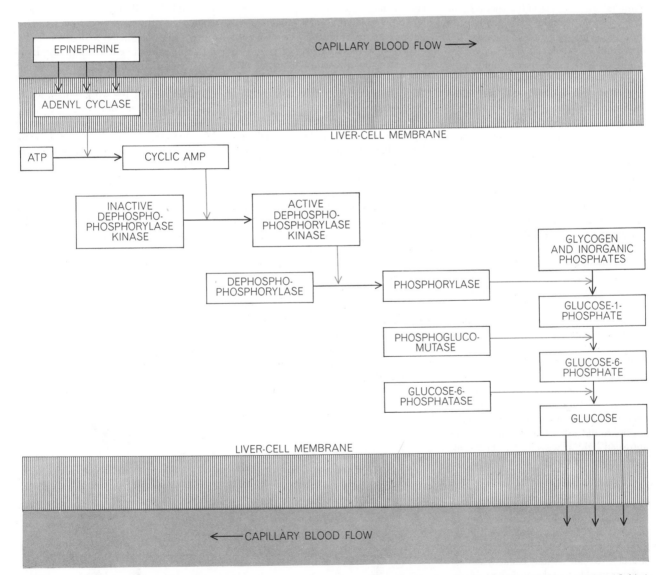

ACTION OF SEVERAL HORMONES in mammals utilizes cyclic AMP as a chemical messenger; illustrated here is the role of the substance in the production of blood sugar by a liver cell following stimulation by the presence of epinephrine in the bloodstream. The enzyme adenyl cyclase, attached to the cell membrane (*top left*), is stimulated by epinephrine and changes ATP inside the cell into cyclic AMP. The messenger substance activates a second enzyme, beginning a five-step sequence that yields glucose (*bottom right*).

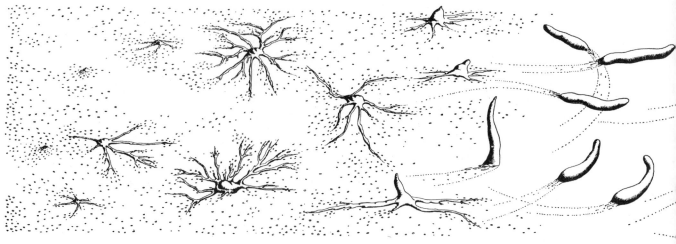

SOCIAL PHASE of the life cycle of the cellular slime molds begins when individual amoebae (*far left*) begin to stream together into a central collection point and gather into a cartridge-shaped mass that moves through the soil like a slug and is attracted to light.

temperature for extended periods of time. The membrane had passed the small molecules of acrasin but barred the large molecules of the destructive enzyme, or acrasinase, that would ordinarily have destroyed the attractant.

Shaffer's experiment with the glass slide and the agar block constituted a test or an assay for the presence or absence of acrasin. The existence of such an assay, even though it was not a quantitative one, led to a flurry of activity as workers in various laboratories attempted to determine the chemical identity of acrasin. Various substances became suspects, and I shall mention a few of the attempts that were clearly going in the right direction. Barbara E. Wright of the Retina Foundation found that urine from pregnant women contained a component that attracted the amoebae of *Dictyostelium* (it does contain such a component), but she thought this had something to do with the steroids in the urine (it does not). She and her co-workers made the interesting discovery, however, that *Dictyostelium* synthesizes rather large amounts of one steroid with an unknown function. At Brandeis University, Maurice Sussman and his collaborators found attractive substances that absorbed light at wavelengths in the ultraviolet part of the spectrum (which was on the right track), and Shaffer, then working in conjunction with organic chemists at Cambridge, went so far as to suspect that acrasin was a member of the class of compounds called purines.

The ultimate result of these various efforts was that a time came when no one was working on the acrasin problem. Simultaneously, and without knowing about each other's activities until much later, Theo M. Konijn of the Hubrecht Laboratory in the Netherlands and I

decided that since all was quiet the time had come to have a calm, noncompetitive look at the problem. We had both decided that the first necessity was to devise a truly quantitative assay, one that would not only show that an attractant was present but also measure just how attractive it was.

My test consisted in placing small cellophane squares on a dish of agar that contained the test substance; amoebae were then put on the cellophane squares and the rate at which they moved off the cellophane was recorded. At the time we thought the test had limitations (a concern that has recently proved groundless), but it taught us several things. It showed, for example, that urine from both pregnant and nonpregnant women was loaded with an active substance, and my colleague Ruth Reisberg found good evidence that the substance was a phosphate compound. It also proved that bacteria were loaded with an attractant of some kind. (This was a fact that a number of workers had

learned before. Barbara Wright had even shown the attractiveness of bacterial extracts using the Shaffer assay.) Undoubtedly the greatest benefit of my test, however, was that when Konijn read our first report, he immediately wrote to tell us of his remarkable work, which led to his coming to Princeton for a year.

Because the Konijn test was the key to later successes I shall describe it in some detail. The basic idea is to put a very small droplet of saline solution that contains amoebae on an agar plate and then to put a water droplet containing the test substance nearby. If after a few hours the amoebae spread out of the confines of their droplet, the test is scored as positive. The assay is made quantitative simply by varying the distance between the two droplets. There are, however, two key factors in making the Konijn test work. One is that the agar must be washed repeatedly and must have exactly the right degree of rigidity. Unless these requirements are met the amoebae may not stay within the confines of their droplet even though

a   *b*

ABILITY OF BACTERIA to secrete the slime mold attractant, named acrasin by the author, was discovered during the search for the substance's chemical identity; one laboratory proof is illustrated here. The collection point (*a*) of a slime mold aggregation is re-

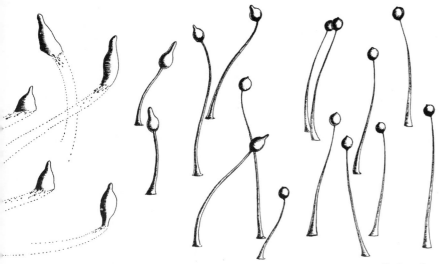

After a period of migration the slug upends itself and some cells form a stalk that rises upward, carrying aloft a spherical mass of other cells that have turned into spores (*right*).

the test droplet contains no attractant.

The other key factor in the Konijn test is that the amoebae must be at just the right stage in their development in order to respond. At first Konijn would stay up until the small hours waiting for that precise moment. Then matrimony intervened, with the result that he made an interesting discovery. When he stored the test plates overnight at five degrees Celsius, within minutes after the plates had returned to room temperature the next day all the amoebae were responsive. In addition, their movements were more precisely synchronized than they had ever been before.

Working with this test, Konijn and a colleague at the University of Utrecht, J. G. C. van de Meene, began an attempt to identify acrasin. We were making a similar attempt with my test, but they carried the matter much further. We had both shown that it was unaffected by heating and that it had a negative electric charge. They were able to add the fact that the attractant had a low molecular weight (between 200 and 400). When he came to Princeton, Konijn brought along some of his purified extract. He and David S. Barkley, who was then a graduate student, determined that the attractant also absorbed light in the ultraviolet part of the spectrum, with an absorption peak at a wavelength of 259 microns.

One evening Konijn and Barkley were sitting together reviewing all they knew about acrasin—the data just noted and the fact that the substance was found in bacteria and urine. Barkley had an inspiration: Why not see if cyclic AMP was an attractant? They quickly obtained some of the substance and in no time the hunch was verified. I was in Canada at the time and received an excited telephone call from the two; they had found that cyclic AMP was amazingly effective at attracting amoebae in the Konijn test. Eventually they were to learn that as little as .01 millimicrogram (10^{-11} gram) was enough to make the amoebae burst out of their droplet.

This important discovery was a wonderful bit of good fortune but a big question remained. Was cyclic AMP a naturally occurring acrasin, synthesized by the amoebae? By chance we did an incredibly stupid thing that turned out to be incredibly foresighted. At the time we happened to have another slime mold species, *Polysphondylium pallidum*, growing well in liquid culture. Without thinking, we decided to see if this species synthesized cyclic AMP. We found that it produced the substance in large quantities and that it evidently produced no other kind of attractant. This took us about a month.

We decided to complete the story by showing that cyclic AMP also attracts *Polysphondylium*. To our horror it did no such thing: the species produces the attractant but does not respond to it. This discovery raised interesting questions that are still unanswered regarding the specificity of slime mold species and also presented a puzzle: We do not know the chemical mechanism whereby *Polysphondylium* amoebae form aggregates.

It seemed as though a whole month had been wasted. To make matters worse, when we attempted to extract cyclic AMP from *Dictyostelium* we not only found that it was totally absent but also could discover no trace of any attractant of any kind. While groping in the depths of depression, one of our collaborators, Ying-Ying Chang, decided to see if our negative results were related to those that had plagued Shaffer when he first tried to "bottle" acrasin. Shaffer had shown that *Dictyostelium* produced the destructive enzyme acrasinase, and he had kept the enzyme from making the acrasin disappear by separating the two.

There is an enzyme (a phosphodiesterase) that changes cyclic AMP into 5′ AMP. In mammals it had been isolated and partially purified by R. W. Butcher of Vanderbilt University, who was kind enough to send us some. In fact, Butch-

d

e

moved (*b*), leaving the stream of converging amoebae disoriented (*c*). A nearby clump of the common bacterium *Escherichia coli* now begins to exercise an attraction (*d*) and within 45 minutes the amoebae are streaming in the opposite direction (*e*). The ancestors of the bacteria-devouring social amoebae probably were helped in their search for food by their prey's secretion of the attractant.

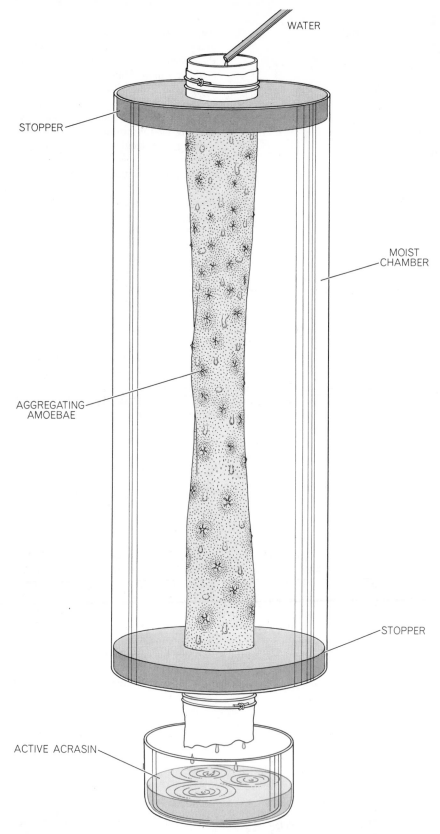

WATER

STOPPER

MOIST
CHAMBER

AGGREGATING
AMOEBAE

STOPPER

ACTIVE ACRASIN

"RUBE GOLDBERG" DEVICE for collecting acrasin, the attractant produced by slime molds, was constructed by Brian M. Shaffer of the University of Cambridge during his visit to Princeton University. A sausage casing was suspended in a moist chamber; aggregating amoebae were placed on the outside of the casing while water was dripped down the inside. The small molecules of acrasin passed through the membrane, entered the water and were collected (*bottom*). The molecules of an enzyme that is produced by the amoebae along with acrasin and that usually destroys the attractant, however, were too large to pass through the membrane. Acrasin collected by this method remained active for long periods.

er's enzyme was one of the substances we had used to prove that the attractant secreted by *Polysphondylium* actually was cyclic AMP; when we incubated the attractant with the enzyme, all the attractant's activity disappeared. If *Dictyostelium* was producing such an enzyme, our failure to find cyclic AMP would be explained.

In a rather rapid series of experiments Chang showed that such was the case. *Dictyostelium* produces large amounts of a phosphodiesterase that has many (although not all) of the properties of Butcher's enzyme. The main similarity is that, like the mammalian enzyme, the slime mold enzyme breaks down cyclic AMP. The main difference is that the mammalian enzyme is found within the cell but the slime mold enzyme is almost entirely extracellular. I shall return to the significance of this fact.

Clearly Chang's work had pulled us out of our difficulties. Now all we needed was some way of proving that cyclic AMP was actively synthesized by *Dictyostelium*. We approached the problem in two ways. The first method was to grow the amoebae on dead bacteria that contained residual amounts of cyclic AMP. After a period the culture was assayed, great pains being taken to denature the destructive enzyme quickly. The assays showed that there was an increase in an attractant that had the characteristics of cyclic AMP.

The second method, devised by Barkley, depended on separating the destructive enzyme from the attractant. One way he did this was to add beads made of a special resin to a solution containing both substances. The resin had the property of adsorbing molecules with a negative charge, such as the molecules of cyclic AMP, thus putting them out of reach of the enzyme. Another approach was a modification of Shaffer's device: amoebae were kept on one side of a semipermeable membrane and water was kept on the other. Since the membrane would not pass the enzyme's big molecules, the attractant that escaped through the membrane and entered the water was stable. The material collected in both ways was subjected to chemical analysis. Once again the only attractant of low molecular weight to be isolated was identical in characteristics with cyclic AMP. Thus we now seem to have an understanding of the basic acrasin-acrasinase system, at least in the case of *Dictyostelium discoideum*.

We have recently examined the question of when during their life cycle *Dictyostelium* amoebae produce cyclic

AMP. We grew the amoebae on one side of a membrane and bathed the other side with water that was collected at two-hour intervals. The water samples were concentrated and tested for the presence of cyclic AMP, using the Konijn assay. The relative strength of each sample was determined by comparing it with solutions containing known amounts of commercial cyclic AMP.

We found that when the amoebae pass through the aggregation phase there is at least a hundredfold increase in their secretion of the attractant. Simultaneously there is a hundredfold increase in the amoebae's sensitivity to cyclic AMP. This means that the whole mechanism of chemical attraction is at least 10,000 times more effective during aggregation than during the earlier stages of the amoebae's development [see illustration on page 293].

We were also able to show that, just as the amoebae continuously produce small amounts of cyclic AMP during the early stages of their development, they also continuously produce the enzyme that destroys the attractant. This mechanism of simultaneous production and destruction evidently serves an important purpose in the amoebae's life cycle. If the individual amoeba's output of attractant were not steadily eliminated, the attractant would keep accumulating until the gradient in attractant concentration that guides the individual during aggregation would be drowned out. An example is the role the destructive enzyme plays in the Konijn assay. The fact that the amoebae are initially concentrated in one spot means that a considerable amount of the enzyme is also concentrated in the same area; it destroys any attractant in the immediate vicinity of the amoebae. The result is that if attractant is applied externally, it is destroyed in the immediate region of the amoebae, producing a steep outward gradient of attractant that stimulates the amoebae to move out radially beyond the confines of their droplet.

I have pointed out that soil amoebae feed on soil bacteria and also that bacteria secrete cyclic AMP. It is quite reasonable to suppose that a positive response to cyclic AMP, among other substances, helped the solitary predators that were the ancestors of the slime mold amoebae to track their prey. Cyclic AMP is a small molecule that diffuses readily and is remarkably stable. As evolution proceeded and slime molds with multicellular fruiting bodies began to enjoy the selective advantages of more effective spore dispersal, it is possible to

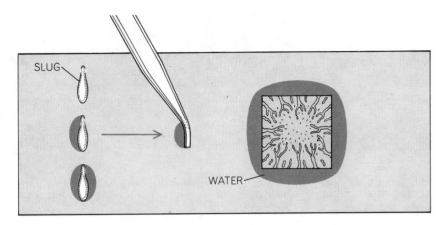

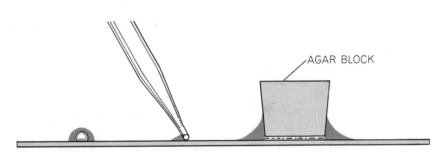

QUALITATIVE ASSAY, proving that slime mold collection centers are rich in acrasin but that the attractant is short-lived, was devised when Shaffer sandwiched amoebae between a glass slide and an agar block. He then dampened the edges of the block every few seconds with water that had been in contact with cell masses that were actively secreting acrasin; the amoebae streamed toward the dampened edges. When the applications of water were minutes apart, however, the amoebae were no longer attracted; the acrasin had vanished.

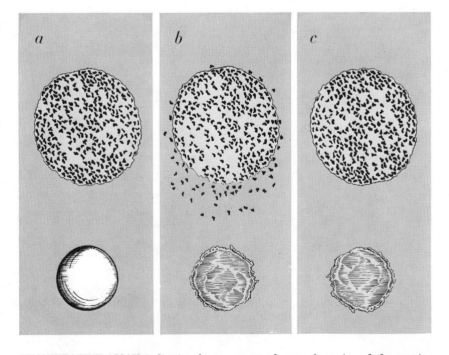

QUANTITATIVE ASSAY, indicating the presence or absence of acrasin and also gauging the amount present, was devised by Theo M. Konijn of the Hubrecht Laboratory in the Netherlands. A droplet containing amoebae in saline solution is put on one part of an agar plate and a droplet of the substance being tested is put nearby (a). If the amoebae emerge from their droplet in an hour or so (b), the test is scored as positive. If the amoebae stay inside the droplet (c), the test substance is deemed lacking in acrasin. The strength of the attractant may be measured by repeating the test with increased distances between droplets.

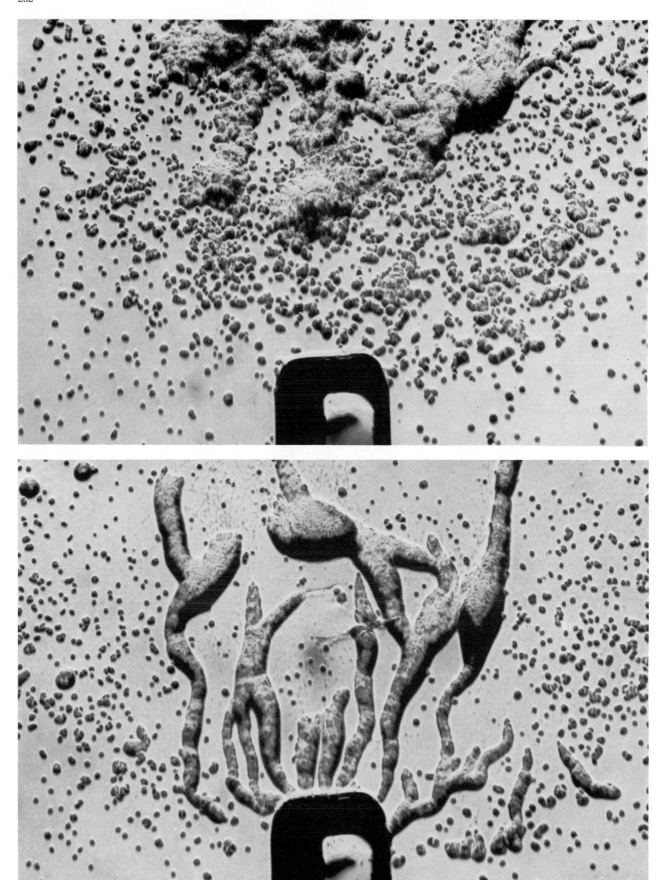

EFFECT OF ATTRACTANT on the social amoebae of the slime mold *Dictyostelium* is shown in these two photographs. In each, a small block of agar (*bottom*) contains the attractant, cyclic-3′,5′-adenosine monophosphate (cyclic AMP). In the upper photograph hundreds of amoebae have been placed near the block. In the lower photograph many have responded to the attractant by streaming toward the block. In normal development secretion of the attractant brings the amoebae together to form a multicellular organism.

assume that social amoebae were helped to aggregate by the same mechanism of chemical attraction that already assisted their quest for food. The only required change would be to increase the sensitivity of the mechanism enormously, so that individuals would ignore their surroundings, stop hunting and swarm together instead.

Thus we find that the question of chemical attractants among social amoebae and the question of hormonal activity in mammals are connected by a single biochemical link, and that the two large pieces of the puzzle that we mentioned in the beginning do indeed fit together. There remains much to be done, however; the rest of the puzzle must be filled in.

If we look at the link, we see that in *Dictyostelium* cyclic AMP appears to be the main hormone responsible for the amoebae's social existence. We should like to know more about how it is controlled within the individual amoeba and how it orients amoebae, but its central role is now established. Like the principal mammalian hormones, it is extracellular and provides communication between cells that are separated from one another in space. In mammals, on the other hand, cyclic AMP is triggered by the extracellular hormones and acts inside the cells. It may possibly act externally as well, but such actions have not yet been elucidated. That the chain of chemical events involving cyclic AMP in mammals is longer and more complex than it is in social amoebae is something that is to be expected of more complicated organisms.

To me, from a biochemical and an evolutionary point of view, one of the most fundamental unsolved questions is the role of cyclic AMP in bacteria. There have been some interesting beginnings on this problem, and when it is fully elucidated, I feel it will provide an important insight into the basic role of the substance in all living systems, an insight that is very much needed at this moment.

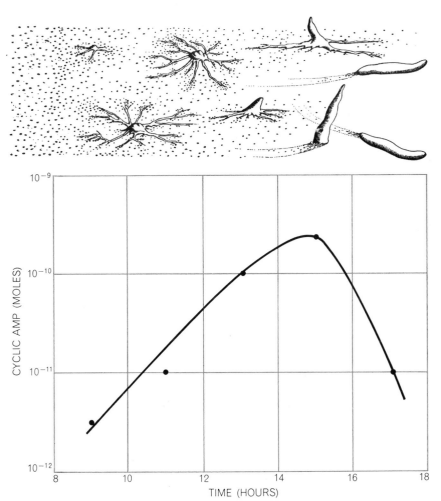

HUNDREDFOLD INCREASE in the amount of attractant secreted by social amoebae takes place during the seven hours in which the individual amoebae begin to stream together at collection centers. The peak is reached as aggregation is completed (*top, right of center*); production falls sharply as the fully formed multicellular slug starts to migrate.

31

Communication between Ants and Their Guests

by Bert Hölldobler
March 1971

Ants feed and shelter many other species of arthropods. The key to this hospitality lies in the guests' ability to communicate in the same chemical and mechanical language used by their hosts

The world of the arthropods is marked by a curious phenomenon, discovered nearly a century ago, that has never ceased to intrigue investigators. Many species of insects and other arthropods live with ants and have developed a thriving parasitic relationship with them. A number of these myrmecophiles make their home in the ants' nests and enjoy all the benefits. Although the interlopers in some cases eat the host ants' young, the ants treat the guests with astonishing cordiality: they not only admit the invading species to the nest but feed, groom and rear the guest larvae as if they were the ants' own brood.

How do the myrmecophiles manage to gain this acceptance? Ants, as highly social animals, possess a complex system of internal communication that enables the colony to carry out its collaborative activities in nest-building, food-gathering, care of the young and defense against enemies. The fact that the ants do not treat their alien guests as strangers suggests that the guests must somehow have broken the ants' code, that is, attained the ability to "speak" the ants' language, which involves a diversity of visual, mechanical and chemical cues.

Studies of the social behavior and communication of ants over the past 10 years have produced information that now provides a basis for well-grounded investigation of the relations between myrmecophiles and their hosts. Among the thousands of myrmecophilous animals (which include not only such arachnids as mites but also collembolans, flies, wasps and many other insect groups) the staphylinid beetles, commonly known as the rove beetles, demonstrate the parasitic relationships in a particularly clear-cut fashion. Focusing mainly on this family of myrmecophiles, I have been looking into the details of their communication and relations with certain species of ants.

The relations vary considerably with the beetle species. Some species live along the ants' food-gathering trails, others at the garbage dumps outside the nest, others in outer chambers within the nest and still others all the way inside the brood chambers. Let us consider first the beetles that attain the brood chamber.

A well-known example of such a beetle is *Atemeles pubicollis,* a European species. During its larval stage it lives in the nest of the mound-making wood ant *Formica polyctena*. I found that the ant's adoption of the beetle larva depends in the first instance on chemical communication. The larva secretes from glandular cells in its integument a substance that apparently acts as an attractant for the ant. This substance may be an imitation of a pheromone that ant larvae themselves emit to release brood-keeping behavior in the adults [see "Pheromones," by Edward O. Wilson; Scientific American Offprint 157]. The brood-tending ants respond to the chemical signal from the *Atemeles* beetle larvae with intense grooming of the larvae. I was able to verify the existence of chemical communication by two kinds of experimental evidence. Experiments with radioactive tracers demonstrated that substances were transferred from the beetle larvae to the ants. When larvae coated with shellac to prevent liberation of their secretion were placed at the entrance to the nest, the ants either ignored the larvae or carried them off to the garbage dump. If at least one segment of a larva's body was left uncovered by shellac, however, the ants would take the larva into the nest and adopt it. They even carried in dummies of filter paper soaked with secretions extracted from the beetle larvae.

A different form of communication comes into play to elicit the ants' feeding of larvae. The beetle larvae imitate certain begging behavior of ant larvae involving mechanical stimulation of the brood-keeping adults. When a larva is touched by the adult ant's mouthparts or antennae, it promptly rears up and tries to make contact with the ant's head. If the larva succeeds in tapping the ant's lip with its own mouthparts, the ant regurgitates a droplet of food. The beetle larvae perform the begging behavior more intensely than the ant larvae do, and apparently for this reason they receive more food. In order to trace and measure the distribution of food to the larvae in a brood chamber I gave the ants food labeled with radioactive sodium phosphate. The experiment showed that in a mixed population of beetle and ant larvae the beetles obtained a disproportionately large share of the food. The presence of beetle larvae reduced the normal flow of food to the ant larvae; on the other hand, the presence of ant larvae did not affect the food flow to the beetle larvae.

The question arises: How does the ant colony manage to survive the beetle larvae's competition for the food and their intense predation on the ant larvae in the brood chamber? This turns out to have a simple answer. The beetle larvae are cannibalistic and unable to distinguish their fellow larvae from ant larvae by odor. They therefore cut down their

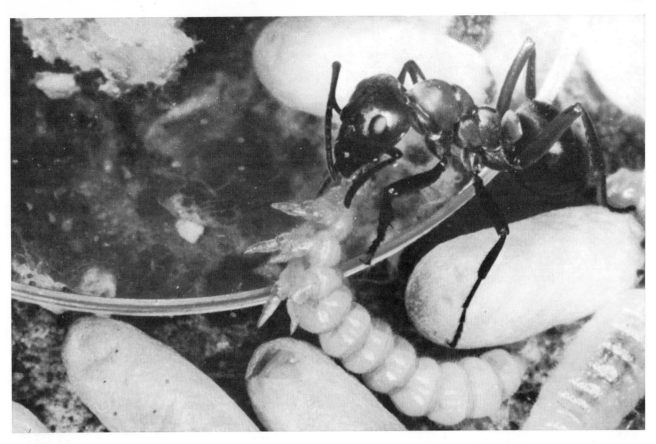

YOUNG GUEST, a beetle larva of the genus *Atemeles,* is given a droplet of liquid food by an ant attendant in the brood chamber of a nest of *Formica* ants. Not only beetles but also wasps, flies and many other arthropods are fed and sheltered by various ant hosts.

MATURING LARVA obtains added nourishment in the *Formica* brood chamber by eating the small ant larvae sheltered there. The beetle larvae also prey on one another. The nest in this photograph and the one at top of page is a man-made laboratory structure.

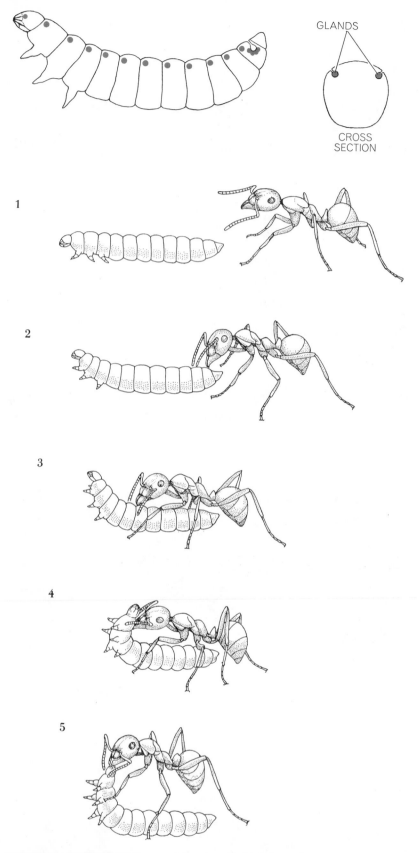

GLANDS

CROSS
SECTION

1

2

3

4

5

CHEMICAL ATTRACTANT secreted by a row of glands on each side of the beetle larva (*top*) causes the ant host to groom the larva intensively (*1–4*). Experiments with tracer substances indicate that the grooming process transfers secretions from larva to ant. Contact between ant and larva also makes the larva rear up. If the larva can make mouth-to-mouth contact with the ant (*5*), the stimulus causes the ant to regurgitate a droplet of liquid food. Except where noted, the representations of ant hosts and their guests, including the illustration on the cover, are based on a series of original illustrations by the author's wife.

own population, whereas the ant larvae do not. Typically one finds that in a brood chamber the ant larvae are found in clusters but beetle larvae (particularly those belonging to the genus *Lomechusa*) show up as loners, having devoured their neighbors.

The *Atemeles* beetles have two homes with ants—one for the summer, the other for winter. After the larvae have pupated and hatched in a *Formica* nest, the adult beetles migrate in the fall to nests of dark brown, insect-eating ants of the genus *Myrmica*. The reason for the beetles' migration is that brood-keeping and the food supply are maintained in *Myrmica* colonies throughout the winter, whereas *Formica* ants suspend their raising of young. In the *Myrmica* nests the beetles, still sexually immature, can be fed and ripen to maturity by the spring, at which time they return to *Formica* nests for mating and the laying of eggs. Thus the life cycles and behavior of the *Atemeles* beetles and the *Formica* and *Myrmica* ants are so synchronized that the beetles can take maximum advantage of the social life of each of the two ant species that serve as hosts. In this respect *Atemeles* shows a remarkably advanced form of evolutionary adaptation. The *Lomechusa* beetles, also co-dwellers with *Formica* ants, do not change their environment for the winter; after hatching they simply move on to another *Formica* colony of the same species and share the food shortage. It appears that *Atemeles* evolved myrmecophilous relations with *Formica* to begin with and then "discovered" and adapted to a winter home with *Myrmica*, developing proficiency in a second language for that purpose.

Before leaving the *Formica* nest at the beginning of the fall the *Atemeles* beetle obtains a supply of food for its migration by begging from its *Formica* hosts. For this it employs a technique of tactile stimulation. The beetle first drums rapidly on an ant with its antennae to attract attention. It then induces the ant to regurgitate food by touching the ant's mouthparts with its maxillae and forelegs [*see bottom illustration on page 299*]. High-speed motion pictures show that ants themselves obtain food from one another by similar mechanical signals.

How does the migrating beetle find its way to a *Myrmica* nest? *Formica* nests are normally found in woodland, whereas *Myrmica* nests are found in the grassland beyond the woods. It can be shown experimentally that when *Atemeles* beetles leave the *Formica* nest, they generally move in the direction of increasing light. This may explain how the beetles manage to reach the relatively open grasslands where the *Myrmica* ants live. After reaching the open grassland the beetle has to use other cues to find a *Myrmica* nest. By means of laboratory experiments I have ascertained that it is guided to the nest by the odor of the host species of ant. The odor must be windborne; the beetle is not drawn to it in still air. Curiously, the beetle possesses only a temporary sensitivity to this

odor; it is limited to the two weeks after leaving the *Formica* nest. (In the spring the beetle locates a *Formica* nest in the same way, by its odor.)

On finding a *Myrmica* nest, the beetle obtains recognition and adoption with a ritual involving chemical communication. The beetle first taps the ant lightly with its antennae and raises the tip of its abdomen toward the host. The ant responds by licking secretions from glands on the abdomen's tip that I call "ap-

peasement glands," because their secretion apparently suppresses aggressive behavior in the ant. The ant is next attracted to a series of glands along the sides of the beetle's abdomen; I call these the "adoption glands," as the ant will not welcome or adopt the beetle unless it senses their secretion. Presumably the odor of this secretion mimics the odor of the ant species. Finally the beetle lowers its abdomen so that the ant can approach, and the ant then grasps

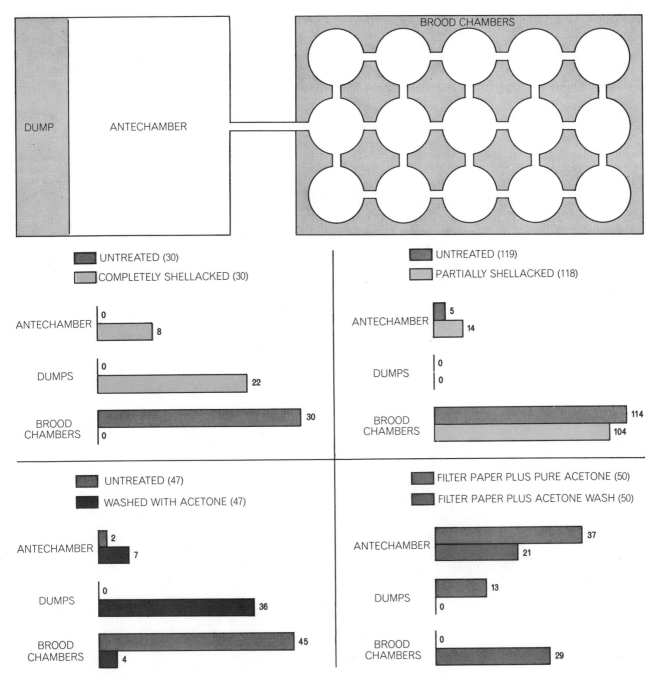

POWER OF ATTRACTANT was shown in four experiments with an artificial ant nest (*top*). The controls in three cases were normal beetle larvae. When normal larvae and others covered with shellac were placed in the antechamber, the ants moved all the normal larvae but none of the shellacked ones into the brood chambers. Most of the shellacked larvae were taken to a dump. Presented

with larvae that were coated except for one body segment, the ants accepted most of them. Faced with a choice between normal larvae and others whose attractant had been washed off in an acetone bath, the ants accepted most normal larvae but dumped most deodorized ones. In a choice between attractant-scented and unscented dummy larvae made of paper, the ants accepted only scented ones.

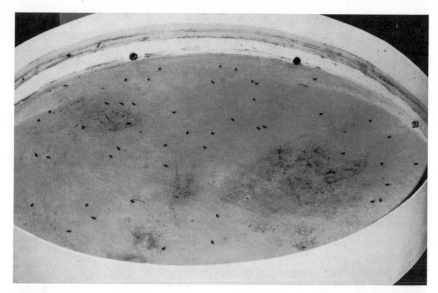

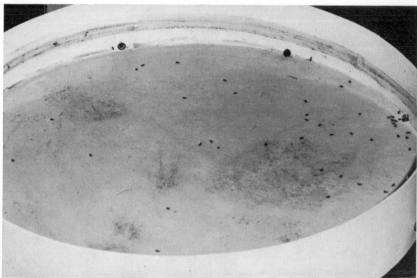

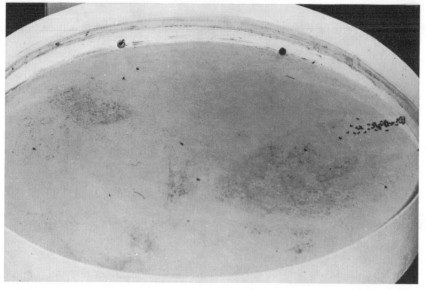

ROLE OF SCENT in directing the migration of *Atemeles* beetles from their initial residence in *Formica* nests to winter quarters in *Myrmica* nests is demonstrated in these
laboratory photographs. At the start of the experiment (*top*) the beetles are distributed at
random in a circular enclosure. Air scented with the odor of *Myrmica* ants is then blown
into the enclosure through one of the eight holes around its rim (*center*). Ten minutes later
(*bottom*) most of the beetles have collected in a cluster near the scent-emitting hole.

some bristles around the beetle's lateral
glands and carries the guest into the
brood chamber.

The *Atemeles* beetles are not the only
myrmecophiles capable of making themselves at home with more than one kind
of ant. More than 50 years ago the Harvard University entomologist William
Morton Wheeler discovered that staphylinid beetles of the *Xenodusa* genus
change their domicile with the seasons.
The larvae live in *Formica* nests through
the summer and the adults overwinter in
nests of the carpenter ants of the genus
Camponotus. It is interesting to note that
the carpenter ants also maintain larvae
throughout the winter. It may well be
that the evolutionary history of the
Xenodusa beetles parallels that of *Atemeles* in selecting and adapting to a
winter home.

Unlike the *Atemeles*, *Lomechusa* and
Xenodusa beetles, various other genera
of staphylinid beetles do not possess the
command of ant language that is required to gain entry into the brood chamber, the optimum niche for obtaining
food. Staphylinids of the European
genus *Dinarda*, for example, are limited
to the peripheral chambers of the nests
of their host (*Formica sanguinea*). *Dinarda* offers secretions from glands similar to *Atemeles*' appeasement glands,
but these secretions only induce the ant
to tolerate the beetle, not to adopt it and
take it into the brood chamber. *Dinarda*
is therefore reduced to living on such
food as it can find or scrounge in the
peripheral chambers. There it feeds on
dead ants that have not yet been taken
to the garbage dump and on food that it
may steal or wheedle from the worker
ants. Occasionally *Dinarda* snatches a
food droplet from the mouth of a forager
that is about to pass the food to another
worker. Or the beetle may surreptitious-
ly approach a food-laden forager and, by
touching the forager's lip, induce the
regurgitation of a small food droplet.
The ant immediately recognizes the beetle as an alien, however, and starts to
attack it. The beetle staves off the attack
by raising its abdomen and offering the
ant its appeasement secretion; while the
ant is savoring the substance it has licked
up, the beetle makes its escape.

Other groups of myrmecophilous
staphylinid beetles (for example the
genus *Myrmedonia*) have a bare minimum of communication with their ant
hosts, sufficient only to allow the beetle to feed at the ants' garbage dumps.
At the dump the foraging beetle can
avert attack by an ant from the nest by
offering it the appeasement secretion

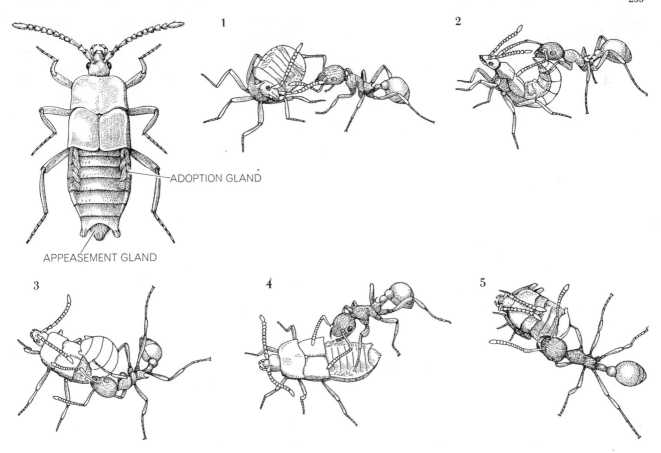

ADOPTION GLAND

APPEASEMENT GLAND

ADOPTION RITUAL, which gains entry for *Atemeles* beetles to nests of the ant genus *Myrmica,* depends on secretions from glands at the tip and along the sides of the beetle's abdomen (*top left*). On encountering a potential *Myrmica* host in the nest antechamber, the beetle attracts its attention (*1*) by tapping the ant lightly with its antennae and raising the tip of its abdomen. The response of the ant (*2*) is to taste the secretion from the glands there, known as "appeasement glands" because they apparently suppress the ant's aggressive behavior toward intruders. The next attractant is the secretion from the "adoption glands" (*3, 4*), so named because an ant will not welcome an intruder before it senses this secretion. The tightly curled beetle is then carried to a brood chamber (*5*).

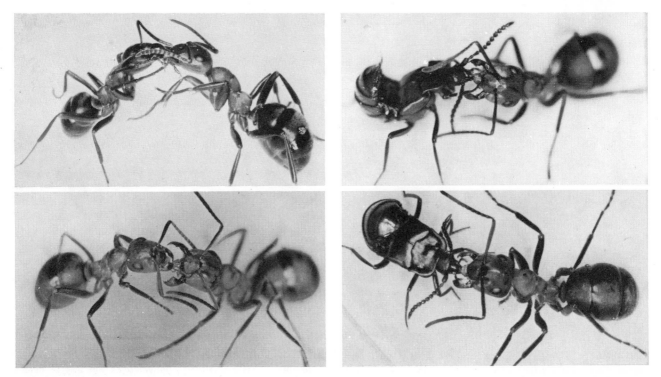

IDENTICAL MEANS is used by ants (*left*) and beetles (*right*) to make a food-laden forager regurgitate. A tapping of the forager's mouthparts with the food seeker's forelegs is the key stimulus (*top photographs*). Regurgitation is immediate (*bottom photographs*).

and thus winning time to escape. If the beetle is placed anywhere inside the nest, however, the appeasement does not avail; the ants promptly kill the beetle as an intruder.

There are myrmecophiles that possess only an elementary, one-way form of chemical communication with the ants they depend on for food. This is simply the ability to recognize the odor of the trail laid down between the nest and a food source by foragers of the particular ant species. For example, a small nitidulid beetle in Europe (*Amphotis marginata*) can thus identify the trails of a shiny black wood ant (*Lasius fuliginosus*), and in many localities these beetles abound along the trails. They act as begging highwaymen, intercepting food-laden ants on the trail and inducing them to regurgitate food droplets by tapping the ant's labium. The ant soon realizes it has been tricked and attacks the beetle. Although the beetle has no appeasement mechanism, it avoids injury by retracting its appendages and flattening itself on the ground.

Many myrmecophilous beetles closely resemble their ant hosts in appearance. This is particularly true of guests of the

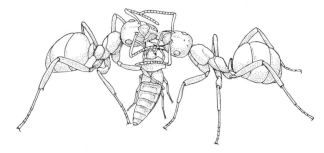

FOOD THIEVES, beetles of the genus *Dinarda*, are tolerated in the outer chambers but not the brood chambers of the *Formica* nest. By touching a forager ant's mouthparts (*left*) a *Dinarda* beetle may induce regurgitation of a food droplet. The beetle may also intercept a droplet (*right*) as it is being passed between two ants. If attacked, the beetle raises its abdomen, drawing attention to the glands that, like the *Atemeles* "appeasement glands," secrete a substance that the ants savor. While the ant tastes, the beetle flees.

HIGHWAYMAN BEETLES of the genus *Amphotis* locate foraging trails of the wood ant *Lasius* by the scent and ambush food-laden workers. By stimulating the ant's mouthparts (*top left*) the beetle causes it to regurgitate its cropful of food (*top right*). The robbed ant then reacts aggressively (*bottom left*), but passive defense (*bottom right*) enables the armored beetle to weather the attack.

army ants, and some investigators concluded that the factor inducing the ants
to accept beetles as their nestmates was
the beetles morphological resemblance
to themselves. It was even thought to be
the case with *Atemeles* and *Lomechusa*,
although they do not particularly resemble their hosts. I have been able to show,
by artificially altering the shape and col-
or of these beetles, that morphological
features do not contribute to the success
of their relationship with their hosts. Instead it appears that communicative behavior remains the essential requirement
for acceptance. I believe this is probably
generally the case, even for the guests of
the army ants. Very likely the guests'
mimicry of their hosts' appearance has
evolved as a protection against predation
by birds. Predatory birds that follow
army ants as they travel over open
ground do not attack the ants themselves; they feed only on other insects
that are stirred up by the ants' march.
Presumably insects that resemble ants
are ignored by the birds. With carefully
designed experiments we should be able
to resolve this question.

There remains for further exploration
the fascinating question of how the
extraordinarily effective system of communication between myrmecophiles and
their hosts evolved. We can think of this
evolution as a two-part process. First,
viewing the myrmecophile as a signal-
receiver, we can suppose the potential
guest of a specific ant underwent gradual evolutionary modification of its receptor system that endowed it with the
ability to recognize the ant's odor, the
distinctions between the ant larva and
the adult and other signals opening the
way to a parasitic relationship. Second,
regarding the myrmecophile as a transmitter of signals, we can see that through
natural selection it must have evolved
the set of secretions and behavioral acts
that induces the ant to accept the guest
in the nest and nurture it. Thus the development of the guest-host relationship
would involve adaptive changes only in
the guest, accommodating itself to the
nature of the specific host. In all probability this is the way *Atemeles* and *Lomechusa* won their welcome and support
in the homes of their respective species
of ant hosts.

By careful analysis of various related
species of beetles that enjoy differing degrees of intimacy with their hosts we
can expect to learn more about the details of the evolution of the social associations between myrmecophiles and
ants and their communication systems.

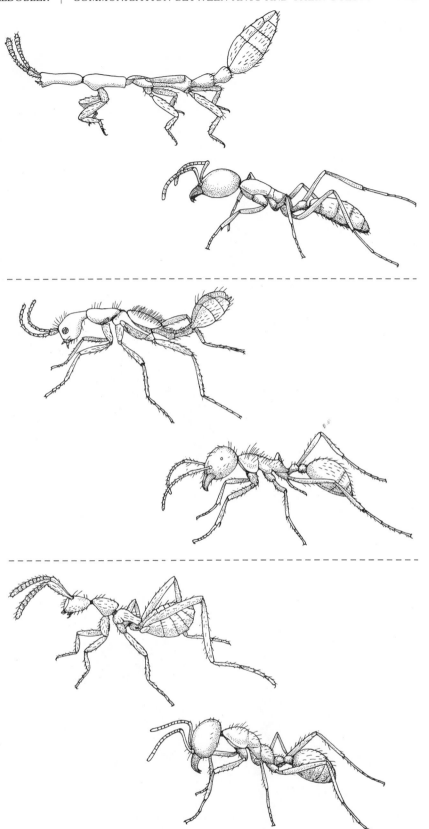

BEETLE MIMICS that resemble their army-ant hosts include a seemingly headless species,
Mimanomma spectrum (*top left*). Its host is the African driver ant *Dorylus nigricoms* (*top
right*). Other army-ant mimics are *Crematoxenus aenigma* (*center left*) and *Mimeciton
antennatum* (*bottom left*), shown with their respective New World hosts, *Neivamyrmex
melanocephalum* and *Labidus praedator*. The drawings are based on original sketches by
the late Charles H. Seevers of Roosevelt College. There is no evidence that mimicry affects
guest-host relations. It may, however, protect the beetles from attack by other predators.

BEES ARE PAINTED with colored dots so that they can be identified during an experiment at the author's station near Munich. In this way the feeding station of a bee can be associated with its dance within the hive. The dish contains sugar water.

TWO VARIETIES OF BEE, the yellow Italian bee *Apis mellifera ligustica* and the black Austrian bee *A. mellifera carnica*, feed together. These two bees can live together in the same hive, but their dances do not have quite the same meaning. Accordingly one variety cannot accurately follow the feeding "instructions" of the other. Both of these photographs were made by Max Renner.

32

Dialects in the Language of the Bees

by Karl von Frisch

August 1962

The dances that a honeybee does to direct its fellows to a source of nectar vary from one kind of bee to another. These variations clarify the evolution of this remarkable system of communication

For almost two decades my colleagues and I have been studying one of the most remarkable systems of communication that nature has evolved. This is the "language" of the bees: the dancing movements by which forager bees direct their hivemates, with great precision, to a source of food. In our earliest work we had to look for the means by which the insects communicate and, once we had found it, to learn to read the language [see "The Language of the Bees," by August Krogh; SCIENTIFIC AMERICAN Offprint 21]. Then we discovered that different varieties of the honeybee use the same basic patterns in slightly different ways; that they speak different dialects, as it were. This led us to examine the dances of other species in the hope of discovering the evolution of this marvelously complex behavior. Our investigation has thus taken us into the field of comparative linguistics.

Before beginning the story I should like to emphasize the limitations of the language metaphor. The true comparative linguist is concerned with one of the subtlest products of man's powerfully developed thought processes. The brain of a bee is the size of a grass seed and is not made for thinking. The actions of bees are mainly governed by instinct. Therefore the student of even so complicated and purposeful an activity as the communication dance must remember that he is dealing with innate patterns, impressed on the nervous system of the insects over the immense reaches of time of their phylogenetic development.

We made our initial observations on the black Austrian honeybee (*Apis mellifera carnica*). An extremely simple experiment suffices to demonstrate that these insects do communicate. If one puts a small dish of sugar water near a beehive, the dish may not be discovered for several days. But as soon as one bee has found the dish and returned to the hive, more foragers come from the same hive. In an hour hundreds may be there.

To discover how the message is passed on we conducted a large number of experiments, marking individual bees with colored dots so that we could recognize them in the milling crowds of their fellows and building a hive with glass walls through which we could watch what was happening inside. Briefly, this is what we learned. A bee that has discovered a rich source of food near the hive performs on her return a "round dance." (Like all the other work of the colony, food-foraging is carried out by females.) She turns in circles, alternately to the left and to the right [see top illustration on next page]. This dance excites the neighboring bees; they start to troop behind the dancer and soon fly off to look for the food. They seek the kind of flower whose scent they detected on the original forager.

The richer the source of food, the more vigorous and the longer the dance. And the livelier the dance, the more strongly it arouses the other bees. If several kinds of plants are in bloom at the same time, those with the most and the sweetest nectar cause the liveliest dances. Therefore the largest number of bees fly to the blossoms where collecting is currently most rewarding. When the newly recruited helpers get home, they dance too, and so the number of foragers increases until they have drained most of the nectar from the blossoms. Then the dances slow down or stop altogether. The stream of workers now turns to other blossoms for which the dancing is livelier. The scheme provides a simple and purposeful regulation of supply and demand.

The round dance works well for flowers close to the beehive. Bees collect their nourishment from a large circuit, however, and frequently fly several miles from the hive. To search at such distances in all directions from the hive for blossoms known only by scent would be a hopeless task. For sources farther away than about 275 feet the round dance is replaced by the "tail-wagging dance." Here again the scent of the dancer points to the specific blossoms to be sought, and the liveliness of the dance indicates the richness of the source. In addition the wagging dance transmits an exact description of the direction and distance of the goal. The amount and precision of the information far exceeds that carried by any other known communication system among animals other than man.

The bee starts the wagging dance by running a short distance in a straight line and wagging her abdomen from side to side. Then she returns in a semicircle to the starting point. Then she repeats the straight run and comes back in a semicircle on the opposite side. The cycle is repeated many times [see middle illustration on next page]. By altering the tempo of the dance the bee indicates the distance of the source. For example, an experimental feeding dish 1,000 feet away is indicated by 15 complete runs through the pattern in 30 seconds; when the dish is moved to 2,000 feet, the number drops to 11.

There is no doubt that the bees understand the message of the dance. When they fly out, they search only in the neighborhood of the indicated range, ignoring dishes set closer in or farther away. Not only that, they search only in the direction in which the original feeding dish is located.

The directional information contained in the wagging dance can be followed most easily by observing a forager's per-

ROUND DANCE, performed by moving in alternating circles to the left and to the right, is used by honeybees to indicate the presence of a nectar source near the hive.

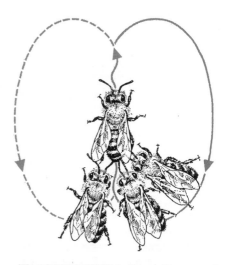

WAGGING DANCE indicates distance and direction of a nectar source farther away. Bee moves in a straight line, wagging her abdomen, then returns to her starting point.

SICKLE DANCE is used by the Italian bee. She moves in a figure-eight-shaped pattern to show intermediate distance. A dancer is always followed by her hivemates.

formance when it takes place out in the open, on the small horizontal landing platform in front of the entrance to the hive. The bees dance there in hot weather, when many of them gather in front of the entrance. Under these conditions the straight portion of the dance points directly toward the goal. A variety of experiments have established that the pointing is done with respect to the sun. While flying to the feeding place, the bee observes the sun. During her dance she orients herself so that, on the straight run, she sees the sun on the same side and at the same angle. The bees trooping behind note the position of the sun during the straight run and position themselves at the same relative angle when they fly off.

The composite eye of the insect is an excellent compass for this purpose. Moreover, the bee is equipped with the second navigational requisite: a chronometer. It has a built-in time sense that enables it to compensate for the changes in the sun's position during long flights.

Usually the wagging dance is performed not on a horizontal, exposed platform but in the dark interior of the hive on the vertical surface of the honeycomb. Here the dancer uses a remarkable method of informing her mates of the correct angle with respect to the sun. She transposes from the ability to see the sun to the ability to sense gravity and thereby to recognize a vertical line. The direction to the sun is now represented by the straight upward direction along the wall. If the dancer runs straight up, this means that the feeding place is in the same direction as the sun. If the goal lies at an angle 40 degrees to the left of the sun, the wagging run points 40 degrees to the left of the vertical. The angle to the sun is represented by an equal angle with respect to the upright. The bees that follow the dancer watch her position with respect to the vertical, and when they fly off, they translate it back into orientation with respect to the light.

We have taken honeycombs from the hive and raised the young bees out of contact with older bees. Then we have brought the young bees back into the colony. They were immediately able to indicate the direction of a food source with respect to the position of the sun, to transpose directional information to the vertical and to interpret correctly the dances of the other bees. The language is genuinely innate.

When we extended our experiments to the Italian variety of honeybee (*Apis mellifera ligustica*), we found that its innate system had developed somewhat

differently. The Italian bee restricts her round dance to representing distances of only 30 feet. For sources beyond this radius she begins to point, but in a new manner that we call the sickle dance. The pattern is roughly that of a flattened figure eight bent into a semicircle [*see bottom illustration at left*]. The opening of the "sickle" faces the source of food; the vigorousness of the dance, as usual, indicates the quality of the source.

At about 120 feet the Italian bee switches to the tail-wagging dance. Even then she does not use exactly the same language as the Austrian bee does. The Italian variety dances somewhat more slowly for a given distance. We have put the two varieties together in a colony, and they work together peacefully. But as might be expected, confusion arises when they communicate. An Austrian bee aroused by the wagging dance of an Italian bee will search for the feeding place too far away.

Since they are members of the same species, the Austrian and Italian bees can interbreed. Offspring that bear the Italian bee's yellow body markings often do the sickle dance. In one experiment 16 hybrids strongly resembling their Italian parent used the sickle dance to represent intermediate distances 65 out of 66 times, whereas 15 hybrids that resembled their Austrian parent used the round dance 47 out of 49 times. On the other two occasions they did a rather dubious sickle dance: they followed the pattern but did not orient it to indicate direction.

Other strains of honeybee also exhibited variations in dialect. On the other hand, members of the same variety have proved to understand each other perfectly no matter where they come from.

Our next step was to study the language of related species. The only three known species of *Apis* in addition to our honeybee live in the Indo-Malayan region, which is thought to be the cradle of the honeybee. The Asian species are the Indian honeybee *Apis indica*, the giant bee *Apis dorsata* and the dwarf bee *Apis florea*. Under a grant from the Rockefeller Foundation my associate Martin Lindauer was able to observe them in their native habitat.

The Indian honeybee, which is so closely related to ours that it was for a long time believed to be a member of the same species, has also been domesticated for honey production. Like the European bees, it builds its hive in a dark, protected place such as the hollow of a tree. Its language is also much like that of the European bees. It employs the round dance for distances up to 10 feet, then switches directly to the tail-wagging dance. Within its dark hive the Indian bee also transposes from the visual to the gravity sense. The rhythm of the dance, however, is much slower than that of the European bees.

The giant bee also exhibits considerable similarity to its European cousins and to the Indian species in its communications. It changes from the round dance to the wagging dance at 15 feet. In its rhythm it moves at about the same rate as the Italian bee does. The

hive of the giant bee, however, is built on tree branches or other light, exposed places. The inhabitants dance on the vertical surface of the comb, converting the angle with respect to the sun correctly into an angle from the vertical. But since the comb is out in the open, the dancers can always find a spot that commands a clear view of the sky. The fact that they do this indicates that the following bees can understand the instructions better when they have direct information about the position of the sun.

In the case of the dwarf bee, Lindauer found a clearly more primitive social organization and a correspondingly less

highly developed language. The dwarf bees, which are so small that a layman would probably mistake them for winged ants, build a single comb about the size of a man's palm. It dangles from an upper branch of a small tree. When the dwarf bees return from feeding, they always alight on the upper rim of the comb, where their mates are sitting in a closely packed mass that forms a horizontal landing place for the little flyers. Here they perform their dances. They too use a round dance for distances up to 15 feet, then a wagging dance. Their rhythm is slow, like that of the Indian bee.

DIALECTS in the language of the bees are charted. The dwarf bee dances on a horizontal surface. All others dance on a vertical surface. The speed of the wagging dance carries distance instructions. The more rapidly the bee performs its wagging runs, the shorter is the distance. The figures in the squares represent the number of wagging runs in 15 seconds for each distance and kind of bee.

The dwarf bee can dance only on a horizontal platform. Lindauer obtained striking proof of this on his field trip. When he cut off the branch to which a comb was attached and turned the comb so that the dancing platform was shifted to a vertical position, all the dancers stopped, ran up to the new top and tried to stamp out a dancing platform by running about through the mass of bees. When he left the hive in its normal position but placed an open notebook over its top, the foragers became confused and stopped dancing. In time, however, a few bees assembled on the upper surface of the notebook; then the foragers landed there and were able to perform their dances. Then, to remove every possible horizontal surface, Lindauer put a ridged, gable-shaped glass tile on top of the comb and closed the tile at both ends. In this situation the bees could not dance at all. After three days in this unnatural environment the urge to dance had become so great that a few bees tried to dance on the vertical surface. But they continued to depend on vision for their orientation and did not transpose the horizontal angle to a vertical one. Instead they looked for a dancing surface on which there was a line parallel to the direction of their flight. They tried to make a narrow horizontal path in the vertical curtain of bees, keeping their straight runs at the same angle to the sun as the angle at which they had flown when they found food. Under these circumstances only a very few bees were able to dance. Obviously the dwarf bee represents a far more primitive stage of evolution than the other species. She cannot transpose from light to gravity at all.

In trying to follow the dancing instinct farther back on the evolutionary scale, we must be satisfied with what hints we can get by observing more primitive living insects. Whereas a modern fossil rec-ord gives some of the physical development of insects, their mental past has left no trace in the petrified samples.

The use of sunlight as a means of orientation is common to many insects. It was first observed among desert ants about 50 years ago. When the ants creep out of the holes of their subterranean dwellings onto the sandy and barren desert surface, they cannot depend on landmarks for orientation because the wind constantly changes the markings of the desert sands. Yet they keep to a straight course, and when they turn around they find their way home along the same straight line. Even the changing position of the sun does not disturb them. Like the bee, the desert ant can take the shift into account and use the sun as a compass at any hour, compensating correctly for the movement of the sun in the sky.

Perhaps even more remarkable is the fact that many insects have developed an ability to transpose from sight to gravity. If a dung beetle in a dark room is placed on a horizontal surface illuminated from one side by a lamp, the beetle will creep along a straight line, maintaining the same angle to the light source for as long as it moves. If the light is turned off and the surface is tilted 90 degrees so that it is vertical, the beetle will continue to crawl along a straight line in the dark; it now maintains the same angle with respect to gravity that it earlier maintained with respect to light. This transposition is apparently an automatic process, determined by the arrangement of the nervous system. Some insects transpose less accurately, keeping the same angle but placing it sometimes to the right and sometimes to the left of the vertical without regard to the original direction with respect to the light. Some are also impartial as to up and down, so that an angle is transposed in any of four ways. Since the patterns do not transmit in-formation, their exact form makes no difference. Among the ancestors of the bees transposition behavior was probably once as meaningless as it is in the dung beetle and other insects today. In the course of evolution, however, the bee learned to make meaningful use of this central nervous mechanism in its communication system.

Both navigation by the sun and transposition, then, have evolved in a number of insects. Only the bees can use these abilities for their own orientation and for showing their mates the way to food. The straight run in the wagging dance, when performed on a horizontal surface, indicates the direction in which the bees will soon fly toward their goal. Birds do something like this; when a bird is ready to take off, it stretches its neck in the direction of its flight. Such intention movements, as they are called, sometimes influence other animals. In a flock of birds the movements can become infectious and spread until all the birds are making them. It is possible that among the honeybees the strict system of the wagging dance gradually developed out of such intention movements, performed by forager bees before they flew off toward their goal.

The most primitive communication system we have found among the bees does not contain information about distance or direction. It is used by the tiny stingless bee *Trigona iridipennis*, a distant relative of the honeybee. Lindauer observed this insect in its native Ceylon. Its colonies are less highly organized, resembling bumblebee colonies rather than those of honeybees.

When a foraging *Trigona* has found a rich source of nectar, she also communicates with her nestmates. But she does not dance. She simply runs about in great excitement on the comb, knocking against her mates, not by chance but intentionally. In this somewhat rude manner she attracts their attention to the fragrance of blossoms on her body. They fly out and search for the scent, first in the nearby surroundings, then farther away. Since they have learned neither the distance nor the direction of the goal, they make their way to the food source one by one and quite slowly.

We probably find ourselves here at the root of the language of the bees. Which way the development went in detail we do not know. But we have learned enough so that our imagination can fill in the evolutionary gaps in a general way.

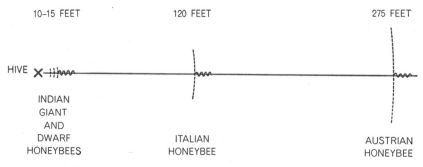

10–15 FEET   120 FEET   275 FEET

HIVE

INDIAN
GIANT
AND
DWARF
HONEYBEES   ITALIAN   AUSTRIAN
                    HONEYBEE   HONEYBEE

CHANGE FROM ROUND TO WAGGING DANCE occurs when nectar source lies beyond a certain radius of the hive. Change occurs at different distances among different bees. Because the wagging dance shows direction as well as distance, the Indian, giant and dwarf bees can give more precise information about a nearby source than the European bees can.

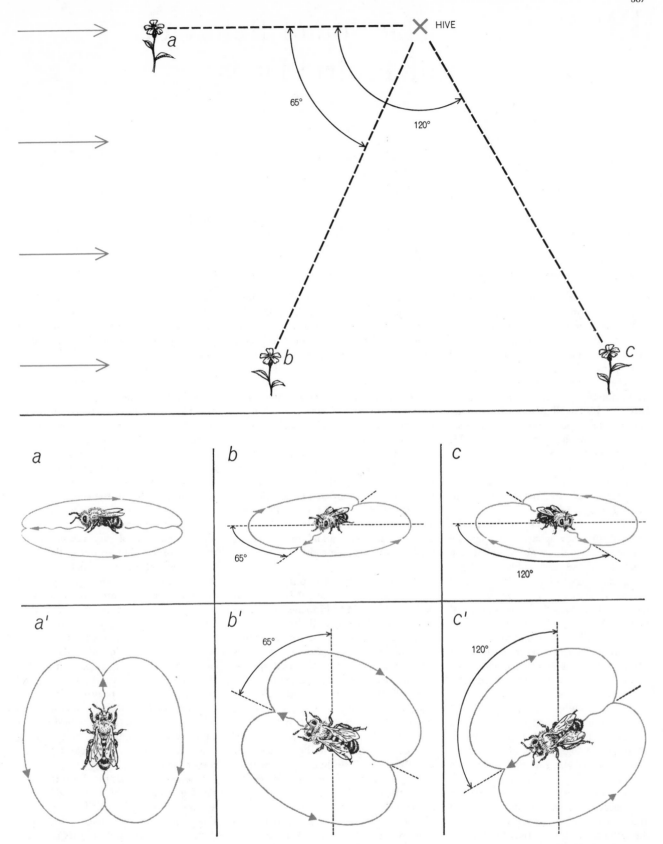

DIRECTION of a nectar source from the hive is shown by the direction in which a bee performs the straight portion of the wagging dance. The top section of the drawing shows flowers in three directions from the hive. The colored arrows represent the sun's rays. The middle section shows the dwarf bee, which dances on a horizontal surface. Her dance points directly to the goal: she orients herself to see the sun at the same angle as she saw it while flying to her food. The bottom section shows the bees that dance on a vertical surface. They transpose the visual to the gravitational sense. Movement straight up corresponds to movement toward the sun (a'). Movement at an angle to the vertical (b', c') signifies that the food lies at that same angle with respect to the sun.

33

The Social Behavior of Prairie Dogs

by John A. King
October 1959

In building towns the prairie dog protects itself against predators and encourages the plants on which it feeds. Its social behavior is passed from generation to generation by both heredity and learning

The Great Plains of the North American West seemed endless and uninviting to many pioneers, who brought memories of the forests and hills of their homes in the East. Rainfall was sparse, the streams were shallow and there was little timber for building. The Indians, the deer and the great herds of bison—all perpetually on the move—seemed to share the pioneer's restlessness as he pressed across this open wilderness toward a homestead. The only settlers here were the town-dwelling prairie dogs; they sat upon their thresholds and barked at the intruding wagon trains.

When pioneers at last settled in the Great Plains, they opened war upon the prairie dogs and sought to eliminate them from the now-cultivated land. Long campaigns of poisoning were so successful that three decades ago some naturalists predicted the prairie dogs' imminent extinction. The prediction has not been fulfilled, but the busy rodents have all but disappeared from the Great Plains that once belonged to them; they flourish chiefly in the refuges afforded by the national parks.

It was in one of these refuges, Wind Cave National Park in the Black Hills of South Dakota, that I undertook to study prairie dogs during three summers and a winter. Not far from the park's main road is Shirttail Canyon, the lower terraces and floor of which are occupied by a large prairie-dog town of 75 acres and nearly 1,000 inhabitants. On one of the terraces, separated from the town by steep slopes, I built two burlap-screened platforms on six-foot towers. Thus concealed, I was able to observe the little animals without upsetting their normal way of life.

As befits a town-dweller, the prairie dog has developed well-defined patterns of social behavior that set it apart from other prairie rodents, such as the dispersed and secretive deer mice and ground squirrels. Moreover, the prairie dog's social behavior is central to its survival in the Great Plains environment.

By building towns that occupy acres and even square miles, prairie dogs modify their environment and make it more suitable to their needs. Within the towns their social organization protects them from predators, and their own regulation of population density minimizes the risk of famine. This highly effective mode of behavior is transmitted to each prairie-dog generation by the preceding one. From my vantage point in Shirttail Canyon, this "education" of the young was a most engaging sight.

The prairie dog is a plump, tawny rodent, weighing about two pounds and measuring 14 to 17 inches from nose to tail-tip. It belongs to the squirrel family and is related to the marmots. Of the five species of the prairie-dog genus *Cynomys*, it is the black-tailed species *ludovicianus* that builds its towns in the Great Plains. Its front feet are equipped with long digging claws, its ears are small and its eyes—which are especially adapted to the detection of aerial predators—are so high on its head that they are about the first things to appear as the animal emerges from its burrow.

On the Great Plains this species lives between the temperature extremes of the hot summers of Texas and the intense winters of North Dakota. Precipitation seems to be an important factor in its distribution. It occupies a belt that gets some 20 inches a year, which supports a dominant vegetation of prairie grasses. The relationship between the prairie dogs and the vegetation upon which they feed is an intricate one, and

much remains to be learned about it. Apparently their feeding habits eliminate the taller vegetation growing in a prairie-dog town. This yields a twofold advantage: it deprives predators of dense cover and at the same time encourages fast-growing weeds with abundant fruits and seeds. The latter probably furnish a more varied and nutritious diet than the original vegetation. Some weeds such as the thistle seem to provide the prairie dogs, which rarely drink any free water, with moisture. Those non-food plants that still grow are clipped and left to wither in the sun.

A feature of the prairie-dog town that is perhaps even more striking than the altered pattern of vegetation is the mounds of soil that surround the 20 to 50 burrow entrances that dot each acre. These are a foot or two high and five or six feet across, and are often built after a rain, when the soil is soft and pliable. The prairie dogs dig and push the soil to the burrow and then tamp it into place with their noses. The mounds serve a dual purpose. During rains, water often rises several inches on the flat soil, and the mounds prevent swamping of the labyrinth of burrows. The mounds also serve as lookout stations; the prairie dogs sit atop them and from that vantage watch for predators and invading prairie dogs of other clans.

The many animals that hunt the prairie dog make watchfulness and readily accessible refuges vitally important. Eagles and hawks swoop upon prairie dogs from above, and bobcats and coyotes stalk them from any vegetative cover not cleared from the towns. In addition to being abundant the burrow-refuges must be deep enough to permit escape from fast-digging badgers, and they must have at least one additional exit in case the prairie dogs are followed through their burrows by the black-footed ferret.

Not all trespassers prey on the prairie dogs. At times deer mice and cottontail rabbits use less-frequented burrows for temporary refuge. Ground squirrels may, when hard-pressed, go into a prairie-dog burrow. The prairie dogs react to these small mammals with indifference or mild curiosity. On one occasion, as I watched, a prairie-dog pup crept up to a rabbit, sniffed it and then attempted to chase it, but soon gave up as the rabbit bounded easily away.

Frequently I saw antelope spend most of a day in the prairie-dog town, feeding and lying in the shade of scattered pine trees. When the antelope first approached the town, the prairie dogs would bark at them and scurry to their burrows. But once the visitors proved harmless, the prairie dogs would peer at them from their burrows and even come out and feed within a few feet of them.

Occasionally a herd of some 200 bison passed through the town, sometimes staying for a week and sometimes only for an hour. A herd of that size spending the afternoon on a 75-acre town soon makes a shambles of it. A bison would approach a well-formed crater, paw it, dig at it with its horns and then, seeing that it made a nice tub for a dust bath, lie down to wallow, destroying both crater and burrow.

The prairie dogs reacted to the bison much as they did to the less robustious antelope. While the bison were active—wallowing or sparring with one another—the prairie dogs would remain in their burrows. But when the bison turned to grazing or quietly ruminating, the prairie dogs came out and fearlessly fed nearby. After the bison had left, the prairie dogs diligently repaired the damage. But some craters seemed to be favorite wallowing sites, and in the face of constant destruction the prairie dogs relinquished these to the bison and left them unrepaired.

Some prairie-dog towns are divided by topographic features into wards, but the most significant divisions within towns or wards are invisible and are imposed by the prairie dogs themselves. These are the territories of the prairie-dog clans, which we call coteries. The coterie is the basic unit of the town's social organization. Its identity is established and maintained by frequent friendly contacts among its members and hostile contacts with members of other coteries. Each coterie member is known to the others; all share the burrows and

TERRITORIAL CALL is sounded by prairie dogs of all ages and both sexes to signify territorial proprietorship, departure of a predator or just general well-being. Animal throws itself into upright posture to deliver the call, which is almost always echoed by others. When it indicates "all clear," the town may resound as other prairie dogs repeat call.

resources of their territory. Members of a coterie frequently groom each other, play and "kiss." They rarely dominate or antagonize one another. A coterie territory usually covers about seven tenths of an acre, though its extent may vary, particularly at the edge of a town, where the animals may move into adjacent uninhabited regions.

The coteries vary in composition. The elusive "average" consists of an adult male and three adult females together with about six offspring. Some coteries consist of two males and five females. One coterie I observed had 31 young, another failed to produce any young one year, and still another consisted of only two old males and a barren female.

While both the membership and size of the coterie change from season to season, its territory is substantially permanent. Even a complete change of individuals, as the old die or emigrate and the young replace them, does not alter the boundaries between territories. Males that invade a territory from another section of the town quickly learn from the members of the coterie where the territorial boundaries lie. And the prairie-dog pups soon learn the extent of their home territory. Probably the main feature of a territory's stability is the network of burrows that are so essential to the prairie dog's survival. Each burrow entrance is jealously guarded, since the loss of one entrance to an invader may open the entire system to trespass. Consequently the inhabitants carry their defense to the boundaries of their territory. Thus delimited and maintained, the territory endures as a heritage handed down through the generations.

Most of a prairie dog's travels are confined to its coterie territory, and these typically add up to more than a mile a day. Any part of the territory that is not visited on one day may become a center of activity the next. One adult male that I watched for an entire day traveled almost three miles within his three-quarter-acre domain. Early in the morning he drove an invader from the territory, and for the rest of the day he remained alert against further invasions—circling the periphery of the territory and examining other animals apparently to make sure that they were bona fide coterie members.

Many prairie-dog activities are directed toward making social contacts, and these seem to serve at least two functions. First, they provide a con-

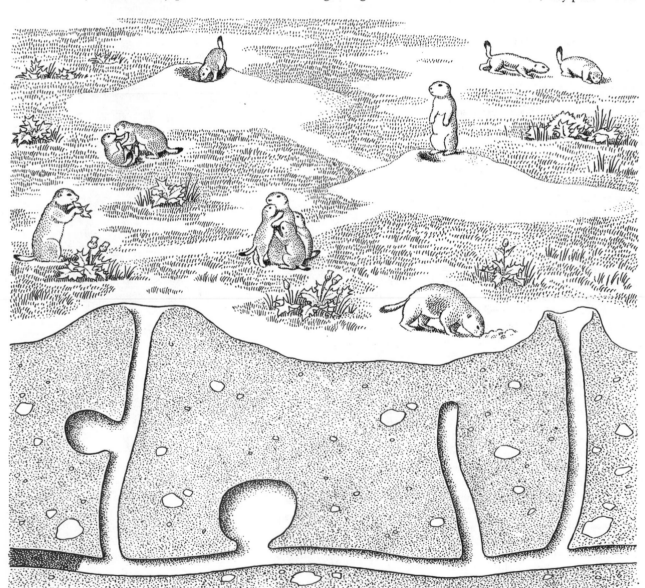

LABYRINTH OF BURROWS honeycombs a prairie-dog town. The mounds surrounding the burrow entrances keep surface water out of the system. The mounds also serve as lookout vantage-points. Many burrows have at least two exits. Bulbous excavations are nesting chambers. The shaded area at bottom left is a temporary plug of dung. The passages are about three feet below the surface.

stant check against trespassers, and second, they help maintain relations among the members of a coterie and thus preserve the group. Two individuals feeding some distance apart may, upon seeing each other, run forward to meet; or one, seeing another on a crater, will run and sit beside him. Whatever the occasion, when two individuals meet they turn their heads toward each other, open their mouths, bare their teeth and "kiss."

When familiars pass in a flight to a burrow, the kiss is hastily given. At other times it is more leisurely and prolonged. Two animals will meet and kiss; then one will roll over on its back, still maintaining oral contact. Often the kiss ends with both animals rapturously stretched out side by side; they then rise and move off to feed with their bodies pressed together.

The kiss seems primarily a means of distinguishing friend from foe. If two animals sight each other near the territorial boundary, or in any other circumstances in which they are uncertain of each other's identity, they lie down on their bellies and, flicking their tails, slowly creep toward each other until contact is made and the identification kiss exchanged. The open mouth which characterizes the kiss probably serves as a threat rather than an expression of af-

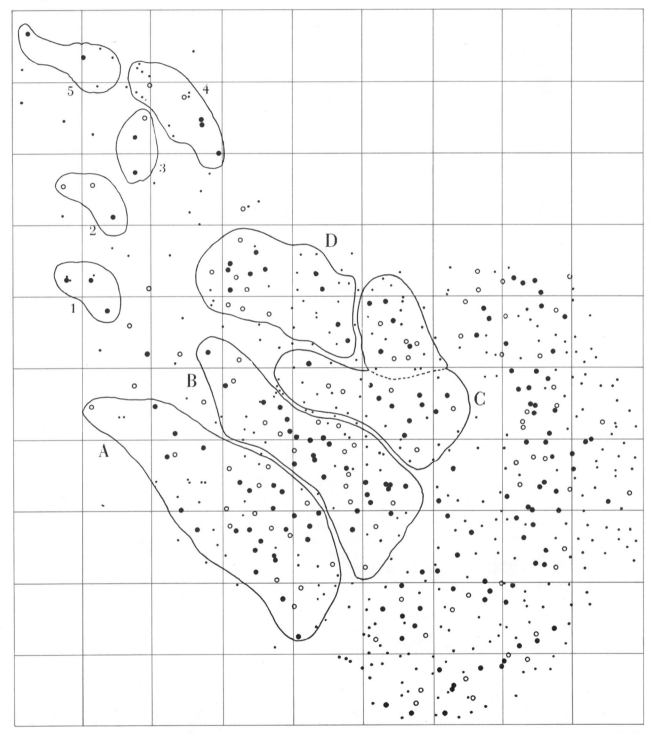

PRAIRIE DOGS DIVIDE THEIR TOWNS into coterie territories (A, B, C *and* D), the borders of which they zealously guard. Numbered areas at upper left are new territories established by emigrating adults. Territory C is in process of being split. Solid circles indicate large, active burrows; open circles are smaller burrows; dots are holes without craters. Each square is 50 feet wide.

fection. Each, seeing the other's mouth open and ready to bite, is probably made aware that the other is defending the integrity of the same territory and is hence a coterie member. Faced with the bared teeth and open mouth, an interloper runs off, while a resident stays and osculates.

The identification kiss is frequently a preliminary to a more elaborate ritual called grooming. After kissing, one prairie dog will begin to nibble and paw the other. The passive animal may roll over on its back better to expose itself to the stimulating ministrations of its partner; it crawls under the other's muzzle and encourages it to continue or to start again if it has stopped. All members of a coterie groom one another; males groom females and vice versa, adults groom the young and the young groom the adults and one another. The pups are particularly responsive to grooming by the adults and will chase and crawl under them in efforts to attract attention.

Perhaps the most significant sensory cue for uniting the coterie, and even the whole town, is the prairie-dog vocalizations. Prairie dogs were named for their bark, which is their only resemblance to dogs and which differs appreciably from a dog's bark. Prairie-dog barks may convey a particular meaning or several different ones. Generally the prairie-dog bark consists of a short nasal yip which varies in intensity and frequency with the stimulus. Some have a quite specific sound, others have many variations and, probably, many interpretations. The approach of danger provokes a warning call. Upon hearing it the other prairie dogs of the town interrupt their activities and sit up to see the cause of the alarm. If the bark is high-pitched and rapid, they stop whatever they are doing and dash immediately to their burrows.

The most distinctive vocalization is the two-syllable territorial call, which is delivered in a loud, clear series of two or three. The prairie dog throws its body upward, rising on its hind legs with its nose pointed straight up and its forefeet thrust out, and cries out with such force that it sometimes leaps from the ground. The pups also throw their whole bodies into the call, sometimes tumbling over backward in the effort. The territorial call is almost the antithesis of the bark. The bark communicates alarm, while most territorial calls are given in home territory where the animal feels secure. When delivered by a foraging prairie dog, the call may merely proclaim ownership of a territory. But it serves also to challenge an invader or to announce victory after a dispute. Among these gregarious creatures the territorial call sel-

GROOMING ENCOUNTERS punctuate the prairie-dog day. All coterie members, young and old, male and female, groom each other. The practice probably serves to maintain friendly relations and provide precondition for social organization. Animals in top drawing rise after recognition kiss and begin grooming; then one lies down better to expose itself to the other. Later animals often go off to graze with their bodies pressed together.

dom goes unanswered. When an animal utters it after the departure of a predator, the others join in, and what was simply an all-clear signal becomes a cacophony of togetherness.

With the exception of the vocalizations, which can be heard by the whole town, cooperative behavior does not extend to prairie dogs of other coteries; indeed, relations between coteries are quite hostile. Most conflicts between coteries arise over boundaries. When a coterie member, foraging at the limits of its own territory, passes into an adjoining territory, a resident of the area rushes up to drive it away. The invader may be only a few feet outside its own territory and, failing to recognize that it is trespassing, may refuse to yield. The

ensuing struggle is a stereotyped ritual that consists more of threat than fight. The animals rush toward each other, stop short and freeze face to face. Then, in a kind of reverse kissing encounter, one of the disputants turns, raises and spreads its tail, exposing its anal glands, and waits for the other to approach. The latter cautiously draws near and sniffs. Then they exchange roles; they alternate in this way until the stalemate is broken by an attempt of one to bite the rump of the other. The "bitten" contestant (prairie dogs are rarely scarred) backs away a few feet and then returns to the fray. The dispute is often accompanied by much rushing back and forth and repetition of the smelling encounters. Finally some arbitrary boundary is established and the antagonists return

to their foraging. Such disputes rarely result in boundary changes of more than a few feet and rarely in injury to the contenders.

In contrast to such innocuous and accidental trespasses, one coterie member may undertake a purposeful invasion of another territory. These invasions may simply be exploratory, as when an invader enters an unguarded territory and takes the opportunity to examine the strange burrows and terrain. If the residents do not discourage these explorations, the invader may grow arrogant and even sound his territorial call at the burrow entrances. At such times, however, the mere appearance of a resident is usually enough to send the explorer scurrying home. But some invasions are made by more aggressive males

IDENTIFICATION KISS is exchanged whenever prairie dogs meet (*upper drawings*). If animals recognize each other, grooming encounters follow (*see drawings on page 312*); if they do not, the interloper usually turns and runs off. Here nonrecognition is followed by a tail-raising ritual in which the animals alternately sniff each other's anal glands (*lower left*), attempt to bite each other's rump, stalk off (*lower right*) and return to the ritual until one retreats a few feet. Rarely does either suffer more than a nip.

who are willing to fight for the territory. Their tactics are to remain just inside the invaded borders, sniffing at burrows and acquainting themselves with the various coterie members. This may go on for days or weeks. If the residents threaten, the invaders run. But they do not run directly out of the territory; rather, they circle it while being chased and leave only when hard pressed, to return later. Each such encounter stops just short of combat and seems designed merely to test the residents' resistance. But ultimately the showdown comes and the invader fights the resident male. The victor takes, or reaffirms, possession of the territory and the vanquished is permanently driven away.

These and other trespasses impose a constant vigilance upon coterie members, especially upon the coterie's dominant male. He is constantly running about the territory and identifying the individuals in it, climbing atop mounds and rocks and surveying the territorial boundaries, and keeping a wary eye on the activities of neighboring coteries. Through zealous guarding of boundaries, the coterie maintains a discrete area in a large town, and with its epimeletic (care-giving) behavior achieves a stable, cohesive and amiable society.

High on the list of benefits that accrue to the coterie and the individual prairie dog through such cooperative behavior is control of vegetation, with its encouragement of bountiful weeds and elimination of grass cover for predators. Equally high is the more effective defense afforded by the common watch for danger and, with this, the protective advantage of common access to all burrows in a coterie territory. The coterie system has another decisive, though not so immediately apparent, role in the prairie dog's adaptation to its environment. The division of the town into coterie territories secures a uniform distribution of population that minimizes overgrazing. Within the coterie, another social mechanism protects the group from the disaster of overpopulation.

In an area of some five acres in the Shirttail Canyon town the July population in three successive years went from 44 to 28 to 82. Much of this fluctuation can be accounted for by breeding habits. In contrast to other rodents, which reproduce throughout a long breeding season, prairie dogs deliver only one litter of about five pups each year of their three or four years of life. Indeed, some females, at least in the northern ranges, may not breed at all during their first year. As a result the age composition of

a coterie, its reproduction rate and its total numbers will vary greatly from year to year. This was the case in the town I observed, where the young constituted only 10 per cent of the population one year and about 70 per cent the next; as a result the population density increased from four per acre to 15 in three months. Such a population leap—almost four-fold in three months—might well seem a weakness in the prairie dog's social order; it could threaten both the food supply and the social order itself. But prairie dogs control population density by a singular mechanism: the emigration of the older generation from the coterie territory.

Though individuals may permanently depart from the coterie at any time, group emigration takes place less capriciously. Between March and May, the period of gestation and lactation, the coterie system breaks down and is replaced by individual nest-territories. The parturient females vehemently defend their nests against all intruders. They may associate with coterie members on neutral ground away from the nests, but the free access to all burrows that normally prevails in the coterie is now suspended. At this juncture some adults and yearlings begin to forage and construct burrows beyond the town's edge. They do this at intervals during the day, and at night return to their established homes. The daily migrations to the new suburbs continue through June. Then, as the young prairie dogs emerge from their burrows and require less care, the migrant adults spend more and more of the day at the new burrows and finally remain all night. They still come back into the old town occasionally, but the center of their activities has permanently changed. Not all such expansions are into unsettled areas. Emigrating adults may invade a neighboring coterie territory and take it over if its tenants are too few in number or fail to defend their rights. The expansion in population that I observed not only increased the size of the ward under my surveillance from five acres to more than seven; it also created new breeding assemblages as other prairie dogs, mostly yearlings, were attracted from the town's more remote sections. And the redistribution of population reduced the over-all density of the town from 15 to 11 prairie dogs per acre.

Population pressure was doubtless a factor in this emigration of coterie adults to new suburbs, though all the motivating factors are not yet clear. It may also be that the behavior of the young—their peculiar fondness for grooming and their tireless pursuit of attention from the

adults—played a part. Very rarely were the pups rebuffed. If disinclined to groom them, the adults walked away. But this was scarcely enough to discourage the overtures of the pups. Finally, with something that must border on exasperation, the adults started their emigration.

Such forbearance on the part of adults toward the young is an important feature of the social milieu that the young prairie dog enters when he leaves his burrow. Certainly it helps to account for the extremely low mortality-rate among the young (in one section of the town that I observed only one pup out of 58 died). Of equal consequence is the perpetuation of the prairie dogs' social habits. After leaving his birthplace the emergent pup meets his father and other members of the coterie and enters a pup paradise. He plays with his siblings and the other young. All the adults kiss and groom him as his mother does, and he responds to them as he does to her. He readily accepts foster mothers and may spend the night with their broods. He attempts to suckle adults indiscriminately—males as well as females. A female will submit quietly; the male gently thwarts him and grooms him instead, rolling him over on his back and running his teeth through the pup's belly fur. The pup's demands for this treatment increase as he grows. He follows the adults about, climbing on them, crawling under them, doing everything he can to entice an adult into a grooming session. Sometimes, if he fails to win attention, he may playfully jump at them, and they may enter into the game. Only on the rarest of occasions is he rebuffed by an importuned adult; seldom is he kicked or bitten or drubbed.

During these first pleasant weeks the pup may even meander into adjacent coterie territories with impunity. But as he begins to mature he wanders farther and his invasions meet with less forbearance. At first the receptions are mildly hostile, but they gradually grow more severe. Soon he comes to recognize the territorial boundaries and learns that not all prairie dogs treat him alike. At about this time he begins to originate and respond to territorial calls. When he utters them in foreign territory, the immediate reprisals soon teach him to confine them to his own territory. He learns further to associate the calls with safety and his own well-being. By this time he is more cautious when he enters strange territory, and if he is approached by a resident, he retreats to his own area. He begins to use the identification kiss to discriminate between coterie members and strangers. If his kiss is not returned by another animal in his territory, he treats it suspiciously, barks at it, runs up to smell it and then dashes off. As he grows older this behavior elaborates into the tail-spreading ritual.

By the end of summer the young prairie dog has become mature in the behavior that is essential to the social organization of the town. His fighting is less playful and more often resembles real hostility. His associations with adults are fewer because of their emigration to other areas. When he does encounter them, he may groom them as often as they do him. He keeps within his territory and invades only uninhabited areas. He guards his territory against invaders, particularly those of his own age, and when older prairie dogs invade, he barks as though calling for help. He has matured now, except in sexual behavior.

A yearling male that has not emigrated the year before may show interest in females during the breeding season, but he is probably checked by the adults. He may join with the migrating adults and leave his territory, or even try to go alone, and this may explain the high mortality of yearling males; their ratio to females, about 14 to 10 at birth, now drops to about six to 10. His sisters, probably not yet in heat, do not engage in sexual activities.

If the yearling male remains in the coterie until summer, his disputes with other coterie yearlings may increase, though he shows no antagonism to the adults or to the young pups now coming from the burrows. His restlessness grows and he invades other territories, constantly testing the combativeness of strange males. If he finds a weak neighbor, he may contest with him for the territory. Should he succeed, he dominates the other males in the new coterie, or they leave and try for new territories of their own.

In a newly settled area other prairie dogs may be attracted to his site. Yearling females, which have been crowded out of their old homes by the new crop of pups, or adult females departing from their young, may come to live with him. Other yearling males may immigrate. All bring the social order learned in their youth to the new territory. Strangers are repelled and acquaintances are recognized and cared for. By the next breeding season the male is sexually mature, and he mates with the females of his coterie. The social behavior and organization communicated to him as a pup he now passes on to the next generation.

The Social Life of Baboons

by S. L. Washburn and Irven DeVore
June 1961

*A study of "troops" of baboons in their natural
environment in East Africa has revealed patterns of
interdependence that may shed light on the evolution
of the human species*

The behavior of monkeys and apes has always held great fascination for men. In recent years plain curiosity about their behavior has been reinforced by the desire to understand human behavior. Anthropologists have come to understand that the evolution of man's behavior, particularly his social behavior, has played an integral role in his biological evolution. In the attempt to reconstruct the life of man as it was shaped through the ages, many studies of primate behavior are now under way in the laboratory and in the field. As the contrasts and similarities between the behavior of primates and man—especially preagricultural, primitive man—become clearer, they should give useful insights into the kind of social behavior that characterized the ancestors of man a million years ago.

With these objectives in mind we decided to undertake a study of the baboon. We chose this animal because it is a ground-living primate and as such is confronted with the same kind of problem that faced our ancestors when they left the trees. Our observations of some 30 troops of baboons, ranging in average membership from 40 to 80 individuals, in their natural setting in Africa show that the social behavior of the baboon is one of the species' principal adaptations for survival. Most of a baboon's life is spent within a few feet of other baboons. The troop affords protection from predators and an intimate group knowledge of the territory it occupies. Viewed from the inside, the troop is composed not of neutral creatures but of strongly emotional, highly motivated members. Our data offer little support for the theory that sexuality provides the primary bond of the primate troop. It is the intensely social nature of the baboon, expressed in a diversity of inter-

individual relationships, that keeps the troop together. This conclusion calls for further observation and experimental investigation of the different social bonds. It is clear, however, that these bonds are essential to compact group living and that for a baboon life in the

troop is the only way of life that is feasible.

Many game reserves in Africa support baboon populations but not all were suited to our purpose. We had to be able to locate and recognize particular troops and their individual members

GROOMING to remove dirt and parasites from the hair is a major social activity among baboons. Here one adult female grooms another while the second suckles a year-old infant.

and to follow them in their peregrinations day after day. In some reserves the brush is so thick that such systematic observation is impossible. A small park near Nairobi, in Kenya, offered most of the conditions we needed. Here 12 troops of baboons, consisting of more than 450 members, ranged the open savanna. The animals were quite tame; they clambered onto our car and even allowed us to walk beside them. In only 10 months of study, one of us (DeVore) was able to recognize most of the members of four troops and to become moderately familiar with many more. The Nairobi park, however, is small and so close to the city that the pattern of baboon life is somewhat altered. To carry on our work in an area less disturbed by humans and large enough to contain elephants, rhinoceroses, buffaloes and other ungulates as well as larger and less tame troops of baboons, we went to the Amboseli game reserve and spent two months camped at the foot of Mount Kilimanjaro. In the small part of Am-

boseli that we studied intensively there were 15 troops with a total of 1,200 members, the troops ranging in size from 13 to 185 members. The fact that the average size of the troops in Amboseli (80) is twice that of the troops in Nairobi shows the need to study the animals in several localities before generalizing.

A baboon troop may range an area of three to six square miles but it utilizes only parts of its range intensively. When water and food are widely distributed, troops rarely come within sight of each other. The ranges of neighboring troops overlap nonetheless, often extensively. This could be seen best in Amboseli at the end of the dry season. Water was concentrated in certain areas, and several troops often came to the same water hole, both to drink and to eat the lush vegetation near the water. We spent many days near these water holes, watching the baboons and the numerous other animals that came there. On one occasion we counted more

than 400 baboons around a single water hole at one time. To the casual observer they would have appeared to be one troop, but actually three large troops were feeding side by side. The troops came and went without mixing, even though members of different troops sat or foraged within a few feet of each other. Once we saw a juvenile baboon cross over to the next troop, play briefly and return to his own troop. But such behavior is rare, even in troops that come together at the same water hole day after day. At the water hole we saw no fighting between troops, but small troops slowly gave way before large ones. Troops that did not see each other frequently showed great interest in each other.

When one first sees a troop of baboons, it appears to have little order, but this is a superficial impression. The basic structure of the troop is most apparent when a large troop moves away from the safety of trees and out onto open plains. As the troop moves the less dominant

MARCHING baboon troop has a definite structure, with females and their young protected by dominant males in the center of the formation. This group in the Amboseli reserve in Kenya includes a female (*left*), followed by two males and a female with juvenile.

adult males and perhaps a large juvenile or two occupy the van. Females and more of the older juveniles follow, and in the center of the troop are the females with infants, the young juveniles and the most dominant males. The back of the troop is a mirror image of its front, with less dominant males at the rear. Thus, without any fixed or formal order, the arrangement of the troop is such that the females and young are protected at the center. No matter from what direction a predator approaches the troop, it must first encounter the adult males.

When a predator is sighted, the adult males play an even more active role in defense of the troop. One day we saw two dogs run barking at a troop. The females and juveniles hurried, but the males continued to walk slowly. In a moment an irregular group of some 20 adult males was interposed between the dogs and the rest of the troop. When a male turned on the dogs, they ran off. We saw baboons close to hyenas, cheetahs and jackals, and usually the baboons seemed unconcerned—the other animals kept their distance. Lions were the only animals we saw putting a troop of baboons to flight. Twice we saw lions near baboons, whereupon the baboons climbed trees. From the safety of the trees the baboons barked and threatened the lions, but they offered no resistance to them on the ground.

With nonpredators the baboons' relations are largely neutral. It is common to see baboons walking among topi, eland, sable and roan antelopes, gazelles, zebras, hartebeests, gnus, giraffes and buffaloes, depending on which ungulates are common locally. When elephants or rhinoceroses walk through an area where the baboons are feeding, the baboons move out of the way at the last moment. We have seen wart hogs chasing each other, and a running rhinoceros go right through a troop, with the baboons merely stepping out of the way. We have seen male impalas fighting while baboons fed beside them. Once we saw a baboon chase a giraffe, but it seemed to be more in play than aggression.

Only rarely did we see baboons engage in hostilities against other species. On one occasion, however, we saw a baboon kill a small vervet monkey and eat it. The vervets frequented the same water holes as the baboons and usually they moved near them or even among them without incident. But one troop of baboons we observed at Victoria Falls pursued vervets on sight and attempted, without success, to keep

BABOON EATS A POTATO tossed to him by a member of the authors' party. Baboons are primarily herbivores but occasionally they will eat birds' eggs and even other animals.

INFANT BABOON rides on its mother's back through a park outside Nairobi. A newborn infant first travels by clinging to its mother's chest, but soon learns to ride pickaback.

BABOONS AND THREE OTHER SPECIES gather near a water
hole (*out of picture to right*). Water holes and the relatively lush

vegetation that surrounds them are common meeting places for a
wide variety of herbivores. In this scene of the open savanna of

them out of certain fruit trees. The
vervets easily escaped in the small
branches of the trees.

The baboons' food is almost entirely
vegetable, although they do eat meat on
rare occasions. We saw dominant males
kill and eat two newborn Thomson's
gazelles. Baboons are said to be fond of
fledglings and birds' eggs and have even
been reported digging up crocodile
eggs. They also eat insects. But their diet
consists principally of grass, fruit, buds
and plant shoots of many kinds; in the
Nairobi area alone they consume more
than 50 species of plant.

For baboons, as for many herbivores,
association with other species on the

range often provides mutual protection.
In open country their closest relations
are with impalas, while in forest areas
the bushbucks play a similar role. The
ungulates have a keen sense of smell,
and baboons have keen eyesight. Baboons are visually alert, constantly looking in all directions as they feed. If they
see predators, they utter warning barks
that alert not only the other baboons but
also any other animals that may be in
the vicinity. Similarly, a warning bark
by a bushbuck or an impala will put a
baboon troop to flight. A mixed herd of
impalas and baboons is almost impossible to take by surprise.

Impalas are a favorite prey of

cheetahs. Yet once we saw impalas, grazing in the company of baboons, make
no effort to escape from a trio of approaching cheetahs. The impalas just
watched as an adult male baboon
stepped toward the cheetahs, uttered a
cry of defiance and sent them trotting
away.

The interdependence of the different
species is plainly evident at a water
hole, particularly where the bush is thick
and visibility poor. If giraffes are drinking, zebras will run to the water. But the
first animals to arrive at the water hole
approach with extreme caution. In the
Wankie reserve, where we also observed
baboons, there are large water holes

the Amboseli reserve there are baboons in the foreground and middle distance. An impala moves across the foreground just left of center. A number of zebras are present; groups of gnus graze together at right center and move off toward the water hole (*right*).

surrounded by wide areas of open sand between the water and the bushes. The baboons approached the water with great care, often resting and playing for some time in the bushes before making a hurried trip for a drink. Clearly, many animals know each other's behavior and alarm signals.

A baboon troop finds its ultimate safety, however, in the trees. It is no exaggeration to say that trees limit the distribution of baboons as much as the availability of food and water. We observed an area by a marsh in Amboseli where there was water and plenty of food. But there were lions and no trees

and so there were no baboons. Only a quarter of a mile away, where lions were seen even more frequently, there were trees. Here baboons were numerous; three large troops frequented the area.

At night, when the carnivores and snakes are most active, baboons sleep high up in big trees. This is one of the baboon's primary behavioral adaptations. Diurnal living, together with an arboreal refuge at night, is an extremely effective way for them to avoid danger. The callused areas on a baboon's haunches allow it to sleep sitting up, even on small branches; a large troop can thus find sleeping places in a few trees. It is known that Colobus monkeys

have a cycle of sleeping and waking throughout the night; baboons probably have a similar pattern. In any case, baboons are terrified of the dark. They arrive at the trees before night falls and stay in the branches until it is fully light. Fear of the dark, fear of falling and fear of snakes seem to be basic parts of the primate heritage.

Whether by day or night, individual baboons do not wander away from the troop, even for a few hours. The importance of the troop in ensuring the survival of its members is dramatized by the fate of those that are badly injured or too sick to keep up with their fellows. Each day the troop travels on a circuit

LIONESS LEAPS AT A THORN TREE into which a group of baboons has fled for safety. Lions appear to be among the few animals that successfully prey on baboons. The car in the background drove up as the authors' party was observing the scene.

of two to four miles; it moves from the sleeping trees to a feeding area, feeds, rests and moves again. The pace is not rapid, but the troop does not wait for sick or injured members. A baby baboon rides its mother, but all other members of the troop must keep up on their own. Once an animal is separated from the troop the chances of death are high. Sickness and injuries severe enough to be easily seen are frequent. For example, we saw a baboon with a broken forearm. The hand swung uselessly, and blood showed that the injury was recent. This baboon was gone the next morning and was not seen again. A sickness was widespread in the Amboseli troops, and we saw individuals dragging themselves along, making tremendous efforts to stay with the troop but falling behind. Some of these may have rejoined their troops; we are sure that at least five did not. One sick little juvenile lagged for four days and then apparently recovered. In the somewhat less natural setting of Nairobi park we saw some baboons that

had lost a leg. So even severe injury does not mean inevitable death. Nonetheless, it must greatly decrease the chance of survival.

Thus, viewed from the outside, the troop is seen to be an effective way of life and one that is essential to the survival of its individual members. What do the internal events of troop life reveal about the drives and motivations that cause individual baboons to "seek safety in numbers"? One of the best ways to approach an understanding of the behavior patterns within the troop is to watch the baboons when they are resting and feeding quietly.

Most of the troop will be gathered in small groups, grooming each other's fur or simply sitting. A typical group will contain two females with their young offspring, or an adult male with one or more females and juveniles grooming him. Many of these groups tend to persist, with the same animals that have been grooming each other walking together when the troop moves. The nucleus of

such a "grooming cluster" is most often a dominant male or a mother with a very young infant. The most powerful males are highly attractive to the other troop members and are actively sought by them. In marked contrast, the males in many ungulate species, such as impalas, must constantly herd the members of their group together. But baboon males have no need to force the other troop members to stay with them. On the contrary, their presence alone ensures that the troop will stay with them at all times.

Young infants are equally important in the formation of grooming clusters. The newborn infant is the center of social attraction. The most dominant adult males sit by the mother and walk close beside her. When the troop is resting, adult females and juveniles come to the mother, groom her and attempt to groom the infant. Other members of the troop are drawn toward the center thus formed, both by the presence of the pro-

BABOONS AND IMPALAS cluster together around a water hole. The two species form a mutual alarm system. The baboons have keen eyesight and the impalas a good sense of smell. Between them they quickly sense the presence of predators and take flight.

tective adult males and by their intense interest in the young infants.

In addition, many baboons, especially adult females, form preference pairs, and juvenile baboons come together in play groups that persist for several years. The general desire to stay in the troop is strengthened by these "friendships," which express themselves in the daily pattern of troop activity.

Our field observations, which so strongly suggest a high social motivation, are backed up by controlled experiment in the laboratory. Robert A. Butler of Walter Reed Army Hospital has shown that an isolated monkey will work hard when the only reward for his labor is the sight of another monkey [see "Curiosity in Monkeys," by Robert A. Butler; SCIENTIFIC AMERICAN Offprint 426]. In the troop this social drive is expressed in strong individual prefer-

ences, by "friendship," by interest in the infant members of the troop and by the attraction of the dominant males. Field studies show the adaptive value of these social ties. Solitary animals are far more likely to be killed, and over the generations natural selection must have favored all those factors which make learning to be sociable easy.

The learning that brings the individual baboon into full identity and participation in the baboon social system begins with the mother-child relationship. The newborn baboon rides by clinging to the hair on its mother's chest. The mother may scoop the infant on with her hand, but the infant must cling to its mother, even when she runs, from the day it is born. There is no time for this behavior to be learned. Harry F. Harlow of the University of Wisconsin has shown that an infant monkey will auto-

matically cling to an object and much prefers objects with texture more like that of a real mother [see the article "Love in Infant Monkeys," by Harry F. Harlow, beginning on page 250]. Experimental studies demonstrate this clinging reflex; field observations show why it is so important.

In the beginning the baboon mother and infant are in contact 24 hours a day. The attractiveness of the young infant, moreover, assures that he and his mother will always be surrounded by attentive troop members. Experiments show that an isolated infant brought up in a laboratory does not develop normal social patterns. Beyond the first reflexive clinging, the development of social behavior requires learning. Behavior characteristic of the species depends therefore both on the baboon's biology and on the social situations that are present in the troop.

As the infant matures it learns to ride on its mother's back, first clinging and then sitting upright. It begins to eat solid foods and to leave the mother for longer and longer periods to play with other infants. Eventually it plays with the other juveniles many hours a day, and its orientation shifts from the mother to this play group. It is in these play groups that the skills and behavior patterns of adult life are learned and practiced. Adult gestures, such as mounting, are frequent, but most play is a mixture of chasing, tail-pulling and mock fighting. If a juvenile is hurt and cries out, adults come running and stop the play. The presence of an adult male prevents small juveniles from being hurt. In the protected atmosphere of the play group the social bonds of the infant are widely extended.

Grooming, a significant biological function in itself, helps greatly to establish social bonds. The mother begins grooming her infant the day it is born, and the infant will be occupied with grooming for several hours a day for the rest of its life. All the older baboons do a certain amount of grooming, but it is the adult females who do most. They groom the infants, juveniles, adult males and other females. The baboons go to each other and "present" themselves for grooming. The grooming animal picks through the hair, parting it with its hands, removing dirt and parasites, usually by nibbling. Grooming is most often reciprocal, with one animal doing it for a while and then presenting itself for grooming. The animal being groomed relaxes, closes its eyes and gives every indication of complete pleasure. In addition to being pleasurable, grooming serves the important function of keeping the fur clean. Ticks are common in this area and can be seen on many animals such as dogs and lions; a baboon's skin, however, is free of them. Seen in this light, the enormous amount of time baboons spend in grooming each other is understandable. Grooming is pleasurable to the individual, it is the most important expression of close social bonds and it is biologically adaptive.

The adults in a troop are arranged in a dominance hierarchy, explicitly revealed in their relations with other members of the troop. The most dominant males will be more frequently groomed and they occupy feeding and resting positions of their choice. When a dominant animal approaches a subordinate one, the lesser animal moves out of the way. The observer can determine the order of dominance simply by watching the reactions of the baboons as they move past each other. In the tamer troops these observations can be tested by feeding. If food is tossed between two baboons, the more dominant one will take it, whereas the other may not even look at it directly.

The status of a baboon male in the dominance hierarchy depends not only on his physical condition and fighting ability but also on his relationships with other males. Some adult males in every large troop stay together much of the time, and if one of them is threatened, the others are likely to back him up. A group of such males outranks any individual, even though another male outside the group might be able to defeat any member of it separately. The hierarchy has considerable stability and this is due in large part to its dependence on clusters of males rather than the fighting ability of individuals. In troops where the rank order is clearly defined, fighting is rare. We observed frequent bickering or severe fighting in only about 15 per cent of the troops. The usual effect of the hierarchy, once relations among the males are settled, is to decrease disruptions in the troop. The dominant animals, the males in particular, will not let others fight. When bickering breaks out, they usually run to the scene and stop it. Dominant males thus protect the weaker animals against harm from inside as well as outside. Females and juveniles come to the males to groom them or just to sit beside them. So although dominance depends ultimately on force, it leads to peace, order and popularity.

	ECOLOGY			ECONOMIC SYSTEM	
	GROUP SIZE, DENSITY AND RANGE	HOME BASE	POPULATION STRUCTURE	FOOD HABITS	ECONOMIC DEPENDENCE
	GROUPS OF 50–60 COMMON BUT VARY WIDELY. ONE INDIVIDUAL PER 5–10 SQUARE MILES. RANGE 200–600 SQUARE MILES. TERRITORIAL RIGHTS; DEFEND BOUNDARIES AGAINST STRANGERS.	OCCUPY IMPROVED SITES FOR VARIABLE TIMES WHERE SICK ARE CARED FOR AND STORES KEPT.	TRIBAL ORGANIZATION OF LOCAL, EXOGAMOUS GROUPS.	OMNIVOROUS. FOOD SHARING. MEN SPECIALIZE IN HUNTING, WOMEN AND CHILDREN IN GATHERING.	INFANTS ARE DEPENDENT ON ADULTS FOR MANY YEARS. MATURITY OF MALE DELAYED BIOLOGICALLY AND CULTURALLY. HUNTING, STORAGE AND SHARING OF FOOD.
	10–200 IN GROUP. 10 INDIVIDUALS PER SQUARE MILE. RANGE 3–6 SQUARE MILES; NO TERRITORIAL DEFENSE.	NONE; SICK AND INJURED MUST KEEP UP WITH TROOP.	SMALL, INBREEDING GROUPS.	ALMOST ENTIRELY VEGETARIAN. NO FOOD SHARING, NO DIVISION OF LABOR.	INFANT ECONOMICALLY INDEPENDENT AFTER WEANING. FULL MATURITY BIOLOGICALLY DELAYED. NO HUNTING, STORAGE OR SHARING OF FOOD.

APES AND MEN are contrasted in this chart, which indicates that although apes often seem remarkably "human," there are fundamental differences in behavior. Baboon characteristics, which may be taken as representative of ape and monkey behavior in

Much has been written about the importance of sex in uniting the troop, it has been said, for example, that "the powerful social magnet of sex was the major impetus to subhuman primate sociability" [see "The Origin of Society," by Marshall D. Sahlins; SCIENTIFIC AMERICAN Offprint 602]. Our observations lead us to assign to sexuality a much lesser, and even at times a contrary, role. The sexual behavior of baboons depends on the biological cycle of the female. She is receptive for approximately one week out of every month, when she is in estrus. When first receptive, she leaves her infant and her friendship group and goes to the males, mating first with the subordinate males and older juveniles. Later in the period of receptivity she goes to the dominant males and "presents." If a male is not interested, the female is likely to groom him and then present again. Near the end of estrus the dominant males become very interested, and the female and a male form a consort pair. They may stay together for as little as an hour or for as long as several days. Estrus disrupts all other social relationships, and consort pairs usually move to the edge of the troop. It is at this time that fighting may take place, if the dominance order is not clearly established among the males. Normally there is no fighting over females, and a male, no matter how dominant, does not monopolize a female for long. No male is ever associated with more than one estrus female; there is nothing resembling a family or a harem among baboons.

Much the same seems to be true of other species of monkey. Sexual behavior appears to contribute little to the cohesion of the troop. Some monkeys have breeding seasons, with all mating taking place within less than half the year. But even in these species the troop continues its normal existence during the months when there is no mating. It must be remembered that among baboons a female is not sexually receptive for most of her life. She is juvenile, pregnant or lactating; estrus is a rare event in her life. Yet she does not leave the troop even for a few minutes. In baboon troops, particularly small ones, many months may pass when no female member comes into estrus; yet no animals leave the troop, and the highly structured relationships within it continue without disorganization.

The sociableness of baboons is expressed in a wide variety of behavior patterns that reinforce each other and give the troop cohesion. As the infant matures the nature of the social bonds changes continually, but the bonds are always strong. The ties between mother and infant, between a juvenile and its peers in a play group, and between a mother and an adult male are quite different from one another. Similarly, the bond between two females in a friendship group, between the male and female in a consort pair or among the members of a cluster of males in the dominance hierarchy is based on diverse biological and behavioral factors, which offer a rich field for experimental investigation.

In addition, the troop shares a considerable social tradition. Each troop has its own range and a secure familiarity with the food and water sources, escape routes, safe refuges and sleeping places inside it. The counterpart of the intensely social life within the troop is the coordination of the activities of all the troop's members throughout their lives. Seen against the background of evolution, it is clear that in the long run only the social baboons have survived.

When comparing the social behavior of baboons with that of man, there is little to be gained from laboring the obvious differences between modern civilization and the society of baboons. The comparison must be drawn against the fundamental social behavior patterns that lie behind the vast variety of human ways of life. For this purpose we have charted the salient features of baboon life in a native habitat alongside those of human life in preagricultural society [see chart below]. Cursory inspection shows that the differences are more numerous and significant than are the

SOCIAL SYSTEM					COMMUNICATION
ORGANIZATION	SOCIAL CONTROL	SEXUAL BEHAVIOR	MOTHER-CHILD RELATIONSHIP	PLAY	
BANDS ARE DEPENDENT ON AND AFFILIATED WITH ONE ANOTHER IN A SEMIOPEN SYSTEM. SUBGROUPS BASED ON KINSHIP.	BASED ON CUSTOM.	FEMALE CONTINUOUSLY RECEPTIVE. FAMILY BASED ON PROLONGED MALE-FEMALE RELATIONSHIP AND INCEST TABOOS.	PROLONGED; INFANT HELPLESS AND ENTIRELY DEPENDENT ON ADULTS.	INTERPERSONAL BUT ALSO CONSIDERABLE USE OF INANIMATE OBJECTS.	LINGUISTIC COMMUNITY. LANGUAGE CRUCIAL IN THE EVOLUTION OF RELIGION, ART, TECHNOLOGY AND THE CO-OPERATION OF MANY INDIVIDUALS.
TROOP SELF-SUFFICIENT, CLOSED TO OUTSIDERS. TEMPORARY SUBGROUPS ARE FORMED BASED ON AGE AND INDIVIDUAL PREFERENCES.	BASED ON PHYSICAL DOMINANCE.	FEMALE ESTRUS. MULTIPLE MATES. NO PROLONGED MALE-FEMALE RELATIONSHIP.	INTENSE BUT BRIEF; INFANT WELL DEVELOPED AND IN PARTIAL CONTROL.	MAINLY INTERPERSONAL AND EXPLORATORY.	SPECIES-SPECIFIC, LARGELY GESTURAL AND CONCERNED WITH IMMEDIATE SITUATIONS.

general, are based on laboratory and field studies; human characteristics are what is known of preagricultural Homo sapiens. The chart suggests that there was a considerable gap between primate behavior and the behavior of the most primitive men known.

BABOONS AND ELEPHANTS have a relationship that is neutral rather than co-operative, as in the case of baboons and impalas. If an elephant or another large herbivore such as a rhinoceros moves through a troop, the baboons merely step out of the way.

similarities.

The size of the local group is the only category in which there is not a major contrast. The degree to which these contrasts are helpful in understanding the evolution of human behavior depends, of course, on the degree to which baboon behavior is characteristic of monkeys and apes in general and therefore probably characteristic of the apes that evolved into men. Different kinds of monkey do behave differently, and many more field studies will have to be made before the precise degree of difference can be understood.

For example, many arboreal monkeys have a much smaller geographical range than baboons do. In fact, there are important differences between the size and type of range for many monkey species. But there is no suggestion that a troop of any species of monkey or ape occupies the hundreds of square miles ordinarily occupied by preagricultural human societies. Some kinds of monkey may resent intruders in their range more than baboons do, but there is no evidence that any species fights for complete control of a territory. Baboons are certainly less vocal than some other monkeys, but no nonhuman primate has even the most rudimentary language. We believe that the fundamental contrasts in our chart would hold for the vast majority of monkeys and apes as compared with the ancestors of man. Further study of primate behavior will sharpen these contrasts and define more clearly the gap that had to be traversed from ape to human behavior. But already we can see that man is as unique in his sharing, co-operation and play patterns as he is in his locomotion, brain and language.

The basis for most of these differences may lie in hunting. Certainly the hunting of large animals must have involved co-operation among the hunters and sharing of the food within the tribe. Similarly, hunting requires an enormous extension of range and the protection of a hunting territory. If this speculation proves to be correct, much of the evolution of human behavior can be reconstructed, because the men of 500,000 years ago were skilled hunters. In locations such as Choukoutien in China and Olduvai Gorge in Africa there is evidence of both the hunters and their campsites [see "Olduvai Gorge," by L. S. B. Leakey; SCIENTIFIC AMERICAN, January, 1954]. We are confident that the study of the living primates, together with the archaeological record, will eventually make possible a much richer understanding of the evolution of human behavior.

BIBLIOGRAPHIES

I ENVIRONMENT TO ACTION

1. Predatory Fungi

THE FRIENDLY FUNGI: A NEW APPROACH TO THE EEL-WORM PROBLEM. C. L. Duddington. Faber and Faber, 1957.

PREDACEOUS FUNGI. Charles Drechsler in *Biological Reviews*, Vol. 16., No. 4, pages 265–290; October, 1941.

THE PREDACIOUS FUNGI AND THEIR PLACE IN MICROBIAL ECOLOGY. C. L. Duddington in *Microbial Ecology*, pages 218–237; 1957.

THE PREDACIOUS FUNGI: ZOOPAGALES AND MONILIALES. C. L. Duddington in *Biological Reviews*, Vol. 31, No. 2, pages 152–193; May, 1956.

SOME HYPHOMYCETES THAT PREY ON FREE-LIVING TERRICOLOUS NEMATODES. Charles Drechsler in *Mycologia*, Vol. 29, No. 4, pages 447–552; July–August, 1937.

STIMULATED ACTIVITY OF NATURAL ENEMIES OF NEMATODES. M. B. Linford in *Science*, Vol. 85, No. 2, 196, pages 123–124; January 29, 1937.

2. Life and Light

PHOTOSYNTHESIS AND RELATED PROCESSES. Eugene I. Rabinowitch. Interscience Publishers, Inc., 1945–1956.

RADIATION BIOLOGY. VOLUME III: VISIBLE AND NEAR-VISIBLE LIGHT. Edited by Alexander Hollaender. McGraw-Hill Book Company, Inc., 1956.

VISION AND THE EYE. M. H. Pirenne. The Pilot Press Ltd., 1948.

THE VISUAL PIGMENTS. H. J. A. Dartnall. Methuen & Co. Ltd., 1957.

3. Small Systems of Nerve Cells

HETEROSYNAPTIC FACILITATION IN NEURONS OF THE ABDOMINAL GANGLION OF APLYSIA DEPILANS. E. R. Kandel and L. Tauc in *Journal of Physiology* (London), Vol. 181, No. 1, pages 1–27; November, 1965.

ON THE FUNCTIONAL ANATOMY OF NEURONAL UNITS IN THE ABDOMINAL CORD OF THE CRAYFISH, PROCAMBARUS CLARKII (GIRARD). C. A. G. Wiersma and G. M. Hughes in *The Journal of Comparative Neurology*, Vol. 116, No. 2, pages 209–228; April, 1961.

RELEASE OF COORDINATED BEHAVIOR IN CRAYFISH BY SINGLE CENTRAL NEURONS. Donald Kennedy, W. H. Evoy, and J. T. Hanawalt in *Science*, Vol. 154, No. 3751, pages 917–919; November 18, 1966.

TYPES OF INFORMATION STORED IN SINGLE NEURONS. Felix Strumwasser in *Invertebrate Nervous Systems: Their Significance for Mammalian Neurophysiology*, edited by Cornelius A. G. Wiersma. The University of Chicago Press, 1967.

4. Moths and Ultrasound

THE DETECTION AND EVASION OF BATS BY MOTHS. Kenneth D. Roeder and Asher E. Treat in *American Scientist*, Vol. 49, No. 2, pages 135–148; June, 1961.

MOTH SOUNDS AND THE INSECT-CATCHING BEHAVIOR OF BATS. Dorothy C. Dunning and Kenneth D. Roeder in *Science*, Vol. 147, No. 3654, pages 173–174; January 8, 1965.

NERVE CELLS AND INSECT BEHAVIOR. Kenneth D. Roeder. Harvard University Press, 1963.

5. The Sex-Attractant Receptors of Moths

CHEMICALS CONTROLLING INSECT BEHAVIOR: SYMPOSIUM OF THE AMERICAN CHEMICAL SOCIETY. Edited by M. Beroza. Academic Press, 1970.

OLFACTORY RECEPTORS FOR THE SEXUAL ATTRACTANT (BOMBYKOL) OF THE SILK MOTH. D. Schneider in The Neurosciences: Second Study Program, edited by F. O. Schmitt. Rockefeller University Press, 1970.

INSECT OLFACTION. Karl-Ernst Kaissling in Handbook of Sensory Physiology: Vol. IV/1, edited by L. M. Beidler. Springer-Verlag, 1971.

INSECT SEX PHEROMONES. M. Jacobson. Academic Press, 1972.

GYPSY MOTH CONTROL WITH THE SEX ATTRACTANT PHEROMONE. M. Beroza and E. F. Knipling in Science, Vol. 177, No. 4043, pages 19–27; July 7, 1972.

6. Eye Movements and Visual Perception

PATTERN RECOGNITION. Edited by Leonard M. Uhr. John Wiley & Sons, Inc., 1966.

CONTEMPORARY THEORY AND RESEARCH IN VISUAL PERCEPTION. Edited by Ralph Norman Haber. Holt, Rinehart & Winston, Inc., 1968.

A THEORY OF VISUAL PATTERN PERCEPTION. David Noton in IEEE Transactions on Systems Science and Cybernetics, Vol. SSC-6, No. 4, pages 349–357; October, 1970.

SCANPATHS IN EYE MOVEMENTS DURING PATTERN PERCEPTION. David Noton and Lawrence Stark in Science, Vol. 171, No. 3968, pages 308–311; January 22, 1971.

7. The Flight-Control System of the Locust

THE CENTRAL NERVOUS CONTROL OF FLIGHT IN A LOCUST. Donald M. Wilson in The Journal of Experimental Biology, Vol. 38, No. 2, pages 471–490; June, 1961.

EXPLORATION OF NEURONAL MECHANISMS UNDERLYING BEHAVIOR IN INSECTS. Graham Hoyle in Neural Theory and Modeling: Proceedings of the 1962 Ojai Symposium, edited by Richard F. Reiss. Stanford University Press, 1964.

8. Brains and Cocoons

COCOON CONSTRUCTION BY THE CECROPIA SILKWORM. William G. Van der Kloot and C. M. Williams in Behaviour, Vol. 5, Parts 2 and 3, pages 141–174, and Vol. 6, Part 4, pages 233–255; 1953 and 1954.

9. The Neurobiology of Cricket Song

THE ROLE OF THE CENTRAL NERVOUS SYSTEM IN ORTHOPTERA DURING THE COORDINATION AND CONTROL OF STRIDULATION. Franz Huber in Acoustic Behaviour of Animals, edited by R. G. Busnel. American Elsevier Publishing Company, 1964.

INTRACELLULAR ACTIVITY IN CRICKET NEUROSIS DURING THE GENERATION OF SONG PATTERNS. David Bentley in Zeitschrift für vergleichende Physiologie, Vol. 62, pages 267–283; 1969.

NEUROMUSKULÄRE AKTIVITÄT BEI VERSCHIEDENEN VERHALTENSWEISEN VON DREI GRILLENARTEN. Wolfram Kutsch in Zeitschrift für vergleichende Physiologie, Vol. 63, pages 335–378; 1969.

POSTEMBRYONIC DEVELOPMENT OF ADULT MOTOR PATTERNS IN CRICKETS: A NEURAL ANALYSIS. David R. Bentley and Ronald R. Hoy in Science, Vol. 170, No. 3965, pages 1409–1411; December 25, 1970.

GENETIC CONTROL OF AN INSECT NEURONAL NETWORK. David R. Bentley in Science, Vol. 174, No. 4014 pages 1139–1141; December 10, 1971.

10. How We Control the Contraction of Our Muscles

TREATISE ON PHYSIOLOGICAL OPTICS: VOL. III. Hermann L. F. Helmholtz. English translation by James P. C. Southall. Dover Publications, Inc., 1962.

POSITION SENSE AND SENSE OF EFFORT. P. A. Merton in Homeostasis and Feedback Mechanisms. Symposia of the Society for Experimental Biology, Vol. 18, pages 387–400; 1964.

THE INNERVATION OF MAMMALIAN SKELETAL MUSCLE. D. Barker in Ciba Foundation Symposium: Myotatic, Kinesthetic and Vestibular Mechanisms, edited by A. V. S. de Reuck and Julie Knight. Little, Brown and Company, 1967.

SERVO ACTION AND STRETCH REFLEX IN HUMAN MUSCLE AND ITS APPARENT DEPENDENCE ON PERIPHERAL SENSATION. C. D. Marsden, P. A. Merton, and H. B. Morton in The Journal of Physiology, Vol. 216, pages 21–22P; July, 1971.

PROJECTION FROM LOW THRESHOLD MUSCLE AFFERENTS OF HAND AND FOREARM TO AREA 3a OF BABOON'S CORTEX. C. G. Phillips, T. P. S. Powell, and M. Wiesendanger in The Journal of Physiology, Vol. 217, pages 419–446; September, 1971.

11. Annual Biological Clocks

THE EFFECT OF TEMPERATURE AND PHOTOPERIOD ON THE YEARLY HIBERNATING BEHAVIOR OF CAPTIVE GOLDEN-MANTLED GROUND SQUIRRELS (Citellus Lateralis Tescorum). Eric T. Pengelley and Kenneth C. Fisher in Canadian Journal of Zoology, Vol. 41, No. 6, pages 1103–1120; September, 1963.

CIRCANNUALE PERIODIK ALS GRUNDLAGE DES JAHRES-ZEITLICHEN FUNKTIONS-WANDELS BEI ZUGVÖGELN. Eberhard Gwinner in *Journal für Ornithologie*, Vol. 109, No. 1, pages 70–95; January, 1968.

CIRCANNUAL RHYTHMICITY—EVIDENCE AND THEORY. E. T. Pengelley and Sally J. Asmundson in *Life Sciences and Space Research: Vol. VIII*. North-Holland Publishing Company, 1970.

II THE ADAPTIVENSSS OF BEHAVIOR

12. The Evolution of Behavior

THE STUDY OF INSTINCT. N. Tinbergen. Clarendon Press, 1952.

13. The Evolution of Behavior in Gulls

BEHAVIOR AND EVOLUTION. Edited by Anne Roe and George Gaylord Simpson. Yale University Press, 1958.

THE HERRING GULL'S WORLD. Nikolaas Tinbergen. William Collins Sons & Co., 1953.

THE ORIGIN AND EVOLUTION OF COURTSHIP AND THREAT DISPLAY. N. Tinbergen by Julian Huxley, A. C. Hardy, and E. B. Ford, pages 233–250. George Allen & Unwin Ltd., 1954.

14. The Behavior of Lovebirds

THE COMPARATIVE ETHOLOGY OF THE AFRICAN PARROT GENUS AGAPORNIS. William C. Dilger in *Zeitschrift für Tierpsychologie*, Vol. 17, No. 6, pages 649–685; 1960.

THE EVOLUTION OF BEHAVIOR IN GULLS. N. Tinbergen in *Scientific American*, Vol. 203, No. 6, pages 118–130; December, 1960.

SOME RECENT TRENDS IN ETHOLOGY. R. A. Hinde in *Psychology: A Study of a Science*, Vol. 2, edited by Sigmund Koch, pages 561–610. McGraw-Hill Book Company, Inc., 1959.

THE STUDY OF INSTINCT. N. Tinbergen. Oxford University Press, 1951.

15. Genetic Dissection of Behavior

BIOLOGY OF DROSOPHILA. Edited by M. Demerec. Hafner Publishing Co., Inc., 1965.

BEHAVIORAL MUTANTS OF DROSOPHILA ISOLATED BY COUNTERCURRENT DISTRIBUTION. Seymour Benzer in *Proceedings of the National Academy of Sciences of the United States of America*, Vol. 58, No. 3, pages 1112–1119; September, 1967.

CLOCK MUTANTS OF DROSOPHILA MELANOGASTER. Ronald J. Konopka and Seymour Benzer in *Proceedings of the National Academy of Sciences of the United States of America*, Vol. 68, No. 9, pages 2112–2116; September, 1971.

MAPPING OF BEHAVIOR IN DROSOPHILA MOSAICS. Yoshiki Hotta and Seymour Benzer in *Nature*, Vol. 240, pages 527–535; December 29, 1972.

16. Habitat Selection

THE DISTRIBUTION AND ABUNDANCE OF ANIMALS. H. G. Andrewartha and L. C. Birch. The University of Chicago Press, 1954.

THE ROLE OF EXPERIENCE IN HABITAT SELECTION BY THE PRAIRIE DEER MOUSE, PEROMYSCUS MANICULATUS BAIRDI. Stanley C. Wecker in *Ecological Monographs*, Vol. 33, No. 4, pages 307–325; Autumn, 1963.

THE STRATEGY OF THE GENES. C. H. Waddington, The Macmillan Company, 1957.

17. Visual Isolation in Gulls

ANIMAL SPECIES AND EVOLUTION. Ernst Mayr. Harvard University Press, 1963.

THE EVOLUTION OF BEHAVIOR IN GULLS. N. Tinbergen in *Scientific American*, Vol. 203, No. 6, pages 118–130; December, 1960.

THE STUDY OF INSTINCT. Nikolaas Tinbergen. Oxford University Press, 1950.

18. Electric Location by Fishes

ECOLOGICAL STUDIES ON GYMNOTIDS. H. W. Lissmann in *Bioelectrogenesis: A Comparative Survey of its Mechanisms with Particular Emphasis on Electric Fishes*. American Elsevier Publishing Co., Inc., 1961.

ON THE FUNCTION AND EVOLUTION OF ELECTRIC ORGANS IN FISH. H. W. Lissmann in *Journal of Experimental Biology*, Vol. 35, No. 1, pages 156–191; March, 1958.

THE MECHANISM OF OBJECT LOCATION IN GYMNARCHUS NILOTICUS AND SIMILAR FISH. H. W. Lissmann and K. E. Machin in *Journal of Experimental Biology*, Vol. 35, No. 2, pages 451–486; June, 1958.

THE MODE OF OPERATION OF THE ELECTRIC RECEPTORS IN GYMNARCHUS NILOTICUS. K. E. Machin and H. W. Lissmann in *Journal of Experimental Biology*, Vol. 37, No. 4, pages 801–811; December, 1960.

19. The Homing Salmon

THE CHEMICAL SENSE. R. W. Moncrieff. John Wiley & Sons, Inc., 1946.

HOMING INSTINCT IN SALMON. Bradley T. Scheer in *The Quarterly Review of Biology*, Vol. 14, No. 4, pages 408–430; December, 1939.

THE MIGRATION AND THE CONSERVATION OF SALMON. Edited by Forest R. Moulton. American Association for the Advancement of Science, 1939.

SENSORY PHYSIOLOGY AND THE ORIENTATION OF ANIMALS. Donald R. Griffin in *American Scientist*, Vol. 41, No. 2, pages 209–244; April, 1953.

20. The Mystery of Pigeon Homing

ORIENTATION BY PIGEONS: IS THE SUN NECESSARY? William T. Keeton in *Science*, Vol. 165, pages 922–928; 1969.

MAGNETS INTERFERE WITH PIGEON HOMING. William T. Keeton in *Proceedings of the National Academy of Sciences of the United States of America*, Vol. 68, No. 1, pages 102–106; January, 1971.

HOMING IN PIGEONS WITH IMPAIRED VISION. K. Schmidt-Koenig and H. J. Schlichte in *Proceedings of the National Academy of Sciences of the United States of America*, Vol. 69, No. 9, pages 2446–2447; September, 1972.

RELEASE-SITE BIAS AS A POSSIBLE GUIDE TO THE "MAP" COMPONENT IN PIGEON HOMING. William T. Keeton in *Journal of Comparative Physiology*, Vol. 86, pages 1–16; 1973.

ORIENTATION OF HOMING PIGEONS ALTERED BY A CHANGE IN THE DIRECTION OF AN APPLIED MAGNETIC FIELD. Charles Walcott and Robert P. Green in *Science*, Vol. 184, No. 4133, pages 180–182; 1974.

THE ORIENTATIONAL AND NAVIGATIONAL BASIS OF HOMING IN BIRDS. William T. Keeton in *Advances in the Study of Behavior: Vol. 5*, edited by D. S. Lehrman, R. Hinde, and E. Shaw. Academic Press, 1974.

III THE ADAPTABILITY OF BEHAVIOR

21. The Reproductive Behavior of Ring Doves

CONTROL OF BEHAVIOR CYCLES IN REPRODUCTION. Daniel S. Lehrman in *Social Behavior and Organization among Vertebrates*, edited by William Etkin. The University of Chicago Press, 1964.

HORMONAL REGULATION OF PARENTAL BEHAVIOR IN BIRDS AND INFRAHUMAN MAMMALS. Daniel S. Lehrman in *Sex and Internal Secretions*, edited by William C. Young. Williams & Wilkins Company, 1961.

INTERACTION OF HORMONAL AND EXPERIENTIAL INFLUENCES ON DEVELOPMENT OF BEHAVIOR. Daniel S. Lehrman in *Roots of Behavior*, edited by E. L. Bliss. Harper & Row, Publishers, 1962.

22. Stress and Behavior

ADRENOCORTICAL ACTIVITY AND AVOIDANCE LEARNING AS A FUNCTION OF TIME AFTER AVOIDANCE TRAINING. Seymour Levine and F. Robert Brush in *Psysiology & Behavior*, Vol. 2, No. 4, pages 385–388; October, 1967.

HORMONES AND CONDITIONING. Seymour Levine in *Nebraska Symposium on Motivation: 1968*, edited by William J. Arnold. University of Nebraska Press, 1968.

EFFECTS OF PEPTIDE HORMONES ON BEHAVIOR. David de Wied in *Frontiers in Neuroendocrinology*, edited by William F. Ganong and Luciano Martini. Oxford University Press, 1969.

THE NEUROENDOCRINE CONTROL OF PERCEPTION. R. I. Henkin in *Perception and Its Disorders: Proceedings of the Association for Research in Nervous Mental Disease, 32*, edited by D. Hamburg. The Williams & Wilkins Co., 1970.

23. Learning in the Octopus

BRAIN AND BEHAVIOUR IN CEPHALOPODS. M. J. Wells. Stanford University Press, 1962.

THE FUNCTIONAL ORGANIZATION OF THE BRAIN OF THE CUTTLEFISH SEPIA OFFICINALIS. B. B. Boycott in *Proceedings of the Royal Society of London*, Series B, Vol. 153, No. 953, pages 503–534; February, 1961.

IN SEARCH OF THE ENGRAM. K. S. Lashley in *Physiological Mechanisms in Animal Behaviour*. Symposia of the Society for Experimental Biology, No. 4, 1950.

A MODEL OF THE BRAIN. J. Z. Young. Oxford University Press, 1964.

SOME ESSENTIALS OF NEURAL MEMORY SYSTEMS: PAIRED CENTRES THAT REGULATE AND ADDRESS THE SIGNALS OF THE RESULTS OF ACTION. J. Z. Young in *Nature*, Vol. 198, No. 4881, pages 626–632; May, 1963.

24. "Imprinting" in a Natural Laboratory

"Imprinting" in Animals. Eckhard H. Hess in *Scientific American*, Vol. 198, No. 3, pages 81–90; March, 1958.

Imprinting in Birds. Eckhard H. Hess in *Science*, Vol. 146, No. 3648, pages 1128–1139; November 27, 1964.

Innate Factors in Imprinting. Eckhard H. Hess and Dorle B. Hess in *Psychonomic Science*, Vol. 14, No. 3, pages 129–130; February 10, 1969.

Development of Species Identification in Birds: An Inquiry into the Prenatal Determinants of Perception. Gilbert Gottlieb. University of Chicago Press, 1971.

Natural History of Imprinting. Eckhard H. Hess in *Integrative Events in Life Processes: Annals of the New York Academy of Sciences*, Vol. 193, in press.

25. How an Instinct is Learned

On the Stimulus Situation Releasing the Begging Response in the Newly Hatched Herring Gull Chick (Larus argentatus argentatus Pont).

N. Tinbergen and A. C. Perdeck in *Behaviour*, Vol. 3, pages 1–39; 1950.

A Critique of Konrad Lorenz's Theory of Instinctive Behavior. Daniel S. Lehrman in *The Quarterly Review of Biology*, Vol. 28, No. 4, pages 337–363; December, 1953.

An Introduction to Animal Behavior: Ethology's First Century. Peter H. Klopfer and Jack P. Hailman. Prentice-Hall, Inc., 1967.

The Ontogeny of an Instinct: The Pecking Response in Chicks of the Laughing Gull (Larus atricilla L.) and Related Species. Jack P. Hailman in *Behaviour*, Supplement 15; 1967.

26. Love in Infant Monkeys

The Development of Affectional Responses in Infant Monkeys. Harry F. Harlow and Robert R. Zimmermann in *Proceedings of the American Philosophical Society*, Vol. 102, pages 501–509; 1958.

The Nature of Love. Harry F. Harlow in *American Psychologist*, Vol. 12, No. 13, pages 673–685; 1958.

IV SOCIAL BEHAVIOR

27. Animal Communication

Mechanisms of Animal Behavior. P. R. Marler and W. J. Hamilton III. John Wiley & Sons, Inc., 1966.

Animal Communication: Techniques of Study and Results of Research. Edited by T. A. Sebeok. Indiana University Press, 1968.

Sex Pheromone Specificity: Taxonomic and Evolutionary Aspects in Lepidoptera. Wendell L. Roelofs and Andre Comeau in *Science*, Vol. 165, No. 3891, pages 398–400; July 25, 1969.

The Insect Societies. Edward O. Wilson. The Belknap Press of Harvard University Press, 1971.

Language in Chimpanzee? David Premack in *Science*, Vol. 172, No. 3985, pages 802–822; May 21, 1971.

Non-Verbal Communication. Edited by R. A. Hinde. Cambridge University Press, 1972.

28. The Language of Birds

Characteristics of Some Animal Calls. P. Marler in *Nature*, Vol. 176, No. 4470, pages 6–8; July, 1955.

Comments on 'The Bird Fancyer's Delight': Together with Notes on Imitation in the Sub-Song of the Chaffinch. W. H. Thorpe in *Ibis*, Vol. 97, No. 2, pages 247–251; April, 1955.

The Process of Song Learning in the Chaffinch as Studied by Means of the Sound Spectograph. W. H. Thorpe in *Nature*, Vol. 173, No. 4402, pages 465–469; March, 1954.

29. The Fighting Behavior of Animals

Aggression. J. P. Scott. University of Chicago Press, 1958.

Kampf und Paarbildung einiger Cichliden. Beatrice Oehlert in *Zeitschrift für Tierpsychologie*, Vol. 15, No. 2, pages 141–174; August, 1958.

Studies on the Basic Factors in Animal Fighting. Parts I–IV. Zing Yang Kuo in *The Journal of Genetic Psychology*, Vol. 96, Second Half, pages 201–239; June, 1960.

Studies on the Basic Factors in Animal Fighting. Parts V–VII. Zing Yang Kuo in *The Journal of Genetic Psychology*, Vol. 97, Second Half, pages 181–225; December, 1960.

Zum Kampf- und Paarungsverhalten einiger Antilopen. Fritz Walther in *Zeitschrift für Tierpsychologie*, Vol. 15, No. 3, pages 340–380; October, 1958.

30. Hormones in Social Amoebae and Mammals

THE CELLULAR SLIME MOLDS. John Tyler Bonner. Princeton University Press, 1967.

CYCLIC AMP. G. A. Robison, R. W. Butcher, and E. W. Sutherland in *Annual Review of Biochemistry*, Vol. 37, pages 149–174; 1968.

CYCLIC AMP: A NATURALLY OCCURRING ACRASIN IN THE CELLULAR SLIME MOLDS. T. M. Konijn, D. S. Barkley, Y. Y. Chang, and J. T. Bonner in *The American Naturalist*, Vol. 102, No. 925, pages 225–233; May–June, 1968.

31. Communication Between Ants and Their Guests

THE SYSTEMATICS, EVOLUTION AND ZOOGEOGRAPHY OF STAPHYLINID BEETLES ASSOCIATED WITH ARMY ANTS (COLEOPTERA, STAPHYLINIDAE). Charles H. Seevers in *Fieldiana: Zoology*, Vol. 47, No. 2, pages 139–351; March 22, 1965.

BEHAVIOR OF STAPHYLINIDAE ASSOCIATED WITH ARMY ANTS (FORMICIDAE: ECITONINI). Roger D. Akre and Carl W. Rettenmeyer in *Journal of the Kansas Entomological Society*, Vol. 39, No. 4, pages 745–782; October, 1966.

CONTRIBUTION TO THE PHYSIOLOGY OF GUEST-HOST RELATIONS (MYRMECOPHILY) IN ANTS, II: THE RELATION BETWEEN THE IMAGOS OF ATEMELES PUBICOLLIS AND FORMICA AND MYRMICA. B. Hölldobler in *Zeitschrift für Vergleichend Physiologie*, Vol. 66, pages 215–250; 1970.

32. Dialects in the Language of the Bees

COMMUNICATION AMONG SOCIAL BEES. Martin Lindauer. Harvard University Press, 1961.

THE DANCING BEES. Karl von Frisch. Harcourt, Brace & Company, 1955.

"SPRACHE" UND ORIENTIERUNG DER BIENEN. Karl von Frisch. Verlag Hans Huber, 1960.

33. The Social Behavior of Prairie Dogs

THE ANALYSIS OF SOCIAL ORGANIZATION IN ANIMALS. J. P. Scott in *Ecology*, Vol. 37, No. 2, pages 213–220; April, 1956.

PRAIRIE DOGS, WHITEFACES AND BLUE GRAMA. Carl B. Koford in *Wildlife Monographs*, No. 3, pages 1–78; December, 1958.

SOCIAL BEHAVIOR, SOCIAL ORGANIZATION AND POPULATION DYNAMICS IN A BLACK-TAILED PRAIRIEDOG TOWN IN THE BLACK HILLS OF SOUTH DAKOTA. John A. King in *Contributions from the Laboratory of Vertebrate Biology*, No. 67, pages 1–123; April, 1955.

34. The Social Life of Baboons

BEHAVIOR AND EVOLUTION. Edited by Anne Roe and George Gaylord Simpson. Yale University Press, 1958.

A STUDY OF BHAVIOUR OF THE CHACMA BABOON, PAPIO URSINUS. Niels Bolwig in *Behaviour*, Vol. 14, No. 1–2, pages 136–163; 1959.

INDEX